Extensions of Minimal Transformation Groups

Extensions of Minimal Transformation Groups

I. U. Bronšteĭn

Academy of Sciences of the Moldavian
Soviet Socialist Republic
Institute of Mathematics and Computer Center

SIJTHOFF & NOORDHOFF 1979
Alphen aan den Rijn The Netherlands
Germantown, Maryland USA

ISBN 90 286 0368 9

Original title:
"Rassirenija minimal'nyh grupp preobrazovanii"
published in 1975 in Kisihev

Edited by Dr. S. Swierczkowski

Printed in The Netherlands

CONTENTS

CHAPTER III

CHAPTER IV

PREFACE TO THE ENGLISH EDITION

This edition is an almost exact translation of the original Russian text. A few improvements have been made in the presentation. The list of references has been enlarged to include some papers published more recently, and the latter are marked with an asterisk.

THE AUTHOR

LIST OF SYMBOLS

INTRODUCTION

1. It is well known that an autonomous system of ordinary differential equations satisfying conditions that ensure uniqueness and extendibility of solutions determines a flow, *i.e.* a one-parameter transformation group. G.D. Birkhoff was the first to note that many notions and results obtained in the theory of autonomous differential equations can be extended to the case of flows in abstract spaces. Introducing the very important concepts of minimal sets and of recurrent and central motions, he laid the foundation of the general theory of flows or, as it is usually called, the topological theory of dynamical systems. This theor-was further developed by A.A. Markov, G.F. Hilmy, V.V. Stepanov, A.N. Tikhonov, M.V. Bebutov, A.G. Maier, V.V. Nemytskiĭ, E.A. Barbas in and others (see Chapter V of the book by V.V. Nemytskii and V.V. Stepanov [*1*], the survey article of V.V. Nemytskiĭ [*2*], and the monograph by K.S. Sibirskiĭ [*1*]). At the end of the 1940's W.H. Gottschalk and G.A. Hedlund [*2*], V.V. Nemytskiĭ [*1*] and E.A. Barbašin [*1*,*2*] proposed a broad, and at the same time, a very natural generalization of classical dynamical systems, namely, the notion of a topological transformation group. The precise definition is as follows. Let X be a topological space, T be a topological group and let $\pi : X \times T \to X$ be a continuous map, satisfying the following identities (where the image of $(x,t) \in X \times T$ under π is denoted by $x\pi^t$):

(1) $x\pi^e = x$ ($x \in X$; e is the identity of the group T);

(2) $(x\pi^t)\pi^s = x\pi^{ts}$ ($x \in X; t,s \in T$).

Then the triple (X,T,π) is called a *TOPOLOGICAL TRANSFORMATION GROUP*. The part of the theory of transformation groups devoted to the study of those notions whose prototype occurred previously in classical dynamics (such as various types of motions, minimal and limit sets, recurrence, stability, *etc.*) is called *TOPOLOGICAL DYNAMICS*.

Let (X,T,π) and (Y,T,ρ) be two topological transformation groups. A continuous map $\varphi : X \to Y$ is said to be a homomorphism and is denoted by $\varphi : (X,T,\pi) \to (Y,T,\rho)$ if for each $t \in T$ the diagram:

$$\begin{array}{ccc} X & \xrightarrow{\varphi} & Y \\ \pi^t \downarrow & & \downarrow \rho^t \\ X & \xrightarrow{\varphi} & Y \end{array}$$

is commutative. If φ is onto, then one also says that (X,T,π) is an *EXTENSION* of (Y,T,ρ) by the homomorphism φ. The map φ: $(X,T,\pi) \to (Y,T,\rho)$ itself is often called an extension.

The class of all transformation groups (X,T,π) with a fixed group T forms a category Cat(T) whose homomorphisms are the morphisms defined above and whose objects are these transformation groups. Extensions may also be considered as objects of some category, namely, the category whose objects are the morphisms of Cat(T). Thus, extensions of topological transformation groups are analogous to fiber bundles in topology and to group extensions in algebra.

During the last ten to fifteen years topological dynamics has been a rapidly developing area of research. R. Ellis introduced the important concept of the *ENVELOPING SEMIGROUP* which is defined to be the closure of the set $\{\pi^t : X \to X \mid t \in T\}$ in the space of all maps of X into X provided with the topology of pointwise convergence. This notion proved to be essential and fruitful in the study of minimal sets. A set $M \subset X$ is called *MINIMAL SET* if M coincides with the orbit closure of each point $x \in M$. Minimal sets are now the central theme in topological dynamics. Their classification is probably a very difficult problem whose complete solution seems to lie beyond the reach of the present methods. In this connection we mention an interesting result due to Jewett [*1*]: for any weakly mixing automorphism of a Lebesgue space there exists a compact zero-dimensional minimal having a unique invariant measure and isomorphic with respect to this measure to the given automorphism. Hence it follows that there exist, for example, minimal sets with completely positive entropy. Thus the structure of minimal sets may be very complicated even from the measure-theoretic viewpoint, whence a description of minimal sets up to a topological isomorphism seems to be a hopeless task. Therefore, it is of interest to study some special 'simple' types of minimal sets.

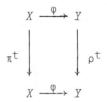

The structure of the almost periodic minimal sets has been known for a long time. L. Auslander, F. Hahn and L. Markus constructed interesting examples of distal, but not almost periodic minimal sets, the so-called nil-flows (if (X,d) is a metric space, then distality means that $\inf_{t \in T} d(x\pi^t, y\pi^t) > 0$ for any two distinct points $x,y \in X$). H. Furstenberg [*1*] discovered the structure of the distal minimal sets. He introduced the notion of an isometric (equicontinuous) extension, and he proved that every distal minimal set can be built up from a point by taking equicontinuous extensions and inverse limits.

Figuratively speaking, equicontinuous extensions are building blocks with the help of which we can construct, step by step, any distal minimal set.

Since Furstenberg's paper [*1*] on distal flows appeared, it became clear that not only iminmal sets, but also (and to a marked degree) minimal extensions are of interest. (A minimal extension is a pair of minimal sets together with a homomorphism of one onto the other). R. Ellis and the author (independently and by different methods) proved a structure theorem for distal minimal extensions which is a generalization of the above-mentioned Furstenberg theorem. Some other types of extensions were investigated by Ellis, Veech, Petersen, Shapiro, and the author.

It should be noted that extensions of transformation groups occur not only when considering minimal sets. There are other sources of extensions no less important than the theory of minimal sets; it is possible to associate extensions with broad classes of differential, integral, algebraic and some other types of equations. Let us consider, for example, differential equations of the form:

$$\frac{dx}{dt} = f(x,t) \qquad (x \in W \subset \mathbb{R}^n,\ t \in \mathbb{R}). \tag{A}$$

Let $Y = C(W \times \mathbb{R}, \mathbb{R}^n)$ be the space of all continuous functions f: $W \times \mathbb{R} \to \mathbb{R}^n$ provided with the compact open topology and let $(Y, \mathbb{R}, \sigma)$ be the shift transformation group. Let X denote the subspace of $Y \times C(\mathbb{R}, W)$ consisting of all pairs (f, φ) where $f \in Y$ and $\varphi: \mathbb{R} \to W$ is a solution of (A). Then X is invariant under the shift transformation group and the canonical projection $\rho: X \to Y$ is a homomorphism. This construction is very useful in solving many problems occurring in the qualitative theory of differential equations. For example, let the right side of (A) be periodic (or almost periodic); does there exists a periodic (almost periodic, respectively) solution of (A)? This question can be reformulated as a problem of 'lifting' some properties of $(Y, \mathbb{R})$ to $(X, \mathbb{R})$ by means of the extension $\rho: (X, \mathbb{R}) \to (Y, \mathbb{R})$.

The consideration of various problems in the qualitative theory of differential equations from the viewpoint of extensions seems to be advantageous for a variety of reasons. Firstly, this is a unifying approach which embraces in one framework various types of equations. Secondly, extensions enable us to bring to light the essential features and to remove details of minor importance. Thirdly, in this way it becomes possible to use the technique and the results of topological dynamics for solving many problems, for example, the above-mentioned problem on the existence of almost periodic solutions (see Chapter IV). To give one more example, let us remark that the recent results on genericity of some properties of dynamical systems obtained by V.A. Dobrynskiĭ and A.N. Šarkovskiĭ are also contained within the scheme of the extensions theory.

The origins of the above construction can be found in J. Favard's work on almost periodic solutions of linear differential equations with almost periodic coefficients, published as early as 1927. The shift dynamical system, which is the basic 'technical' device binding equations with extensions, was introduced by M.V. Bebutov in 1941, but extensions associated with equations began to appear (in a more or less explicit form) only a quarter of a century later in the papers by R.K. Miller, V.M. Millionsčikov, G. Seifert, G.R. Sell, B.A. Ščerbakov, V.V. Žikov, and others. Moreover, all the authors mentioned above (with the exception of V.V. Žikov) considered only relationships between equations and some dynamical systems constructed by means of these equations, but the above extension $p:(X,\mathbb{R},\sigma) \to (Y,\mathbb{R},\sigma)$ was altogether ignored. However, it is just this dynamical system associated with the equation (A) (for example, the shift transformation group in the space of solutions) considered apart from this extension.

It is also to be noted that V.V. Nemytskiĭ pointed out on numerous occasions that it would be disirable to generalize the concept of a dynamical system in such a way that it would reflect the properties of non-autonomous differential eqations to the same extent that a dynamical system does for autonomous equations. Such a generalization was proposed by V.I. Zubov [*1*]. These are the so-called *general systems*, or *two-parameter families of transformations* defined as follows: let X be a metric space, let $\mathbb{R}$ be the group of real numbers, and let $f:X\times\mathbb{R}\times\mathbb{R}\to X$ be a continuous map, satisfying the following identities:

(1) $f(x,t,t) = x, \quad (x \in X, t \in \mathbb{R})$;

(2) $f(x,t_1 + t_2,\tau) = f(f(x,t_1,\tau),t_2,t_1 + \rho)$.

Then we say that f defines a two-parameter family of transformations of the space X, or that $[X,\mathbb{R},f]$ is a general system. Some authors call $[X,\mathbb{R},f]$ a non-autonomous dynamical system. The theory of general systems has not developed enough to become a branch of mathematics on its own footing.

In our opinion, the only abstract notion which adquately reflects the properties of non-autonomous differential equations is that of an extension of dynamical systems. We note in this respect that many non-autonomous equations arise simply from extensions. For example, linear homogeneous equations with variable coefficients often arise by linearization of a smooth flow, hence, from a special extension of the form $p:(TM,\mathbb{R}) \to (M,\mathbb{R})$ where M is a smooth manifold, TM is its tangent space and p is the canonical projection. The recent attempts to extend the deep theory of smooth dynamical systems to the case of time-dependent vector fields on manifolds indicate that the task will most probably be accomplished within the framework of the extension theory.

2. This book presents an exposition of both old and new results on minimal transformation groups and minimal extensions together with some applications to non-autonomous differential equations. It consists of four chapters.

In Chapter I (§(1·1-6)) the basic properties of topological transformation groups are presented. The material in the first two sections is adapted from the book by W.H. Gottschalk and G.A. Hedlund [2]. The reader interested in those parts of topological dynamics, which we have omitted (for example, symbolic dynamics), is referred to this book. In §(1·3) several proximal relations concerning transformation groups defined by W.H. Gottschalk, R. Ellis, and J.P. Clay are considered. §(1·4) is devoted to the concepts of the enveloping semigroup introduced by R. Ellis. In §(1·5) we consider transformation groups with the property that every orbit closure is minimal. The results established in §(1·5) are used in §(1·6) to describe equicontinuous and distal transformation groups.

Chapter II (§(2·7-11)) is concerned with minimal transformation groups (X,T,π), *i.e.* groups such that X is a minimal set. In §(2·7) some relations between a minimal set and its enveloping semigroup are described, endomorphisms and automorphisms of minimal transformation groups are considered and the existence of universal minimal sets is established. These results are due to J. Auslander, R. Ellis and H. Chu. §(2·8) is devoted to almost periodic and locally almost periodic minimal sets. We also present here the well-known results due to V.V. Stepanov, A.N. Tikhonov and A.A. Markov and investigate the structure of minimal sets containing Levitan almost periodic points. The main result of §(2·9) is the Furstenberg Structure Theorem . §(2·10) is devoted to the study of transitive distal transformation groups and nil-flows. A theorem due to Green on the minimality of such flows is also presented. In §(2·11) some topological properties of minimal sets and their location in the phase space are investigated, various concepts of homogeneity are introduced and relations between homogeneity, distality and equicontinuity are considered. In particular, we present theorems of A.A. Markov and G.F. Hilmy on finite-dimensional minimal sets and a theorem due to W.H. Gottschalk which solves Birkhoff's problem concerning homogeneous minimal sets. Some conditions guaranteeing the nontriviality of the fundamental group of a minimal set are also given.

In Chapter III (§(3·12-19)), which is central for the book, various types of minimal extensions are investigated. In §(3·12) we present some basic properties of extensions and we investigate relationships between the most important classes of extensions. We prove some auxiliary statements including, in particular, a very useful lemma due to Beech on the existence of an open homomorphism. The purpose of §(3·13) is to investigate equicontinuous and stable extensions. Much attention is paid to describing the equicontinuous structure relation, which corresponds to the maximal equicontinuous extension subordinated to the given extension. The 'machinery' of §(3·13) is used in §(3·14) to investigate the structure of distal and almost distal extensions. The main results contained in §(3·14) are due to R. Ellis, W.A. Veech and the author. In §(3·15) the algebraic theory of minimal sets developed by R. Ellis is treated from a new viewpoint. In particular, the Galois theory of distal extensions is presented and a structure theorem for a class of minimal extensions including point-distal extensions is proved. §(3·16) contains a complete description of locally trivial minimal group extensions in the context of homological algebra. The next section (§(3·17)) is devoted to a detailed study of equicontinuous extensions. In this way we obtain an improved version of the structure theorem for distal extensions and we give a complete description of two- and three-dimensional locally arcwise connected distal minimal sets. §(3·18) presents a constructive method due to R. Ellis of producing minimal group extensions, which is illustrated by some interesting examples invented by Ellis and Shapiro. §(3·19) is devoted to the notion of disjointness introduced by Furstenberg, which related minimal sets to sets with opposite properties, namely, the distal and weakly mixing sets. We note, in passing, that the second class has not yet been sufficiently studied.

Finally, Chapter IV is devoted to several applications of the above results concerning extensions of dynamical systems to various equations (primarily to non-autonomous differential equations). Our purpose is to demonstrate, by considering a number of examples, that the concept of an extension enables one to pose within the framework of topological dynamics and solve there many problems arising in the qualitative theory of differential equations. We hope that after mastering the methods and the main results presented in this book, the reader will be able to deduce even more consequences and to obtain new results for other classes of equations. We also note that the comparatively slow tempo of presentation adopted in the first three chapters gives place in the last chapter to a more rapid one, and some proofs presented there may possibly be looked upon as too laconic. The author apologizes in advance for this; he has been motivated by the modest intention pursued in this chapter. In our opinion, the application of extensions to various equations is a potential research area which is presently at the beginning stage of development. The multicolored moasic drawn in Chapter IV just testifies to the fact that the investigation in this area is far from being complete.

Chapter IV consists of five sections. In §(4·20) we consider shift transformation groups in various functional spaces, which we use in §(4·21) to describe extensions associated with some classes of equations and to obtain thereby some existence theorems for recurrent and Poisson stable solutions of non-autonomous differential equations. In §(4·22) we show that synchronous solutions of differential equations, introduced by B.A. Ščerbakov, exist if and only if the associated extension is almost automorphic. Some simple sufficient conditions ensuring the existence of synchronous solutions are given. In particular, we present some results due to B.A. Ščerbakov on the existence of a unique bounded synchronous solution for some linear and quasi-linear equations. Towards the end of that section scalar (one-dimensional) differential equations are considered. The purpose of §(4·23) is to investigate almost periodic solutions of almost periodic differential equations. This difficult problem is reduced to that of investigating the existence of equicontinuous minimal extensions with zero-dimensional fibers. The last question depends, in turn, on the existence of a common fixed point for a compact connected transformation group. In this way it is possible to prove the existence of almost periodic solutions in the two- and three-dimensional cases when assuming uniform positive Liapunov stability of the solutions. These results are due to V.V. Žikov. By means of extensions it is proved that uniformly asymptotically stable solutions are asymptotically almost periodic. We also consider algebraic equations with recurrent or almost periodic coefficients and we present the well known Bohr-Flanders-Walter Theorem. In §(4·24) synchronous solutions of linear differential equations are investigated. Generalizing the properties of extensions associated with linear differential equations, we introduce the concept of a linear extension and we extend Favard's minimax method to this case. We prove some general theorems and deduce a number of corollaries. Finally, we treat some theorems due to Bohl, Bohr and Kadets on the primitive of an almost periodic function.

3. Let us mention some books closely related to the subject matter of this monograph. As we have already remarked, the basic topics of topological dynamics are considered by Gottschalk and and Hedlund [*2*]. Many results concerning minimal sets are presented in the monograph by Ellis [*13*]. The material of Chapter IV is in part contained in the following books: Miller and Sell [*1*], Sell [*1*] and Ščerbakov [*6*]. For related topics see Amerio and Prouse [*1*], Krasnoselskiĭ, Burd and Kolesov [*1*], Massera and Schäffer [*1*], Kharasakhal [*1*].

4. It is assumed that the reader is familiar with the basic facts about topological and uniform spaces and topological groups. Occasionally we shall use fundamental groups and covering spaces (see, for example, Hu [*1*]). With few exceptions, all statements will be provided with complete proofs. Otherwise the reader will be referred to the original papers.

5. Following Gottschalk and Hedlund [*2*], we shall usually place the function sign on the right, that is, the image of a point $x \in X$ under a map $f: X \to Y$ will be denoted by xf and the map f will be denoted by $x \mapsto xf$ $(x \in X)$. By using this notation the

composition of the maps $f:X \to Y$ and $g:Y \to Z$ is written in the form $fog:X \to Z$. That is precisely the reason why we adopt this notation. But sometimes (especially in Chapter IV) we keep to the customary usage and we write the function sign on the left. We hope that the 'coexistence' of both systems of notation will not cause much inconvenience for the reader.

All topological and uniform spaces are assumed to be Hausdorff (separated) unless the contrary is specified. Each compact topological space is considered to be provided with the uniquely defined uniformity structure compatible with its topology. Given a subset A of a topological space, the closure of A will be denoted by $\bar{A}$ and the interior of A will be denoted by int$A$. We adopt also the following notations: $\emptyset$ is the empty set, $\mathbb{Z}$ is the group of integers, $\mathbb{R}$ is the group of real numbers, $\mathbb{C}$ is the space of all complex numbers. Given a family $\{A_i | i \in I\}$ of subsets of a set A, the intersection of this family is denoted by $\cap [A_i | i \in I]$ and the union is denoted by $\cup [A_i | i \in I]$. If X is a uniform space, $\mathfrak{U}[X]$ (or simply $\mathfrak{U}$) will denote the uniformity structure of the space X (*i.e.* the filter of indices of the uniform structure). If $x \in X$ and $\alpha \in \mathfrak{U}$, then $x\alpha$ denotes the set $\{y | y \in X, (x,y) \in \alpha\}$. The end of a proof is denoted by ▮.

6. The sections are divided into subsections, each numbered (within a chapter) by a pair of numbers, where the first one refers to the number of the section and the second one refers to the number of the subsection in this section. At the conclusion of each section there are remarks and bibliographical notes. The Bibliography at the end of the book contains only papers and books quoted in the text. A more complete bibliography on topological dynamics and related topics containing approximately 2,500 items was complied by Gottschalk [*10*].

The author is deeply indebted to K.S. Sibirskiĭ, Corresponding Member of the Academy of Sciences of the Moldavian Soviet Socialist Republic, for his encouragement and continuing interest in this work. The author is also grateful to A.I. Gerko who read the manuscript and made many useful remarks, and to V.F. Černiĭ for help rendered in the preparation of the manuscript.

CHAPTER I

TOPOLOGICAL TRANSFORMATION GROUPS

§(1·1): BASIC DEFINITIONS

1·1.1 Let X be a topological space, T be a topological group, π be a map of the Cartesian product $X \times T$ into X. The triple (X,T,π) is called a *TOPOLOGICAL TRANSFORMATION GROUP*, or more briefly, a *TRANSFORMATION GROUP* if the following three conditions are satisfied:

(1): Identity Axiom:

$$(x,e)\pi = x \qquad (x \in X, e \text{ is the identity of the group } T);$$

(2): Homomorphism Axiom:

$$((x,t)\pi,s)\pi = (x,ts)\pi \qquad (x \in X; t,s \in T);$$

(3): Continuity Axiom:

The function $\pi: X \times T \to X$ is continuous.

Let (X,T,π) be a topological transformation group. Then X is called the *PHASE SPACE* and T is called the *PHASE GROUP*. If $x \in X$, $t \in T$, then we shall often write xt instead of $(x,t)\pi$, when there is no danger of ambiguity.

The identity and homomorphism axioms may be rewritten in the following form:

$$(1) \quad xe = x \quad (x \in X); \qquad (2) \quad (xt)s = x(ts) \quad (x \in X; t,s \in T).$$

Instead of (X,T,π) we shall sometimes write (X,T) or even simply T.

A statement such as "(X,T,π) has some property $\mathcal{P}$" will often be paraphrased as "T has the property $\mathcal{P}$ on X" or "X has the property $\mathcal{P}$ under the action of T". Analogously, for a point $x \in X$, the assetion that "(X,T,π) has a certain property $\mathcal{P}$ at x" may be paraphrased as "T has the property $\mathcal{P}$ at the point x" or "x has the property $\mathcal{P}$ under the action of T"

Let $t \in T$. The map $\pi^t : X \to X$ defined by $x\pi^t = (x,t)\pi$ $(x \in X)$ is called the *t-SHIFT*. Let $G = \{\pi^t | t \in T\}$. If $x \in X$, then the *MOTION of the point* x is the map $\pi_x : T \to X$, where $t\pi_x = (x,t)\pi$ $(t \in T)$.

It follows from the definition of a topological transformation group that:

(1) π^e is the identity map of X;

(2) if $t,s \in T$, then $\pi^t \circ \pi^s = \pi^{ts}$;

(3) if $t \in T$, then π^t is a homeomorphism of X onto X and

$$(\pi^t)^{-1} = \pi^{t^{-1}};$$

(4) G is a group of homeomorphisms of X onto X;

(5) the map $\lambda : T \to G$ defined by $t\lambda = \pi^t$ $(t \in T)$ is a homomorphism of the group T onto G;

(6) if $x \in X$, then π_x is a continuous function.

1·1.2 If $A \subset X$ and $B \subset T$, then the set $(A \times B)\pi = \{xt | x \in A, t \in B\}$ will be, for brevity denoted by AB. Similarly, if $t \in T$, then instead of $\{a\pi^t | a \in A\}$, we shall write At. If $x \in X$, then xB denotes the set $\{x\pi^t | t \in B\}$.

1·1.3 LEMMA: *The following assertions hold:*

(1): *If* $A \subset X$, $t \in T$, *then* $\overline{At} = \bar{A}t$;

(2): *If* $A \subset X$ *and* $B \subset T$ *are compact subsets, then* AB *is also compact;*

(3): *If* A *and* B *are compact subsets of* X *and* T, *respectively, and* W *is a neighbourhood of* AB, *then there exist neighbourhoods* U *of* A *and* V *of* B *such that* $UV \subset W$.

Proof: (1): This assertion is true, since $\pi^t : X \to X$ is a homeomorphism. (2): The set AB is compact as an image of the compact set $A \times B$ under the continuous map π. (3): Let $a \in A$ and $t \in B$. There exist neighbourhoods $U_{a,t}$ of a and $V_{a,t}$ of t such that $U_{a,t}V_{a,t} \subset W$. Choose a finite set $F \subset A$ with $A \subset \cup [U_{a,t} | a \in F] \equiv U_t$. Let $V_t = \cap [V_{a,t} | a \in F]$, then $U_tV_t \subset W$ $(t \in B)$. Thus, for each element $t \in B$ there exist neighbourhoods V_t of t and U_t of the set A, such that $U_tV_t \subset W$. Since the set B is compact, we may choose a finite set $E \subset B$ such that $B \subset \cup [V_t | t \in E]$. Put $V = \cup [V_t | t \in E]$ and $U = \cap [U_t | t \in E]$. Then U is a neighbourhood of A, V is a neighbourhood of B and $UV \subset W$. ▌

1·1.4 LEMMA: *Let* X,Y *be uniform spaces, let* $\varphi : X \to Y$ *be a continuous map and let* A *be a compact subset of* X. *Then for each* $\beta \in \mathfrak{U}[Y]$ *there exists an index* $\alpha \in \mathfrak{U}[X]$ *such that* $x\alpha\varphi \subset x\varphi\beta$ *for all* $x \in A$.

The proof is left to the reader.

1·1.5 LEMMA: *Let* X *be a uniform space, let* $\alpha \in \mathfrak{U}[X]$*, and let* A *and* B *be compact subsets of* X *and* T*, respectively. Then:*

(1): *If* $t \in T$*, then there exist a* $\beta \in \mathfrak{U}[X]$ *and a neighbourhood* V *of the identity* $e \in T$ *such that* $x\beta tV \subset xt\alpha$ *for all* $x \in A$;

(2): *There exists an index* $\beta \in \mathfrak{U}[X]$ *such that* $x\beta t \subset xt\alpha$ *for all* $x \in A$, $t \in B$.

Proof: (1) follows immediately from Lemma 1·1.4. The second assertion follows from (1).∎

The property expressed in (2) is sometimes called *UNIFORM INTEGRAL CONTINUITY*.

1·1.6 Let $A \subset X$ and $S \subset T$. The set A is called an *S-INVARIANT SET* if $AS \subset A$. T-invariant sets are, for brevity, called *INVARIANT SETS*. By definition, the empty set is invariant. It is easy to verify the following propositions: (1) if the set A is invariant, then $X \setminus A$, $\bar{A}$, int$A$ are also invariant; (2) the intersection (union) of a family of invariant subsets of X is invariant.

Let $x \in X$ and $S \subset T$. The set xS is called the *S-ORBIT of* x. The set xT will be simply called the *ORBIT* (or *TRAJECTORY*) *of the point* x. It is clear that the orbit xT is the smallest invariant subset of X, containing the point x. If $y \in xT$, then $xT = yT$. The set $\overline{xT}$ is the smallest closed invariant set containing x. If Y is an invariant subset of X, then the transformation group (X,T,π) induces on Y a topological transformation group, which will be denoted by (Y,T,π).

A non-empty closed invariant subset $M \subset X$ is called a *MINIMAL SUBSET* if M does not contain a proper closed invariant subset. Note that a set M is minimal iff $M = \overline{xT}$ for every point $x \in M$.

1·1.7 THEOREM: *Any non-empty compact invariant set* $A \subset X$ *contains a minimal subset.*

Proof: Let I be the family of all non-empty closed invariant subsets of A. $I \neq \emptyset$, since $A \in I$. Further, I is partially ordered by inclusion: $A_1 \leqslant A_2$, iff $A_1 \subset A_2$. If $\mathfrak{I}$ is a linearly ordered sub-family of I, then the intersection B of all sets from $\mathfrak{I}$ is non-empty, because A is compact. Clearly, B is invariant. Hence, $B \in I$. By Zorn's Lemma, I contains at least one minimal element M. It is clear that M is a minimal subset of A.∎

1·1.8 Let $\{(X_i,T,\pi_i) | i \in I\}$ be a family of transformation groups indexed by a set I and let $X = \Pi[X_i | i \in I]$ be the (Cartesian) topological product of the family $\{X_i | i \in I\}$. Let us define a map $\pi: X \times T \to X$ as follows: if $x = \{x_i | i \in I|$ and $t \in T$, then

$(x,t)\pi = \{x_i\pi_i^t \mid i \in I\}$. It is easy to verify that (X,T,π) is a topological transformation group. We shall say that (X,T,π) is the *CARTESIAN PRODUCT OF THE FAMILY* and denote it as:

$$(X,T,\pi) = \prod[(X_i,T,\pi_i) \mid i \in I].$$

1·1.9 Let (X,T,π) and (Y,T,ρ) be topological transformation groups. A continuous map φ of X onto (into) Y is called a *HOMOMORPHISM* of (X,T,π) *ONTO* (*INTO*) (Y,T,ρ) if $x\pi^t\varphi = x\varphi\rho^t$ $(t \in T, x \in X)$. A homomorphism φ of a transformation group (X,T,π) onto (into) (Y,T,ρ) is said to be an *ISOMORPHISM of* (X,T,π) *INTO* (Y,T,ρ) if $\varphi : X \to Y$ is a homeomorphism of X onto $X\varphi$. If φ is a homomorphism of (X,T,π) onto (Y,T,ρ), then one says that (Y,T,ρ) *is a FACTOR of* (X,T,π) and that (X,T,π) *is an EXTENSION of* (Y,T,ρ) *BY THE HOMOMORPHISM* φ.

1·1.10 A transformation group (X,T,π) is called a *FLOW* if $T = \mathbb{R}$, and a *CASCADE* if $T = \mathbb{Z}$.

REMARKS AND BIBLIOGRAPHICAL NOTES

The notion of a topological transformation group originated in the latter part of the nineteenth century in the theory of local Lie groups, but precise definitions were given relatively recently. Presently two directions of research exist in the theory of topological transformation groups. The first one is represented by the book of Montgomery and Zippin [*1*], which is devoted mainly to *compact* transformation groups. The second direction (topological dynamics) is represented by the book of Gottschalk and Hedlund [*2*]. Topological dynamics considers transformation groups (X,T,π) with a *non-compact* phase group T, in particular, flows and cascades.

As was pointed out in the Introduction, flows arise in the study of those systems of autonomous differential equations whose solutions are defined on the whole real line and are uniquely determined by the initial data. The notion of a flow can be extended to more general situations. Thus, if the solutions are defined only for non-negative moments of time, we get 'one-sided' (semigroup) dynamical systems. Equations whose solutions are defined on intervals depending on the initial point, lead to local dynamical systems. The reader who is interested in such generalizations is referred to the books of Hajek [*1*] and of Bhatia and Hajek [*1*].

Many notions and results given in the first two chapters can be translated into the context of topological transformation semigroups (this generalization is left to the reader as a useful exercise). The generalization of results presented in Chapter III to the case of semigroup dynamical systems seems to be a more difficult task. Some results in this direction were obtained by Gerko [*1*,*2*]. In Chapter IV we shall consider not only transformation groups, but also semigroup dynamical systems, which is primarily motivated by the requirements of the applications of topological dynamics to the study of equations.

§(1·2): RECURSION

1·2.1 We begin in this section the investigation of 'recurrence' properties. This is of great importance for topological dynamics. Classical dynamics long ago discovered the essential role of periodic motions. H. Poincaré showed the first examples of motions which are non-periodic but are such that the moving point returns to every neighbourhood of the initial state at arbitrarily distant moments of time. Such motions are called *POISSON STABLE MOTIONs* (*cf.* Birkhoff [2]).

Let (X,T,π) be a topological transformation group, let $x \in X$, U be a neighbourhood of x and

$$A(x,U) = \{t \mid t \in T \; x\pi^t \in U\}.$$

It is clear that the properties of $A(x,U)$, as U ranges over the neighbourhood filter of the point x, tell about the recurrence character of the motion π_x. Therefore various 'recurrence' properties can be simultaneously investigated by means of the following approach.

Suppose that certain subsets of T are distinguished and called *ADMISSIBLE SUBSETS*. Denote the class of all admissible subsets by $\mathbf{A} = \mathbf{A}(T)$.

Let $x \in X$. The transformation group (X,T,π), or simply T, is said to *RECURSIVE AT THE POINT* x and the point x is called a *POINT RECURSIVE UNDER* T if for each neighbourhood U of x there exists an admissible subset $A \subset T$ such that $xA \subset U$ or, equivalently, $A \subset A(x,U)$. The transformation group T is said to be *POINTWISE RECURSIVE* if T is recursive at each point of X. The transformation group T is called *LOCALLY RECURSIVE AT* $x \in X$ and the point x is called *LOCALLY RECURSIVE UNDER* T, if for each neighbourhood U of x we can choose a neighbourhood V of x and an admissible subset $A \in \mathbf{A}$ so that $VA \subset U$. The transformation group T is said to be *LOCALLY RECURSIVE* if t is locally recursive at each point $x \in X$.

Let $(X,\mathfrak{U})$ be a uniform space. The transformation group (X,T,π) is called a *RECURSIVE TRANSFORMATION GROUP* if for every index (*i.e.*, entourage) $\alpha \in \mathfrak{U}[X]$ there exists an admissible set $A \in \mathbf{A}$ such that $xA \subset x\alpha$ for each $x \in X$.

The main logical hierarchy between the notions introduced above is indicated in the following table (where arrows denote implications):

T is recursive
$\Downarrow$
T is locally recursive $\Rightarrow$ T is locally recursive at a point $x \in X$
$\Downarrow$ $\Downarrow$
T is pointwise recursive $\Rightarrow$ T is recursive at $x \in X$.

1·2.2 THEOREM: *If*

$$tAt^{-1} \in \mathbf{A} \qquad (A \in \mathbf{A}, t \in T), \tag{2.1}$$

then the set R of all recursive (locally recursive) points $x \in X$ is invariant under T.

Proof: Let $x \in R$, $t \in T$. We shall show that $xt \in R$. Let U be any neighbourhood of xt. Since $\pi^{t^{-1}} X \to X$ is a homeomorphism, Ut^{-1} is a neighbourhood of x. Since $x \in R$, there exists an admissible set $A \in \mathbf{A}$ such that $xA \subset Ut^{-1}$. Put $B = t^{-1}At$, then $xtB \subset U$ and $B \in \mathbf{A}$ by (2.1). Hence $xt \in R$. The proof of the other assertion (in brackets) is analogous.▌

1·2.3 We introduce now two important classes of subsets of a topological group T which will serve as admissible classes.

A subset $A \subset T$ is called *LEFT-SYNDETIC* or, simply, *SYNDETIC*, if there exists a compact subset $K \subset T$ such that $T = AK$. It is easy to verify that a set A is syndetic iff there exists a compact subset $K \subset T$ such that $tK \cap A \neq \emptyset$ for each element $t \in T$. This implies, in particular, that for $T = \mathbb{R}$ the notions of a syndetic set and of a relatively dense set coincide (Nemytskiĭ and Stepanov [*1*], Ch. V). If A is a syndetic set and $t, s \in T$, then the set sAt is also syndetic.

Let us now define the second class of admissible subsets. A subset $S \subset T$ is called a *SUBSEMIGROUP* of the group T if $SS \subset S$. A subset $P \subset T$ is said to be *REPLETE IN* T if for each compact subset $K \subset T$ there exist elements $t, s \in T$ such that $tKs \subset P$. Let P be a replete subsemigroup of the group T. A subset $B \subset T$ is called *P-EXTENSIVE* if $pP \cap B \neq \emptyset$ for every $p \in P$. The class of all P-extensive subsets of T is denoted by $E_P(T)$.

If the class of admissible subsets $\mathbf{A}$ is chosen to be the class $S(T)$ of all syndetic subsets (the class $E_P(T)$ of all P-extensive subsets), then the term 'recursive' is replaced by 'almost periodic' ('P-recurrent').

REMARKS AND BIBLIOGRAPHICAL NOTES

The general approach to the investigation of various 'recurrence 'properties, based on the concept of an 'admissible' set, was first proposed by Gottschalk and Hedlund [*1*] (see also their book [*2*]). The notion of an almost periodic point is a generalization of recurrent points in the sense of Birkhoff [*1*] (to be more precise, of almost recurrent points in the sense of Bebutov [*1*]). If $T = \mathbb{R}$ and $P = \mathbb{R}^+ \equiv \{t | t \in \mathbb{R}, t > 0\}$, then a set $A \subset \mathbb{R}$ is P-extensive iff A contains a sequence, which tends to $+\infty$. Therefor in the case of $P = \mathbb{R}^+$ the notion of P-recurrence and the notion of Poisson stability in the positive direction coincide.

§(1·3): RELATIONS

We shall establish in this section a connection between invariant equivalence relations defined on the phase space of a transformation group and homomorphisms of this transformation group. Moreover, we shall introduce and study four relations which will be repeatedly used in the following.

1·3.1 Let X be a set and let $X \times X$ denote the Cartesian product, *i.e.*, the set of all ordered pairs (x,y) $(x \in X, y \in X)$. A non-empty subset $R \subset X \times X$ is said to be a (*BINARY*) *RELATION*. Let $\Delta \equiv \Delta(X) = \{(x,x) | x \in X\}$ be the *DIAGONAL of* $X \times X$. A relation is called

(1) A *REFLEXIVE RELATION* if $\Delta \subset R$;

(2) A *SYMMETRIC RELATION* if $(x,y) \in R$ implies $(y,x) \in R$;

(3) A *TRANSITIVE RELATION* if $(x,y) \in R$ and $(y,z) \in R$ imply $(x,z) \in R$.

A reflexive symmetric transitive relation is called an *EQUIVALENCE RELATION*. An equivalence relation $R \subset X \times X$ defines a *PARTITION* of X into subsets called *EQUIVALENCE CLASSES*. If $x \in X$, then the equivalence class xR of the point x is defined to be the set $\{y | y \in X, (x,y) \in R\}$. Conversely, each partition of X (*i.e.*, a family of non-empty disjoint subsets, whose union is X) defines the equivalence relation on X for which two point x and y are said to be equivalent if they belong to the same element of the partition.

If X is a set and R is an equivalence relation in X, then X/R *denotes the set* $\{xR | x \in X\}$.

Let X be a set, D be a partition of X and $E \subset X$. The set $\cup\{A | A \in D, A \cap E \neq \emptyset\}$ *is denoted by ED and called the STAR* (or *the SATURATION*) *of* E *RELATIVE TO* D. A set A is called *SATURATED with respect to* D, or *D-SATURATED*, if $E = ED$.

Let X be a topological space, let R be an equivalence relation on X and let D_R be the partition of X into equivalence classes. The relation R is called an *OPEN relation* (and the partition D_R is called a *STAR-OPEN partition*), if the star of each open subset $E \subset X$ is open in X. The relation R is called a *CLOSED relation* (and the partition D is called a *STAR-CLOSED partition*), if the star of each closed subset is closed.

Let X/R be the set $\{xR|x \in X\}$ provided with the quotient topology. The relation R is open (closed) iff the canonical projection of X onto X/R is open (closed).

1·3.2 Let (X,T,π) be a topological transformation group and let T be a relation in X. Then R is called a *RELATION INVARIANT UNDER* (X,T,π), if $(x,y) \in R$ implies that $(x\pi^t,y\pi^t) \in T$ for all $t \in T$. The notion of an invariant equivalence relation is intimately related to that of a homomorphism, introduced in §(1·1). Let (Y,T,ρ) be a transformation group and let $\varphi:X \to Y$ be a homomorphism of (X,T,π) onto (Y,T,ρ). Then the set

$$R \equiv R_\varphi \equiv R(X \to Y) = \{(a,b)|a,b \in X, a\varphi = b\varphi\}$$

is an invariant equivalence relation.

Conversely, suppose that R is an invariant open equivalence relation and X/R is the quotient space. Define a map $\rho:(X/R) \times T \to X/R$ by:

$$(A,t)\rho = A\pi^t \qquad (A \in X/R, t \in T).$$

It is easy to verify that $(X/R,T,\rho)$ is a topological transformation group and that the canonical projection $\varphi:X \to X/R$ is a homomorphism of (X,T,π) onto $(X/R,T,\rho)$. In fact, it is evident that the identity axiom and the homomorphism axiom are satisfied. To verify the third axiom, let $A \in X/R$, $t \in T$ and let U be a neighbourhood of $(A,t)\rho = A\pi^t$ in X/R. The set $U\varphi^{-1}$ is open in X, since φ is continuous. If $x \in A$, then $x\pi^t \in A\pi^t \subset U\varphi^{-1}$. We can choose a neighbourhood V of x and a neighbourhood W of the element t such that $(V,W)\pi \subset U\varphi^{-1}$. Since the map φ is open, the set $V\varphi = \{B|B \in X/R, B \cap V \neq \emptyset\}$ is a neighbourhood of the point $A \in X/R$. Furthermore, if $B \in V\varphi$ and $s \in W$, then $B\pi^s \subset U\varphi^{-1}$, and therefore $(B,s)\rho \equiv B\pi^s \in U$. Thus $(X/R,T,\rho)$ is a transformation group. Let $A \in X/R$ and $x \in A$. Then $x\varphi = A$ and $x\pi^t\varphi = A\pi^t$, since $x\pi^t \in A\pi^t$. Hence $x\pi^t\varphi = x\varphi\rho^t$ $(x \in X, t \in T)$, *i.e.*, φ is a homomorphism of (X,T,π) onto $(X/R,T,\rho)$.

Suppose now that X is compact, R is a closed invariant equivalence relation on X and φ is the canonical projection of X onto X/R. We shall show that $(X/R,T,\rho)$, where ρ is defined as above, is a topologcal transformation group. It will suffice to verify the continuity axiom. Let $A \in X/R$, $t \in T$ and let U be a neighbourhood of the point $(A,t)\rho = A\pi^t$ in X/R. The set $U\varphi^{-1}$ is open in X and $A\pi^t \subset U\varphi^{-1}$. The set A is closed and therefore compact. By Lemma 1·1.3(3) there exist neighbourhoods V of the set $A \subset X$ and W of the element $t \in T$ such that $(V,W)\pi \subset U\varphi^{-1}$. Since the relation R is closed, V contains an R-saturated neighbourhood V_1 of the set A. By definition of the quotient topology, $V_1\varphi$ is a neighbourhood of the point $A \in X/R$ and $(V_1\varphi,W)\rho \subset U$. Thus $(X/R,T,\rho)$ is a topological transformation group. The above considerations show that $\varphi:X \to X/R$ is a homomorphism of (X,T,π) onto $(X/R,T,\rho)$.

Conversely, let X be a compact space and let $\varphi: X \to Y$ be a homomorphism of (X,T,π) onto a transformation group (Y,T,σ). The set $R = \{(x_1,x_2) \mid x_1,x_2 \in X, x_1\varphi = x_2\varphi\}$ is a closed invariant equivalence relation and, moreover, (Y,T,σ) is isomorphic to $(X/R,T,\rho)$.

Thus, in the class of transformation groups with compact phase spaces, the notion of a homomorphism is in fact equivalent to the notion of a closed invariant equivalence relation.

Let (X,T,π) be a topological transformation group and let $(X \times X,T,\pi)$ be the Cartesian product of the transformation group (X,T,π) with itself. By definition, the map $\pi:(X \times X) \times T \to (X \times X)$ is given by $((x,y),t)\pi = (x,t)(\pi^t \times \pi^t) \equiv (x\pi^t, y\pi^t)$ $((x,y) \in X \times X,$ $t \in T)$. A relation R is invariant iff the set $R \subset X \times X$ is invariant under $(X \times X,T,\pi)$.

Let X be a compact space, let (X,T,π) be a transformation group and $R \subset X \times X$ be a relation. By R^* we will denote the smallest closed invariant equivalence relation containing R.

1·3.3 Let $(X,\mathfrak{U})$ be a uniform space and let (X,T,π) be a transformation group. Let us introduce the following definitions.

Two points x and y of X are called *PROXIMAL under* (X,T,π) if for every index $\alpha \in \mathfrak{U}[X]$ there exists an element $t \in T$ such that $(xt,yt) \in \alpha$. Two points, which are not proximal are called *DISTAL*. The set of all proximal pairs of points is called the *PROXIMAL RELATION* and is *denoted by* $P(X,T,\pi)$ or, simply, *by* P.

Two points x and y of X are said to be *REGIONALLY PROXIMAL under* (X,T,π) if for each index $\alpha \in \mathfrak{U}[X]$ and for each neighbourhood U of x and each neighbourhood V of y we can choose points $x_1 \in U$, $y_1 \in V$ and an element $t \in T$ in such a way that $(x_1t, y_1t) \in \alpha$. The set of all regionally proximal pairs of points is called the *REGIONALLY PROXIMAL RELATION* and is *denoted by* $Q(X,T,\pi)$ *or*, simply, *by* Q.

Let $L = L(X,T,\pi)$ denote the set of all pairs $(x,y) \in X \times X$ such that for every index $\alpha \in \mathfrak{U}[X]$ there exists a syndetic subset $A \subset T$, satisfying the condition $(xt,yt) \in \alpha$ for all $t \in A$. The set $L(X,T,\pi)$ is called the *SYNDETICALLY PROXIMAL RELATION*.

By $M = M(X,T,\pi)$ we denote the set of all pairs $(x,y) \in X \times X$ such that for every index $\alpha \in \mathfrak{U}$ and for every neighbourhood U of x and every neighbourhood V of y there exist points $x_1 \in U$, $y_1 \in V$ and a syndetic subset $A \subset T$ such that $(x_1t, y_1t) \in \alpha$ for all $t \in A$. The set $M(X,T,\pi)$ is called the *REGIONALLY SYNDETICALLY PROXIMAL RELATION*.

By K^* we denote the set of all compact subsets of the group T. The following equalities are valid:

$$P = \bigcap [\alpha T \mid \alpha \in \mathfrak{U}]; \tag{3.1}$$

$$Q = \bigcap [\overline{\alpha T} \mid \alpha \in \mathfrak{U}]; \tag{3.2}$$

$$L = \bigcap_{\alpha \in \mathfrak{U}} \bigcup_{K \in K^*} \bigcap_{t \in T} \alpha K t; \tag{3.3}$$

$$M = \bigcap_{\alpha\in\mathfrak{U}} \overline{\bigcup_{K\in K^*} \bigcap_{t\in T} \alpha K t}. \tag{3.4}$$

We prove, for example, the second formula. If $(x,y) \in Q$ and W is a neighbourhood of $(x,y) \in X \times X$, then there exist neighbourhoods U of x and V of y such that $U \times V \subset W$. For these neighbourhoods and each index $\alpha \in \mathfrak{U}$ there exist points $x_1 \in U$, $y_1 \in V$ and an element $t \in T$ such that $(x_1t, y_1t) \in \alpha$. Put $x_2 = x_1t$, $y_2 = y_1t$; then $(x_1,y_1) \in U \times V \subset W$ and $(x_1,y_1) = (x_2t^{-1}, y_2t^{-1}) \in \alpha T$, *i.e.*, $W \cap \alpha T \neq \emptyset$ for every neighbourhood W of (x,y) and each index $\alpha \in \mathfrak{U}$. But this just means that

$$(x,y) \in \bigcap [\alpha T \mid \alpha \in \mathfrak{U}].$$

Conversely, suppose that the last formula holds. If U is a neighbourhood of x, V is a neighbourhood of y and $\alpha \in \mathfrak{U}[X]$, then there exists a pair $(x_1,y_1) \in \alpha T$ such that $x_1 \in U$, $y_1 \in V$. Since $(x_1,y_1) \in \alpha T$, there exist a pair $(x_2,y_2) \in \alpha$ and an element $t \in T$ for which $x_1 = x_2t$, $y_1 = y_2t$. Hence $(x_1t^{-1}, y_1t^{-1}) \in \alpha$, from which it follows that $(x,y) \in Q$. The proof of the other three formulae is analogous.

It is immediate that

$$\Delta \subset L \subset M \subset Q \subset X \times X, \qquad \Delta \subset L \subset P \subset Q \subset X \times X, \tag{3.5}$$

where $\Delta = \{(x,x) \mid x \in X\}$. The relations P,Q,L,M are invariant, reflexive and symmetric.

Let us investigate some properties of these relations. We prove first two lemmas.

1·3.4 LEMMA: *Let T be a group and let A_1,A_2,K_1,K_2 be subsets of T such that $T = A_1K_1 = A_2K_2$. Then*

$$T = (A_1 \cap A_2K_1^{-1})K_1K_2.$$

Proof: For each $t \in T$ there exist elements $a_2 \in A_2$ and $k_2 \in K_2$ such that $t = a_2k_2$. Choose elements $a_1 \in A_1$ and $k_1 \in K_1$ for which $a_2 = a_1k_1$. Then $t = a_1k_1k_2$ and $a_1 = a_2k_1^{-1} \in A_1 \cap A_2K_1^{-1}$. ∎

1·3.5 LEMMA: *Let T be a group, n be a positive integer and $A_1,\dots,A_n,K_1,\dots,K_n$ be subsets of T such that $T = A_iK_i$ ($i = 1,\dots,n$). Then*

$$T = \left(\bigcap_{i=1}^{n} \left(A_i \left(\prod_{j=0}^{i-1} K_j\right)^{-1}\right)\right) \prod_{i=1}^{n} K_i,$$

where $K_0 = \{e\}$ and e is the identity of T.

Proof: The statement is true for $n = 1$. Suppose that it holds for $n = p$. We shall prove that it is true also for $n = p + 1$. Since $T = A_{p+1}K_{p+1}$, we get by our assumption that

$$T = \left(\bigcap_{i=1}^{p}\left(A_i\left(\prod_{j=0}^{i-1} K_j\right)^{-1}\right)\right)\prod_{i=1}^{p} K_i.$$

It follows by the preceding lemma that

$$T = \left(\left(\bigcap_{i=1}^{p}\left(A_i\left(\prod_{j=0}^{i-1} K_j\right)^{-1}\right)\right)\cap A_{p+1}\left(\prod_{j=0}^{p} K_j\right)^{-1}\right)\left(\prod_{i=0}^{p} K_i\right)K_{p+1}$$

$$= \left(\bigcap_{i=1}^{p+1}\left(A_i\left(\prod_{j=0}^{i-1} K_j\right)^{-1}\right)\right)\prod_{i=1}^{p+1} K_i. \blacksquare$$

1·3.6 THEOREM: *If the space X is compact, then L is an equivalence relation on X.*

Proof: We need only to show that L is transitive. Let $(x,y) \in L$ and $(y,z) \in L$. We shall prove that $(x,z) \in L$. Let $\alpha \in \mathcal{U}[X]$. Choose $\beta \in \mathcal{U}[X]$ so that $\beta^2 \subset \alpha$. Then there exist subsets $A_1 \subset T$ and $K_1 \subset T$ such that $K_1 \in K^*$, $T = A_1K_1$ and $(x,y)A_1 \equiv \{(xt,yt) | t \in A_1\} \subset \beta$. Since the set K is compact, the set K_1^{-1} is also compact. Since the space X is compact, for $\beta \in \mathcal{U}[X]$ we can choose an index $\gamma \in \mathcal{U}[X]$ such that $\gamma K_1^{-1} \subset \beta$ (see Lemma 1·1.4(2)). From $(y,z) \in L$ it follows that there exist sets $A_2 \subset T$ and $K_2 \subset T$ for which $K_2 \in K^*$, $T = A_2K_2$ and $(y,z)A_2 \subset \gamma$. Then $(y,z)A_2K_1^{-1} \subset \beta$. Let $A = A_1 \cap A_2K_1^{-1}$; then $(x,y)A \subset \beta$ and $(y,z)A \subset \beta$; therefore $(x,z)A \subset \alpha$. By Lemma 1·3.4, $T = AK_1K_2$. Since K_1 and K_2 are compact sets, the set K_1K_2 is also compact. Hence A is a syndetic set. Thus $(x,z) \in L$. ∎

1·3.7 LEMMA: *If $u \in L$, then $\overline{uT} \subset L$.*

Proof: Let $w \in uT$ and $\alpha \in \mathcal{U}[X]$. Choose an index $\beta \in U[X]$ such that $\bar{\beta} \subset \alpha$. There exist sets $A \subset T$ and $K \subset T$ such that $K \in K^*$, $T = AK$ and $uA \subset \beta$. We shall prove that $w \in L$. It will suffice to prove that $wtK^{-1} \cap \alpha \neq \emptyset$ for all $t \in T$. Suppose that to the contrary $wtK^{-1} \cap \alpha = \emptyset$ for some $t \in T$. Then $wtK^{-1} \cap \bar{\beta} = \emptyset$, *i.e.*, $wtK^{-1} \subset (X \times X) \setminus \bar{\beta}$. Since the set $(X \times X) \setminus \bar{\beta}$ is open in $X \times X$ and tK^{-1} is compact, there exists a neighbourhood U of w such that $UtK^{-1} \cap \bar{\beta} = \emptyset$. Since $w \in \overline{uT}$, there exists an element $s \in T$ with $us \in U$. But $st \in T = AK$, and hence $stK^{-1} \cap A \neq \emptyset$; therefore $ustK^{-1} \cap uA \neq \emptyset$ and moreover $ustK^{-1} \cap \beta \neq \emptyset$, since $uA \subset \beta$. But this contradicts the definition of U. ∎

1·3.8 LEMMA: *If X is compact, $z \in X \times X$ and $zT \subset P$, then $z \in L$.*

Proof: Let $\alpha \in \mathcal{U}[X]$. For each point $w \in \overline{zT} \subset P$ there exists an element t_w such that $wt_w \in \alpha$, and therefore $U_wt_w \subset \alpha$ for some neighbourhood U_w of the point w in $X \times X$. Since X is compact, $X \times X$ is also compact; hence we can choose a finite subset F of $\overline{zT}$ so that $\overline{zT} \subset \bigcup [U_w | w \in F]$. Let us define $K = \{t_w^{-1} | w \in F\}$, $A = \{t | t \in T, zt \in \alpha\}$; we shall prove that $T = AK$. Let $t \in T$. Since

$zt \in \overline{zT} \subset \cup [U_w | w \in F]$, there exists a point $f \in F$ such that $zt \in U_f$. Hence $zttf \in \alpha$; therefore $ttf \in A$ and $t \in AK$. ∎

1·3.9 THEOREM: *Let X be a compact space and let (X,T,π) be a transformation group. Then*

(1) $L = \{z | z \in X \times X \; zT \subset P\}$;

(2) $P = L$ *iff* $z \in P$ *implies* $zT \subset P$;

(3) *If P is closed in $X \times X$, then $P = L$ and P is a closed invariant equivalence relation.*

Proof: This follows immediately from the preceding two lemmas. ∎

1·3.10 LEMMA: *Let n be a positive integer, let $X_1,\ldots,X_n$ be compact spaces, $(X_1,T),\ldots,(X_n,T)$ transformation groups, let $(x_i,y_i) \in L(X_i,T)$ and let $\alpha_i \in \mathfrak{U}[X_i]$ $(i = 1,\ldots,n)$. Then there is a syndetic subset $A \subset T$ such that $(x_i,y_i)A \subset \alpha_i$ for each i $(i = 1,\ldots,n)$.*

Proof: There are subsets $A_1 \subset T$ and $K_1 \in K^*(T)$ which satisfy the conditions $T = A_1K_1$ and $(x_1,y_1)A \subset \alpha_1$. Since X_2 and K_1^{-1} are compact and $(x_2,y_2) \in L(X_2,T)$, there exist subsets $A_2 \subset T$ and $K_2 \in K^*(T)$ such that $T = A_2K_2$, and $(x_2,y_2) \in A_2K_1^{-1} \subset \alpha_2$. Proceeding by induction, we get subsets $A_1,\ldots,A_n,K_1,\ldots,K_n$ of the group T which satisfy the conditions $K_i \in K^*(T)$, $T = A_iK_i$, $K_0 = \{e\}$ and

$$(x_i,y_i)A_i\left(\prod_{j=0}^{i-1} K_j\right)^{-1} \subset \alpha_i \qquad (i = 1,\ldots,n).$$

Denote

$$A = \bigcap_{i=1}^{n}\left(A_i\left(\prod_{j=0}^{i-1} K_j\right)^{-1}\right).$$

Then $(x_i,y_i)A \subset \alpha_i$ $(i = 1,\ldots,n)$. Since $K_1,\ldots,K_n$ are compact, it follows from Lemma 1·3.5 that A is a syndetic subset of T. ∎

1·3.11 LEMMA: *Let n be a positive integer, let $X_1,\ldots,X_n$ be compact spaces and let $(X_1,T),\ldots,(X_n,T)$ be transformation groups. Further, let $(x_i,y_i) \in M(X_i,T)$, $\alpha_i \in \mathfrak{U}[X_i]$, and let U_i be a neighbourhood of x_i and V_i a neighbourhood of y_i $(i = 1,\ldots,n)$. Then there exist points $p_i \in U_i$, $q_i \in V_i$ $(i = 1,\ldots,n)$ and a syndetic subset $A \subset T$ for which $(p_i,q_i)A \subset \alpha_i$ $(i = 1,\ldots,n)$.*

Proof: Since $(x_1,y_1) \in M(X_1,T)$, there exist subsets $A_1 \subset T$, $K_1 \subset T$ and points $p_1 \in U_1$, $q_1 \in V_1$ such that $K_1 \in K^*(T)$, $T = A_1K_1$ and $(p_1,q_1)A_1 \subset \alpha_1$. Further, since $(x_2,y_2) \in M(X_2,T)$, we can choose subsets $A_2 \subset T$, $K_2 \subset T$ and points $p_2 \in U_2$, $q_2 \in U_2$ such that $K_2 \in K^*$, $T = A_2K_2$, $(p_2,q_2)A_2K_1^{-1} \subset \alpha_2$. Proceeding by induction, we get points $p_i \in U_i$, $q_i \in V_i$ $(i = 1,\ldots,n)$ and subsets $A_1,\ldots,A_n$, $K_1,\ldots,K_n$ of T for which $K_i \in K^*(T)$, $T = A_iK_i$,

$$(p_i,q_i)A_i\left(\prod_{j=0}^{i-1} K_j\right)^{-1} \subset \alpha_i \qquad (i = 1,\ldots,n),$$

and $K_0 = \{e\}$. Let us set

$$A = \bigcap_{i=1}^{n}\left(A_i\left(\prod_{j=0}^{i-1} K_j\right)^{-1}\right).$$

Then $(p_i,q_i)A \subset \alpha_i$ $(i = 1,\ldots,n)$. Lemma 1·3.5 shows that A is a syndetic subset. ∎

1·3.12 Let us introduce some definitions.

Let $\{(X_i,T)|i \in I\}$ be a family of transformation groups and let $\Pi[(X_i,T)|i \in I]$ be their Cartesian product. Define a map

$$\theta:(\prod[X_i|i \in I] \times \prod[X_i|i \in I]) \to \prod[(X_i \times X_i)|i \in I]$$

by $(\{x_i\},\{y_i\})\theta = \{(x_i,y_i)|i \in I\}$ Evidently θ is an isomorphism of the corresponding transformation groups.

Let $R = R(X,T)$ be a relation defined for all transformation groups from a class closed under the formation of Cartesian products. The relation R is called a *MULTIPLICATIVE RELATION IN THIS CLASS* if $\Pi[R(X_i,T)|i \in I] = (R(\Pi[(X_i,T)|i \in I]))\theta$ for every family $\{(X_i,T)|i \in I\}$ of transformation groups belonging to the class under consideration.

Before we state the next theorem we note that by Tikhonov's Theorem the class of all transformation groups with compact phase spaces is closed under Cartesian products.

1·3.13 THEOREM: *The relations $L = L(X,T)$ and $M = M(X,T)$ are multiplicative in the class of transformation groups with compact phase spaces.*

Proof: It is easy to see that

$$(L(\prod[(X_i,T)|i \in I]))\theta \subset \prod[L(X_i,T)|i \in I].$$

To prove the reverse inclusion, let

$$\{(x_i,y_i)|i \in I\} \in \prod[L(X_i,T)|i \in I];$$

then $(x_i,y_i) \in L(X_i,T)$ for each element $i \in I$. We need to show that $(\{x_i|i \in I\},\{y_i|i \in I\}) \in L(\Pi[(X_i,T)|i \in I])$. Let $\mathcal{J}$ be a non-empty finite subset of I and let $\alpha_j \in \mathcal{U}[X_j]$ $(j \in \mathcal{J})$. It suffices to show that there exists a syndetic subset $A \subset T$ such that $(x_j, y_j)A \subset \alpha_j$ for all $j \in \mathcal{J}$. But this follows from Lemma 1·3.10. Thus L is multiplicative. Analogously, Lemma 1·3.11 implies that M is multiplicative. ∎

1·3.14 THEOREM: *The following statements are equivalent:*

(1) $P = Q = L = M$;

(2) $P \supset Q$.

Proof: Clearly (1) implies (2). Let $P \supset Q$. Then $P = Q$, since always $P \subset Q$, and therefore P is closed in $X \times X$. By Theorem 1·3.9, $P = L$. Therefore the equalities (1) follow from the formulae (3.5). ∎

1·3.15 Suppose that X is compact and let R_1, R_2 be closed invariant equivalence relations. If $R_1 \subset R_2$, then the canonical projection λ of X/R_1 onto X/R_2 is a homomorphism of $(X/R_1, T)$ onto $(X/R_2, T)$.

Let $\{(X_i, T) | i \in I\}$ be a family of transformation groups with compact phase spaces and for each $i \in I$ let λ_i be a homomorphism of (X,T) into (X_i, T). Suppose that the maps λ_i $(i \in I)$ separates the points of X, *i.e.*, for each two distinct points x_1 and x_2 of X there is a λ_i such that $x_1\lambda_i \neq x_2\lambda_i$. Define a map $\lambda : X \to \Pi[X_i | i \in I]$ by $x\lambda = \{x\lambda_i | i \in I\}$. It is easy to verify that λ is an isomorphism of (X,T) onto an invariant subset of the Cartesian product $\Pi[(X_i, T) | i \in I]$.

A transformation group (X_0, T) is called a ***TRIVIAL TRANSFORMATION GROUP*** if X_0 consists of a single point.

Let $\mathcal{P} = \mathcal{P}(X,T)$ be a property of transformation groups that is preserved under isomorphisms. The property $\mathcal{P}$ is called an ***ADMISSIBLE PROPERTY*** if it satisfies the following three conditions:

CONDITION (1): *The trivial transformation group (X_0, T) has the property $\mathcal{P}$;*

CONDITION (2): *If (X,T) has the property $\mathcal{P}$ and Y is an invariant subset of X, then (Y,T) also has this property;*

CONDITION (3): *If $\{(X_i, T) | i \in I\}$ is a family of transformation groups, each of which has the property $\mathcal{P}$, then the Cartesian product $\Pi[(X_i, T) | i \in I]$ has the property $\mathcal{P}$.*

Let A be a set and let $\{A_i | i \in I\}$ be a family of non-empty subsets of A. An element A_j, $j \in I$, is called the smallest element of the system $\{A_i | i \in I\}$ iff $A_j \subset A_i$ for every $i \in I$. Each family $\{A_i | i \in I\}$ has no more than one smallest element.

1·3.16 THEOREM: *If $\mathcal{P} = \mathcal{P}(X,T)$ is an admissible property, then for each transformation group (X,T) with a compact phase space, there exists a smallest closed invariant equivalence relation $R \subset X \times X$ such that the transformation group $(X/R, T)$ has the property $\mathcal{P}$.*

Proof: Let $\sigma = \{R_i | i \in I\}$ be the family of all closed invariant equivalence relations $R_i \subset X \times X$ such that $(X/R_i, T)$ has the prcperty $\mathcal{P}$. Note that $\sigma \neq \emptyset$, since $(X \times X) \in \sigma$. Denote $\cap[R_i | i \in I]$ by R. Then R is a closed invariant equivalence relation. It remains to prove that the transformation group $(X/R, T)$ has the property $\mathcal{P}$. Since $R \subset R_i$ $(i \in I)$, there exists a canoncial projection λ_i: $X/R \to X/R_i$ and, moreover, λ_i is a homomorphism of $(X/R, T)$ onto

$(X/R_i,T)$. By the definition of R, the family $\{\lambda_i | i \in I\}$ separates points of X/R. Hence, there exists an isomorphism of $(X/R,T)$ into the transformation group $\Pi[(X/R_i,T)\ i \in I]$. Since the property $\mathcal{P}$ is admissible, this implies that $(X/R,T)$ has this property.

Certain important examples of admissible properties will be given in §(1•6).

REMARKS AND BIBLIOGRAPHICAL NOTES

The material in 1•3.1 and 1•3.2 is taken from the book by Gottschalk and Hedlund [*2*]. The relations P and Q were first studied by Ellis and Gottschalk [*1*]. The relations L and M were introduced by Clay [*1*]. The statements 1•3.10-14 are due to Clay [*1*]. Theorem 1•3.16 is contained in the paper by Ellis and Gottschalk [*1*] (see note 8 of that paper). The proximal relation was investigated by Wu [*1*], Wu and McMahon [*1*], Keynes [*2-4*] and Shapiro [*1*]. In particular, the last paper contains an example of a compact minimal set for which the proximal relation is transitive but not closed. In Subsection 3•18.22 it will be shown in an example that the proximal relation of a minimal set can be non-transitive.

§(1•4): THE ELLIS SEMIGROUP

We introduce in this section the enveloping semigroup (defined by Ellis) of a transformation group and we investigate some algebraic properties of this semigroup and various relations between a transformation group and its enveloping semigroup.

1•4.1 To begin with we recall some definitions from the theory of semigroups (see, for example, Ljapin [*1*]).

Let S be a set. Suppose that for each two elements $a,b \in S$ a third element of S (called the *PRODUCT of a and b* and *denoted by ab*) is defined in such a way that the associativity axiom is satisfied:

$$a(bc) = (ab)c \qquad (a,b,c \in S).$$

Then S is said to be a *SEMIGROUP*. An element $e \in S$ is called the *IDENTITY of the semigroup S* if $ae = ea = a$ for each $a \in S$. A subset $P \subset S$ is called a *SUBSEMIGROUP* of the semigroup S if $PP \subset P$, *i.e.*, the product of each two elements of P belongs to P. A nonempty subset $P \subset S$ is called a *RIGHT IDEAL of the semigroup S* if $PS \subset P$. A right ideal $P \subset S$ is said to be a *MINIMAL RIGHT IDEAL* if it does not contain proper subsets which are right ideals of the semigroup.

We list now some well known facts (see Ljapin [*1*]). For the sake of completeness we shall provide them with proofs.

1·4.2 LEMMA: *Let I be a semigroup without proper right ideals and let* $\mathcal{J}(I)$ *be the set of all idempotents of the semigroup I. If* $\mathcal{J}(I) \neq \emptyset$, *then*

(1) $pI = I$ *for all* $p \in I$;

(2) $up = p$ $(p \in I, u \in \mathcal{J}(I))$;

(3) *If* $u \in \mathcal{J}(I)$, $p \in I$ *and* $pu = p$, *then* $p \in \mathcal{J}(I)$;

(4) *If* $u \in \mathcal{J}(I)$, *then* Iu *is a group with identity* u;

(5) *For each element* $p \in I$ *there is a unique element* $u \in \mathcal{J}(I)$ *such that* $pu = p$;

(6) $I = \bigcap [Iu \mid u \in \mathcal{J}(I)]$;

(7) *If* $u, v \in \mathcal{J}(I)$ *and* $u \neq v$, *then* $Iu \cap Iv = \emptyset$.

Proof: (1) Let $p \in I$. Then $pI \subset I$ and $(pI)I \subset pI$. Since I contains no proper right ideals, $pI = I$. (2) Let $u \in \mathcal{J}(I)$, $p \in I$. By (1), $uI = I$, and hence there exists an element $q \in I$ such that $up = q$. Then $up = u^2q = uq = p$. (3) If $u \in \mathcal{J}(I)$, $p \in I$ and $pu = p$, then $p^2 = pup = up = p$ by (2). (4) If $p, q \in Iu$, then $pq \in IuIu \subset Iu$. It is clear that u is a right identity for Iu and by (2) u is a left identity. Since $p \in Iu \subset I$, we have $pI = I$, hence there exists an element $t \in I$ such that $pt = u$. Then $r \equiv tu \in Iu$ and $pr = u$. Analogously, we get that $rs = u$ for an element $s \in Iu$. Then $p = pu = prs = us = s$. Consequently r is the inverse of p in Iu. (5) Let $v \in \mathcal{J}(I)$. If $p \in I$, then $pI = I$ by (1). Hence there is an element $q \in I$ such that $pq = v$. Denote $u = qp$. Then $pu = pqp = vp = p$ by (2). Moreover, $u^2 = qpqp = qvp = qp = u$. Thus, $u \in \mathcal{J}(I)$ and $pu = p$. If $w \in \mathcal{J}(I)$ and $pq = p$, then $w = uw = qpw = qp = u$. (6) If $p \in I$, then by (5) there exists an element $u \in \mathcal{J}(I)$ such that $pu = p$. Then $p \in Iu$. (7) If $u, v \in \mathcal{J}(I)$, $u \neq v$ and $pu = qv$ for some elements p and q of I, then $pv = puv$ by (2); hence $pv = qvv = qv$. Thus, $(pu)u = pu = qv = pv = (pu)v$. By (5) we get $u = v$, which leads to a contradiction. ▮

1·4.3 COROLLARY: *Let I be a semigroup without proper right ideals. If I contains exactly one idempotent u, then I is a group with identity u. If I contains an identity element, then I is a group.*

1·4.4 LEMMA: *If I is a minimal right ideal of the semigroup S, then I contains no proper right ideals of the semigroup I.*

Proof: Since I is a minimal right ideal of S, $IS \subset I$. If K is a right ideal of the semigroup I, then $K \subset I$ and $KI \subset K$; hence $(KI)S \subset KI$. Thus KI is a right ideal of X and, moreover, $KI \subset I$. This is possible only when $KI = I$. Therefore $K \subset I = KI \subset K$, *i.e.* $K \subset I$. ▮

1·4.5 LEMMA: *If I is a minimal right ideal of a semigroup S with* $\mathcal{J}(I) = \emptyset$ *and the elements* $p \in S$, $q \in I$, $r \in I$ *satisfy the equality* $pq = pr$, *then* $q = r$.

Proof: Denote $s = qp$; then $s \in IS \subset I$ and $sq = qpq = qpr = sr$. By statements (6) and (4) of Lemma 1·4.2 there exist elements $a \in I$ and $v \in \mathcal{J}(I)$ such that $as = v$. Then $q = vq = asq = asr = vr = r$ by Lemma 1·4.2(2).∎

Let u and v be two idempotents of a semigroup S. We say that u and v are ***EQUIVALENT IDEMPOTENTS***, writing $u \sim v$, if $uv = v$ and $vu = u$. If $u \sim v$, $v \sim w$, then $uw = uvw = uv = u$, $wu = wvu = wv = w$. Thus $u \sim w$. Hence the above relation is indeed an equivalence relation. By statement (2) of Lemma 1·4.2, if $u \sim v$ and there is a minimal right ideal of S containing both u and v, then $u = v$.

1·4.6 LEMMA: *Let I_1, I_2 be minimal right ideals of a semigroup S and let u_1 be an idempotent belonging to I_1. Then there exists a unique idempotent $u_2 \in I_2$ such that $u_1 \sim u_2$.*

Proof: The set $u_1 I_2$ is a right ideal of the semigroup S, since $u_1 I_2 S \subset u_1 I_2$. Further, $u_1 I_2 \subset I_1 S \subset I_1$ and since I_1 is a minimal right ideal, we get $u_1 I_2 = I_1$. Consequently there exists an element $u_2 \in I_2$ such that $u_1 u_2 = u_1$. By Lemma 1·4.5, the element u_2 is uniquely defined. Now $u_1 u_2 = u_1$ implies that $u_1 u_2^2 = u_1 u_2 = u_1$ and therefore $u_2 = u_2^2$. Analogously, there exists an element $v_1 \in I_2$ for which $u_2 v_1 = u_2$ and $v_1^2 = v_1$. Then $u_1 = u_1 u_2 = u_1 u_2 v_1 = u_1 v_1$. By Lemma 1·4.2(2), $u_1 v_1 = v_1$. Thus $u_1 = v_1$, and hence $u_2 u_1 = u_2$. This means that $u_1 \sim u_2$.∎

1·4.7 Throughout the remainder of this section (X,T,π) denotes a topological transformation group with the property that each orbit closure $\overline{xT}$ $(x \in X)$ is compact.

Let X^X denote the Cartesian product of X copies of the space X. To be more precise, X^X is the set of all maps $\xi : X \to X$ provided with a topology which is defined as follows. Let $U_1, \ldots, U_n$ be a finite family of non-empty open subsets of the space X and let $x_i \in X$ $(i = 1, \ldots, n)$. Let $S(U_1, \ldots, U_n, x_1, \ldots, x_n)$ denote the set of all elements $\xi \in X^X$ with $x_i \xi \in U_i$ $(i = 1, \ldots, n)$. The family of all sets of the form $S(U_1, \ldots, U_n, x_1, \ldots, x_n)$ is a basis of some topology on X^X. The set X^X provided with this topology is said to be the Cartesian product of X copies of the space X.

Let $\{\xi_i | i \in I\}$ be a net (Kelley [1]) of elements $\xi_i \in X^X$. It follows directly from the above definition that the net $\{\xi_i | i \in I\}$ converges to some element $\eta \in X^X$ iff $\{x\xi_i | i \in I\} \to x\eta$ for each point $x \in X$. Therefore X^X is sometimes called the *SPACE OF ALL MAPS* ξ: $X \to X$ *WITH the TOPOLOGY OF POINTWISE CONVERGENCE*.

The set X can be provided with a semigroup structure in the following way. Let $\xi, \eta \in X$. By $\xi \circ \eta$ we denote the composition of the maps ξ and η, that is, $x(\xi \circ \eta) = (x\xi)\eta$ $(x \in X)$. Clearly, this operation is associative. The identity map is the identity element of the semigroup X. An element $\xi \in X^X$ is invertible iff ξ: $X \to X$ is bijective. In what follows we shall write $\xi\eta$ instead of $\xi \circ \eta$.

Let ξ be a fixed element of X^X. Let us show that left multiplication by ξ is a continuous mapping of X^X into X^X. Let $\{\eta_i | i \in I\}$ be a net of elements $\eta_i \in X^X$, which converges to an element

$\eta \in X^X$, *i.e.*, $\{x\eta_i | i \in I\} \to x\eta$ for all $x \in X$. Then $\{x\xi\eta_i | i \in I\} \to x\xi\eta$ for all $x \in X$, since $x\xi \in X$. But this just means that $\{\xi\eta_i | i \in I\} \to \xi\eta$.

Now let ξ be a continuous map of X into X. Let us show that right multiplication by ξ is a continuous map of X^X into X^X. Suppose that a net $\{\eta_i | i \in I\}$ converges to an element $\eta \in X^X$, *i.e.*, $\{x\eta_i | i \in I\} \to x\eta$ for all $x \in X$. Since $\xi : X \to X$ is continuous, it follows that $\{x\eta_i\xi | i \in I\} \to x\eta\xi$, *i.e.*, $\{\eta_i\xi | i \in I\} \to \eta\xi$. It should be noted that, in general, right multiplication is not continuous and thus, *a fortiori*, multiplication is not jointly continuous.

Denote $\{\pi^t : X \to X | t \in T\}$ by G; then $G \subset X^X$. By $E = E(X,T,\pi)$ we denote the closure of the set G in X^X. Since $\overline{xT}$ is compact for each $x \in X$, it follows from Tikhonov's Theorem (see Bourbaki [2]) that the set E is compact. Let us show that E is a subsemigroup of X^X. Let $\eta \in G$. Then $E\eta = \bar{G}\eta = \overline{G\eta}$, since right multiplication by a continuous map $\eta \in G$ is continuous. But $G\eta = G$, and hence $E\eta = \bar{G} = E$. Thus, $EG = E$. If $\xi \in E$, then $\xi E = \xi\bar{G} \subset \overline{\xi G}$, since left multiplication by ξ is continuous. From $EG = E$ it follows that $\overline{\xi G} \subset \bar{E}$. But $\bar{E} = E$, since E is closed in X^X; therefore $\xi E \subset E$, and hence $EE \subset E$. Thus E is a compact subsemigroup of the semigroup X^X. The semigroup $E = E(X,T,\pi)$ is called the *ENVELOPING SEMIGROUP OF THE TOPOLOGICAL TRANSFORMATION GROUP* (X,T,π), or briefly, the *ELLIS SEMIGROUP*.

Define a map $\bar{\pi} : E \times T \to E$ by $(\xi,t)\bar{\pi} = \xi\circ\pi^t \equiv \xi\pi^t$ $(\xi \in E, t \in T)$. Then $(E,T,\bar{\pi})$ is a topological transformation group, because it is a restriction of the Cartesian product of X copies of the topological group (X,T,π) to the invariant subspace $E \subset X^X$. Since we have agreed to write $\xi\bar{\pi}^t$ instead of $(\xi,t)\bar{\pi}$, we get $\xi\bar{\pi}^t = \xi\pi^t$, hence we may remove the bar over ξ without any danger of ambiguity.

For $x \in X$ we define a map $\varphi : E \to X$ by $\xi\varphi = x\xi$ $(\xi \in E)$. It follows from the definition of the topology in $E \subset X^X$ that φ_x is continuous. Since $\xi\pi\ \varphi_x = \xi\bar{\pi}^t\varphi_x = x\xi\pi^t = \xi\varphi_x\pi^t$, φ_x is a homorphism of (E,T,π) onto $(\overline{xT},T,\pi)$.

Let I be a minimal right ideal of the semigroup E. Since $IT \subset IE \subset I$, I is an invariant subset under (E,T,π), and hence we get a transformation group (I,T,π).

1·4.8 LEMMA: *A set $I \subset E$ is a minimal right ideal of the semigroup E iff I is a minimal set of (I,T). Therefore each minimal right ideal is closed.*

Proof: Let I be a minimal right ideal and let $p \in I$. Then $pE \subset IE \subset I$, but $(pE)E \subset pE$ implies that pE is a right ideal. Hence $I = pE$. Since left multiplication is continuous in E, the set I is closed. Thus, each minimal right ideal is closed. Conversely, let I be a minimal subset of E. Then I is closed and $IG \subset I$; therefore $IE \subset E$, *i.e.*, I is a right ideal. Let K be a non-empty right ideal contained in I and take $p \in K$. Then $pE \subset KE \subset K \subset I$. The set pE is closed and $pEG = pE$. Since I is a minimal set, $pE = I$ and, consequently, $K = I$. ∎

1·4.9 COROLLARY: *Every closed right ideal of the semigroup E contains at least one minimal right ideal.*

Proof: This follows from Theorem 1·1.7 and the preceding lemma, since each ideal of E is an invariant set of (E,T,π). ∎

1·4.10 LEMMA: *Let I be a minimal right ideal of the semigroup E and let $x \in X$. Then xI is a minimal subset of X. Conversely, if M is a minimal set and $x \in M$, then $M = xI$ for each minimal right ideal I of E.*

Proof: The set xI is the image of the compact minimal set I under the homomorphism $\varphi_X:(E,T) \to (X,T)$, and hence xI is minimal. Let M be a minimal set, let $x \in M$ and let I be any minimal right ideal of E. Then xI is a minimal set. Since $xI \subset xE = \overline{xT} = M$ and M is minimal, we get $M = xI$. ∎

1·4.11 LEMMA: *Let S be a compact space provided with a semigroup structure such that the left multiplication by each element is continuous. Then S contains an idempotent r.*

Proof: Let S^* denote the set of all non-empty compact subsets $A \subset S$ such that $AA \subset A$. Then $S^* \neq \emptyset$, as $S \in S^*$. When partially ordered by inclusion, S^* is inductive. Hence, by Zorn's Lemma, there exists a minimal element $M \in S^*$. If $R \in M$, then rM is a non-empty compact set with $(rM)(rM) \subset rMM \subset rM$. Hence, $rM \in S$ and $rM = M$. Since M is a minimal element, we get $rM = M$. Therefore $r = rp$ for some element $p \in M$. Denote $\{a|a \in M, ra = r\}$ by L; then $p \in L$. Since left multiplication is continuous in S, the set L is closed and therefore compact. If $k \in L$, $\ell \in L$, then $(rk)\ell = r\ell = r$. Hence $LL \subset L$. Thus $L \in S^*$ and $L \subset M$ and hence $L = M$. Thus $r \in L$, which means that $r^2 = r$. ∎

1·4.12 COROLLARY: *Each minimal right ideal of I contains at least one idempotent.*

1·4.13 COROLLARY: *If a minimal right ideal I contains no more than one idempotent or contains the identity element of E, then I is a group.*

1·4.14 THEOREM: *The following statements are equivalent:*

(1) *$(x,y) \in P$, where P denotes the proximal relation in X;*

(2) *There exists an element $p \in E$, for which $xp = yp$;*

(3) *There exists a minimal right ideal I of the semigroup E such that $xp = yp$ for all $p \in I$.*

Proof: Let $(x,y) \in P$. For each index $\alpha \in \mathfrak{U}[X]$ there exists an element $t_\alpha \in T$, such that $(xt_\alpha, yt_\alpha) \in \alpha$. The set $\mathfrak{U}[X]$ is directed, when partially ordered by inclusion. Since the space E is compact, the net $\{\pi^{t_\alpha}|\alpha \in \mathfrak{U}[X]\}$ has a limit point $p \in E$. Then $xp = yp$.

If $xp = yp$ for some element $p \in E$ and K is a minimal right ideal of E, then $I \equiv pK$ is a minimal right ideal and $xr = yr$ for all elements $r \in I$. Conversely (3) implies (2).

Suppose that $\alpha \in \mathfrak{U}$ and $xp = yp$ for some element $p \in E$. Choose an index $\beta \in \mathfrak{U}[X]$ so that $\beta = \beta^{-1}$ and $\beta^2 \subset \alpha$. Since $p \in E = \bar{G}$, there exists an element $t \in T$ for which $(x\pi^t, xp) \in \beta$ and $(y\pi^t, yp)$

$\in \beta$. Then $(x\pi^t, y\pi^t) \in \beta^2 \subset \alpha$. Therefore $(x,y) \in P$. ▮

1·4.15 THEOREM: *The following statements are equivalent:*

(1) *P is an equivalence relation;*

(2) *E contains exactly one minimal right ideal.*

Proof: Suppose that P is an equivalence relation. It has already been pointed out above that E contains at least one minimal right ideal. We shall show that it is unique. Suppose that, to the contrary, I_1, I_2 are two distinct minimal right ideals. For each idempotent $u_1 \in I_1$, there exists, by Lemma 1·4.6, an idempotent $u_2 \in I_2$ such that $u_1 \sim u_2$. Let $x \in X$, $y = xu_1$, $z = xu_2$. Then $yu_1 = xu_1^2 = xu_1$, $zu_2 = xu_2^2 = xu_2$. Therefore $(x,y) \in P$ and $(x,z) \in P$. Since P is an equivalence relation, $(y,z) \in P$. By Theorem 1·4.14, there is a minimal right ideal I_3 such that $yr = zy$ for all $r \in I_3$. By Lemma 1·4.6, we can choose an idempotent $u_3 \in I_3$ in such a way that $u_1 \sim u_2 \sim u_3$. Then $yu_3 = zu_3$, $y = xu_1 = xu_1u_3 = yu_3 = zu_3 = xu_2u_3 = xu_2 = z$, *i.e.*, $xu_1 = xu_2$. Since x is an arbitrary point of X, this means that $u_1 = u_2$. Thus $I_1 \cap I_2 \neq \emptyset$ and since I_1 and I_2 are minimal right ideals, this can happen only when $I_1 = I_2$, which leads to a contradiction. Hence (1) implies (2).

Conversely, let (2) be satisfied and let $(x,y) \in P$, $(y,z) \in P$. We shall show that $(x,z) \in P$. By Theorem 1·4.14, there exist minimal right ideals I_1 and I_2 such that $xr = yr$ $(z \in I_1)$ and $yp = zp$ $(p \in I_2)$. But $I_1 = I_2$ by (2), and hence $xr = yr = zr$ $(r \in I_1)$, whence $(x,z) \in P$. Thus, the relation $P(X,T,\pi)$ is transitive. Consequently (1) holds. ▮

1·4.16 LEMMA: *Let $x \in X$ and let M be a minimal set contained in xT. Then $M = xI$ for some minimal right ideal I of E and there exists a point $y \in M$ such that x and y are proximal.*

Proof: Denote $\{p | p \in E, xp \in M\}$ by F. If $p \in F$, $q \in E$, then $xpq \in Mq \subset M$, and hence $pq \in F$. Thus $FE \subset F$. Since M is closed in X, the set F is closed in E and therefore F is a closed right ideal. By Corollary 1·4.9, F contains a minimal right ideal I. Then $xI \subset xF \subset M$ and $xI = M$, since M and xI are minimal sets. Let u be an idempotent belonging to I. Then $y \equiv xu \in xI = M$ and $yu = xu^2 = xu$. Hence $(x,y) \in P$ by Theorem 1·4.14.

Let $\{M_\alpha | \alpha \in A\}$ denote the class of all minimal subsets of X. Define $N_\alpha = \{x | x \in X, \overline{xT} \supset M_\alpha\}$.

1·4.17 THEOREM: *If $P = P(X,T,\pi)$ is an equivalence relation, then*

(1) *$\{N_\alpha | \alpha \in A\}$ is a partition of X, each set N_α is invariant and M_α is the only minimal set contained in N_α;*

(2) *If $x \in N_\alpha$, then there is a point $y \in M_\alpha$ such that $(x,y) \in P$;*

(3) *If P is closed in $X \times X$, then all the sets M_α are closed.*

Proof: By Theorem 1·4.15, the enveloping semigroup E contains exactly one minimal right ideal. Let $x \in X$. From the preceding Lemma it follows that $\overline{xT}$ contains no more than one minimal set. Since $\overline{xT}$ is compact, we conclude by Theorem 1·1.7 that $\overline{xT}$ contains exactly one minimal set. Therefore (1) holds. (2) If $x \in N_\alpha$, then $M_\alpha \subset \overline{xT}$ and our statement follows from Lemma 1·4.16. (3) Suppose that the relation P is closed, $\{x_i | i \in I\}$ is a net of points $x_i \in N_\alpha$ and $\{x_i | i \in I\} \to x$. There exist points $y_i \in M_\alpha$ $(i \in I)$ such that $(x_i, y_i) \in P$. Since M_α is compact, we may assume without loss of generality that $\{y_i | i \in I\} \to y \in M_\alpha$. Then $(x,y) \in P$, since the relation P is closed. By Theorem 1·4.14, there exists an element $p \in E$ such that $xp = yp$. Since M_α is a minimal set, we get $xp = yp \in M_\alpha$, and hence $M_\alpha \subset xE = \overline{xT}$. This means that $x \in N_\alpha$. Therefore N_α is closed.▌

1·4.18 LEMMA: *If $P(X) \equiv P(X,T,\pi)$ is an equivalence relation on X, then $P(X \times X) \equiv P(X \times X, T, \hat{\pi})$ is an equivalence relation on $X \times X$.*

Proof: Since $P(X)$ is an equivalence relation, the semigroup $E = E(X,T,\pi)$ contains, by Theorem 1·4.15, exactly one minimal right ideal I. Suppose that:

$$((x_1,x_2),(x_3,x_4)) \in P(X \times X), \quad ((x_3,x_4),(x_5,x_6)) \in P(X \times X). \quad (4.1)$$

We shall prove that:

$$((x_1,x_2),(x_5,x_6)) \in P(X \times X). \quad (4.2)$$

It follows from (4.1) that $(x_1,x_3) \in P(X)$, $(x_2,x_4) \in P(X)$, $(x_3,x_5) \in P(X)$ and $(x_4,x_6) \in P(X)$. By Theorem 1·4.14, this implies $x_1p = x_3p = x_5p$ and $x_2p = x_4p = x_6p$ for all $p \in I$. Consequently (4.2) holds.▌

1·4.19 THEOREM: *The following statements are mutually equivalent:*

(1) *P is an equivalence relation on X;*

(2) *The orbit closure of each point in $(X \times X, T, \hat{\pi})$ contains exactly on minimal set;*

(3) $P(X,T,\pi) = L(X,T,\pi)$.

Proof: Suppose (1) holds. By the preceding lemma, $P(X \times X)$is an equivalence relation on $X \times X$, and Theorem 1·4.17 implies (2).

Let (1) be satisfied and let $(x,y) \in P(X)$. We shall show that $(x,y) \in L(X)$. By Theorem 1·3.9(2), it suffices to show that $\overline{(x,y)T}$ $P(X)$. Suppose to the contrary that there exists an $(a,b) \in \overline{(x,y)T} \setminus P(X)$. Let M be a minimal set contained in $\overline{(a,b)T}$. Then there are two possibilities:

(a) $M \cap P(X) = \emptyset$. Since $M \subset \overline{(a,b)T} \subset \overline{(x,y)T}$, by Lemma 1·4.16 there is a point $(u,v) \in M$ such that $((x,y),(u,v)) \in P(X \times X)$. Then it follows that $(x,u) \in P(X)$, $(y,v) \in P(X)$ and since $P(X)$ is assumed to be an equivalence relation and $(x,y) \in P(X)$, we get $(u,v) \in P(X)$.

But this leads to a contradiction, since $(u,v) \in M$.

(b) $M \cap P(X) \neq \emptyset$. Let $(x',y') \in X \times X$; then $\overline{(x',y')T}$ contains exactly one minimal set N, since (1) implies (2).

If $(x',y') \in P(X)$, then $\overline{(x',y')T} \cap \Delta \neq \emptyset$, where $\Delta = \{(x,x) \mid x \in X\}$. Since $\overline{(x',y')T} \cap \Delta$ is closed and invariant, it follows that $N \subset \overline{(x',y')T} \cap \Delta$, and hence $N \subset \Delta$. If, in particular, $(x',y') \in P(X) \cap M$, then $M \subset \Delta$. Consequently $(a,b)T \cap \Delta \neq \emptyset$. Therefore $(a,b) \in P(X)$, but this contradicts the choice of (a,b).

Thus $\overline{(x,y)T} \subset P(X)$, and therefore $(x,y) \in L(X)$, *i.e.*, (3) holds. By Theorem 1·3.6, (3) implies (1).

It remains to prove that (3) follows from (2). Assume (2) to be fulfilled, and let $(x,y) \in P$ and $q \in E$. By Theorem 1·3.9(2) it suffices to show that $(xq,yq) \in P$, which is equivalent to $\overline{(xq,yq)T} \cap \Delta \neq \emptyset$. Let M be a minimal set contained in $\overline{(xq,yq)T}$. Since $(xq,yq) \in \overline{(x,y)T}$, we have $\overline{(xq,yq)T} \subset \overline{(x,y)T}$, and hence $M \subset \overline{(x,y)T}$. From $(x,y) \in P$ it follows that $\overline{(x,y)T} \cap \Delta \neq \emptyset$. Since $\overline{(x,y)T} \cap \Delta$ is a closed non-empty invariant set, it contains a minimal set M'. But $\overline{(x,y)T}$ contains exactly one minimal set, hence $M = M'$, and therefore $M \subset \overline{(xq,yq)T} \cap \Delta$. Thus $\overline{(xq,yq)T} \cap \Delta \neq \emptyset$ and $(xq,yq) \in P$. ∎

1·4.20 Let (X,T,π) and (Y,T,ρ) be two transformation groups with the property that each orbit closure in both X and Y is compact. Set $G(X) = \{\pi^t \mid t \in T\}$, $G(Y) = \{\rho^t \mid t \in T\}$, $E(X) = E(X,T,\pi)$, $E(Y) = E(Y,T,\rho)$.

1·4.21 THEOREM: *Let $\varphi: X \to Y$ be a homomorphism of (X,T,π) onto (Y,T,ρ). Then*

(1) *There exists a unique mapping ψ of the set $G(X)$ into $G(Y)$ such that $\pi^t\psi = \rho^t$ $(t \in T)$ and moreover ψ is a homomorphism of the group $G(X)$ onto $G(Y)$ with $p \circ \varphi = \varphi \circ (p\psi)$, $(p \in G(X))$;*

(2) *There exists a uniquely defined mapping $\vartheta: E(X) \to E(Y)$ such that ϑ is the extension of ψ by continuity and moreover ϑ is a uniformly continuous homomorphism of the semigroup $E(X)$ onto $E(Y)$ and $p \circ \varphi = \varphi \circ (p\vartheta)$, $(p \in E(X))$;*

(3) *ϑ is a homomorphism of $(E(X),T,\bar{\pi})$ onto $(E(Y),T,\bar{\rho})$;*

(4) *If I is a right ideal of $E(X)$, then $I\vartheta$ is a right ideal of $E(Y)$;*

(5) *If K is a right ideal of $E(Y)$, then $K\vartheta^{-1} = \{p \mid p \in E(X), p\vartheta \in K\}$ is a right ideal of $E(X)$;*

(6) *If I is a minimal right ideal of the semigroup $E(X)$, then $I\vartheta$ is a minimal right ideal of $E(Y)$;*

(7) *If (Y,T,ρ) coincides with (X,T,π), then ϑ is the identity map;*

(8) *If K is a minimal right ideal of the semigroup $E(Y)$, then there exists a minimal right ideal I of $E(X)$ such that $I\vartheta = K$;*

(9) *If K is a minimal right ideal of $E(Y)$ and $v \in K$ is an idempotent, then there exists an idempotent $u \in E(X)$ belonging to some minimal right ideal of the semigroup $E(X)$ and such that $u\vartheta = v$.*

Proof: (1) Suppose $t, s \in T$ are such that $\pi^t = \pi^s$. We shall prove that $\rho^t = \rho^s$. If $x \in X$, then $x\varphi\rho^t = x\pi^t\varphi = x\pi^s\varphi = x\varphi\rho^s$. Since $X\varphi = Y$, it follows that $y\rho^t = y\rho^s$ $(y \in Y)$, *i.e.*, $\rho^t = \rho^s$. Thus the existence and the uniqueness of the required map are proved. Clearly, $G(X)\psi = G(Y)$. If $t, s \in T$, then $(\pi^t o \pi^s)\psi = \pi^{ts}\psi = \rho^{ts} = \rho^t o \rho^s = (\pi^t\psi)o(\pi^s\psi)$. Hence ψ is a homomorphism. If $t \in T$, then $\pi^t o\varphi = \varphi o \pi^t\psi$, hence $po\varphi = \varphi o(p\psi)$ for all $p \in G(X)$.

(2) Let $p \in E(X)$ and let $\{\pi^{t_\alpha}\}$, $\{\pi^{t_\alpha}\}$ be two nets which converge in $E(X)$ to p. Given $y \in Y$, we choose a point $x \in X$ so that $y = x\varphi$. Then $\{y\rho^{t_\alpha}\} \to (xp)\varphi$ and $\{y\rho^{s_\alpha}\} \to (xp)\varphi$. This allows us to define a map $\vartheta: E(X) \to E(Y)$ by $p\vartheta = \lim\{\rho^{t_\alpha}\}$. We note that $y(p\vartheta) = \lim\{y\rho^{t_\alpha}\} = (xp)\varphi$; hence $po\varphi = \varphi o(p\vartheta)$ $(p \in E(X))$. We shall now prove that $\psi: E(X) \to E(Y)$ is continuous. Let $\{p_\alpha\} \to p$ in $E(X)$; then $\lim\{y(p_\alpha\vartheta)\} = \lim\{(xp_\alpha)\varphi\} = (xp)\varphi = y(p\vartheta)$, that is, $\{p_\alpha\vartheta\} \to p\vartheta$. Since $E(X)$ is compact, the mapping ϑ is uniformly continuous.

Let us prove that $\vartheta: E(X) \to E(Y)$ is a semigroup homomorphism. Let $p \in E(X)$ and $q \in G(X)$. We shall show that $(poq)\vartheta = (p\vartheta)o(q\vartheta)$. Choose a net $\{p_i | i \in I\}$ $(p_i \in G(X))$ such that $\{p_i | i \in I\} \to p$. Then $\{p_i oq | i \in I\} \to poq$ by the continuity of $q: X \to X$. Since ϑ is continuous, we get $\{(p_i oq)\vartheta | i \in I\} \to (poq)\vartheta$. On the other hand, $\{(p_i oq)\psi\} = \{(p_i\psi)o(q\psi)\} = \{(p_i\vartheta)o(q\vartheta)\} \to p\vartheta o q\vartheta$. Thus $(poq)\vartheta = p\vartheta o q\vartheta$ $(p \in E(X),\ q \in G(X))$. Now let $p \in E(X)$ and $q \in E(X)$. We shall show that $(poq)\vartheta = p\vartheta o q\vartheta$. There exists a net $\{q_i | i \in I\}$, $q_i \in G(X)$, such that $\{q_i | i \in I\} \to q$. Then $\{poq_i\} \to poq$ and $\{(poq_i)\vartheta\} \to (poq)\vartheta$. On the other hand, $(poq_i)\vartheta = p\vartheta o q_i\vartheta$ by the above. But $\{q_i | i \in I\} \to q$ implies $\{qo\vartheta\} \to q\vartheta$, hence $\{p\vartheta o q_i\vartheta\} \to p\vartheta o q\vartheta$. Thus $(poq)\vartheta = p\vartheta o q\vartheta$ $(p, q \in E(X))$. This completes the proof of (2).

(3) As was stated above, the map ϑ is continuous. Moreover, $(po\pi^t)\vartheta = p\vartheta o \pi^t\vartheta = p\vartheta o \rho^t$ $(p \in E(X),\ t \in T)$.

(4) Let $q \in I\vartheta$ and $r \in E(Y)$. We shall prove that $qr \in I\vartheta$. There is an element $p \in I$ such that $q = p\vartheta$. Since $E(X)\vartheta = E(Y)$, an element $s \in E(X)$ can be chosen with $s\vartheta = r$. Then $ps \in I$, since I is a right ideal, and $qr = p\vartheta o s\vartheta = (ps)\vartheta \in I\vartheta$.

(5) Let $p \in K\vartheta^{-1}$ and $s \in E(X)$. Then $p\vartheta \in K$ and $(ps)\vartheta = p\vartheta\ s\vartheta \in K(s\vartheta) \subset K$; hence $ps \in K\vartheta^{-1}$.

(6) Let K be a right ideal of $E(Y)$ and assume that $K \subset I\vartheta$. Then $K\vartheta^{-1}$ is a right ideal in $E(X)$, by (5). Since $K\vartheta^{-1} \cap I \neq \emptyset$, the set I is contained in $K\vartheta^{-1}$. Thus $I\vartheta \subset K$, and hence $I\vartheta = K$.

(7) Since $\pi^t\vartheta = \pi^t\psi = \rho^t = \pi^t$ and $G(X)$ is dense in $E(X)$, ϑ is the identity map.

(8) By (5), $K\vartheta^{-1}$ is a right ideal of $E(X)$. Let I be a minimal right ideal of $E(X)$ contained in $K\vartheta^{-1}$. Then $I\vartheta \subset K$ and $I\vartheta = K$, since K is a minimal right ideal.

(9) By (8), there exists a minimal right ideal I of $E(X)$ such

that $I\vartheta = K$. The set $S = \{\xi \mid \xi \in I, \xi\vartheta = v\}$ is a closed subsemigroup of $E(X)$. By Lemma 1·4.11, there is an idempotent $u \in S$, which is just the required one. ▮

1·4.22 THEOREM: *Let* φ *be a homomorphism of* (X,T,π) *onto* (Y,T,ρ). *If* $P(X) = L(X)$, *then* $P(Y) = L(Y)$.

Proof: By Theorems 1·3.6 and 1·4.15 it suffices to show that $E(Y)$ contains only one minimal right ideal. If I_1, I_2 are minimal right ideals in $E(Y)$, then $I_1\vartheta^{-1}$ and $I_2\vartheta^{-1}$ are right ideals of $E(X)$. Since $E(X)$ contains exactly one minimal right ideal, say I, by Corollary 1·4.9 we obtain $I \subset I_1\vartheta^{-1} \cap I_2\vartheta^{-1}$. Thus $I_1 \cap I_2 \neq \emptyset$, and hence $I_1 = I_2$. ▮

REMARKS AND BIBLIOGRAPHICAL NOTES

The results in 1·4.7-15 are due to Ellis [*5*]. Theorem 1·4.17 was proved by J. Auslander [*2*]. Theorem 1·4.19 follows from results obtained by J. Auslander [*2*] and Wu [*1*]. Theorem 1·4.21 is contained in the paper by Ellis and Gottschalk [*1*].

§(1·5): POINTWISE ALMOST PERIODIC TRANSFORMATION GROUPS

In this section we shall consider relationships between compact minimal sets, almost periodic points and minimal right ideals of the enveloping semigroup. We shall investigate the proximal relation in a transformation group with the property that all points of the phase space are almost periodic.

1·5.1 Let (X,T,π) be a topological transformation group. As in 1·2.3, a point $x \in X$ is called *ALMOST PERIODIC UNDER* (X,T,π), if for each neighbourhood U of x there exists a syndetic subset $A \subset T$ such that $xA \subset U$.

1·5.2 LEMMA: *Let* $x \in X$. *The following statements are mutually equivalent:*

(1) *The point* x *is almost periodic;*

(2) *For each neighbourhood* U *of* x *there is a compact subset* $K \subset T$ *such that* $xtK \cap U \neq \emptyset$ *for all* $t \in T$;

(3) *Given a neighbourhood* U *of the point* x, *we can choose a compact subset* $K \subset T$ *in such a way that* $xT \subset UK$.

The proof is left to the reader.

1·5.3 THEOREM: *Let* X *be a regular space and let* $x \in X$ *be almost periodic under* (X,T,π). *Then* $\overline{xT}$ *is a minimal set.*

Proof: Suppose that the set $\overline{xT}$ is not minimal. Then there exists a point $y \in \overline{xT}$ such that $x \notin \overline{yT}$. Since the space X is regular, there exist a neighbourhood U of x and a neighbourhood V of the set $\overline{yT}$ such that $U \cap V = \emptyset$. By Lemma 1·5.2, there is a compact set $K \subset T$ with $xT \subset UK$. Since $yK^{-1} \subset V$ and the set K^{-1} is compact, Lemma 1·1.3(3) implies that there is a neighbourhood W of y such that $WK^{-1} \subset V$; hence $WK^{-1} \cap U = \emptyset$ and $W \cap UK = \emptyset$. But $y \in \overline{xT}$, whence $xT \cap W \neq \emptyset$. Moreover, $UK \cap W \neq \emptyset$, since $xT \subset UK$. This is a contradiction. ▮

1·5.4 THEOREM: *Let M be a compact minimal set. Then each point $x \in M$ is almost periodic.*

Proof: Let $x \in M$ and let U be a neighbourhood of x. For each point $y \in M$, the equality $\overline{yT} = M$ holds, and hence $yT \cap U \neq \emptyset$. It follows that $M \subset UT$. Since M is compact, there exists a finite set $K \subset T$ with $M \subset UK$, and consequently $xT \subset UK$. By Lemma 1·5.2, this means that x is almost periodic. ▮

1·5.5 THEOREM: *Let $x \in X$ be almost periodic and suppose that there exists a neighbourhood U of x such that its closure $\bar{U}$ is compact. Then the set $\overline{xT}$ is compact and the group T is pointwise almost periodic on $\overline{xT}$.*

Proof: Let U be a neighbourhood of x with $\bar{U}$ compact. Since x is almost periodic, there exists a compact set $K \subset T$ such that $xT \subset UK$. Then $\overline{xT} \subset \overline{UK} = \bar{U}K$, and hence $\overline{xT}$ is a compact set. Since a compact Hausdorff space is regular, to conclude the proof it suffices now to refer to the Theorems 1·5.3,4. ▮

1·5.6 THEOREM: *Let X be a locally compact topological space. Then the following statements are equivalent to each other:*

(1) *The transformation group (X,T,π) is pointwise almost periodic;*

(2) *For each point $x \in X$ the set $\overline{xT}$ is compact and minimal.*

Proof: This follows from Theorems 1·5.4,5. ▮

1·5.7 LEMMA: *Let X be a compact space and take $x \in X$. Then the following assertions are equivalent to each other:*

(1) *$\overline{xT}$ is a minimal set of (X,T,π);*

(2) *For each element $p \in E = E(X,T,\pi)$ there is an element $q \in E$ such that $xpq = x$.*

Proof: Suppose that (1) holds and $p \in E$. Then $xp \in xE = \overline{xT}$. Since $\overline{xT}$ is a minimal set, we have $\overline{xpT} = \overline{xT}$. But $\overline{xpT} = xpE$; hence $x \in xpE$, and therefore $x = xpq$ for some element $q \in E$. Thus (1) implies (2). Conversely, let (2) be true and let $y \in \overline{xT}$. We shall prove that $\overline{xT}$ is a minimal set. If suffices to show that $x \in \overline{yT}$. Since $y \in \overline{xT} = xE$, we get $y = xp$ for some element $p \in E$. By (2), there exists an element $q \in E$ such that $yq = xpq = x$. This means that $x \in yE = \overline{yT}$. ▮

Let $(X \times X,T,\hat{\pi})$ be the Cartesian product of two copies of $(X,T,\hat{\pi})$. The map $\varphi: X \times X \to X$ defined by $(x,y)\varphi = x$ $((x,y) \in X \times X)$ is a homomorphism of $(X \times X,T,\hat{\pi})$ onto (X,T,π). Let ϑ denote the semigroup homomorphism of $E(X \times X, T, \pi)$ onto $E(X, T, \pi)$ induced by φ.

1·5.8 LEMMA: *The homomorphism* $\vartheta: E(X \times X, T, \pi) \to E(X, T, \pi)$ *is in fact an isomorphism.*

Proof: This follows immediately from the definition of the enveloping semigroup. If $p \in E(X \times X, T)$, then there is an element $q \in E(X, T)$ such that $(x,y)p = (xq, yq)$ $((x,y) \in X \times X)$.

Conversely, if $q \in E(X,T)$, then the map p defined as above is an element of $E(X \times X, T)$. ▮

In what follows we shall identify the semigroups $E(X,T)$ and $E(X \times X, T)$.

We recall that two points x_1 and x_2 of X are called distal if $(x_1, x_2) \notin P(X,T,\pi)$. Let M be any invariant subset of X. We say that x is distal in M if $(x,y) \notin P$ for every point $y \in M$, $y \neq x$.

1·5.9 THEOREM: *Let X be a compact space, let $x \in X$ and let (X,T,π) be a pointwise almost periodic transformation group. The following statements are mutually equivalent:*

(1) *The point x is distal in $\overline{xT}$;*

(2) *The point x is distal in X;*

(3) *For each point $y \in X$ the orbit-closure of $(x,y) \in X \times X$ in $(X \times X, T, \hat{\pi})$ is a minimal set.*

Proof: We first show that (1) implies (3). Let $y \in X$ and let $p \in E(X,T,\pi)$. By Lemmas 1·5.7,8, it suffices to show that there exists an element $r \in E(X,T,\pi)$ with $xpr = x$, $ypr = y$. Put $F = \{q \mid q \in E(X,T,\pi), ypr = y\}$. Since $\overline{yT}$ is a minimal set, $F \neq \emptyset$ by Lemma 1·5.7. Obviously F is closed in $E(X,T,\pi) \subset X^X$ and therefore compact. Since the left multiplication by p is continuous in the semigroup $E(X,T,\pi)$, the set pF is also compact. Let us show that pF is a subsemigroup of E. In fact, if $upq_1 = y$ and $ypq_2 = y$, then $ypq_1pq_2 = y$, *i.e.*, $pFpF \subset pF$. By Lemma 1·4.11, the semigroup pF contains an idempotent pr. Thus it follows that $(xpr, x) \in P$. But $xpr \in \overline{xT}$ and hence $xpr = x$, by (1). By the definition of F, $r \in F$ implies $ypr = y$. Thus (1) implies (3).

Let us show now that (3) implies (2). Let $y \in X$ and $y \neq x$. We shall prove that $(x,y) \notin P$. Suppose that to the contrary $xp = yp$ for some element $p \in E(X,T,\pi)$. By (3) the orbit closure of $(x,y) \in X \times X$ is a minimal set. Hence there is an element $q \in E(X,T,\pi)$ such that $(xpq, ypq) = (x,y)$ (this follows from Lemmas 1·5.7,8). Then $y = ypq = xpq = x$, which contradicts $x \neq y$.

Since (2) obviously implies (1), the proof is complete. ▮

By xP we *denote the set of all points* $y \in X$ *such that* $(x,y) \in P$. For a minimal right ideal I of the semigroup $E(X,T,\pi)$ let $\mathfrak{J}(I)$ be the *set of all idempotents belonging to* I. We shall *denote by* $\mathfrak{J}$ *the union of the sets* $\mathfrak{J}(I)$ *over all minimal right ideals* I *of* E.

1·5.10 LEMMA: *Let X be a compact space, let $\overline{xT}$ $(x \in X)$ be a minimal set of (X,T,π) and let I be a minimal right ideal of the semigroup E. Then there is an idempotent $u \in \mathfrak{J}(I)$ such that $x = xu$.*

Proof: By Lemma 1·4.10, $xt = xI$. Denote $\{r | r \in I, xr = r\}$ by F. It is easy to verify that F is non-empty, closed and that $FF \subset F$. By Lemma 1·4.11, F contains an idempotent u. Then $u \in \mathcal{J}(I)$ and $xu = x$. ▮

1·5.11 THEOREM: *Let X be a compact space and let $x \in X$. If (X,T,π) is a pointwise almost periodic transformation group, then:*

(1) $xP = x\mathcal{J}$;

(2) *If $u,v \in \mathcal{J}(I)$, then $(xu,xv) \in P$ and, conversely, if $(y,z) \in P$ for each point $y \in x\mathcal{J}(I)$, then $z \in x\mathcal{J}(I)$;*

(3) *If $q \in I$ and $(x,xq) \in P$ for all $x \in X$, then $q \in \mathcal{J}(I)$.*

Proof: (1) If u is idempotent, then $(xu)u = xu^2 = xu$, and therefore $(xu,x) \in P$. Conversely, let $y \in xP$. By Theorem 1·4.14, there is a minimal right ideal $I \subset E$ such that $xr = yr$ for all $r \in I$. By the preceding lemma, $yu = y$ for some idempotent $u \in \mathcal{J}(I)$. Then $xu = yu = y$, and hence $y \in x\mathcal{J}(I)$. (2) Let $u,v \in \mathcal{J}(I)$. By Lemma 1·4.2(2), $uv = v$, hence $xuv = xv^2 = xvv$, therefore $(xu,xv) \in P$. If $(y,z) \in P$ for each point $y \in x\mathcal{J}(I)$, then $(x,z) \in P$, because $x \in x\mathcal{J}(I)$ by Lemma 1·5.10. By (1) there is a minimal right ideal $I' \subset E$ such that $z \in x\mathcal{J}(I')$, *i.e.*, $z = xu'$ for some element $u' \in \mathcal{J}(I')$. Choose an element $u \in \mathcal{J}(I)$ so that $u \sim u'$. Then $(xu', xu) = (z,xu) \in P$, because $xu \in x\mathcal{J}(I)$. By Theorem 1·4.14, there is a minimal right ideal $I'' \subset E$ such that $xu'p'' = xup''$ for all $p'' \in I''$. By Lemma 1·4.6, we can choose an element $u'' \in \mathcal{J}(I'')$ with $u \sim u' \sim u''$. Then $u'u'' = u'$, $uu'' = u$, $xu'u'' = xuu''$. Therefore $z = xu' = xu'u'' = xu \in x\mathcal{J}(I)$. (3) Let $q \in I$ and $(x,xq) \in P$ for all $x \in X$. If $u \in \mathcal{J}(I)$, then $uq = q$ by Lemma 1·4.2(2). Since $(xu,xuq) \in P$, by the choice of q, we get $(xu,xq) \in P$. Denote $y = xq$. Then $(y,xu) \in P$ for all $u \in \mathcal{J}(I)$. By (2) there is an element $v \in \mathcal{J}(I)$ such that $xq = y = xv$. By Lemma 1·4.2(3), there is an element $w \in \mathcal{J}(I)$ with $qw = q$. Then $xq = xqw = xvw = xw$, because $vw = w$. Hence $xq = xw$ for each point $x \in X$, and thus $q = w$ and $q \in \mathcal{J}(I)$. ▮

REMARKS AND BIBLIOGRAPHICAL NOTES

Properties of motions in compact minimal sets (in the case of a flow) were first investigated by Birkhoff [*1*] and later by Bebutov [*1*]. Theorems 1·5.3-6 are generalizations of results obtained by Birkhoff and Bebutov. The material in Subsections 1·5.1-6 is adapted from the book by Gottschalk and Hedlund [*2*]. There are many papers devoted to the study of minimal sets, we mention here only the three survey articles by Gottschalk [*4*,*8*,*9*]. Theorem 1·5.9 was proved by Flor [*1*]. Theorem 1·5.11 is due to J. Auslander [*2*].

§(1·6): DISTAL AND EQUICONTINUOUS TRANSFORMATION GROUPS

We shall investigate in this section the distal and equicontinuous transformation groups and their relations with almost periodic and locally almost periodic transformation groups. We shall also discuss some further properties of the relations P, Q and M.

1·6.1 A transformation group (X,T,π) is called a ***DISTAL TRANSFORMATION GROUP*** if every two distinct points of X are distal, *i.e.*, if $P(X,T,\pi) = \Delta(X)$. We shall present some important properties of distal transformation groups. We shall first prove the following propositions.

1·6.2 LEMMA: *Let X be a compact space, and let (X,T) be a transformation group. If $a,b \in X$, $a \neq b$, and (a,b) is an almost periodic point of the transformation group $(X,T)\times(X,T)$, then a and b are distal. Conversely, if a and b are distal, then there exist points $a',b' \in X$ such that $a' \neq b'$, $(a',b') \in \overline{(a,b)T}$ and (a',b') is almost periodic under $(X,T)\times(X,T)$.*

Proof: Let $a,b \in X$, $a \neq b$ and assume that (a,b) is an almost periodic point. Then $\overline{(a,b)T}$ is a minimal set by Theorem 1·5.3. We shall prove that a and b are distal. Suppose that the contrary is true. Then $\overline{(a,b)T}\cap\Delta \neq \emptyset$, hence $\overline{(a,b)T} \subset \Delta$, because $\overline{(a,b)T}$ is a minimal set. But this contradicts $a \neq b$.

Conversely, suppose a and b are distal. Since $X\times X$ is a compact space, $\overline{(a,b)T}$ contains a minimal set M. Let $(a',b') \in M$. Since a and b are distal, $M\cap\Delta = \emptyset$, and hence $a' \neq b'$. By Theorem 1·5.4 the point (a',b') is almost periodic.∎

1·6.3 THEOREM: *Let X be a compact space. The following statements are mutually equivalent:*

(1) *The transformation group (X,T,π) is distal;*

(2) *$E \equiv E(X,T,\pi)$ is a group;*

(3) *The orbit closure of each point $(x,y)\in X\times X$ under $(X\times X,T,\hat{\pi})$ is a minimal set.*

Proof: (1) implies (2). Let $x \in X$ and let u be an idempotent in E. The points x and xu are proximal because $xu = xu^2 = (xu)u$. But $P = \Delta$, and hence $x = xu$ $(x \in X)$, *i.e.*, $u = e$. Thus the identity $e \in E$ is the only idempotent of E and therefore E is a group (Corollary 1·4.3).

(2) implies (3). Let $x,y \in X$ and $p \in E$. By Lemma 1·5.7 it suffices to show that there is an element $q \in E$ for which $xpq = x$ and $ypq = y$. Since E is a group, we can take $q = p^{-1} \in E$.

(3) implies (1) by Lemma 1·6.2.∎

1·6.4 THEOREM: *If X is compact and (X,T,π) is a distal transformation group, then the orbit closure of each point $x \in X$ is a minimal set.*

Proof: By the preceding theorem, $E = E(X,T,\pi)$ is a group. The assertion follows now from Lemma 1·5.7.∎

Since (X,T,π) is a homomorphic image of the direct product $(X\times X,T,\hat{\pi})$, the statement of Theorem 1·6.4 is in fact a corollary of Theorem 1·6.3.

Let (X,T), (Y,T) be transformation groups and let $\varphi:X\to Y$ be a homomorphism of (X,T) onto (Y,T). Let $\hat{\varphi}:X\times X\to Y\times Y$ denote the map defined by $(x_1,x_2)\hat{\varphi} = (x_1\varphi,x_2\varphi)$ $(x_1,x_2\in X)$.

1·6.5 LEMMA: *If the space X is compact, then the transformation group (Y,T) is distal iff $P(X,T)\hat{\varphi} = \Delta(Y) \equiv \{(y,y)\,|\,y\in Y\}$.*

Proof: Suppose that $P(X,T)\hat{\varphi} = \Delta(Y)$; we shall prove that (Y,T) is distal. By Theorem 1·6.3, it suffices to show that $E(Y,T)$ is a group, but this, in turn, will be so, if we could show that $E(Y,T)$ has only one idempotent. Let u be any idempotent of $E(Y,T)$. We shall prove that $v:Y\to Y$ is the identity mapping. Let ϑ denote the canonical homomorphism of $E(X,T)$ onto $E(Y,T)$ induced by φ. By Theorem 1·4.21(2), the set $v\vartheta^{-1}$ is non-empty and compact. It is a subsemigroup of $E(X,T)$, because $vv = v$. By Lemma 1·4.11, there is an idempotent $u\in v\vartheta^{-1}$. It follows from Theorem 1·4.14 that $(x,xu)\in P(X,T)$ for each point $x\in X$, because $(xu)u = xu^2 = xu$. The condition $(x,xu)\hat{\varphi}\in\Delta(Y)$ implies that $x\varphi = xu\varphi$ $(x\in X)$. By Theorem 1·4.21(2), $x\varphi v = xu\varphi = x\varphi$ $(x\in X)$. Since $X\varphi = Y$, we get $yv = y$ for every point $y\in Y$. Therefore $v:Y\to Y$ is the identity map.

Conversely, suppose that the transformation group (Y,T) is distal. It is clear that $P(X,T)\hat{\varphi}\subset P(Y,T)$. Hence $\Delta(Y) = \Delta(X)\hat{\varphi} \subset P(X,T)\hat{\varphi}\subset P(Y,T) = \Delta(Y)$, and conseqently $P(X,T)\hat{\varphi} = \Delta(Y)$.∎

1·6.6 THEOREM: *Let X be a compact space. If R is a closed invariant equivalence relation on X, then the transformation group $(X/R,T)$ is distal iff $R\supseteq P(X,T)$.*

Proof: Let R be a closed invariant equivalence relation on X and let $\varphi:X\to X/R$ be the canonical projection. Then φ is a homomorphism of (X,T) onto $(X/R,T)$. By the preceding lemma, $(X/R,T)$ is distal iff $P(X,T)\hat{\varphi} = \Delta(X/R)$; but this is equivalent to $R\supset P(X,T)$.∎

1·6.7 COROLLARY: *If X is a compact space, (X,T) is a distal transformation group and φ is a homomorphism of (X,T) onto a transformation group (Y,T), then (Y,T) is also distal.*

1·6.8 LEMMA: *In the class of all transformation groups (X,T) with compact phase spaces, the property "(X,T) is distal" is admissible.*

Proof: This follows directly from the definitions.∎

1·6.9 THEOREM: *If (X,T) is a transformation group with a compact phase space, then there exists the smallest closed invariant equivalence relation $P^{\#}$ such that $(X/P^{\#},T)$ is distal and, moreover, $P^{\#}$ coincides with the least closed invariant equivalence relation containing the proximal relation P.*

Proof: The assertion follows from Lemma 1·6.8 and the Theorems 1·3.16 and 1·6.6.∎

1·6.10 THEOREM: *If X is compact and (X,T) is distal, then $Q(X,T) = M(X,T)$.*

Proof: Let $(x,y) \in Q(X,T)$. We shall prove that $(x,y) \in M(X,T)$. Let $\alpha \in \mathfrak{U}[X]$, let U be a neighbourhood of x and let V be a neighbourhood of y. Choose an index $\gamma \in \mathfrak{U}[X]$ with $\gamma^3 \subset \alpha$ and $\gamma = \gamma^{-1}$. Since $(x,y) \in Q$, there exist points $x_1 \in U$, $y_1 \in V$ and an element $t \in T$ such that $(x_1t,y_1t) \in \gamma$. By Theorem 1·6.3, the transformation group $(X \times X,T)$ is pointwise almost periodic because (X,T) is distal. Hence there exists a syndetic subset $B \subset T$ such that $(x_1t, y_1t)B \subset x_1t\gamma \times y_1t\gamma$. Consequently $(x_1t,y_1t)B \subset \gamma^3 \subset \alpha$. Denote tB by A; then $(x_1,y_1)A \subset \alpha$. Since B is syndetic, the set A is also syndetic, and hence $(x,y) \in M(X,T)$. Thus $Q(X,T) \subset M(X,T)$. The converse inclusion is obvious.∎

1·6.11 Let X be a uniform space. A transformation group (X,T) is said to be an *EQUICONTINUOUS TRANSFORMATION GROUP* if for each point $x \in X$ and every index $\alpha \in \mathfrak{U}[X]$ there is an index $\beta \in \mathfrak{U}[X]$ such that $x\beta t \subset xt\alpha$ for all $t \in T$. A transformation group (X,T) is called a *UNIFORMLY CONTINUOUS TRANSFORMATION GROUP* if for each index $\alpha \in \mathfrak{U}[X]$ we can choose an index $\beta \in \mathfrak{U}[X]$ such that $x\beta t \subset xt\alpha$ for all $x \in X$ and $t \in T$. This is equivalent to $\beta T \subset \alpha$. If X is compact, then the transformation group (X,T) is uniformly continuous iff (X,T) is equicontinuous. This is also equivalent to the condition $Q(X,T) = \Delta$. In fact, if (X,T) is uniformly equicontinuous, then

$$Q = \bigcap [\overline{\beta T} \mid \beta \in \mathfrak{U}] \subset \bigcap [\bar{\alpha} \mid \alpha \in \mathfrak{U}] = \Delta.$$

Conversely, let $Q = \Delta$ and $\alpha \in \mathfrak{U}[X]$. We shall prove that $\beta T \subset \alpha$ for some index $\beta \in \mathfrak{U}[X]$. Suppose that, to the contrary, for each $\beta \in \mathfrak{U}[X]$ there is a pair $(x_\beta,y_\beta) \in \beta$ and an element $t_\beta \in T$ with

$$(x_\beta t_\beta, y_\beta t_\beta) \notin \alpha. \tag{6.1}$$

Since X is compact, we may suppose without loss of generality that

$$\{x_\beta t_\beta \mid \beta \in \mathfrak{U}\} \to a \in X, \qquad \{y_\beta t_\beta \mid \beta \in \mathfrak{U}\} \to b \in X.$$

It follows from (6.1) that $a \neq b$. Since $(x_\beta t_\beta, y_\beta t_\beta)t_\beta^{-1} = (x_\beta, y_\beta) \in \beta$, the pair (a,b) belongs to $Q(X,T)$; but this contradicts $Q = \Delta$.

Since $P(X,T) \subset Q(X,T)$, each uniformly equicontinuous transformation group is distal. The converse is not true in general, as we shall show in Example 2·9.19.

1·6.12 We recall some concepts and facts from the theory of function spaces (see, for example, Kelley [*1*], Chapter 7).

Let X be a compact space, let $C(X,X)$ be the set of all continuous maps $\varphi:X \to X$, let $\alpha \in \mathcal{U}[X]$ and put $\tilde{\alpha} = \{(\varphi,\psi)|\varphi,\psi \in C(X,X), (x\varphi,x\psi) \in \alpha, (x \in X)\}$. The family $\{\tilde{\alpha}|\alpha \in \mathcal{U}[X]\}$ forms a basis for some uniformity in $C(X,X)$. The set $C(X,X)$ provided with this uniformity will also be denoted by $C(X,X)$. The uniform space $C(X,X)$ is complete.

Let $\Phi \subset C(X,X)$. By Ascoli's Theorem, the closure $\bar{\Phi}$ of the set Φ in $C(X,X)$ is compact iff the family Φ is equicontinuous, *i.e.*, for each $x \in X$ and each index $\alpha \in \mathcal{U}[X]$ there exists an index $\beta \in \mathcal{U}[X]$ such that $x\beta\varphi \subset x\varphi\alpha$ for all $\varphi \in \Phi$.

If the family $\Phi \subset C(X,X)$ is equicontinuous, then the topology induced on Φ by X^X coincides with the topology of uniform convergence, that is, with the topology of a subspace of $C(X,X)$.

Let F be a group of homeomorphisms of X onto X. Define a map $\xi:X \times F \to X$ by

$$(x,p)\xi = xp \qquad (x \in X, p \in F), \tag{6.2}$$

where xp denotes, as usual, the image of the point x under the homeomorphism $p:X \to X$.

1·6.13 LEMMA: *Let X be a compact space and let F be a group of homeomorphisms of X onto X. Provide F with the subspace topology induced by $F \subset X^X$. If F is compact, then ξ is continuous.*

We omit the difficult proof of this profound result. The reader will find it in the reference Ellis [*1*].

1·6.14 THEOREM: *Let the space X be compact. The following statements are equivalent:*

(1) *(X,T,π) is an equicontinuous transformation group;*

(2) *The enveloping semigroup $E = E(X,T,\pi)$ is a group of homeomorphisms of X onto itself.*

Proof: (1) implies (2). By Ascoli's Theorem, the closure E_0 of the set $G = \{\pi^t|t \in T\}$ in $C(X,X)$ is compact. Since the family G is equicontinuous, the family E_0 is also equicontinuous, and therefore E_0 coincides with the closure of G in X^X, that is, $E_0 = E(X,T,\pi)$. Since the transformation group (X,T,π) is equicontinuous, it is distal. By Theorem 1·6.3, E is a group. Consequently each element $p \in E$ is a continuous bijective map of X onto X, and hence (2) is fulfilled.

(2) implies (1). By Lemma 1·6.13, the map $\xi:X \times E \to X$ defined by (6.2) is continuous, and since X and E are compact spaces, the family E is equicontinuous, and consequently (1) is true. ∎

1·6.15 LEMMA: *Let X be a compact space, and let φ be a homomorphism of a transformation group (X,T) onto a transformation group (Y,T). Then (Y,T) is equicontinuous iff*

$$Q(X,T)\hat{\varphi} = \Delta(Y). \tag{6.3}$$

Proof: Since φ is a homomorphism, $Q(X,T)\hat{\varphi} \subset Q(Y,T)$. If the transformation group (Y,T) is equicontinuous, then $Q(Y,T) = \Delta(Y)$ and (6.3) holds. By Lemma 1·6.5, (Y,T) is distal, and hence $E(Y,T)$ is a group (Theorem 1·6.3).

We shall now prove that (Y,T) is equicontinuous. By Theorem 1·6.14 it suffices to show that each element $q \in E(Y,T)$ is a continuous map of Y onto Y. Let $q \in E(Y,T)$, $y \in Y$ and let $\{y_i | i \in I\}$ be a net which converges to y. We shall prove that yq is a limit point of the net $\{y\ q | i \in I\}$. This will imply that the map $q: Y \to Y$ is continuous at the point y. By Theorem 1·4.21 the homomorphism φ induces a homomorphism $\vartheta: E(X,T) \to E(Y,T)$. Since $E(X,T)\vartheta = E(Y,T)$, there is an element $p \in E(X,T)$ such that $p\vartheta = q$. Choose points $x_i \in X$ with $x_i\varphi = y_i$ $(i \in I)$. Since X is compact, there is a subnet $\{x_j | j \in J\}$ of the net $\{x_i | i \in I\}$ such that $\{x_j | j \in J\}$ converges to a point $a \in X$ and $\{x_jp | j \in J\}$ converges to a point $b \in X$. Let us prove that $(b,ap) \in Q(X,T)$. Let U be a neighbourhood of ap, let V be a neighbourhood of b and let $\alpha \in \mathcal{U}[X]$. There exists an element $k \in J$, for which $x_k \in a\alpha$, $x_kp \in V$. Since the group $G(X,T)$ is dense in $E(X,T)$, there is an element $t \in T$ such that $at \in U$ and $x_kt \in V$. Then

$$(x_kt, at)t^{-1} = (x_k, a) \in \alpha.$$

This means that $(b,ap) \in Q(X,T)$. Since $Q(X,T)\hat{\varphi} = \Delta(Y)$, this yields $b\varphi = ap\varphi$. Further, $\{y_jq | j \in J\} = \{x_j\varphi q | j \in J\} \to b\varphi = ap\varphi = a\varphi q$. Since $\{x_j | j \in J\} \to a$ and the map $\varphi: X \to Y$ is continuous, $\{x_j\varphi | j \in J\} \to a\varphi$. Recall that $\{y_i | i \in I\} \to y$. Since $\{y_j | j \in J\}$ is a subnet of $\{y_i | i \in I\}$, we get $\{x_j\varphi | j \in J\} \to y$. Thus $a\varphi = y$. Therefore $\{y_jq | j \in J\} \to a\varphi q = yq$, and hence yq is a limit point of the net $\{y\ q | i \in I\}$. ▯

1·6.16 COROLLARY: *If X is a compact space and φ is a homomorphism of an equicontinuous transformation group (X,T) onto (Y,T), then (Y,T) is also equicontinuous.*

1·6.17 THEOREM: *Let X be a compact space, let (X,T) be a transformation group and let R be a closed invariant equivalence relation on X. Then the transformation group $(X/R, T)$ is equicontinuous iff $R \supset Q(X,T)$.*

Proof: Let φ denote the canonical homomorphism of (X,T) onto $(X/R,T)$. By the preceding Lemma, $(X/R,T)$ is equicontinuous iff $Q(X,T)\varphi = \Delta(X/R)$; but this is equivalent to the inclusion $R \supset Q(X,T)$. ▮

It is easy to see that the property "(X,T) is equicontinuous" is admissible in the class of transformation groups (X,T) with compact phase spaces.

1·6.18 THEOREM: *For any transformation group (X,T) with a compact phase space there is a smallest closed invariant equivalence relation $Q^\#$ such that $(X/Q^\#,T)$ is equicontinuous. The relation $Q^\#$ coincides with the smallest closed invariant equivalence relation, containing $Q(X,T)$,* i.e., *$Q^\# = Q^*$.*

Proof: This is a consequence of Theorems 1·3.16 and 1·6.17.∎

1·6.19 A subset S of the topological group T is called a ***RIGHT REPLETE SUBSET*** if for each compact subset $K \subset T$ there is an element $t \in T$ such that $Kt \subset S$.

1·6.20 LEMMA: *Let S be a right replete subset of the group T. Then:*

(1) $tS \cap S \neq \emptyset$ *for all* $t \in T$;

(2) $T = AS$ *for each syndetic subset* $A \subset T$.

Proof: (1) Let $t \in T$ and $K = \{e, t^{-1}\}$. There is an element $s \in T$ such that $Ks \subset S$. Then $s \in tS \cap S$. (2) Since A is syndetic, there is a compact subset $K \subset T$ with $T = AK$. Select an element $s \in T$ such that $Ks \subset S$. Then $T \equiv AKs \subset AS \subset T$, and hence $T = AS$.∎

1·6.21 THEOREM: *Let X be a compact space. Then the following statements are equivalent:*

(1) (X,T) *is an equicontinuous transformation group;*

(2) *For* $\alpha \in \mathfrak{U}[X]$ *we can choose an index* $\beta \in \mathfrak{U}[X]$ *and a right replete subset* $S \subset T$ *so that* $\beta S \subset \alpha$.

Proof: It will suffice to show that (2) implies (1).

Let (2) be true. We first show that $Q(X,T) \subset P(X,T)$. Let $(x,y) \in Q$ and $\alpha \in \mathfrak{U}[X]$. Choose an index $\beta \in \mathfrak{U}[X]$ such that $\beta^3 \subset \alpha$. For $\beta \in \mathfrak{U}[X]$ we select $\gamma \in \mathfrak{U}[X]$ with $\gamma = \gamma^{-1}$ and a right replete subset $S \subset T$, for which $\gamma S \subset \beta$. Since $(x,y) \in Q$, there exist points $x_1 \in x\gamma$, $y_1 \in y\gamma$ and an element $t \in T$ such that $(x_1t, y_1t) \in \gamma$. By the preceding lemma we can find an element $s \in tS \cap S$ for which

$$(xs, ys) = (xs, x_1s)\circ(x_1s, y_1s)\circ(y_1s, ys) \in \beta^3 \subset \alpha.$$

Therefore $(x,y) \in P$. Thus $Q \subset P$.

By Theorem 1·3.14, we have $P = L = Q = M$. To complete the proof, it is enough to show that $L = \Delta$. Let $(x,y) \in L(X,T)$ and $\alpha \in \mathfrak{U}[X]$. Select a right replete subset $S \subset T$ and an index $\beta \in \mathfrak{U}[X]$ in such a way that $\beta S \subset \alpha$. There exists a syndetic subset $A \subset T$ with $(x,y)A \subset \beta$. By the preceding lemma, $(x,y)T = (x,y)AS \subset \beta S \subset \alpha$. Hence $(x,y) \in \alpha$ for each $\alpha \in \mathfrak{U}[X]$. Therefore $x = y$, that is, $L = \Delta$.∎

1·6.22 Let X be a compact space. Recall that a transformation group (X,T) is called *almost periodic* if for every index $\alpha \in \mathfrak{U}[X]$ there exists a syndetic subset $A \subset T$ such that $xA \subset x\alpha$ for all points $x \in X$.

1·6.23 THEOREM: *Let (X,T) be a transformation group with a compact phase space. The following assertions are equivalent:*

(1) (X,T) *is equicontinuous;*

(2) (X,T) *is almost periodic.*

Proof: (1) implies (2). By hypothesis, the family $G = \{\pi^t | t \in T\}$ is equicontinuous. Hence by Ascoli's Theorem the closure E_0 of G in $C(X,X)$ is compact. Let $\alpha \in \mathcal{U}[X]$, $\alpha = \alpha^{-1}$. From the covering $\{\pi^t\alpha | t \in T\}$ of the set E_0 in $C(X,X)$ we can select a finite subcovering, *i.e.*, there exists a finite set $K \subset T$ such that $E_0 \subset \cup[\pi^k\tilde{\alpha} | k \in K]$. Therefore for each element $t \in T$ we can find an element $k_t \in K$ such that $(\pi^{k_t}, \pi^t) \in \tilde{\alpha}$, *i.e.*, $xk_t \in xt\alpha$ for each point $x \in X$. Then $xt^{-1}k_t \in x\alpha$ $(x \in X)$. Denote $\{t^{-1}k_t | t \in T\}$ by A. Clearly $AK^{-1} = T$ and $xA \subset x\alpha$ $(x \in X)$. Thus (2) holds.

(2) implies (1). Choose $\alpha \in \mathcal{U}[X]$ such that $\beta = \beta^{-1}$ and $\beta^2 \subset \alpha$. By hypothesis, there exist $A \subset T$ and a compact subset $K \subset T$ such that $T = AK$ and $xA \subset x\beta$ for all $x \in X$, *i.e.*, $(x,xa) \in \beta$ for all $x \in X$, $a \in A$. This implies that $(x,xa^{-1}) \in \beta$ $(x \in X,\ a \in X)$. Let $t \in T$; then $t^{-1} = ak$ for some elements $a \in A$ and $k \in K$. Hence $(xk^{-1},xt) = (xk^{-1},xk^{-1}a^{-1}) \in \beta$ $(x \in X)$. Consequently for each $t \in T$ there is an element $\ell \in K^{-1}$ which satisfies the condition $(x\ell,xt) \in \beta$ $(x \in X)$. Since the set K^{-1} is compact, the family $\{\pi^\ell | \ell \in K^{-1}\}$ is equicontinuous and, consequently, compact in $C(X,X)$. Hence there exists a finite subset $F \subset K^{-1}$ with $\{\pi^\ell | \ell \in K^{-1}\} \subset \cup[\pi^f\tilde{\beta} | f \in F]$. Thus for each element $t \in T$ we can find an element $f \in F$ such that $(xf,xt) \in \alpha$ $(x \in X)$. Therefore the family G is totally bounded, and, since the space $C(X,X)$ is complete, the set E_0 is compact. This implies that (1) is true. ∎

1·6.24 A transformation group (X,T) is called a *LOCALLY WEAKLY ALMOST PERIODIC TRANSFORMATION GROUP* iff for each point $x \in X$ and every neighbourhood U of x we can find a finite set $K \subset T$ and a neighbourhood V of x such that $VT \subset UK$.

1·6.25 THEOREM: *Let X be a compact space and (X,T) be a transformation group. The following statements are equivalent:*

(1) *(X,T) is locally weakly almost periodic;*

(2) *$\mathcal{A} \equiv \{\overline{xT} | x \in X\}$ is a star-closed partition of X.*

Proof: (1) implies (2). Let $x \in X$ and let U be an open neighbourhood of the set $A \equiv \overline{xT}$. If E is an open subset of X, then the saturation of E is equal to ET; therefore it will suffice to show that there exists a neighbourhood V of A with $VT \subset U$. Select a neighbourhood W of A in such a way that $\bar{W} \subset U$. For each point $x \in X \setminus U$ there exist a neighbourhood N_x of x and a finite subset $K \subset T$ such that $N_xT \subset (X \setminus \bar{W})K_x$. Select a finite subset $F \subset X \setminus U$ such that:

$$X \setminus U \subset \bigcup [N_x | x \in F],$$

and define $K = \cup[K_x | x \in F]$. Then $(X \setminus U)T \subset (X \setminus \bar{W})K$. There is a neighbourhood V of the set A such that $VK^{-1} \subset W$. Then $V \cap (X \setminus \bar{W})K = \emptyset$, hence $V \cap (X \setminus U)T = \emptyset$, and consequently $VT \subset U$.

(2) implies (1). If $x \in X$ and U is a neighbourhood of x, then $\overline{xT} \subset UT$, because (2) implies that $\overline{xT}$ is a compact minimal set.

Hence we can find a finite set $K \subset T$, for which $\overline{xT} \subset UK$. By hypothesis (2), there exists a neighbourhood V of x such that $V\mathcal{A} \equiv VT \subset UK$. ▮

1·6.26 LEMMA: *Let X be a compact space. The following statements are equivalent:*

(1) *(X,T) is an equicontinuous transformation group;*

(2) *$(X \times X,T)$ is locally weakly almost periodic.*

Proof: Obviously (1) implies (2). Conversely, let (2) be true. By Theorem 1·6.25, for every index $\alpha \in \mathfrak{U}[X]$ and each point $x \in X$ we can choose a neighbourhood V of x such that $(V \times V)\mathcal{A} \subset \alpha$ where $\mathcal{A}$ is the orbit closure partition of $X \times X$. But $(V \times V)\mathcal{A} = (V \times V)T$; hence $(x,y)T \subset \alpha$ for each point $y \in V$. Since X is compact, there exists an index $\alpha \in \mathfrak{U}[X]$ with $\beta T \subset \alpha$. Thus (1) is true. ▮

1·6.27 A transformation group (X,T,π) is called *LOCALLY ALMOST PERIODIC AT A POINT* $x \in X$ iff for each neighbourhood U of x there is a neighbourhood V of x and a syndetic subset $A \subset T$ with $VA \subset U$. It is easy to verify that this definition does not depend on the topology in X; therefore we can suppose in the definition that $AK = T$ for some finite subset $K \subset T$. A transformation group (X,T,π) is called *LOCALLY ALMOST PERIODIC* iff it is locally almost periodic at each point $x \in X$.

1·6.28 LEMMA: *Let X be a compact topological space. A transformation group (X,T) is locally almost periodic at a point $x \in X$ iff for each neighbourhood U of x we can choose a neighbourhood V of the same point and a finite subset $K \subset T$ in such a way that for every element $t \in T$ there exists an element $k \in K$ with $Vt \subset Uk$.*

Proof: If (X,T) is locally almost periodic at x, then for each neighbourhood U of x there exist a neighbourhood V of x and a syndetic subset $S \subset T$ such that $VS \subset U$. As was noted above, we may suppose that $T = SK$ for some finite set $K \subset T$. Therefore each element $t \in T$ can be expressed in the form $t = sk$, where $s \in S$, $k \in K$. Then $Vt = Vsk \subset Uk$.

Conversely, let $x \in X$ and let U be an arbitrary neighbourhood of x. Select a neighbourhood V of x and a finite subset $K \subset T$ such that $Vt \subset Uk$ for each $t \in T$ and a suitable $k = k_t \in K$. Denote $\{tk_t^{-1} | t \in T\}$ by A; then $AK = T$ and $VA \subset U$. ▮

Two subsets $Y \subset X$ and $Z \subset X$ are said to be *DISTAL SUBSETS* (one from another) if there exists an index $\alpha \in \mathfrak{U}[X]$ such that $(yt,zt) \notin \alpha$ for all $y \in Y$, $z \in Z$, $t \in T$.

1·6.29 LEMMA: *Let X be a compact space and let (X,T) be a locally almost periodic transformation group. If the point $x \in X$ is distal from a set $Z \subset X$, then there exists a neighbourhood U of x, such that U and Z are distal from one another.*

Proof: By hypothesis, there exists an index $\alpha \in \mathfrak{U}[X]$ such that

$(xt,zt) \notin \alpha$ for all $z \in Z$ and $t \in T$. Choose an index $\beta \in \mathfrak{U}[X]$, for which $\beta = \beta^{-1}$ and $\beta^3 \subset \alpha$. Since (X,T) is locally almost periodic, there exist a neighbourhood U of x and a syndetic subset $A \subset T$ such that $UA \subset x\beta$. If $y \in U$, $z \in Z$ and $a \in A$, then $(ya,za) \notin \beta$, for otherwise $(xa,za) = (xa,x)\circ(x,ya)\circ(ya,za) \in \beta^3 \subset \alpha$, which is impossible. Let K be a compact subset such that $T = AK$. Since X is compact, Lemma 1·1.5 guarantees the existence of an index $\gamma \in \mathfrak{U}$ with $\gamma K^{-1} \subset \beta$. It is easy to see that $(yt,zt) \notin \gamma$ for all $t \in T$, $y \in U$ and $z \in Z$. ▮

1·6.30 THEOREM: *If X is compact and (X,T) is a locally almost periodic transformation group, then $P = Q = L = M$.*

Proof: By Theorem 1·3.14 it will suffice to show that $Q \subset P$. If $x \in X$ and $y \in X$ are distal, then by Lemma 1·6.29 there exists a neighbourhood U of x such that y and U are distal from one another. Applying Lemma 1·6.29 again, we conclude that there exists a neighbourhood V of y which is distal from U. Thus if $(x,y) \notin P$, then $(x,y) \notin Q$. ▮

1·6.31 THEOREM: *Let X be a compact space and let (X,T) be a transformation group. The following assertions are equivalent:*

(1) *(X,T) is equicontinuous;*

(2) *(X,T) is locally almost periodic and distal.*

Proof: Let (1) be true. Then the transformation group (X,T) is distal. By Theorem 1·6.23, (X,T) is almost periodic, and therefore locally almost periodic. Conversely, let (2) be true. Then $P(X,T) = \Delta$, because (X,T) is distal. By the preceding theorem, $P = Q$. Therefore (1) is satisfied. ▮

REMARKS AND BIBLIOGRAPHICAL NOTES

Theorems 1·6.3,4 are due to Ellis [*2*]. Lemma 1·6.5 and Theorem 1·6.6 are taken from the paper by Ellis and Gottschalk [*1*]. Theorem 1·6.10 was proved by Clay [*1*]. Lemma 1·6.13 and Theorem 1·6.14 are taken from the paper by Ellis [*1*], which contains also some more general results. Lemma 1·6.51 and Theorem 1·6.17 were proved by Ellis and Gottschalk [*1*]. Theorem 1·6.21, which is a generalization of a result due to Baum [*1*], was obtained by Clay [*1*]. Theorem 1·6.23 is due to Gottschalk [*1*]. Theorem 1·6.25 is taken from the book by Gottschalk and Hedlund [*2*]. Lemma 1·6.28 is contained in the paper by McMahon and Wu [*1*]. Theorems 1·6.30,31 follow from the results of Ellis and Gottschalk [*1*], Clay [*1*] and Gottschalk [*3*]. For related results see L. Auslander and Hahn [*1*], Clay [*2*], Ellis [*3*], England [*1*], Markov [*2*] and Wu [*3*].

CHAPTER II

MINIMAL TRANSFORMATION GROUPS

§(2·7): THE ENVELOPING SEMIGROUP OF A MINIMAL TRANSFORMATION GROUP

In this section we shall investigate some relations between a minimal transformation group (X,T,π) and its enveloping semigroup $E = E(X,T,\pi)$ and we shall describe the endomorphisms and automorphisms of (X,T,π). Furthermore, the existence of universal minimal transformation groups is proved. Throughout this section it will be assumed that the phase spaces of all transformation groups are compact.

2·7.1 The notion of a *minimal set* $M \subset X$ was introduced in §(1·1). Recall that M is minimal iff $\overline{xT} = M$ for each point $x \in M$. A transformation group (X,T,π) is said to be a *MINIMAL TRANSFORMATION GROUP* if the whole space X is a minimal set, *i.e.*, $\overline{xT} = X$ for every point $x \in X$.

Let $E = E(X,T,\pi)$ denote, as usual, the *Ellis semigroup of the transformation group* (X,T,π). The semigroup E contains at least one minimal right ideal I. By Lemma 1·4.8, $(I,T,\bar{\pi}) \equiv (I,T,\pi)$ is a minimal transformation group.

2·7.2 LEMMA: *Let I be a minimal right ideal of E, let $u \in I$ be an idempotent and let $q \in I$. If $qu = q$, then $Iq = Iu$.*

Proof: Since $qu = q$ and $IE \subset I$, we have $Iq = Iqu \subset Iu$. By Lemma 1·4.2(1), $qI = I$. Hence there exists an element $r \in I$ with $qr = u$. Set $v = rq$. Then $v \in I$ and $v^2 = rqrq = ruq$. By Lemma 1·4.2(2), $uq = q$, consequently $v^2 = rq = v$, hence v is an idempotent. Further, $v = rq = rqu = vu = u$ by Lemma 1 4.2(2). Thus $u = v = rq \in Iq$, and therefore $Iu \subset I^2q \subset Iq$. We have shown above that $Iq \subset Iu$. Thus $Iq = Iu$. ▮

2·7.3 Let $H = H(X,T)$ denote the *set of all endomorphisms of the transformation group* (X,T), *i.e.*, homomorphisms of (X,T)

into (X,T), and let $A = A(X,T)$ denote the *set of all automorphisms of* (X,T), *i.e.*, isomorphisms of (X,T) into (X,T). It is easy to see that if (X,T) is a minimal transformation group, then each endomorphism maps X onto X; therefore $H(X,T)$ is a semigroup and $A(X,T)$ is a group. By Theorem 1·4.21(7) each endormorphism $\varphi \in H(X,T)$ induces the identity transformation of $E(X,T)$. We shall often use this simple fact in the sequel.

2·7.4 LEMMA: *Let* (X,T) *be a minimal transformation group and let* $\varphi \in H(X,T)$. *If* φ *is not the identity map, then* $(x,x\varphi) \in P$ *for each point* $x \in X$.

Proof: Since X is a minimal set and the homomorphism φ is not the identity map, $x \neq x\varphi$ $(x \in X)$. Since each point $x \in X$ is almost periodic by Theorem 1·5.4 and φ is an endomorphism, the point $(x,x\varphi)$ is almost periodic under $(X \times X,T)$, and hence $(x,x\varphi) \notin P$ by Lemma 1·6.2. ▮

2·7.5 LEMMA: *Let* (X,T) *be a minimal transformation group and let* I *be a minimal right ideal of the semigroup* $E(X,T)$. *If for each point* $x \in X$ *there exists only one element* $u \in I$ *which satisfies the condition* $xu = x$, *then the transformation groups* (X,T) *and* (I,T) *are isomorphic.*

Proof: It follows from Lemma 1·5.10 that u is an idempotent. Let $x \in X$. The map $\varphi_x : I \to X$, defined by $p\varphi_x = xp$ $(p \in I)$ is a homomorphism of (I,T) onto (X,T). We shall show that φ_x is injective. Let $p,q \in I$ and $p\varphi_x = q\varphi_x$, that is, $xp = xq$. By Lemma 1·4.2(1), $pI = I$, hence there is an element $r \in I$ with $pr = q$. Thus $xp = xq = xpr$. Since there always exists an idempotent $w \in I$ for which $xp = xpw$, it follows from the hypothesis that r is an idempotent. By Lemma 1·4.2(3), $pv = p$ for some idempotent $v \in I$. Then $(xp)v = xp$ and $r = v$, by the hypothesis. Therefore $q = pr = pv = p$. ▮

2·7.6 LEMMA: *Let* I *be a minimal right ideal of the semigroup* $E(X,T)$. *Then*

(1): *For each* $p \in I$ *the map* $L_p : I \to I$, *defined by* $qL_p = pq$ $(q \in I)$, *belongs to* $H(I,T)$;

(2): *If* $\varphi \in H(I,T)$, *then* $\varphi = L_p$ *for some element* $p \in I$;

(3): $H(I,T) = A(I,T)$.

Proof: (1): Since left multiplication is continuous in the semigroup $E(X,T)$, L_p is a continuous map. Further, $(q\pi^t)L_p = pq\pi^t = (pq)\pi^t = (pq)\bar{\pi}^t = (qL_p)\bar{\pi}^t$ $(q \in I, t \in T)$, whence $L_p \in H(I,T)$. (2): Let $\varphi \in H(I,T)$, $u \in \mathcal{J}(I)$ and $p = u\varphi$. Then $pu = (u\varphi)u = (uu)\varphi = u\varphi = p$; hence $u\varphi = uL_p$. If $t \in T$, then $u\pi^t\varphi = u\varphi\pi^t = pu\pi^t = (u\pi^t)L_p$. Since $\overline{uT} = I$, it follows that $q\varphi = qL_p$ for each element $q \in I$, that is, $\varphi = L_p$. (3): It will suffice to show that for each element $p \in I$ the mapping $L_p : I \to I$ is bijective. But this follows from Lemma 1·4.5. ▮

2·7.7 Let (X,T) be a minimal transformation group, I be a right ideal of the semigroup $E = E(X,T)$ and let $x \in X$. It was shown above that the map $\varphi_x : I \to X$, where $p\varphi_x = xp$ $(p \in I)$, is a homomorphism of (I,T) onto (X,T). It is easy to show that each homomorphism ψ of (I,T) onto (X,T) can be obtained in this way. For, if $u \in \mathfrak{J}(I)$ and $u\psi = x$, then $(u\pi^t)\psi = (u\psi)\pi^t = x\pi^t$ $(t \in T)$, but since ψ is continuous and the set $G = \{\pi^t | t \in T\}$ is dense in E, we conclude that $up\psi = xp$ for each $p \in E$. If $p \in I$, then $up = p$ by Lemma 1·4.2(2). Then $p\psi = xp = p\varphi_x$ $(p \in I)$, that is, $\psi = \varphi_x$.

Set $R_x = \{(q_1,q_2) | q_1,q_2 \in I, xq_1 = xq_2\}$ for each $x \in X$. Then R_x is a closed invariant equivalence relation, and, moreover, the transformation groups (X,T) and $(I/R_x,T)$ are isomorphic. Let p be an element of I with $pR_x \subset R_x$, where pR_x denotes the set $\{(pq_1,pq_2) | (q_1,q_2) \in R_x\}$. Define a map $\psi_x{}^p : I/R_x \to I/R_x$ as follows: if $A \in I/R_x$, then $A\psi_x{}^p = pA$. We shall say that $\psi_x{}^p$ is induced by L_p. It is easy to show that $\psi_x{}^p$ is continuous. Since $pI = I$ by Lemma 1·4.2(2), there exist elements $q,q' \in I$, for which $r = pq$, $r' = pq'$. Then $(q,q') \notin R_x$, that is, $xq \neq xq'$. Since $pR_x \subset R_x$, the map $\psi_x{}^p \in H(X,T)$ is well defined. From $(xq)\psi_x{}^p = xpq = xr = xr' = xpq' = (xq')\psi_x{}^p$ it follows that $\psi_x{}^p \notin A(X,T)$. ▮

2·7.9 THEOREM: *Let (X,T) be a minimal transformation group. Then the following statements are mutually equivalent:*

(1): *There exists a minimal right ideal $I \subset E$ such that (X,T) and (I,T) are isomorphic;*

(2): *(X,T) and (I,T) are isomorphic for each minimal right ideal $I \subset E$;*

(3): *If $x,y \in X$, then there exists an endomorphism $\varphi \in H(X,T)$ such that $x\varphi$ and y are proximal;*

(4): *If (x,y) is almost periodic under $(X \times X,T)$, then there exists an endormorphism $\varphi \in H(X,T)$ such that $x\varphi = y$.*

(A transformation group (X,T) which satisfies the latter condition is called a *REGULAR TRANSFORMATION GROUP*).

Proof: Clearly, (2) implies (1). Let (1) be true and let $p,q \in I$. By Lemma 1·4.2(6), there is an idempotent $u \in I$ with $qu = p$. By Lemma 2·7.2, $rq = pu$ for some element $r \in I$. If $s \in I$, then $(qL_r)s = rqs = pus = ps$, because Lemma 1·4.2(2) yields $us = s$. By Theorem 1·4.14 the points qL_r and p are proximal, and therefore (3) is satisfied. We show next that (3) implies (2). Let $x \in X$ and let I be a minimal right ideal. There exists an element $u \in \mathfrak{J}(I)$ such that $x = xu$. Let $p \in I$ and $x = xp$. We shall show that $u = p$. Let $y \in X$. By hypothesis there exists an endomorphism $\varphi \in H(X,T)$ such that $x\varphi$ and y are proximal. By Theorem 1·5.11(1), $x\varphi = yu'$ for some idempotent u' belonging to a minimal right ideal $I' \subset E$. Lemma 1·4.6 guarantees the existence of an idempotent $v' \in I'$ with $u \sim v'$, that is, $v'u = v'$, $uv' = u$. Then $xv' = xuv' = xu = x$; hence $x\varphi = xv'\varphi = x\varphi v' = yu'v' = yv'$. Therefore we may assume that $u' = v'$ and $yu' = x\varphi = xp\varphi = x\varphi p = yu'p$. Thus we have proved that for each point $y \in X$ there exists a minimal right ideal I' such that $yu' = yu'p$ for some element $u' \in \mathcal{J}(I')$ with $u \sim u'$. Analogously, for $z = yu$ there exist a minimal right ideal I'' and an element $u'' \in \mathcal{J}(I'')$ for which $u \sim u''$ and $zu'' = zu''p$. Then $zu'' = yuu'' = yu$ and $zu''p = yuu''p = yup = yp$, and hence $yu = yp$ $(y \in X)$, that is, $u = p$. By Lemma 2·7.5 this implies (2).

It remains to prove that (3) is equivalent to (4). If (x,y) is almost periodic under $(X \times X,T)$ and $x \neq y$, then x and y are distal by Lemma 1·6.2. Therefore (3) implies (4). Conversely, let (4) be true and take $x,y \in X$. Since x is almost periodic, $xu = x$ for some idempotent $u \in E$ (see Theorem 1·5.11(1)). Since $(x,yu)u = (xu,yu^2) = (x,yu)$, Lemma 1·4.10 and Theorem 1·5.4 show that (x,yu) is almost periodic under $(X \times X,T)$. Then there exists an endomorphism φ, such that $x\varphi = yu$. Since $(y,yu) \in P$, (3) is fulfilled.∎

2·7.10 COROLLARY: *If (X,T) is a minimal transformation group and I_1,I_2 are minimal right ideals of the semigroup E, then the transformation groups (I_1,T) and (I_2,T) are isomorphic.*

2·7.11 Let T be a topological group and let $B = B(T)$ be a class of minimal transformation groups (X,T). The class B is said to be an *ADMISSIBLE CLASS OF MINIMAL TRANSFORMATION GROUPS* if the following conditions are satisfied:

(1) $B \neq \emptyset$;

(2) If $(X_i,T) \in B$ $(i \in I)$ and M is a minimal set of $\Pi[(X_i,T) \mid i \in I]$, then $(M,T) \in B$.

Let B be an admissible class of minimal transformation groups. A minimal transformation group (X,T) is called a *B-UNIVERSAL MINIMAL TRANSFORMATION GROUP* if $(X,T) \in B$ and for each $(Y,T) \in B$ there exists a homomorphism of (X,T) into (Y,T). The class $B_0(T)$ of all minimal transformation groups (X,T) is clearly admissible. If $(X,T) \in B$, then X contains a dense subset whose cardinality does not exceed the cardinality of T. It follows that B_0 is a set.

2·7.12 THEOREM: *If $B = B(T)$ is an admissible class of minimal transformation groups, then there exists a B-universal minimal transformation group and, moreover, it is unique up to isomorphism.*

Proof: Let (A,T) be the Cartesian product of all transformation groups of G. Since the phase spaces of all transformation groups belonging to B are compact (as assumed at the beginning of this section), the space A is also compact by Tikhonov's Theorem. Therefore A contains a minimal set M. $(M,T) \in B$ because B is admissible. For each $(X,T) \in B$, the canonical projection of A onto X induces a homomorphism of (M,T) onto (X,T). Hence (M,T) is a B-universal minimal set.

Let I be a minimal right ideal of the semigroup $E(M,T)$. Then (I,T) is a minimal transformation group. Since B is an admissible class and $(M,T) \in B$, it follows directly from the definition of the enveloping semigroup $E(M,T)$ that $(I,T) \in B$. By Lemma 1·4.10, there exists a homomorphism φ of (I,T) onto (M,T). Since (M,T) is a B-universal minimal transformation group, there exists a homomorphism ψ of (M,T) onto (I,T). Then $\varphi\circ\psi$ is a homomorphism of (I,T) onto (I,T). By Lemma 2·7.6(3), $\varphi\circ\psi$ is an isomorphism. Hence φ is also an isomorphism. Therefore $H(M,T) = A(M,T)$.

Let (M',T) be another B-universal minimal transformation group. By definition there exist homomorphisms $\alpha:(M,T) \to (M',T)$ and $\beta:(M',T) \to (M,T)$. Then $\alpha\circ\beta \in H(M,T) = A(M,T)$. This obviously implies that α is an isomorphism. Thus the B-universal minimal transformation group is defined uniquely (up to isomorphism). ∎

REMARKS AND BIBLIOGRAPHICAL NOTES

The material in Subsections 2·7.2-10 is taken from the paper by J. Auslander [*3*]. This paper also contains an example of a minimal transformation group (X,T) with $A(X,T) \neq H(X,T)$. In Subsection 3·17.20 we shall present an example due to Parry and Walters [*1*] of a distal minimal cascade with the same property. Wu [*2*] constructed two non-isomorphic minimal transformation groups (X_1,T) and (X_2,T), but such that there exist homomorphisms $\varphi_1:(X_1,T) \to (X_2,T)$ and $\varphi_2:(X_2,T) \to (X_1,T)$. Homomorphisms of minimal sets were also studied by J. Auslander [*5,6*], Barbašin [*1,2*], Gottschalk [*6*] and Lam [*1*]. Universal minimal sets were first considered by Ellis [*4*] and Chu [*1*]. Theorem 2·7.12 is due to J. Auslander [*4*].

§(2·8): ALMOST PERIODIC AND LOCALLY ALMOST PERIODIC MINIMAL SETS

We shall give in this section a full description of the structure of equicontinuous (almost periodic) minimal transformation groups and we shall study the relationships between almost periodicity, Ljapunov stability and recurrence. Recurrence properties of points in almost periodic minimal sets will be investigated in detail. We shall present some characterizations of the smallest closed invariant equivalence relation containing the regionally proximal relation. Finally, we shall consider locally almost periodic minimal transformation groups and Levitan almost periodic points.

2·8.1 Let X be a compact space. As we have noted in §(1·6), a *transformation group* (X,T,π) is *equicontinuous* iff the closure E_0 of the set $G = \{\pi^t \mid t \in T\}$ in $C(X,X)$ is compact. If E_0 is compact in $C(X,X)$, then it is also compact in X^X. This implies that the topology induced in E_0 by the space $C(X,X)$ coincides with the topology induced by X^0. Therefore E_0 coincides with $E(X,T,\pi)$ and each element $p \in E$ is a continuous map of X onto X (see Section (1·6)).

2·8.2 LEMMA: *Let G be a compact space endowed with a group structure in such a way that multiplication is continuous in each variable separately. Then G is a topological group.*

Proof: For $g \in G$, let R_g denote the map of G onto G defined by $hR_g = hg$ $(h \in G)$. Provide the set $H = \{R_g | g \in G\}$ with the topology of simple convergence. By hypothesis, H is a compact group of homeomorphisms of G onto G. Lemma 1·6.13 asserts that the map $\xi : G \times H \to G$ defined by $(h,R_g)\xi = hg$ $(h,g \in G)$ is continuous. Hence multiplication in G is jointly continuous. It remains to show that the mapping $g \mapsto g^{-1}$ is continuous. Let $\{g_i | i \in I\}$ be a net which converges to some element $g \in G$. It will suffice to prove that $\{g_i^{-1} | i \in I\} \to g^{-1}$. Suppose the contrary holds. Since the space G is compact, there is a subnet $\{g_j^{-1} | j \in \mathfrak{J}\}$ which converges to some element $a \in G$ which is distinct from g^{-1}. It follows that $\{g_j g_j^{-1} | j \in \mathfrak{J}\} \to ga$, consequently $ga = e$ and so $a = g^{-1}$. This is a contradiction. Hence the operation $g \mapsto g^{-1}$ is continuous. ∎

2·8.3 THEOREM: *Let X be a compact space and let (X,T) be a minimal transformation group. Then the following statements are mutually equivalent:*

(1): *(X,T) is equicontinuous;*

(2): *(X,T) is almost periodic;*

(3): *$E(X,T)$ is a compact topological group;*

(4): *$E(X,T)$ is a group of continuous transformationsof X.*

Proof: Statements (1) and (2) are equivalent by Theorem 1·6.23. If (1) is true, then (X,T) is distal, and hence $E(X,T)$ is a group (Theorem 1·6.3). Since the topology in $E(X,T)$ coincides with the one induced by $C(X,X)$, E is a compact topological group. Hence (3) is true. Since $E \subset C(X,X)$, (4) is also true. Thus (1) implies (3) and (4). Let (4) be true. Since all elements $p \in E(X,T)$ are continuous maps of X onto X, multiplication in the group E is continuous in each variable separately. Thus Lemma 2·8.2 implies that (3) is true.

It remains to show that (3) implies (1). If (3) is true, then it is easy to see that (E,T) is a minimal equicontinuous transformation group. There exists a homomorphism of (E,T) onto (X,T). By Corollary 1·6.16 the transformation group (X,T) is also equicontinuous. ∎

2·8.4 Theorem 2·8.3 enables us to describe the structure of any equicontinuous minimal transformation group (X,T,π) with a compact phase space. Let $x \in X$, $H_x = \{p | p \in E, xp = x\}$. Clearly H_x is a closed subgroup of the compact topological group $E = E(X,T,\pi)$. The homomorphism $\varphi_x : (E,T) \to (X,T)$, where $p\varphi_x = xp$ $(p \in E)$, induces an isomorphism of the transformation group $(E/H_x,T,\rho)$, where $E/H_x = \{H_x p \mid p \in E\}$ and $(H_x p,t)\rho = H_x p\pi^t$ $(H_x p \in E/H_x,\ t \in T)$, onto the transformation group (X,T,π). The group $E(X,T,\pi)$ contains a dense subgroup which is a homomorphic image of the group T, namely $G = \{\pi^t \mid t \in T\}$.

Conversely, let E be a compact topological group and let φ be a continuous homomorphism of the group T into the group E such that $\overline{T\varphi} = E$. Let N be a closed subgroup of E and let $E/N = \{Np \mid p \in E\}$ be the quotient space. Define a map $\rho:(E/N)\times T \to E/N$ by $(Np,t)\rho = Np(t\varphi)$ $(Np \in E/N,\ t \in T)$. It is easy to verify that $(E/N,T,\rho)$ is an equicontinuous minimal transformation group.

Let us now consider the case when the group T is commutative.

2·8.5 THEOREM: *Let X be a compact space, let T be a commutative group and let $x \in X$. Then the following statements are pairwise equivalent:*

(1): *(X,T,π) is an equicontinuous minimal transformation group;*

(2): *There exists a group structure on X which makes X into a topological group so that π_x is a homomorphism of the group T onto a dense subgroup of the group X;*

(3): *There exist a compact commutative topological group S and a homomorphism $\varphi:T \to S$ such that $\overline{T\varphi} = S$ and the transformation group (S,T,ρ), where $(s,t)\rho = s(t\varphi)$ $(s \in S,\ t \in T)$ is isomorphic to (X,T,π).*

Proof: Let (1) be true. By the preceding theorem, $E = E(X,T,\pi)$ is a compact topological group. Since T is commutative, the group E is also commutative. If $p \in E$ and $xp = x$, then $x\pi^t p = xp\pi^t = x\pi^t$. Since the map $p:X \to X$ is continuous by the preceding theorem, $yp = y$ for each point $y \in \overline{xT} = X$, *i.e.*, p is the identity map. Therefore $\varphi_x:E \to X$ is an isomorphism of (E,T) onto (X,T). Thus (1) implies (2). Obviously (2) implies (3) and (3) implies (1).▌

2·8.6 We shall now study the relations between almost periodicity and Liapunov stability in a topological transformation group (X,T,π).

Let $S \subset T$ and let $B \subset X$. A point $x \in (X,\mathfrak{U})$ is said to be *LJAPUNOV S-STABLE WITH RESPECT TO A SUBSET B* if for each index $\alpha \in \mathfrak{U}[X]$ there exists an index $\beta \in \mathfrak{U}[X]$ such that $ys \in xs\alpha$ for all $s \in S$ and $y \in x\beta \cap B$.

Recall that a *replete subset* $S \subset T$ is one where for each compact subset $K \subset T$ there are elements $s,t \in T$ such that $sKt \subset S$.

2·8.7 LEMMA: *A sufficient condition for a subset $S \subset T$ to be replete is that S intersects each syndetic subset $A \subset T$. If the group T is commutative, then this condition is also necessary.*

Proof: Suppose that $S \cap A \neq \emptyset$ for each syndetic subset $A \subset T$.

Let us prove that S is replete. Suppose that, to the contrary, there exists a compact subset $K \subset T$ such that $t_1Kt_2 \not\subset S$ for all $t_1, t_2 \in T$. Set $A(t_1,t_2) = t_1Kt_2 \setminus S$ and $A = \cup[A(t_1,t_2) \mid (t_1,t_2) \in T \times T]$. Clearly $A = T \setminus S$. We may always assume that the identity element $e \in T$ belongs to K. We show that A is syndetic. Since $e \in K^{-1}$, we have $T \setminus S \subset AK^{-1}$. Let $s \in S$. If $sK \cap A = \emptyset$, then $sKe \subset T \setminus A = S$, which contradicts our assumption. Hence $sK \cap A \neq \emptyset$, and therefore there exist elements $k \in K$ and $a \in A$ with $sk = a$. Then $s = ak^{-1}$, *i.e.*, $s \in AK^{-1}$. Since s is an arbitrary element of S, this yields $S \subset AK^{-1}$. Now $e \in K^{-1}$ implies $A \subset AK^{-1}$, hence $T = A \cup S \subset AK^{-1} \subset T$, whence $T = AK^{-1}$. This means that A is a syndetic subset. But $A \cap S = \emptyset$, which contradicts the hypothesis.

Let T be a commutative group and let S be a replete set. If A is syndetic, then $T = AK$ for a compact subset $K \subset T$. Since S is replete, there exists an element $t_1 \in T$ such that $K^{-1}t_1 \subset S$. But $e \in K$ implies $t_1 \in S$. Since $T = AK$, there exist elements $a \in A$ and $k \in K$ for which $t_1 = ak$ and therefore $a = t_1k^{-1} = k^{-1}t_1 \in K^{-1}t_1 \subset S$. Thus $a \in A \cap S$. ∎

2·8.8 THEOREM: *Let T be commutative and let S be a replete subset of T. If $x \in X$ is almost periodic and Liapunov S-stable with respect to xT, then $\overline{xT}$ is an almost periodic minimal set.*

Proof: Since a Hausdorff uniform space is regular, it follows from Theorem 1·5.3 that $\overline{xT}$ is a minimal set. Let $\alpha, \beta \in \mathfrak{U}[X]$ be such that $\beta = \beta^{-1}$ and $\beta^2 \subset \alpha$. Since the point x is Ljapunov S-stable with respect to xT, there exists an index $\gamma \in \mathfrak{U}[X]$ such that $ys \in xs\beta$ for all $s \in S$ whenever $y \in x\gamma \cap xT$. Select $\delta \in \mathfrak{U}[X]$ with $\delta = \delta^{-1}$ and $\delta^2 \subset \gamma$. Since x is almost periodic, there exists a syndetic subset $A \subset T$, for which $xA \subset x\delta$. To conclude the proof it suffices to show that $xta \in xt\alpha$ for all $t \in T$ and $a \in A$.

Let $t \in T$, $a \in A$. Since the map $\pi^a: X \to X$ is continuous, there exists an index $\sigma \in \mathfrak{U}[X]$ such that $ya \in xa\delta$ whenever $y \in x\sigma$. Since x is an almost periodic point, we can find a syndetic subset $B \subset T$ with $xB \subset x(\sigma \cap \delta)$. By hypothesis, S is a replete set. Then it follows that S^{-1} is also replete. Since the set Bt^{-1} is syndetic, $Bt^{-1} \cap S^{-1} \neq \emptyset$ by Lemma 2·8.7. Hence there are elements $b \in B$ and $s \in S$ such that $bt^{-1} = s^{-1}$. Since $xb \in xB \subset x\sigma$, it follows that $xba \in xa\delta$. But $xa \in x\delta$; hence $xba \in x\delta^2 \subset x\gamma$. Since $tb^{-1} = s \in S$, we get $xbatb^{-1} \in (xtb^{-1})\beta$ by the choice of γ. Since the group T is commutative, $xta \in (xtb^{-1})\beta$. But $b \in B$ implies $xb \in x\delta \subset x\gamma$; hence $xb(tb^{-1}) \in (xtb^{-1})\beta$, *i.e.*, $xt \in (xtb^{-1})\beta$. Recalling that $\beta = \beta^{-1}$ and $\beta^2 \subset \alpha$ we get $xta \in xt\alpha$ $(t \in T, a \in A)$. ∎

2·8.9 Let $x \in X$ and $P \subset T$. The set

$$P_X = \cap[\overline{xtP} \mid t \in T]$$

is called the *P-LIMIT of* x.

2·8.10 LEMMA: *If P is a replete subsemigroup of a commutative group T, then*

(1): *The set P_x is closed and invariant;*

(2): *If $\overline{xP}$ is compact, then $P_x \neq \emptyset$;*

(3): $P_x = \bigcap [\overline{xpP} \mid p \in P]$.

Proof: Assertion (1) is obvious. Let $t_1,\dots,t_n \in T$. For the set $\{e,t_1,\dots,t_n\}$ there is an element $s \in T$ such that $\{s,t_1s,\dots,t_ns\} \subset P$. Then $t_1\cdots t_ns^n \in t_iP$ $(i = 1,\dots,n)$; hence

$$\overline{xt_1\cdot\ \dots\ \cdot t_ns^nP} \subset \bigcap_{i=1}^{n} \overline{xt_iP}.$$

Thus each finite family of subsets of the form $\overline{xtP}$ has a non-empty intersection. Therefore (2) holds. Let us verify (3). Clearly $P_x \subset \bigcap [\overline{xpP} \mid p \in P]$. Let $t \in T$ and take $y \in \bigcap [\overline{xpP} \mid p \in P]$. Let us prove that $y \in \overline{xtP}$. Since P is a replete semigroup, there exists an element $s \in T$ such that $\{e,t\}s \subset P$, *i.e.*, $s \in P$ and $ts \in P$. Then $y \in \overline{xtsP} \subset \overline{xtP}$, and therefore (3) is true. ∎

2·8.11 Let $S \subset T$ and $A,B \subset (X,\mathcal{U})$. We say that the *SET A is UNIFORMLY LJAPUNOV S-STABLE WITH RESPECT TO THE SUBSET B* if for each index $\alpha \in \mathcal{U}$ there exists an index $\beta \in \mathcal{U}$ such that $(as,bs) \in \alpha$ for all $s \in S$, whenever $a \in A$, $b \in B$ and $(a,b) \in \beta$.

2·8.12 THEOREM: *Let S be a replete subsemigroup of a commutative group T. If the orbit xT is uniformly Ljapunov S-stable with respect to xT and the set $\overline{xS^{-1}}$ is compact, then $\overline{xT}$ is a compact almost periodic minimal set.*

Proof: By Theorem 2·8.8 it will suffice to show that x is an almost periodic point and that $\overline{xT}$ is compact. Since S is a replete subsemigroup, the set $P \equiv S^{-1}$is also a replete subsemigroup. Therefore Lemma 2·8.10 implies that P_x is a non-empty compact set invariant under (X,T,π). By Theorem 1·1.7, P_x contains a compact minimal set M. We shall prove that $x \in M$. Then Theorem 1·5.4 will imply that x is almost periodic and from the compactness of the set P_x it will follow that $\overline{xT}$ is compact.

Suppose that $x \notin M$. Then there exists an index $\alpha \in \mathcal{U}[X]$ such that $x\alpha \cap M\alpha = \emptyset$. Select an index $\beta \in \mathcal{U}[X]$ with $\beta = \beta^{-1}$ and $\beta^2 \subsetneq \alpha$. Since the set xT is uniformly Ljapunov X-stable relative to xT, there is an index $\gamma \in \mathcal{U}$, $\gamma = \gamma^{-1}$ such that $ps \in qs\beta$ for all $s \in S$, whenever $(p,q) \in \gamma$ and $p,q \in xT$. Let $y \in M$, $\delta \in \mathcal{U}$, $\delta = \delta^{-1}$ and $\delta^2 \subset \gamma$. Since $y \in P_x$, $y \in \overline{xS^{-1}}$. Hence there exists an element $s_1 \in S$ such that $xs_1^{-1} \in y\delta$. Choose an index $\sigma \in \mathcal{U}[X]$ such that $ps_1 \in ys_1\beta$ for all $p \in y\sigma$. Since $y \in \overline{xS^{-1}}$, there is an element $s_2 \in S$ satisfying the condition $xs_2^{-1} \in y\sigma \cap y\delta$. Thus $xs_2^{-1}s_1 \in ys_1\beta$. Since $xs_1^{-1} \in y\delta$, we get that

$$xs_1^{-1} \in y\delta \subset xs_2^{-1}\delta^2 \subset xs_2^{-1}.$$

Then $x \equiv xs_1^{-1}s_1 \in xs_2^{-1}s_1\beta$, by the choice of γ. Hence $x \in xs_2^{-1}$. $s_1\beta \subset ys_1\beta^2 \subset \overline{yT}\alpha = M\alpha$, which contradicts the choice of α.▌

2·8.13 Let $x \in X$ and let P be a replete subsemigroup of a commutative group T. Recall that the point x is called a *P-recurrent point* if for every index $\alpha \in \mathfrak{U}[X]$ there exists a P-extensive subset $B \subset T$ such that $xB \subset x\alpha$. In other words, x is P-recurrent if $xpP \cap x\alpha \neq \emptyset$ for all $\alpha \in \mathfrak{U}$, $p \in P$. Therefore Lemma 2·8.10(3) shows that $x \in X$ is P-recurrent iff $x \in P_X$.

2·8.14 THEOREM: *Let P be a replete subsemigroup of a commutative group T and let $x \in X$. Then the following assertions are equivalent:*

(1): *$\overline{xT}$ is a compact almost periodic minimal set;*

(2): *The point x is P-recurrent, P_X is compact and the orbit xT is uniformly Ljapunov P-stable with respect to P_X.*

Proof: Let (1) be fulfilled. Then x is an almost periodic point (by Theorem 1·5.4), *i.e.*, for each index $\alpha \in \mathfrak{U}[X]$ there exists a syndetic set $A \subset T$ with $xA \subset x\alpha$. Whenever $s \in P$, the set sP is a replete subsemigroup. It follows from Lemma 2·8.7 that $sP \cap A \neq \emptyset$, *i.e.*, the set A is P-extensive. Hence the point x is P-recurrent. Since P_X is closed and $P_X \subset \overline{xT}$, the set P_X is compact. By hypothesis, the transformation group $(\overline{xT},T)$ is almost periodic. Therefore it follows from Theorem 1·6.23 that the set $\overline{xT}$ is uniformly Ljapunov T-stable with respect to $\overline{xT} \supset P_X$. Thus (1) implies (2).

Conversely, let (2) be true. Lemma 2·8.10(3) implies that $x \in P_X$, hence $\overline{xT} \subset P_X \subset \overline{xT}$, *i.e.*, $\overline{xT} = P_X$. Since the orbit xT is uniformly Ljapunov P-stable with respect to $P_X = \overline{xT}$ and the set P_X is compact, it follows from Theorem 2·8.12 that $(\overline{xT},T)$ is a minimal almost periodic transformation group.▌

2·8.15 Given a topological group T, it is easy to verify that the class of all equicontinuous minimal transformation groups (X,T) with compact phase spaces is admissible (see section (2·7)). Therefore it follows from Theorem 2·7.12 that the following theorem holds:

2·8.16 THEOREM: *In the class of all equicontinuous minimal transformation groups with compact phase spaces there exists a universal equicontinuous minimal transformation group which is uniquely defined, up to isomorphism.*

2·8.17 LEMMA: *If S is a closed syndetic subsemigroup of the group T, then S is a subgroup of T.*

Proof: Let $s \in S$ and let U be a neighbourhood of the identity

$e \in T$. It will suffice to show that $s^{-1}U \cap S \neq \emptyset$. Let V be a neighbourhood of e with $VV^{-1} \subset U$. Since S is a syndetic subset, $T = SK$ for some compact subset $K \subset T$. Select a finite subset $F \subset T$ such that $K \subset VF$. Let $k_0 \in K$. Then $s^{-1}k_0 = s_1k_1$ for suitable elements $s_1 \in S$, $k_1 \in K$. Further, $s^{-1}k_1 \in T = SK$, hence there exist elements $s_2 \in S$ and $k_2 \in K$ such that $s^{-1}k_1 = s_2k_2$. Thus there are sequences $k_0, k_1, \ldots$ in K and $s_1, s_2, \ldots$ in S such that $s^{-1}k_i = s_{i+1}k_{i+1}$ $(i = 0, 1, \ldots)$. Since $K \subset VF$, we can find k_m, k_n $(0 \leqslant m < n)$ and an element $f \in F$ such that $k_m \in Vf$, $k_n \in Vf$. Then

$$s^{-1}k_mk_n^{-1} = (s^{-1}k_mk_{m+1}^{-1})(k_{m+1}k_{m+1}^{-1}) \cdot \ldots \cdot (k_{n-1}k_n^{-1})$$

$$= s_{m+1}ss_{m+2} \cdot \ldots \cdot ss_n \in S,$$

because S is a subsemigroup. Hence

$$s^{-1}k_mk_n^{-1} \in s^{-1}(Vf)(Vf)^{-1} = s^{-1}VV^{-1} \subset s^{-1}U.$$

Thus $S \cap s^{-1}U \neq \emptyset$. ∎

2·8.18 Let $A^*(T)$ denote the class of all closed syndetic normal subgroups of the topological group T. If $A \in A^*(T)$, then $T = AK$ for some compact subset $K \subset T$. Therefore the quotient group T/A is compact. Conversely, if T is a locally compact group, A is a subgroup of T and the quotient space T/A is compact, then A is a syndetic subset. Indeed, the identity of the group T has a neighbourhood U whose closure is compact. Let $t \in T$ and let $V(At) = \{As \mid s \in tU\}$. Then $V(At)$ is a neighbourhood of the point $At \in T/A$. Since the space T/A is compact, the open covering $\{V(At) \mid t \in T\}$ of the space T/A contains a finite subcovering $\{V(At_1), \ldots, V(At_n)\}$. Set $K = \bigcup_{i=1}^{n} t_i\bar{U}$. It is clear that K is compact, and $T = AK$.

Let $A \in A^*(T)$ and $x \in X$. Set

$$S_x(A) = \{t \mid t \in T, xt \in \overline{xA}\}.$$

2·8.19 LEMMA: *Let X be a compact space, $x \in X$, $A \in A^*(T)$. Then:*

(1): *The set $S_x(A)$ is closed, $S_x(A) \supset A$ and $\overline{xS_x(A)} = \overline{xA}$;*

(2): *$S_x(A)$ is a subgroup of the group T.*

Proof: Statement (1) is obvious. Since (1) implies that $S_x(A)$ is a closed syndetic subset, by Lemma 2·8.17 it remains to show that $S_x(A)$ is a subsemigroup of T, *i.e.*, $S_x(A)S_x(A) \subset S_x(A)$, but this follows from

$$xS_x(A)S_x(A) \subset \overline{xA}S_x(A) \subset \overline{xAS_x(A)} = \overline{xS_x(A)A} = \overline{\overline{xA}A} = \overline{xAA} = \overline{xA}. \blacksquare$$

2·8.20 LEMMA: *If X is a compact space, (X,T,π) is a minimal transformation group and $A \in A^*(T)$, then $\{\overline{xA} | x \in X\}$ is a partition of X.*

Proof: Let $y \in \overline{xA}$. We shall prove that $\overline{yA} = \overline{xA}$. Clearly $\overline{yA} \subset \overline{xA}$. Since X is a minimal set, $x \in \overline{yT}$. Since $A \in A^*(T)$, there exists a compact set $K \subset T$ such that $T = AK$. Hence $x \in \overline{yAK} = \overline{yA}K$, *i.e.*, $x \in \overline{yA}k$ for some element $k \in K$. Then $\overline{xA} \subset \overline{yA}k \subset \overline{xA}k$ and consequently $xk^{-1} \in \overline{xA}$. This means that $k^{-1} \in S_X(A)$; but since $S_X(A)$ is a group, we get $k \in S_X(A)$. Therefore $\overline{xA}k \subset \overline{xA}S_X(A) \subset \overline{xAS_X(A)} = \overline{xS_X(A)} = \overline{xA}$. Hence $\overline{xA} = \overline{yA}k = \overline{xA}k$, and therefore $\overline{yA} = \overline{xA}$. ▮

2·8.21 Let us, for brevity, *denote by C^* the class of all minimal transformation groups with compact phase spaces and a commutative phase group.*

2·8.22 LEMMA: *If $(X,T,\pi) \in C^*$, $A \in A^*(T)$, $x \in X$, $y \in X$, then $S_X(A) = S_Y(A)$.*

Proof: There is an element $k \in T$ such that $y \in \overline{xA}k = \overline{xkA}$. It follows from the preceding lemma that $\overline{yA} = \overline{xA}k$. Hence $\overline{yA} = \overline{xS_X(A)k} = \overline{xkS_X(A)} = \overline{xkAS_X(A)} = \overline{xAkS_X(A)} = \overline{yAS_X(A)} = \overline{yAS_X(A)} = \overline{yS_X(A)}$. This means that $S_X(A) \subset S_Y(A)$. By symmetry $S_Y(A) \subset S_X(A)$, and hence $S_X(A) = S_Y(A)$. ▮

Accordingly, in what follows we shall write simply $S(A)$ instead of $S_X(A)$ $(x \in X,\ A \in A^*(T))$ whenever $(X,T,\pi) \in C^*$.

2·8.23 A minimal transformation group $(X,T,\pi)$is called a *TOTALLY MINIMAL TRANSFORMATION GROUP* if $\overline{xA} = X$ for each point $x \in X$ and each subgroup $A \in A^*(T)$.

2·8.24 LEMMA: *If $(X,T,\pi) \in C^*$ and $A \in A^*(T)$, then $R(A) = \{(x,y) | x \in \overline{yA}, y \in \overline{xA}\}$ is a closed invariant equivalence relation.*

Proof: By Lemma 2·8.20, $R(A)$ is an equivalence relation. Since $A \in A^*(T)$, $T = AK$ for some compact subset $K \subset T$. It was shown above that $\{\overline{yA} | y \in X\} = \{\overline{xA}k | k \in K\}$ for each $x \in X$. Let F be a closed subset of X and let

$$H = \{t | t \in T, \overline{xA}t \cap F \neq \emptyset\}.$$

By Lemma 1·1.3(3) the set $T \setminus H$ is open; hence H is closed in T. Denote $H \cap K$ by M. Then M is a compact set. The saturation of F with respect to the relation $R(A)$ is the closed set $\overline{xA}M$ (see Lemma 1·1.3(2)). It is easy to verify that the relation $R(A)$ is invariant. ▮

2·8.25 It follows from the preceding lemma that the quotient space $X/R(A)$ is Hausdorff and compact, and the transformation group (X,T,π) induces a transformation group $(X/R(A),T,\pi)$.

Let (X,T,π) be a transformation group belonging to C^* and let $A \in A^*(T)$, then $S(A)$ is a closed syndetic subgroup, and therefore the quotient group $T/S(A)$ is compact. Define a map $\rho:(T/S(A))\times T \to T/S(A)$ by

$$(S(A)\tau,t)\rho = S(A)\tau t, \qquad (t,\tau \in T).$$

It is easy to verify that $(T/S(A),T,\rho)$ is a minimal equicontinuous transformation group.

2·8.26 LEMMA: *Let (X,T,π) belong to the class C^*, let $x \in X$ and let $A \in A^*(T)$. Then the map $\psi_x:T/S(A) \to X/R(A)$ defined by*

$$(S(A)\tau)\psi_x = \overline{xS(A)}\tau \equiv \overline{xA}\tau, \qquad (\tau \in T),$$

is an isomorphism of $(T/S(A),T,\rho)$ onto $(X/R(A),T,\pi)$.

Proof: It is easy to see that ψ_x is continuous and $\rho^t \circ \psi_x = \psi_x \circ \pi^t$ $(t \in T)$. We shall show that ψ_x is injective. For if $\overline{xA}\tau_1 = \overline{xA}\tau_2$, then $x\tau_1\tau_2^{-1} \in \overline{xA}$, that is, $\tau_1\tau_2^{-1} \in S(A)$, and therefore $S(A)\tau_1 = S(A)\tau_2$. Since the space $T/S(A)$ is compact and the space $X/R(A)$ is Hausdorff, ψ is a homeomorphism. ▮

In what follows, **F** will *denote the neighbourhood filter at the identity $e \in T$.*

2·8.27 LEMMA: *If the transformation group (X,T,π) belongs to the class C^* and V is a neighbourhood of the identity $e \in T$, then there exists an index $\alpha \in \mathfrak{U}[X]$ such that $(x,y) \in \alpha$ implies*

$$\overline{xA} \subset \overline{yA}V, \qquad \overline{yA} \subset \overline{xA}V.$$

Proof: The set $\{V^* \mid V \in \mathbf{F}\}$, where

$$V^* = \{(S(A)\tau,S(A)t) \mid \tau^{-1}t \in V\}$$

is a uniformity base of the compact topological group $T/S(A)$. Let $z \in X$. Since $\psi_x:T/S(A) \to X/R(A)$ is a homeomorphism, there exists an index $\alpha \in \mathfrak{U}[X]$ such that $\overline{xA} = \overline{zA}$, $\overline{yA} = \overline{zA}t$ and $(S(A), S(A)t) \in V^*$ whenever $(x,y) \in \alpha$. Clearly α is the required index. ▮

2·8.28 THEOREM: *If the transformation group (X,T) belongs to the class C^* and $A \in A^*(T)$, then $R(A)$ is an invariant equivalence relation which is both open and closed.*

Proof: By Lemma 2·8.24, $R(A)$ is a closed invariant equivalence relation. Let $\beta \in \mathfrak{U}[X]$. Since X is compact, there exists a neighbourhood $V \in \mathbf{F}$ such that $(x,xt) \in \beta$ whenever $x \in X$ and $t \in V$. By Lemma 2·8.27, given V we can find an index $\alpha \in \mathfrak{U}[X]$ such that $(x,y) \in \alpha$ implies $\overline{xA} \subset \overline{yA}V$ and $\overline{yA} \subset \overline{xA}V$. Then $\overline{xA} \subset \overline{yA}\beta$, $\overline{yA} \subset \overline{xA}\beta$, whenever $(x,y) \in \alpha$. ▮

2·8.29 LEMMA: *If (X,T,π) belongs to the class C^* and $A_k \in A^*(T)$ $(k = 1,\dots,n)$, then*

$$\bigcap_{V\in \mathbf{F}} \overline{x(\bigcap_{k=1}^{n} S(A_k)V} = \bigcap_{k=1}^{n} \overline{xA_k},$$

for each point $x \in X$.

Proof: By the preceding theorem, $R \equiv \bigcap_{k=1}^{n} R(A_k)$ is both a closed and open invariant equivalence relation and $\{\cap[\overline{xA_k}|k = 1,\dots,n]| x \in X\}$ is the corresponding partition. By Lemma 2·8.26 the transformation group $(X/R(A_k),T,\pi)$ is isomorphic to $(T/S(A_k),T,\rho)$. Hence the transformation group $(X/R,T,\pi)$ can be embedded into the equicontinuous transformation group $\Pi[(T/S(A_k),T,\rho)|k = 1,\dots,n]$. Therefore, given a point $x \in X$ and a neighbourhood $V \in \mathbf{F}$, we can find an index $\alpha_V \in \mathfrak{U}[X]$ such that $\bigcap_{k=1}^{n} \overline{xA_k}\tau \subset (\bigcap_{k=1}^{n} \overline{xA_k})\alpha_V$ implies $\tau \in \bigcap_{k=1}^{n} S(A_k)V$. By Theorem 1·6.23 the transformation group $\Pi[(T/S(A_k),T,\rho)|k = 1,\dots,n]$ is almost periodic, hence the set $\bigcap_{k=1}^{n} S(A_k)V$ is syndetic for each neighbourhood $V \in \mathbf{F}$.

Let $x \in X$, $V \in \mathbf{F}$ and $y \in \cap[\overline{xA_k}|k = 1,\dots,n]$. By Lemma 2·8.27, for $\alpha_V \in \mathfrak{U}[X]$ there exists an index $\beta \in \mathfrak{U}[X]$ such that $(x,z) \in \beta$ implies

$$\bigcap_{k=1}^{n} \overline{xA_k} \subset (\bigcap_{k=1}^{n} \overline{zA_k})\alpha_V, \qquad \bigcap_{k=1}^{n} \overline{zA_k} \subset (\bigcap_{k=1}^{n} \overline{xA_k})\alpha_V.$$

For each index β_1, $\beta_1 \subset \beta$, there is an element $\tau \in T$ with $x\tau \in y\beta_1$ (because X is a minimal set). Thus:

$$(\bigcap_{k=1}^{n} \overline{xA_k})\tau \subset (\bigcap_{k=1}^{n} \overline{yA_k})\alpha = (\bigcap_{k=1}^{n} \overline{xA_k})\alpha .$$

Consequently $\tau \in \cap[S(A_k)V|k = 1,\dots,n]$. This means that $x(\bigcap_{k=1}^{n} S(A_k)V) \cap y\beta_1 \neq \emptyset$ for each index $\beta_1 \in \mathfrak{U}$ which satisfies the condition $\beta_1 \subset \beta$. Therefore $y \in \overline{x(\cap[S(A_k)V|k = 1,\dots,n]}$ and since V is an arbitrary neighbourhood from $\mathbf{F}$, we get that

$$y \in \bigcap_{V\in \mathbf{F}} \overline{x(\bigcap_{k=1}^{n} S(A_k)V)}. \tag{8.1}$$

Conversely, let (8.1) be satisfied. For $\alpha \in \mathfrak{U}[X]$ there is a neighbourhood $V_\alpha \in \mathbf{F}$ such that

$$x(\bigcap_{k=1}^{n} S(A_k)V_\alpha \subset (\bigcap_{k=1}^{n} \overline{xA_k})\alpha.$$

Then $y \in (\cap [\overline{xA_k} | k = 1,\ldots,n])\alpha$. Since α is an arbitrary index, we get that $y \in \cap [\overline{xA_k} | k = 1,\ldots,n]$. ▮

2·8.30 LEMMA: *If X is a compact space, T is a connected commutative locally compact topological group and (X,T,π) is a minimal almost periodic transformation group, then*

$$\cap [xA | A \in A^*(T)] = x,$$

for each point $x \in X$.

Proof: By Theorem 2·8.5, X can be endowed with a group structure in such a way that X becomes a compact commutative connected topological group with identity x and, moreover, $\pi_X : T \to X$ is a homomorphism of T into the group X. For each $y \in X$, $y \neq x$, there exists a continuous homomorphism φ of the group X into the group $\mathbb{K} = \mathbb{R}/\mathbb{Z}$ such that $0 = x\varphi \neq y\varphi$, where 0 is the zero element of (*see:* Pontrjagin [*1*], Chapter VI). Since $X\varphi$ is a connected subgroup of the group $\mathbb{K}$, it follows from $x\varphi \neq y\varphi$ that $X\varphi = \mathbb{K}$. Since the set X is minimal, $\overline{T\pi_X} = \overline{xT} = X$ $(x \in X)$. Since T is a connected group, $(T\pi_X)\varphi \equiv (xT)\varphi$ is a connected dense subgroup of , and hence $(xT)\varphi = \mathbb{K}$. Let H denote the kernel of the homomorphism $\pi_X\varphi : T \to \mathbb{K}$. Clearly H is a closed subgroup of T. Since T is locally compact, the group T/H is isomorphic to $\mathbb{K}$ and consequently it is compact. Therefore H is a closed syndetic subgroup of T, *i.e.*, $H \in A^*(T)$. From $xH\varphi = \{0\}$ it follows that $\overline{xH}\varphi = \{0\}$. Since $x\varphi \neq y\varphi$, the point y does not belong to $\overline{xH}$. Thus, given $y \in X$, $x \neq y$, there exists a syndetic subgroup $H \subset T$ with $y \notin \overline{xH}$. Therefore $\cap [\overline{xA}\ A \in A^*(T)] = x$. ▮

2·8.31 By Lemma 1·6.18, given (X,T,π) there exists the smallest closed invariant equivalence relation $Q^\#$ such that $(X/Q^\#, T,\pi)$ is equicontinuous, and, moreover, $Q^\#$ coincides with the smallest closed invariant equivalence relation containing $Q = Q(X,T,\pi)$, *i.e.*, $Q^\# = Q''$.

Denote $\cap [R(A) | A \in A^*(T)]$ by N. Lemma 2·8.24 shows that N is a closed invariant equivalence relation. By definition, $(x,y) \in N$ iff $\overline{xA} = \overline{yA}$ for each subgroup $A \in A^*(T)$.

2·8.32 THEOREM: *If X is a compact space, T is a connected commutative locally compact topological group and (X,T,π) is a minimal transformation group, then $N = Q^*$.*

Proof: As was shown in Lemma 2·8.26, the transformation group $(X/N,T,\pi)$ can be embedded into the equicontinuous transformation group $\Pi[(T/S(A),T,\rho)\ A \in A^*(T)]$; therefore it follows from Theorem 1·6.18 that $N \supset Q^\# = Q^*$.

Conversely, let $(x_1,x_2) \in N$. Let us prove that $(x_1,x_2) \in Q$. Suppose that the contrary is true. Let ψ denote the natural homomorphism of (X,T,π) onto $(X/Q^*,T,\pi)$ and put $y_1 = x_1\psi$, $y_2 = x_2\psi$. Then $y_1 \neq y_2$, and Lemma 2·8.30 asserts that $\overline{y_1A} \neq \overline{y_2A}$ for some subgroup $A \in A^*(T)$. Since ψ is a homomorphism, it follows that $\overline{x_1A} \neq \overline{x_2A}$, which contradicts $(x_1,x_2) \in N$. ∎

2·8.33 THEOREM: *If the hypotheses of Theorem 2 8.32 are fulfilled, then the following statements are mutually equivalent:*

(1): *(X,T,π) is equicontinuous;*

(2): $\bigcap[\overline{xA} \mid A \in A^*(T)] = x$ *for each point* $x \in X$*;*

(3): *Given $x \in X$ and $\alpha \in \mathfrak{U}[X]$, there exist subgroups $A_k \in A^*(T)$ $(k = 1,\ldots,n)$ and a neighbourhood $V \in \mathbf{F}$ such that*

$$x\left(\bigcap_{k=1}^{n} S(A_k)V\right) \subset x\alpha; \qquad (8.2)$$

(4): *For each index $\alpha \in \mathfrak{U}[X]$ there exist $A_k \in A^*(T)$ $(k = 1,\ldots,n)$ and $V \in \mathbf{F}$ such that* (8.2) *is satisfied for every point $x \in X$.*

Proof: By Theorem 2·8.32, (1) is equivalent to (2). If (2) holds, then (X,T,π) can be isomorphically mapped into the Cartesian product $\Pi[(T/S(A),T,\rho) \mid A \in A^*(T)]$ and hence (3) is fulfilled. If (3) is true, then Lemma 2·8.29 implies (2), and consequently, (1). Since X is a minimal set, (3) together with (1) imply (4). Conversely, (4) implies (3). ∎

Observe that statement (4) generalizes the well known property of Bohr almost periodic functions (Levitan [*1*]).

2·8.34 LEMMA: *Suppose that the conditions of Theorem* 2·8.32 *are fulfilled and the transformation group (X,T,π) is equicontinuous. Then there exists a subset $\mathbf{B} \subset \mathbf{A}^*(T)$ such that*

(a): *Given $B \in \mathbf{B}$ and $V \in \mathbf{F}$, we can find an index $\alpha \in \mathfrak{U}[X]$ such that $xt \in x\alpha$ implies $t \in BV$;*

(b): *For any $\alpha \in \mathfrak{U}[X]$ there exist $B_1,\ldots,B_n \in \mathbf{B}$ and $V \in \mathbf{F}$ such that $xt \in x\alpha$ whenever $t \in \bigcap_{k=1}^{n} B_kV$.*

Proof: Define $\mathbf{B}$ to be the family $\{S(A) \mid A \in A^*(T)\}$. Then (a) follows from the proof of Lemma 2·8.27. Assertion (b) is a consequence of Theorem 2·8.33. ∎

2·8.35 LEMMA: *If $(X,T,\pi) \in C^*$, $\alpha \in \mathfrak{U}[X]$ and $A_1,\ldots,A_n \in \mathbf{A}^*(T)$, then there exist $\beta \in \mathfrak{U}[X]$ and $V_0 \in F$ such that*

$$x\beta\left(\bigcap_{k=1}^{n} S(A_k)V_0\right) \subset x\left(\bigcap_{k=1}^{n} S(A_k)V\right)\alpha;$$

for each point $x \in X$ and every neighbourhood $V \in \mathbf{F}$.

Proof: Let $\gamma \in \mathcal{U}[X]$ be such that $\gamma = \gamma^{-1}$ and $\gamma^3 \subset \alpha$. The relation $R = \bigcap_{k=1}^{n} R(A_k)$ is open and closed by Theorem 2·8.28. Hence we can find an index $\beta \in \mathcal{U}[X]$ such that $(x,y) \in \beta$ implies

$$\bigcap_{k=1}^{n} \overline{yA_k} \subset \Big(\bigcap_{k=1}^{n} \overline{xA_k}\Big)\gamma.$$

By Lemma 2·8.26 the transformation group $(X/R,T,\pi)$ can be isomorphically mapped into the transformation group $\Pi[(T/S(A_k),T,\rho)| k = 1,\ldots,n]$; therefore, for $\gamma \in U[X]$ there exists a neighbourhood $V_0 \in \mathbf{F}$ such that

$$x\Big(\bigcap_{k=1}^{n} S(A_k)V_0\Big) \subset \Big(\bigcap_{k=1}^{n} \overline{xA_k}\Big)\gamma,$$

for every point $x \in X$. Then by Lemma 2·8.29

$$x\beta\Big(\bigcap_{k=1}^{n} (S(A_k)V_0)\Big) \subset \Big(\bigcap_{k=1}^{n} \overline{xA_k}\Big)\gamma^2 = \Big[\bigcap_{V\in\mathbf{F}} \overline{x\Big(\bigcap_{k=1}^{n} S(A_k)V\Big)}\Big]\gamma^2$$

$$\subset \Big[x\Big(\bigcap_{k=1}^{n} (S(A_k)V)\Big)\Big]\gamma^3 \subset \Big[x\Big(\bigcap_{k=1}^{n} S(A_k)V\Big)\Big]\alpha,$$

for each point $x \in X$ and every neighbourhood $V \in \mathbf{F}$. ▮

2·8.36 We recall that a *transformation group* (X,T,π) is called *locally almost periodic at a point* $x \in X$ if for every neighbourhood U of x we can choose a neighbourhood V of x and a syndetic subset $A \subset T$ so that $VA \subset U$. A transformation group (X,T,π) is called a *locally almost periodic transformation group* if it is locally almost periodic at each point $x \in X$.

2·8.37 THEOREM: *If a minimal transformation group (X,T,π) is locally almost periodic at some point $x \in X$, then (X,T,π) is locally almost periodic.*

Proof: Let $y \in X = \overline{xT}$ and let U be a neighbourhood of the point y. Choose an element $t \in T$ such that $xt \in U$; then $x \in Ut^{-1}$. By hypothesis, for the neighbourhood Ut^{-1} of x we can choose a neighbourhood W of x and a syndetic set $B \subset T$ such that $WB \subset Ut^{-1}$. Since $\overline{yT} = X$, there exists an element $s \in T$ with $Vs \subset W$. Thus $VsBt \subset WBt \subset U$. Let $A \equiv sBt$. Since the set B is syndetic, the set A is also syndetic. Thus (X,T,π) is locally almost periodic at the point $y \in X$. ▮

2·8.38 THEOREM: *If a minimal transformation groux (X,T) with* *a compact phase space is locally almost periodic, then for each neighbourhood U of a point $x \in X$ there exists an element $t \in T$ such that $(xt)P \equiv (xP)t \subset U$ (we recall that P denotes the proximal relation in (X,T)).*

Proof: Let $x \in X$ and let U be a neighbourhood of x. By Lemma 1·6.28 we can choose a finite subset $K \subset T$ and a neighbourhood V of the point x such that $Vt \subset Uk$ for each $t \in T$ and a suitable $k = k_t \in K$. Let us prove that $(xP)k^{-1} \subset U$ for some $k \in K$. Suppose that, to the contrary, for each $k \in K$ there exists a point $x_k \in xP \setminus Uk$. By Theorem 1·6.30, P is a closed equivalence relation, which coincides with L. Since the set K is finite and the transformation group (X,T) is minimal, there exists an element $t_0 \in T$ such that $x_k t_0 \in V$ for all $k \in K$. Now select an element $k_0 \in K$ with $Vt_0^{-1} \subset Uk_0$. Then $x_{k_0} = (x_{k_0}t_0)t_0^{-1} \in Vt_0^{-1} \subset Uk_0$; but this contradicts the choice of x_{k_0}.▌

2·8.39 COROLLARY: *Every non-trivial locally almost periodic minimal transformation group with a compact phase space has a non-trivial equicontinuous factor.*

Proof: Let $x \in X$ and let U be a neighbourhood of x with $U \neq X$. Then $xtP = xPt \subset U$ for some element $t \in T$. Hence $P \neq X \times X$. Since $P = Q$ is a closed invariant equivalence relation, $(X/P,T)$ is a non-trivial equicontinuous factor of (X,T).▌

2·8.40 LEMMA: *Let X be a compact metric space, let (X,T) be a minimal transformation group, $\varphi:(X,T) \to (Y,T)$ a homomorphism of (X,T) onto (Y,T) and $R = \{(x_1,x_2) | x_1,x_2 \in X, x_1\varphi = x_2\varphi\}$. If for every $\varepsilon > 0$ there exists a point x X such that the diameter of the set $xR \equiv \{x' | x' \in X,(x,x') \in R\}$ is smaller than ε, then φ is injective at some point $x_0 \in X$.*

Proof: Set

$$A_\varepsilon = \{x | x \in X, \mathrm{diam}(xR) < \varepsilon\}, \qquad (\varepsilon > 0).$$

It is clear that the set A_ε is open and dense in X. Therefore the set $A = \cap [A_{1/n} | n = 1,2,\ldots]$ is also dense in X and, moreover, φ is injective at each point $x_0 \in A$.▌

2·8.41 THEOREM: *Let X be a compact space and (X,T) a minimal transformation group. It the space X is metrizable and (X,T) is locally almost periodic, then there exists a point $x_0 \in X$, which is distal from all other points $x \in X$. Conversely, if there is a point $x_0 \in X$ with $x_0P = \{x_0\}$ and $P(X,T) = Q(X,T)$, then (X,T) is locally almost periodic.*

Proof: By Theorem 1·6.30, $P \equiv P(X,T) = Q(X,T)$ is a closed invariant equivalence relation. Let $\varphi:(X,T) \to (X/P,T)$ be the natural homomorphism. It follows from Theorem 2·8.38 and Lemma 2·8.40 that φ is injective at some point $x_0 \in X$. Since $(X/P,T)$ is a distal transformation group, x_0 is the required point.

Conversely, if $P(X,T) = Q(X,T)$, then P is a closed invariant equivalence relation. If $x_0P = \{x_0\}$, then the canoncial projection $\varphi: X \to X/P$ is open at the point x_0. Since the transformation group $(X/P,T) \equiv (X/Q,T)$ is equicontinuous and consequently almost periodic, (X,T) is locally almost periodic.∎

2·8.42 We show by means of an example that the metrizability condition, which is put on X in the first part of Theorem 2·8.41, cannot be dropped.

Let $\mathbb{K} = \{\xi \mid \xi \in \mathbb{C}, |\xi| = 1\}$, $\mathbb{Z}_2 = \{1,-1\}$ and $X = \mathbb{K} \times \mathbb{Z}_2$. Define a topology in X by assigning a neighbourhood base to each point as follows:

$$U_\varepsilon(\xi,1) = \{(\xi e^{it},1) \mid 0 \leqslant t < \varepsilon\} \cup \{(\xi e^{it},-1) \mid 0 < t \leqslant \varepsilon\},$$

$$V_\varepsilon(\xi,-1) = \{(\xi e^{it},-1) \mid -\varepsilon < t \leqslant 0\} \cup \{(\xi e^{it},1) \mid -\varepsilon \leqslant t < 0\},$$

(i denotes the imaginary number $\sqrt{-1}$).

There exists a topology τ on X such that $\{U_\xi(\xi,1) \mid 0 < \varepsilon < \pi\}$ is a neighbourhood base at $(\xi,1)$ and $\{V_\xi(\xi,-1) \mid 0 < \xi < \pi\}$ is a neighbourhood base at $(\xi,-1)$ $(\xi \in \mathbb{K})$.

It is easy to verify that (X,τ) is a first countable compact separable Hausdorff space. Each of the above neighbourhoods is not only open, but also closed, and therefore (X,τ) is zero-dimensional. The space (X,τ) is not metrizable because is is not second countable.

Define homeomorphisms $\sigma: X \to X$ and $\lambda\ \mathbb{K} \to \mathbb{K}$ by

$$(\xi,\pm 1)\sigma = (\xi e^{i},\pm 1), \qquad \xi\lambda = \xi e^{i}, \qquad (\xi \in \mathbb{K}).$$

Then (X,σ) is a locally almost periodic minimal cascade (the proof is left to the reader). If φ denotes the canonical projection of X onto $\mathbb{K}$, then each fiber $\xi\varphi^{-1}$ $(\varphi \in \mathbb{K})$ consists of two points: $(\xi,1)$ and $(\xi,-1)$. It is easy to see that $P = \{((\xi,1),(\xi,-1)) \mid \xi \in \mathbb{K}\}$, where P denotes the proximal relation.

2·8.43 THEOREM: *A minimal transformation group (X,T) with a compact phase space is locally almost periodic iff for each point $x \in X$ and every neighbourhood U of x we can choose subsets $S_1 \subset T$, $S_2 \subset T$ and finite subsets $K_1 \subset T$, $K_2 \subset T$ in such a way that $xS_1S_2 \subset U$ and $S_1K_1 = S_2K_2 = T$.*

Proof: Let $x \in X$ and let V be an arbitrary neighbourhood of x. We choose a neighbourhood U of x with $\bar{U} \subset V$. Let us now, for U, select the subsets S_1,S_2,K_1,K_2 with the indicated properties. Since K_1 is finite, we have $X = \overline{xT} = \overline{xS_1K_1} = \overline{xS_1}K_1$ and therefore

$$W \equiv \operatorname{int}(\overline{xS_1}k) \neq \emptyset,$$

for some element $k \in K_1$. There exists an element $t \in T$ with $xt \in W$.

Then $(Wt^{-1})tk^{-1}S_2 \subset (\overline{xS_1}kt^{-1})tk^{-1}S_2 = \overline{xS_1}S_2 \subset \overline{xS_1S_2} \subset \bar{U} \subset V$. Since Wt^{-1} is a neighbourhood of x and $tk^{-1}S_2$ is a syndetic subset, the transformation group (X,T) is locally almost periodic. The converse is obvious. ▮

2·8.44 COROLLARY: *For minimal transformation groups the property of being almost periodic is preserved under homomorphisms.*

2·8.45 A *POINT* $x \in X$ is said to be *LEVITAN ALMOST PERIODIC UNDER* (X,T,π) if for each index $\alpha \in \mathfrak{U}[X]$ there exist closed syndetic subgroups $A_1,\ldots,A_n$ of the group T and a neighbourhood $V \in F$ such that

$$x\left(\bigcap_{k=1}^{n} A_k V\right) \subset x\alpha. \tag{8.3}$$

Since the transformation group $\Pi[(T/A_k,T,\rho)|k = 1,\ldots,n]$ is equicontinuous, the sets of the form

$$\bigcap_{k=1}^{n} A_k V, \qquad (A_1,\ldots,A_n \in A^*(T); V \in \mathbf{F}),$$

are syndetic. Therefore each Levitan almost periodic point is almost periodic.

2·8.46 THEOREM: *Let X be a compact space, T a connected commutative locally compact topological group, (X,T,π) a minimal transformation group and let $x \in X$. The following statements are mutually equivalent:*

(1): *x is a Levitan almost periodic point;*

(2): *For each index $\alpha \in \mathfrak{U}[X]$ there are subsets $A_1,\ldots,A_n \in A^*(T)$ and a neighbourhood $V \in \mathbf{F}$ such that $x\left(\bigcap_{k=1}^{n} S(A_k)V\right) \subset x\alpha$;*

(3): $\bigcap [\overline{xA} | A \in A^*(T)] = x$;

(4): *(X,T,π) is locally almost periodic and x is distal from other points $y \in X$.*

Proof: Let (1) be true and put $G_1 = \Pi[T/A|A \in A^*(T)]$. Then G_1 is a compact commutative connected topological group. The map $\varphi:T \to G_1$ defined by $t\varphi \equiv \omega_t = \{At|A \in A^*(T)\}$ $(t \in T)$ is a homomorphism. Denote $\Omega = T\varphi$; then the closure G of the group Ω in G_1 is a compact commutative topological group.

Since T is locally compact, connected and commutative, the map $\varphi:T \to \Omega$ is injective. In fact, given $t \neq e$, there exists a homomorphism $\psi:T \to \mathbb{K}$ such that $t\psi \neq 0$ (*see:* Pontrjagin [*1*], Chapter VI). By the same argument as in the proof of Lemma 2·8.30, we have that $Ht \cap H = \emptyset$ for some closed syndetic subgroup H of the group T. Consequently $t\varphi \neq e\varphi$.

Define a map $f:\Omega \to X$ by $\omega_t f = xt$ $(t \in T)$. Let us prove that f is continuous. Let $t \in T$ and $\alpha \in \mathfrak{U}[X]$. Since x is Levitan almost periodic, the point xt is also Levitan almost periodic. Hence there exist $A_1,\ldots,A_n \in A^*(T)$ and a $V \in \mathbf{F}$ such that

$$xt(\bigcap_{k=1}^{n} A_k V) \subset xt\alpha. \tag{8.4}$$

The set $U = \{\omega_\tau | A_k \subset A_k t V, (k = 1,\ldots,n)\}$ is a neighbourhood of the point ω_t in Ω and, moreover, $Uf \subset xt\alpha$, by (8.4). Indeed, let $\omega_\tau \in U$. Then $\tau \in A_k t V$ $(k = 1,\ldots,n)$, and consequently $\tau \in t(\bigcap_{k=1}^{n} A_k V)$. It follows from (8.4) that $\omega_\tau f = x\tau \in xt\alpha$.

Let us consider the multivalued map $F:G \to X$ which assigns to each element $g \in G$ the non-empty closed compact set

$$gF \equiv \bigcap_{V\in\Phi} \overline{\{xt | t\varphi \in gV\}},$$

where Φ is the neighbourhood filter of the identity $e \in G$. Let us prove that

$$gF = \bigcap_{V\in\Phi} \overline{(gV)F} \equiv \bigcap_{V\in\Phi} \overline{(\cup [hF | h \in gV])}. \tag{8.5}$$

Obviously, $gF \subset \cap [(gV)F | V \in \Phi]$. Let $q \in \bigcap_{V\in\Phi} \overline{(gV)F}$ and let $W(q)$ be any neighbourhood of the point q. It $V_0 \in \Phi$, then there exists a point $p \in W(q) \cap (gV_0)F$, and therefore $p \in hF$ for some element $h \in V_0 g$. Since $p \in hF$ we can choose for the neighbourhood $W(q)$ of p and the neighbourhood $V_0 g$ of h an element $t \in T$ such that $xt \in W(q)$ and $t\varphi \in V_0 g$. But this just means that $q \in gF$. Thus formula (8.5) is proved. Observe that for each $t \in T$ the set $(t\varphi)F$ consists of only one point $t\varphi f = xt$, by the continuity of $f:\Omega \to X$.

Let H denote the set of all elements $g \in G$ satisfying the following condition: for every neighbourhood W of the point $x = ef$ and every neighbourhood $V \in \Phi$ there exists an element $t \in T$ such that $t\varphi \in Vg$ and $t\varphi f \equiv xt \in W$. In other words H is the set of those elements $g \in G$ for which gF contains the point $ef = x$.

Let us investigate the structure of the set H. We shall first show that $h \in H$ iff $gF \subset ghF$ for each element $g \in G$. If $gF \subset ghF$ for all $g \in G$, then letting $g = e$ we get $eF \subset hF$, that is, $x = ef \in hF$. Conversely, let $x \in hF$. We shall prove that $(t\varphi)f \in [(t\varphi)h]F$ for all $t \in T$. Let W_1 be an arbitrary neighbourhood of xt and let $V \in \Phi$. There exists a neighbourhood W of x with $Wt \subset W_1$. Since $x \in hF$, we can find for the neighbourhoods W and V

an element $\tau\varphi \in Vh$ such that $x\tau \equiv \tau\varphi f \in W$. Then $(\tau t)\varphi = (\tau\varphi)(t\varphi) \in Vh(t\varphi)$ and $t\varphi f = x\tau t \in W_1$. But this just means that $t\varphi f \in [h \times (t\varphi)]F$. Thus if $h \in H$ and $t \in T$; then

$$t\varphi f \in [h(t\varphi)]F. \tag{8.6}$$

Now let $g \in G$, $q \in gF$, let W be any neighbourhood of q and let $V \in \Phi$. Choose an element $t\varphi \in Vg$ with $t\varphi f \in W$. It follows from (8.6) that $W \cap [h(t\varphi)]F \neq \emptyset$ and, moreover, $W \cap (Vgh)F \neq \emptyset$. Using (8.6), we conclude that

$$q \in \bigcap_{V \in \Phi} \overline{(Vgh)F} = (gh)F.$$

Thus $h \in H$ iff $gF \subset (gh)F$ for every element $g \in G$. This implies that H is a subsemigroup of the group G. In fact, let $h_1 \in H$ and $h_2 \in H$. Then $x \in h_1F$ and $x \in h_2F$. Using the above property of H, we get $x \in h_1F \subset h_1h_2F$, *i.e.*, $x \in h_1h_2F$. But this just means that $h_1h_2 \in H$.

Let us show now that H is closed. Let $h_0 \in \bar{H}$ and $V \in \Phi$. There exists an element $h \in H \cap Vh_0$. Then $x \in hF \subset Vh_0F$. Therefore

$$x \in \bigcap_{V \in \Phi} \overline{(Vh_0)F} = h_0F,$$

that is, $h_0 \in H$.

Thus H is a closed subsemigroup of the compact commutative topological group G and, moreover, the identity e is contained in H. It follows that H is a subgroup of G. Indeed, Lemma 1·4.11 shows that each closed subsemigroup H_1 of H contains the identity $e \in G$. The semigroup H contains a minimal ideal I, but since $e \in I$, we get $H = I$; whence H is a subgroup of G (by Corollary 1·4.3).

We shall show that $h \in H$ iff $gF = (gh)F$ for each element $g \in G$. In fact, if $h \in H$, then $gF \subset (gh)F$ for all $g \in G$. Since H is a group, $h^{-1} \in H$, therefore $gF \subset (gh^{-1})F$ for each element $g \in G$, and hence $(gh)F \subset gF$. Thus $gF = (gh)F$. The converse is obvious. It follows from the previous property that $\omega f = \omega hf$ for all $\omega \in \Omega$ and $h \in H \cap \Omega$. Define a map $f_1 : \Omega H/H \to xT$ by $[\omega_t H]f_1 = \omega_t f = xt$. Since the map $f : \Omega \to X$ is continuous, the mapping f_1 is also continuous. It is clear that f_1 is injective. Let us show that the inverse map $f_1{}^{-1} : xT \to \Omega H/H$ is continuous. Let $\{t_i \mid i \in I\}$ ($t_i \in T$) be a net with $\{xt_i \mid i \in I\} \to x$. We shall prove that $\{H(t_i\varphi)\} \to H$. Suppose the contrary is true. Since the group G is compact, there exists a subnet $\{t_j \mid j \in \mathcal{J}\}$ such that $\{xt_j\} \to x$ and $\{t_j\varphi\} \to g \in G \setminus H$. Thus gF contains x, *i.e.*, $g \in H$, which contradicts $g \in G \setminus H$. Define a map $\sigma : (G/H) \times T \to G/H$ by $(Hg,t)\sigma = Hg(t\varphi) \equiv Hg\omega_t$ ($g \in G$, $t \in T$). It is easy to verify that $(G/H,T,\sigma)$ is a minimal equicontinuous transformation group. Since f_1 is a homeomorphism, it follows from Lemma 2·8.34 that there exists a subset $B \subset A^*(T)$ satisfying conditions (a) and (b) of this lemma for the point $x \in X$.

Let $B \in \mathbf{B}$. We shall show that $S(B) = B$, where $S(B) = \{t | xt \in \overline{xB}\}$. If $xt \in \overline{xB}$, then $\overline{xtB} = \overline{xB}$, and hence $x \in \overline{xtB}$. Let B be an arbitrary neighbourhood from F. For $B \in \mathbf{B}$ there is an index $\alpha \in \mathfrak{U}[X]$ such that $x\tau \in x\alpha$ implies $\tau \in BV$. Since $x \in \overline{xtB}$, there exists an element $b \in B$ with $xtb \in x\alpha$. Thus $tb \in BV$, and hence $t \in BV$.

As V is an arbitrary neighbourhood of $e \in T$, we get $t \in \bar{B} = B$. Thus $S(B) = B$. Therefore (2) is true.

Since (2) implies (1), the statements (1) and (2) are equivalent. It follows from Lemma 2·8.29 that (2) is equivalent to (3). Statement (3) means that the canonical map of (X,T,π) onto $(X/N, T,\pi)$ is one-to-one at the point $x \in X$. Therefore x is distal from all other points $y \in X$. If (2) is fulfilled, then (X,T,π) is locally almost periodic by Lemma 2·8.35. Hence (3) implies (4). Let (4) be true, then $P = Q = Q^* = N$; therefore (3) is satisfied. Thus the four statements are mutually equivalent. ▮

2·8.47 THEOREM: *If the conditions of Theorem* 2·8.46 *are fulfilled, then the following statements are mutually equivalent:*

(1): (X,T,π) *is equicontinuous;*

(2): *All points* $x \in X$ *are Levitan almost periodic.*

Proof: (1) implies (2) by Theorem 2·8.33. If (2) is true, then (X,T,π) is a locally almost periodic distal transformation group (Theorem 2·8.46), and Theorem 1·6.31 shows that (1) is satisfied. ▮

REMARKS AND BIBLIOGRAPHICAL NOTES

Lemma 2·8.2 is a special case of a theorem due to Ellis [*1*]. Theorem 2·8.3 was proved by Ellis and Gottschalk [*1*]. Theorem 2·8.5 was proved by Stepanov and Tikhonov [*1*] for flows and by Gottschalk and Hedlund [*2*] in the general case. Theorems 2·8.8, 12 represent a generalization of the well known results obtained by Markov [*2*] and are due to England [*1*]. The concept of a P-limit set is contained in the book by Gottschalk and Hedlund [*2*]. If $T = \mathbb{R}$ and P is the semigroup of non-negative real numbers, then P_x coincides with the ω-limit set of x in the sense of Birkhoff [*1*]. The material in Subsections 2·8.17-29 is taken from Gottschalk and Hedlund [*2*]. Theorem 2·8.32 is a consequence of some results of Gottschalk [*7*]. Theorem 2·8.33 was proved by Bronšteĭn [*3*]. Totally minimal sets were studied by Chu [*2*]. The material in 2·8.38-41 and 2·8.43-44 is due to McMahon and Wu [*1*]. Example 2·8.42 was constructed by Ellis and presented by Ellis and Gottschalk in [*1*]. Theorems 2·8.46,47 were obtained by Bronšteĭn [*3*]. Very interesting related results were obtained by Veech [*1*,*2*]. We indicate some other papers that are intimately related to the subject matter of this section. The relation N was studied by Brook [*1*]. An interesting characterization of the relation Q^* is given in Veech [*3*] and in the paper by Ellis and Keynes [*1*]. Locally almost periodic minimal sets were considered by Wu [*3*]. A wider class of minimal transformation groups satisfying the condition $Q = P$ was investigated by J. Auslander and Horelick [*1*], Markley [*1*,*2*], Shapiro [*3*,*4*]. Chu [*4*] studied the structure of almost periodic transformation groups. The universal equicontinuous minimal set was considered by Chu [*1*] and Eisenberg [*1*].

§(2·9): DISTAL MINIMAL TRANSFORMATION GROUPS

The main result of this section is Furstenberg's Theorem on the structure of distal minimal transformation groups (Theorem 2·9.18).

2·9.1 LEMMA: *If (X,T,π) is a distal minimal transformation group with a compact phase space and $x \in X$, then the mapping $\varphi_x : E(X,T) \to X$, defined by $p\varphi_x = xp$ $(p \in E)$, is open.*

Proof: Let $y \in X$ and $H_y = y\varphi_x^{-1} = \{p | p \in E, xp = y\}$. If $p \in H_y$, then $H_y = H_x p$. In fact, if $q \in H_y$, then $xq = y$, $xp = y$, hence $xqp^{-1} = x$, that is, $qp^{-1} \in H_x$ and consequently $q \in H_x p$. Conversely, if $r \in H_x$, then $xr = x$ and $xrp = xp = y$; therefore $rp = H_y$.

Let U be an open non-empty subset of E. The saturation of U with respect to the partition $\{H_y | y \in X\}$ is equal to the union of all those H_y $(y \in X)$ for which $H_y \cap U \neq \emptyset$. If $p \in H_y \cap U$, then $H_y = H_x p \subset H_x U$. Conversely, let $p \in H_x U$. If $p = qr$, where $q \in H_x$, $r \in U$, then $q^{-1}p \in U$. It is easy to see that H_x is a subgroup of the group E. Therefore $q^{-1}p \in H_x p \cap U$. Set $y = xp$. Then $H_y = H_x p$ and $H_y \cap U \neq \emptyset$. Thus the saturation of U is equal to $H_x U$.

The map $L_p : E \to E$, defined for $p \in E$ by $sL_p = ps$ is continuous, and since E is a group, L_p is bijective. Hence L_p is a homeomorphism. Therefore the set $H_x U$ is open. Since the space X is homeomorphic to the quotient space E/H_x, the lemma is proved.▌

2·9.2 THEOREM: *Let X be a compact space, (X,T) a distal minimal transformation group and R a closed invariant relation on X. Let, further, $(x,y) \in R$ and let U be a neighbourhood of x. Then there exists a neighbourhood V of y such that for each point $y_1 \in V$ we can choose a point $x_1 \in U$ with $(x_1,y_1) \in R$.*

Proof: Let $U_1 = \{p | p \in E(X,T,\pi), xp \in U\}$. The set U_1 is open in E, by the definition of the topology in E. By Lemma 2·9.1 the set $V \equiv yU_1$ is open in X. Since $xe \in U$, we have $e \in U_1$. Therefore $y \in V$. Thus V is a neighbourhood of y. Let $y_1 \in V$. Then $y_1 = yp_1$ for some element $p_1 \in U_1$. Since R is a closed invariant relation, $(x,y) \in R$ implies $(xp_1, yp_1) \in R$. Thus $x_1 \equiv xp_1$ is the required point.▌

2·9.3 COROLLARY: *If (X,T) is a distal minimal transformation group, then each closed invariant equivalence relation in X*

is open. In other words, each homomorphism of a distal minimal transformation group (X,T) is an open map.

2·9.4 Let X be a compact uniform space, (X,T) a transformation group and $R \subset X \times X$ a closed relataon invariant under (X,T).

Let $Q(R) \equiv Q(X,R)$ denote the set of all pairs $(x,y) \in R$ such that for every index $\alpha \in \mathfrak{U}[X]$ there are points $x_1 \in x\alpha$, $y_1 \in y\alpha$ and an element $t \in T$ satisfying the conditions $(x_1,y_1) \in T$ and $(x_1t, y_1t) \in \alpha$. It is easy to show that

$$Q(R) = \bigcap_{\alpha \in \mathfrak{U}} \overline{\alpha T \cap R} = \bigcap_{\alpha \in \mathfrak{U}} \overline{(\alpha \cap R)T}.$$

If $R = X \times X$, then $Q(R) = Q$, where Q is the regionally proximal relation (see Subsection 1·3.3).

Let $M(R)$ denote the set of all pairs $(x,y) \in X \times X$ such that for each index $\alpha \in \mathfrak{U}[X]$ there exist points $x_1 \in x\alpha$, $y_1 \in y\alpha$ and a syndetic subset $A \subset T$ such that $(x_1,y_1) \in R$ and $(x_1t,y_1t) \in \alpha$ for all $t \in A$. If $R = X \times X$, then $M(R)$ coincides with the syndetically regionally proximal relation M.

Let us study some properties of the relations $Q(R)$ and $M(R)$. It is easy to show that $Q(R)$ and $M(R)$ are invariant closed relations and $M(R) \subset Q(R) \subset R$.

2·9.5 LEMMA: *If (X,T) is a distal transformation group with a compact phase space and R is a closed invariant relation in X, then $M(R) = Q(R)$.*

Proof: It will suffice to show that $Q(R) \subset M(R)$. The proof is similar to that of Theorem 1·6.10. Let $(x,y) \in Q(R)$ and $\alpha \in \mathfrak{U}[X]$. Choose an index $\gamma \in \mathfrak{U}[X]$ with $\gamma = \gamma^{-1}$ and $\gamma^3 \subset \alpha$. There exist points $x_0 \in x\gamma$, $y_0 \in y\gamma$ and an element $t_0 \in T$ such that $(x_0, y_0) \in R$ and $(x_0t_0,y_0t_0) \in \gamma$. Since (X,T) is distal, it follows from Theorem 1·6.3 that $(X \times X,T)$ is pointwise almost periodic. Hence there exists a syndetic subset $B \subset T$ with $(x_0t_0,y_0t_0)B \subset x_0t_0\gamma \times y_0t_0\gamma$. Then $(x_0,y_0)t_0B \subset \gamma^3 \subset \alpha$. The set $A \equiv t_0B$ is syndetic and $(x_0t,y_0t) \in \alpha$ for all $t \in A$. Hence $(x,y) \in M(R)$. ▌

2·9.6 LEMMA: *Let $\{(X_i,T) | i \in I\}$ be a family of transformation groups with compact phase spaces, $(X,T) = \Pi[(X_i,T) | i \in I]$ and let ϑ be the canonical isomorphism of the transformation group $(\Pi([X_i | i \in I] \times [X_i | i \in I]),T)$ onto $\Pi[(X_i \times X_i,T) | i \in I]$. Further, let R be a closed invariant equivalence relation on X, and R_i closed invariant equivalence relations on X_i $(i \in I)$ such that $R\vartheta = \Pi[R_i | i \in I]$. Then $M(R)\vartheta = \Pi[M(R_i) | i \in I]$.*

The proof is similar to that of Theorem 1·3.13, and therefore will be omitted.

2·9.7 THEOREM: *If (X,T) is a distal minimal transformation group with compact phase space and $R \subset X \times X$ is a closed invariant equivalence relation, then $Q(R)$ is also a closed invariant equivalence relation.*

Proof: We only need to show that the relation $Q(R)$ is transitive. Let $(x,y) \in Q(R)$ and $(y,z) \in Q(R)$. Let us prove that $(x,z) \in Q(R)$. Let $U(x)$ and $W(z)$ be arbitrary neighbourhoods and let $\alpha \in \mathfrak{U}[X]$. By Theorem 2·9.2 we can choose a neighbourhood $V(y)$ such that for each point $y' \in V(y)$ there exists a point $z' \in W(z)$ with $(y',z') \in Q(R)$. Select an index $\beta \in \mathfrak{U}[X]$ so that $\beta^4 \subset \alpha$. For $\beta \in \mathfrak{U}[X]$ and for the neighbourhoods $U(x)$ and $V(y)$ there are points $x_1 \in U(x)$, $y_1 \in V(y)$ and an element $t_1 \in T$ such that $(x_1,y_1) \in T$ and $(x_1t_1,y_1t_1) \in \beta$. Since (X,T) is a distal transformation group, the orbit closure of (x_1,y_1) in $(X \times X,T)$ is a compact minimal set. Hence for $\beta \in \mathfrak{U}$ there is a compact set $K \subset T$ with

$$(x_1\xi,y_1\xi)K \cap (x_1t_1\beta \times y_1t_1\beta) = \emptyset \tag{9.1}$$

for each element $\xi \in E(X,T)$.

Now choose an index $\delta \in \mathfrak{U}$ which satisfies the condition $\delta K \subset \beta$. Let $O(e)$ denote a neighbourhood of the identity e in $E(X,T)$ such that $x_1O(e) \subset U(x)$ and $y_1O(e) \subset V(y)$. By the choice of $V(y)$, there exists a point $z_1 \in W(z)$ with $(y_1,z_1) \in Q(R)$. For the neighbourhoods $y_1O(e)$ of the point y_1, the neighbourhood $W(z)$ of z_1 and the index $\delta \in \mathfrak{U}$ we can find an element $t_2 \in T$ and points $y_2 \in y_1O(e)$ and $z_2 \in W(z)$ so that $(y_2,z_2) \in R$ and $(y_2t_2,z_2t_2) \in \delta$. Therefore

$$(y_2t_2k,z_2t_2k) \in \beta, \qquad (k \in K). \tag{9.2}$$

Since $y_2 \in y_1O(e)$, there exists an element $\xi_0 \in O(e)$ with $y_2 = y_1\xi_0$. Then $x_2 \equiv x_1\xi_0 \in x_1O(e) \subset U(x)$.

By (9.1) we can assign to the element $\xi \equiv \xi_0t_2 \in E(X,T)$ an element $k_0 \in K$ with $(x_1\xi_0t_2,y_1\xi_0t_2)k_0 \in (x_1t_1\beta \times y_1t_1\beta)$. Since $(x_1t_1, y_1t_1) \in \beta$, we have

$$(x_1\xi_0t_2k_0,y_1\xi_0t_2k_0) \in \beta^3. \tag{9.3}$$

From (9.2,3) it follows that

$$(x_1\xi_0t_2k_0,z_2t_2k_0) \in \beta^4 \subset \alpha,$$

that is

$$(x_2t_2k_0,z_2t_2k_0) \in \alpha. \tag{9.4}$$

We recall that $x_2 \in U(x)$, $z_2 \in W(z)$. Since T is a closed invariant relation and $(x_1,y_1) \in T$, it follows that $(x_1\xi_0,y_1\xi_0) \in R$, *i.e.*, $(x_2,y_2) \in R$. Moreover, $(x_2,z_2) \in T$, because T is an equivalence relation and $(y_2,z_2) \in R$. Therefore (9.4) means that $(x,z) \in Q(R)$. ▌

2·9.8 Let X be a compact space and let φ be a homomorphism of a transformation group (X,T) onto a transformation group (Y,T). We shall say that (X,T) is an *EQUICONTINUOUS EXTENSION OF*

(Y,T) *BY THE HOMOMORPHISM* φ if, given an index $\alpha \in \mathcal{U}[X]$, there exists an index $\beta \in \mathcal{U}[X]$ such that $(x_1t, x_2t) \in \alpha$ for all $t \in T$, whenever $(x_1, x_2) \in \beta$ and $x_1\varphi = x_2\varphi$.

Observe that if (X,T) is an equicontinuous extension of a distal transformation group (Y,T), then (X,T) is also distal. (The proof is left to the reader). Equicontinuous extensions will be thoroughly studied in Section (3·13), and therefore we present now only one result which will be used in the proof of Theorem 2·9.18.

2·9.9 LEMMA: *Let (X,T) be a distal minimal transformation group with a compact phase space and let R and S be closed invariant equivalence relations on X such that $S \subset R$. Then the transformation group $(X/S,T)$ is an equicontinuous extension of the transformation group $(X/R,T)$ by the canoncial homomorphism of $(X/S,T)$ onto $(X/R,T)$ iff $S \supset Q(R)$.*

Proof: By Corollary 2·9.3, the relation S is open; therefore the family $\tilde{\mathcal{U}} = \{\tilde{\alpha} \mid \alpha \in \mathcal{U}[X]\}$, where

$$\tilde{\alpha} = \{(xS, yS) \mid xS \subset yS\alpha, yS \subset xS\alpha\},$$

and $xS = \{z \mid (x,z) \in S\}$ $(x \in X)$, is a uniformity-base of the compact space X/S.

Since the canonical projection $f: X \to X/S$ is uniformly continuous, for every index $\alpha \in \mathcal{U}[X]$ there is an index $\delta \in \mathcal{U}[X]$ such that $(xS, yS) \in \tilde{\alpha}$ whenever $(x,y) \in \delta$.

Suppose that $(X/S,T)$ is an equicontinuous extension of $(X/R,T)$ and let $(x,y) \in Q(R) = \bigcap_{\alpha \in \mathcal{U}} \overline{(\alpha \cap R)T} \subset T$. For $\delta \in \mathcal{U}[X]$ there exist points $x_1 \in X$, $y_1 \in X$ and an element $t_1 \in T$ such that $(x_1, y_1) \in R \cap (x\delta \times y\delta)$ and $(x_1t_1, y_1t_1) \in \delta$. Set $\bar{x} = xS$, $\bar{x}_1 = x_1S$, $\bar{y} = yS$, $\bar{y}_1 = y_1S$. Then

$$\bar{x}_1 \in \bar{x}\tilde{\alpha}, \qquad \bar{y}_1 \in \bar{y}\tilde{\alpha}, \qquad (\bar{x}_1t_1, \bar{y}_1t_1) \in \tilde{\alpha}. \tag{9.5}$$

Since $(x_1, y_1) \in R$ and $\tilde{\alpha}$ is an arbitrary index in $\mathcal{U}[X]$, it follows from (9.5) that $\bar{x} = \bar{y}$, and hence $(x,y) \in S$. Thus $Q(R) \subset S$.

Conversely, let $Q(R) \subset S$. We shall prove that $(X/S,T)$ is an equicontinuous extension of $(X/R,T)$. Let $x,y \in X$, $xR = yR$, but $xS \neq yS$. If $u \in xS$, $v \in yS$, then $uS \neq vS$, and hence (u,v) does note belong to $Q(R)$. Then there exists an index $\gamma \in \mathcal{U}[X]$ which satisfies the condition $(u\gamma \times v\gamma) \cap (\gamma \cap R)T = \emptyset$. Since the set $xS \times yS \subset X \times X$ is compact, we can choose an index $\delta \in \mathcal{U}[X]$ with

$$(xS\delta \times yS\delta) \cap (\delta \cap R)T = \emptyset. \tag{9.6}$$

Now select an index $\alpha \in \mathcal{U}[X]$ so that $x\alpha S \subset xS\delta$ holds for each point $x \in X$.

Denote $\bar{x} = xS$, $\bar{y} = yS$. It follows from (9.6) that

$$(\bar{x}_1, \bar{y}_1)T \cap \tilde{\delta} = \emptyset.$$

for every two points $\bar{x}_1 = x_1S$, $\bar{y}_1 = y_1S$ satisfying the conditions $\bar{x}_1 \in \bar{x}\tilde{\alpha}$, $\bar{y}_1 \in \bar{y}\tilde{\alpha}$, $(x_1,y_1) \in R$. Therefore $(X/S,T)$ is an equicontinuous extension of the transformation group $(X/R,T)$. ▮

2·9.10 Let (X,T) be a distal minimal transformation group with a compact phase space. For each ordinal λ we define by induction a relation $Q_\lambda \subset X \times X$ as follows:

For $\lambda = 0$ we set $Q_0 = X \times X$. Suppose that relations Q_λ have already been defined for all ordinals λ, $\lambda < \mu$. If $\mu = \lambda + 1$ for some ordinal λ, then we set

$$Q_\mu \equiv Q_{\lambda+1} = Q(Q_1) \equiv \bigcap_{\alpha \in \mathcal{U}} \overline{(\alpha \cap Q_\lambda)T}.$$

If μ is a limit ordinal, we define

$$Q_\mu = \bigcap_{\lambda<\mu} Q_\lambda.$$

2·9.11 THEOREM: *Let (X,T) be a distal minimal transformation group with a compact phase space. For every ordinal λ, the relation Q_λ is a closed and open invariant multiplicative equivalence relation.*

Proof: This is obvious for $\lambda = 0$. Suppose that the statement is true for all λ, $\lambda < \mu$. Let us prove that it is true for μ also. If $\mu = \lambda + 1$, then $Q_{\lambda+1}$ is a closed invariant equivalence relation by Theorem 2·9.7. It follows from Lemmas 2·9.5,6 that $Q_{\lambda+1}$ is multiplicative. The relation $Q_{\lambda+1}$ is open by Corollary 2·9.3.

Now let μ be a limit ordinal. Since $Q_\mu = \cap [Q_\lambda | \lambda < \mu]$, we see that Q_μ is a closed invariant equivalence relation. Q_μ is open by Corollary 2·9.3. It is not difficult to prove that the relation Q_μ is multiplicative. ▮

2·9.12 LEMMA: *Let X be a compact metrizable space and (S,T) be a distal minimal transformation group. Then there exists an ordinal ϑ such that $Q_\vartheta = \Delta(X) \equiv \{(x,x) | x \in X\}$.*

Proof: For each ordinal λ, the relation Q_λ is a closed subset of $X \times X$ and $\lambda < \mu$ implies $Q_\lambda \supset Q_\nu$. Hence there exists an ordinal ϑ such that

$$Q_{\vartheta+1} \equiv \bigcap_{\alpha \in \mathcal{U}} \overline{(\alpha \cap Q_\vartheta)T} = Q_\vartheta. \tag{9.7}$$

Let d be a compatible metric in X and $\alpha_n = \{(x,y) | d(x,y) < 1/n\}$. It follows from (9.7) that $(\alpha_n \cap Q_\vartheta)T$ is a dense open subset of the subspace Q_ϑ. Therefore the set $\bigcap_{n=1}^{\infty} (\alpha_n \cap Q_\vartheta)T$ is also dense in Q_ϑ, that is

$$Q_\vartheta = \bigcap_{n=1}^{\infty} (\alpha_n \cap Q_\vartheta)T \subset \bigcap_{n=1}^{\infty} \alpha_n T. \tag{9.8}$$

Since (X,T) is a distal transformation group, we get $P(X) \equiv \bigcap_{n=1}^{\infty} \alpha_n T = \Delta(X)$, and hence $Q_\vartheta = \Delta(X)$.∎

2•9.13 LEMMA: *Let T be a σ-compact topological group (that is, T can be expressed as an union $\cup[K_i \mid i = 1,2,\dots]$ of compact subsets $K_i \subset T$). Then every transformation group (X,T,π) with a compact phase space can be isomorphically embedded into a Cartesian product of a family of transformation groups with metrizable phase spaces.*

Proof: Since X is a compact Hausdorff space, there exists a family $\Phi = \{\varphi : X \to \mathbb{R}\}$ of continuous functions which separates the points of X. Given $x \in X$ and $\varphi \in \Phi$, we define a function $\varphi_x : T \to \mathbb{R}$ by $t\varphi_x = (x\pi^t)\varphi$. Set $A_\varphi = \{\varphi_x \mid x \in X\}$ $(\varphi \in \Phi)$. Since X is compact, the family A_φ is equicontinuous. Provide A_φ with the metric:

$$\rho(\varphi_x,\varphi_y) = \sup_n [\min\{ \max_{t \in \bigcup_{i=1}^{n} K_i} |t\varphi_x - t\varphi_y|, 1/n\}], \qquad (x,y \in X)$$

Define a transformation group $(A_\varphi,T,\pi_\varphi)$ by $(\varphi_x,\tau)\pi_\varphi = \varphi_y$, where $y = x\pi^\tau$ $(x \in X,\ \tau \in T)$. It is easy to see that the mapping which assigns to $x \in X$ the function φ_x is a homomorphism of (X,T,π) onto $(A_\varphi,T,\pi_\varphi)$. Since the family Φ separates the points of X, this gives an isomorphism of (X,T,π) into the Cartesian product $\Pi[(A_\varphi,T,\pi_\varphi) \mid \varphi \in \Phi]$.∎

2•9.14 LEMMA: *Let T be a σ-compact topological group, X a compact space and (X,T) a distal minimal transformation group. Then $Q_\vartheta = \Delta$ for some ordinal ϑ.*

Proof: This follows from Lemmas 2•9.12,13 and Theorem 2•9.11.∎

2•9.15 Let $\{(X_\lambda,T,\pi_\lambda)\}$ be a transfinite sequence of transformation groups, defined for all ordinals λ smaller than some limit ordinal μ, and suppose that for each two ordinals λ and ν, $\nu < \lambda < \mu$, a homomorphism φ_ν^λ of (X_λ,T) onto (X_ν,T) is defined in such a way that

$$\varphi_{\lambda_2}^{\lambda_3} \circ \varphi_{\lambda_1}^{\lambda_2} = \varphi_{\lambda_1}^{\lambda_3}, \tag{9.9}$$

whenever $\lambda_1 < \lambda_2 < \lambda_3 < \mu$.

Consider the subspace Y of the Cartesian product $\Pi[X_\lambda \mid \lambda < \mu]$, consisting of all points $y = \{x_\lambda \mid \lambda < \mu\}$ with $x_\nu = x_\lambda\varphi_\nu^\lambda$ whenever $\nu < \lambda < \mu$. The map $\pi : Y \times T \to Y$ defined by $(\{x_\lambda\}t)\pi = \{(x_\lambda,t)\pi_\lambda\}$

determines a transformation group (Y,T,π), which is called the *PROJECTIVE* (or *INVERSE*) *LIMIT TRANSFORMATION GROUP* of the transfinite sequence of transformation groups $\{(X_\lambda,T,\pi_\lambda)|\lambda < \mu\}$ with homomorphisms φ_ν^λ $(\nu < \lambda < \mu)$.

Let $\{(X_\lambda,T,\pi_\lambda)|0 \leqslant \lambda \leqslant \mu\}$ be a transfinite sequence of transformation groups with homomorphisms φ_ν^λ $(0 \leqslant \nu \leqslant \lambda \leqslant \mu)$ satisfying (9.9) whenever $0 \leqslant \lambda_1 < \lambda_2 < \lambda_3 \leqslant \mu$, and such that $\varphi_\lambda^\lambda : X_\lambda \to X_\lambda$ is the identity map. Assume also that for each limit ordinal γ, $\lambda \leqslant \mu$, the transformation group (X_γ,T) is the projective limit of the transfinite sequence $\{(X_\lambda,T)|\lambda < \gamma\}$ with homomorphisms φ_ν^λ $(\nu < \lambda < \gamma)$. Then $\{(X_\lambda,T,\pi_\lambda)|0 \leqslant \lambda \leqslant \mu\}$, $\{\varphi_\nu^\lambda\}$ is called a *PROJECTIVE SYSTEM OF TRANSFORMATION GROUPS*.

2·9.16 LEMMA: *The projective limit of a projective system of minimal transformation groups with compact phase spaces is also a minimal transformation group. If, moreover, all transformation groups from the projective system are equicontinuous, then the projective limit is also equicontinuous.*

The proof is left to the reader as an easy exercise.

2·9.17 A projective system $\{(X_\lambda,T,\pi_\lambda)|0 \leqslant \lambda \leqslant \mu\}$, $\{\varphi_\nu^\lambda|0 \leqslant \nu \leqslant \lambda \leqslant \mu\}$ is said to be an *F-SYSTEM* if for each ordinal λ, satisfying the condition $\lambda + 1 \leqslant \mu$, the transformation group $(X_{\lambda+1},T)$ is an equicontinuous extension of (X_λ,T) by the homomorphism $\varphi_\lambda^{\lambda+1}$. We shall say that (X,T) belongs to the class F if there exists an F-system $\{(X_\lambda,T,\pi_\lambda)|0 \leqslant \lambda \leqslant \mu\}$, $\{\varphi_\nu^\lambda|0 \leqslant \nu \leqslant \lambda \leqslant \mu\}$ such that (X,T) is isomorphic to (X_μ,T) and the transformation group (X_0,T) is trivial, *i.e.*, the space X_0 consists of a single point. It is easy to see that every transformation group $(X,T) \in F$ with a compact phase space is distal.

2·9.18 THEOREM: *Let X be a compact space, let (X,T) be a distal minimal transformation group, and suppose that, in addition, at least on of the following conditions is satisfied:*

(a): *The space X is metrizable;*

(b): *The group T is σ-compact.*

Then (X,T) belongs to the class F.

Proof: The hypotheses assure the existence of an ordinal with $Q_\vartheta = \Delta$. Set $(X_\lambda,T) = (X/Q_\lambda,T)$ $(0 \leqslant \lambda \leqslant \vartheta)$. Since $0 \leqslant \nu < \lambda \leqslant \vartheta$ implies $Q_\nu \supset Q_\lambda$, there exists a canonical homomorphism φ_ν^λ of (X_λ,T) onto (X_ν,T). It follows from Lemma 2·9.9 that $(X_{\lambda+1},T)$ is an equicontinuous extension of (X_λ,T) by the homomorphism $\varphi_\lambda^{\lambda+1}$ $(\lambda + 1 \leqslant \vartheta)$. If μ is a limit ordinal, $\mu \leqslant \vartheta$, then $Q_\mu = \bigcap_{\lambda<\mu} Q_\lambda$.

This implies that the transformation group (X_μ,T) is isomorphic to the projective limit of the transfinite sequence $\{(X_\lambda,T)|\lambda < \mu\}$ with homomorphisms φ_ν^λ $(0 \leqslant \nu < \lambda < \mu)$. Since $Q_0 = X \times X$, the transformation group (X_0,T) is trivial, and as $Q_0 = \Delta$, the transformation group (X_ϑ, T) coincides with (X,T). Thus (X,T) belongs to the class **F**.▌

2·9.19 To conclude this section we give an example of a distal, but not equicontinuous, minimal flow on a three-dimensional torus.

Let $\mathbb{K}$ be the quotient group of the real line $\mathbb{R}$ by the subgroup $\{2\pi n \mid n \in \mathbb{Z}\}$, let $\Phi: \mathbb{K}^2 \to \mathbb{R}$ be a continuous function and let μ be an irrational number.

2·9.20 LEMMA: *The flow $(\mathbb{K}^3, \mathbb{R}, \delta)$ defined on the three-dimensional torus $\mathbb{K}^3$ by the system of differential equations*

$$\frac{d\varphi}{dt} = 1, \qquad \frac{d\vartheta}{dt} = \mu, \qquad \frac{d\psi}{dt} = \Phi(\varphi,\vartheta), \tag{9.10}$$

is distal. There exists an integer k and a real number γ such that the flow on $\mathbb{K}^3$ defined by

$$\frac{d\varphi}{dt} = 1, \qquad \frac{d\vartheta}{dt} = \mu, \qquad \frac{d\psi}{dt} = k\Phi(\varphi,\vartheta) + \gamma, \tag{9.11}$$

is distal and minimal.

Proof: Let $a = (\varphi_1, \vartheta_1, \psi_1)$ and $b = (\varphi_2, \vartheta_2, \psi_2)$ be distinct points of the torus $\mathbb{K}^3$. If either $\varphi_1 \not\equiv \varphi_2 \pmod{2\pi}$ or $\vartheta_1 \not\equiv \vartheta_2 \pmod{2\pi}$, then the points a and b are evidently distal. If $\varphi_1 \equiv \varphi_2 \pmod{2\pi}$ and $\vartheta_1 \equiv \vartheta_2 \pmod{2\pi}$, then $\psi_1 \not\equiv \psi_2 \pmod{2\pi}$. In this case the distance between the points $a\delta^t$ and $b\delta^t$ does not depend on $t \in \mathbb{R}$. Hence $(\mathbb{K}^3, \mathbb{R}, \delta)$ is a distal flow.

By Theorem 1·6.3, the orbit closure of each point $a \in \mathbb{K}^3$ is a minimal set. Let $a_0 = (0,0,0)$. Let us prove that the intersection F of the orbit closure of the point a_0 with the circle $(0,0,\mathbb{K})$ is a subgroup of the group $(0,0,\mathbb{K})$. Observe that each rotation of the circle $(0,0,\mathbb{K})$ defines an autormorphism of the transformation group $(\mathbb{K}^3, \mathbb{R}, \delta)$. Therefore if $a_1 \equiv (0,0,\psi_1) \in \overline{a_0\mathbb{R}}$ and $a_2 \equiv (0,0,\psi_2) \in \overline{a_0\mathbb{R}}$, then $(0,0,\psi_1 + \psi_2) \in \overline{a_2\mathbb{R}}$. Since $\overline{a_0\mathbb{R}}$ is a minimal set and $a_2 \in \overline{a_0\mathbb{R}}$, we get $\overline{a_0\mathbb{R}} = \overline{a_2\mathbb{R}}$. Thus $(0,0,\psi_1) \in F$ and $(0,0,\psi_2) \in F$ imply $(0,0,\psi_1 + \psi_2) \in F$, *i.e.*, F is a subsemigroup of the group $(0,0,\mathbb{K})$. It easily follows that F is a subgroup of $(0,0,\mathbb{K})$.

There are two possibilities: (a) $F = (0,0,\mathbb{K})$. It is easy to see that in this case the flow $(\mathbb{K}^3, \mathbb{R}, \delta)$ is minimal, and therefore we may set $\gamma = 0$ and $k = 1$; (b) F is a proper subgroup of $(0,0,\mathbb{K})$. Hence F is a finite cyclic group. Let k denote the order of the group F. If we take $k\Phi(\varphi,\vartheta)$ in place of $\Phi(\varphi,\vartheta)$ in the system (9.10), then the intersection F' of the orbit closure of a_0 (under the new flow) with the circle $(0,0,\mathbb{K})$ consists only of the point a_0. If we take φ to be a real number, rationally independent of 1 and μ, then the flow on $\mathbb{K}^3$ defined by (9.11) becomes distal and minimal.▌

As shown in the book by Nemytskiĭ and Stepanov [*1*], (p. 424), there exist an irrational number μ and a continuous function Φ: $\mathbb{K}^2 \to \mathbb{R}$ such that given $m > 0$ and $\varepsilon > 0$ we can choose an $a \in \mathbb{R}$ and $t \in \mathbb{R}$ in such a way that $|a - 2k_0\pi| < \varepsilon$ for some $k_0 \in \mathbb{Z}$ and nevertheless

$$\left|\int_0^t \Phi(\tau,\mu\tau)d\tau - \int_0^t \Phi(\tau,a + \mu\tau)d\tau\right| > m.$$

Fix this function Φ and select two numbers γ and k according to Lemma 2·9.20. Then the flow on $\mathbb{K}^3$ defined by the equations (9.11) is distal and minimal. It follows from the abovementioned property of Φ that this flow is not equicontinuous.

In the next section we shall give some more examples of distal, but not equicontinuous, minimal flows.

REMARKS AND BIBLIOGRAPHICAL NOTES

Theorem 2·9.18 is a slight generalization of a theorem obtained by Furstenberg [*1*]. The proof given here is due to Bronštein [*6*]. It is much simpler and shorter than the original proof. This is achieved with the use of Theorem 2·9.7, which takes care of the main difficulties of the proof. The concept of an equicontinuous extension is a generalization of an isometric extension introduced by Furstenberg [*1*]. Lemma 2·9.20 and the example are due to Bronštein [*4*]. Some questions related to Furstenberg's Theorem are considered by Knapp [*1*,*2*] and Namioka [*1*].

§(2·10): TRANSITIVE DISTAL TRANSFORMATION GROUPS AND NIL-FLOWS

2·10.1 Let G be a topological group, H a closed subgroup of G and $G/H = \{Hg \mid g \in G\}$ the quotient space. If H is a syndetic subset of G (*i.e.*, there exists a compact subset $K \subset G$ with $G = HK$), then the space G/H is compact. Let F denote the neighbourhood filter at the identity of the topological group G, and for a closed subgroup H of G with $V \in F$ let

$$\alpha_V = \{(Ha,Hb) \mid a,b \in G, b^{-1}a \in V\}.$$

The set $\{\alpha_V \mid V \in F\}$ is a uniformity base of some Hausdorff uniformity $\mathcal{U}$, which is compatible with the quotient topology in G/H. In what follows, when considering G/H as an uniform space, we shall always mean the uniformity $\mathcal{U}$.

Let H be a closed syndetic subgroup of G. Define a transitive topological transformation group $(G/H,G,\pi)$ by

$$(Ha,g)\pi = Hag, \qquad (Ha \in G/H, g \in G).$$

In this section necessary and sufficient conditions are obtained for $(G/H,G,\pi)$ to be distal. Besides that, an important class of transformation groups on homogeneous spaces, namely, that of nil-flows, is investigated.

2·10.2 A subgroup H of a topological group G is called a *d*-SUBGROUP if there exists a transfinite sequence

$$G \equiv G_0 \supset G_1 \supset \cdots \supset G_\alpha \supset \cdots \supset G_\vartheta \equiv H, \qquad (10.1)$$

of closed subgroups G_α $(1 \leqslant \alpha \leqslant \vartheta)$ of G, which satisfies the following two conditions:

(d_1): If $\alpha < \vartheta$, then the transformation group $(G_\alpha/G_{\alpha+1},G_\alpha, \pi_\alpha)$ defined by $(G_{\alpha+1}a,g)\pi_\alpha = G_{\alpha+1}ag$ $(a,g \in G_\alpha)$ is equicontinuous;

(d_2): If β is a limit ordinal, then:

$$G_\beta = \bigcap [G_\gamma | \gamma < \beta].$$

2·10.3 Let H be a closed syndetic subgroup of G and suppose that there exists a decreasing normal series reaching to H, *i.e.*, a series of the form (10.1), such that $G_{\alpha+1}$ is a normal subgroup of G_α $(\alpha < \vartheta)$ and $G_\beta = \bigcap [G_\gamma | \gamma < \beta]$ whenever β is a limit ordinal $(\beta \leqslant \vartheta)$. Then H is a *d*-subgroup. For, in this case $G_\alpha/G_{\alpha+1}$ is a compact topological group, and hence the transformation group $(G_\alpha/G_{\alpha+1},G_\alpha,\pi\ )$ is equicontinuous $(\alpha < \vartheta)$.

2·10.4 THEOREM: *Let G be a topological group, let H be a syndetic closed subgroup of G and suppose in addition that at least one of the following conditions is satisfied:*

(a): *The space G/H is metrizable;*

(b): *G is a σ-compact group.*

Suppose, further, that the transformation group $(G/H,G,\pi)$ *is distal. Then H is a d-subgroup of G.*

Proof: Since G acts transitively on $X \equiv G/H$, the transformation group $(G/H,G) \equiv (X,G)$ is minimal. The space X is compact because H is a syndetic subgroup of G.

It follows from Theorem 2·9.18 that there exists a transfinite sequence of closed invariant equivalence relations Q_α $(1 \leqslant \alpha \leqslant \vartheta)$ that satisfies the following three conditions:

(1): $Q_1 \supset Q_2 \supset \cdots \supset Q_\alpha \supset \cdots \supset Q_\vartheta = \Delta \equiv \{(x,x) | x \in X\}$;

(2): If β is a limit ordinal, $\beta \leqslant \vartheta$, then $Q_\beta = \bigcap_{\gamma<\beta} Q_\gamma$;

(3): The transformation group $(X/Q_{\alpha+1},G)$ is an equicontinuous extension of the transformation group $(X/Q_\alpha,G)$ by the canonical projection of $X/Q_{\alpha+1}$ onto X/Q_α.

Let φ_α denote the canonical homomorphism of $(X,G) \equiv (G/H,G)$ onto $(X/Q_\alpha,G)$ and let:

$$G_\alpha = \{a | a \in G, Ha\varphi_\alpha = H\varphi_\alpha\}.$$

Since φ_α is a homomorphism, $a \in G_\alpha$ implies $Hag\varphi_\alpha = Hg\varphi_\alpha$ $(g \in G_\alpha)$, *i.e.*, $ag \in G_\alpha$ for every element $g \in G_\alpha$. Therefore G_α is a subgroup of G. Clearly G_α is a closed subset.

The transformation group $(G/H,G)$ induces a transformation group $(G/G_\alpha,G)$. Since $H \subset G_\alpha$, G_α is a syndetic subgroup of G. It follows that the transformation group $(G/G_\alpha,G)$ is isomorphic to the transformation group $(X/Q_\alpha,G)$. Therefore $(G/G_{\alpha+1},G)$ is an equicontinuous extension of the transformation group $(G/G_\alpha,G)$ by the canonical map $\psi_\alpha^{\alpha+1}:G/G_{\alpha+1} \to G/G_\alpha$. Since the set $\{G_{\alpha+1}g | g \in G_\alpha\}$ coincides with the preimage of the point $G_\alpha \in G/G_\alpha$ under $\psi_\alpha^{\alpha+1}$, the transformation group $(G_\alpha/G_{\alpha+1},G_\alpha)$ is equicontinuous. Given a limit ordinal β, the condition $Q_\beta = \bigcap [Q_\gamma | \gamma < \beta]$ implies that $G_\beta = \bigcap [G_\gamma | \gamma < \beta]$, consequently $G_\vartheta = H$. Thus H is a d-subgroup of G.▌

2·10.5 THEOREM: *If H is a closed syndetic d-subgroup of the group G, then the transformation group $(G/H,G,\pi)$ is distal.*

Proof: Since H is a d-subgroup of G, there exists a transfinite sequence (10.1) satisfying the conditions (d_1) and (d_2).

Let us prove that the transformation group $(G/G_\alpha,G)$ is distal for each ordinal α $(0 \leqslant \alpha \leqslant \vartheta)$. Letting $\alpha = \vartheta$, we shall get the required assertion. We proceed by induction. If $\alpha = 0$, the statement is trivial. Suppose that it is true for all α with $\alpha < \beta$. We shall prove that the transformation group $(G/G_\beta,G)$ is distal. Let us consider two cases separately. (1) $\beta = \alpha + 1$. Let

$$a \in G, \qquad b \in G, \qquad G_\beta a \neq G_\beta b. \tag{10.2}$$

We shall prove that the points $G_{\alpha+1}a$ and $G_{\alpha+1}b$ are distal under $(G/G_{\alpha+1},G)$. If $G_\alpha a \neq G_\alpha b$, then by the inductive assumption the points $G_\alpha a$ and $G_\alpha b$ are distal under $(G/G_\alpha,G)$. Since the transformation group $(G/G_{\alpha+1},G)$ is homomorphically mapped onto $(G/G_\alpha,G)$, it follows that $G_{\alpha+1}a$ and $G_{\alpha+1}b$ are also distal.

If $G_\alpha a = G_\alpha b$, then there exists an element $t \in G_\alpha$ with $a = tb$. We show that also in this case $G_{\alpha+1}a$ and $G_{\alpha+1}b$ are distal. Suppose that, to the contrary, there exist an element $c \in G$ and a net $\{g_i | i \in I\}$ of elements $g_i \in G$ such that:

$$\{G_{\alpha+1}ag_i\} \equiv \{G_{\alpha+1}tbg_i\} \to G_{\alpha+1}c, \qquad \{G_{\alpha+1}bg_i\} \to G_{\alpha+1}c.$$

Hence it follows that the points $G_{\alpha+1}$ and $G_{\alpha+1}t$ are proximal under $(G/G_{\alpha+1},G)$. Since $t \in G_\alpha$ and G_α is a syndetic subgroup of G, the points $G_{\alpha+1}$ and $G_{\alpha+1}t$ are proximal also under the transformation group $(G_\alpha/G_{\alpha+1},G_\alpha)$. By hypothesis, this transformation group is equicontinuous, hence $G_{\alpha+1} = G_{\alpha+1}t$. Then $G_{\alpha+1}b = G_{\alpha+1}tb = G_{\alpha+1}a$, but this contradicts (10.2).▌

2·10.6 COROLLARY: *If G is a topological group, H is a closed syndetic subgroup of G and there exists a decreasing normal series containing H, then the transformation group $(G/H,G)$ is distal.*

Proof: In fact, the hypotheses imply, by Remark 2·10.3, that H is a d-subgroup of G. It remains to refer to Theorem 2·10.5. ▮

2·10.7 We recall some definitions and facts from group theory (*see*, for example, Kurosh [*1*]).

Let G be a group and let A and B be subgroups of G. The commutator of A and B is defined to be the subgroup $[A,B]$ of G generated by all commutators $aba^{-1}b^{-1}$ $(a \in A,\ b \in B)$. A sequence of subgroups of G

$$G \equiv G_0' \supset G_1' \supset \cdots \supset G_k' \supset G_{k+1}' \supset \cdots ,$$

where $G_{k+1}' = [G_k',G]$, is called the *LOWER CENTRAL SERIES* of the group G.

A group G is called a *NILPOTENT GROUP* if there exists a number k such that $G_{k+1}' = \{e\}$, where e is the identity of G, that is, the lower central series reaches $\{e\}$ in a finite number of steps. The number k is called the *LENGTH OF THE LOWER CENTRAL SERIES*. It is well known that each subgroup H of a nilpotent group G is *ATTAINABLE*, *i.e.*, there exists a finite normal series of the form:

$$G \equiv G_0 \supset G_1 \supset \cdots \supset G_\ell \equiv H. \qquad (10.3)$$

If G is a nilpotent topological group and H is a closed subgroup of G, then without loss of generality we may suppose that all members of the series (10.3) are closed.

2·10.8 COROLLARY: *Let G be a nilpotent topological group and let H be a closed syndetic subgroup of G. Then the transformation group $(G/H,G,\pi)$ is distal.*

2·10.9 Later we shall use the following obvious fact. Let G and T be topological groups, let H be a closed syndetic subgroup of G and let $\varphi : T \to G$ be a continuous homomorphism. If the transformation group $(G/H,G)$ is distal, then the transformation group $(G/H,T,\rho_\varphi)$, where $(Hg,t)\rho_\varphi = Hg(t\varphi)$, is also distal We shall sometimes write $(G/H,\varphi : T \to G)$ or even $(G/H,\varphi)$ instead of $(G/H,T,\rho_\varphi)$.

2·10.10 Let G be a connected simply connected nilpotent Lie group (Pontrjagin [*1*]) and H a discrete syndetic subgroup of G. The homogeneous space $G/H = \{Hg \mid g \in G\}$ is called a *NILPOTENT MANIFOLD* (or simply a *NILMANIFOLD*). Maltsev [*1*] has proved that whenever G_i' is a member of the lower central series, then $H \cap G_i'$ is a discrete syndetic subgroup of G_i' $(i = 1,\ldots,k)$. It is also known, that, given an element $g \in G$, there exists a one-parameter subgroup passing through g, *i.e.*, there exists a continuous homomorphism φ of the group $\mathbb{R}$ of real numbers into the group G such that $g \in \mathbb{R}\varphi$.

Let $A = [G,G]$. Since the group G/A is commutative, the group $T = G/AH$ is connected commutative compact Lie group, *i.e.*, a finite-dimensional torus (Pontrjagin [*1*]), which is called the *ADJOINT TORUS OF THE NILMANIFOLD* G/H.

Let $\varphi:\mathbb{R} \to G$ be a continuous homomorphism. It was noted above that $(G/H,\varphi)$is a distal flow. The flow $(G/H,\varphi)$ is called a *NIL-FLOW*. It induces a flow $(T,\mathbb{R},\sigma)$ onto the adjoint torus $T = G/AH$. Since T is a compact group, the flow $(T,\mathbb{R},\sigma)$ is equicontinuous.

If $\varphi:\mathbb{R} \to G$ is any one-parameter subgroup, then the map $\psi:\mathbb{R} \to G/AH$ defined by $t\psi = AH(t\varphi)$ is a group homomorphism. Since $T \equiv G/AH$ is a finite-dimensional torus, there exist homomorphisms $\psi:\mathbb{R} \to T$ such that $\overline{\mathbb{R}\psi} = T$. In this case the flow (T,ψ) is minimal.

2·10.11 THEOREM: *Let G be a connected simply connected nilpotent Lie group, $A = [G,G]$, H a discrete syndetic subgroup of G and let $\varphi:\mathbb{R} \to G$ be a continuous homomorphism. The flow $(G/H,\mathbb{R},\rho_\varphi) \equiv (G/H,\varphi)$ is minimal iff the flow $(G/AH,\mathbb{R},\sigma)$ induced by the flow $(G/H,\varphi)$ on the adjoint torus $T \equiv G/AH$ is minimal.*

Proof: The necessity is obvious. To prove the sufficiency, suppose that the flow $(G/AH,\mathbb{R},\sigma)$ is minimal. We have to show that $(G/H,\varphi)$ is also minimal. We shall proceed by induction on the length k of the lower central series of the group G:

$$G \equiv G_0' \supset G_1' \supset \cdots \supset G_k' \supset G_{k+1}' = \{e\}.$$

If $k = 0$, the group G is commutative, $A = \{e\}$, therefore the assertion is trivial in this case.

Suppose that our claim has been proved for all connected simply connected nilpotent Lie groups with the length of the lower central series smaller than k $(k > 0)$, and suppose that G is a group with the length of its lower central series equal to k. Clearly the group $N \equiv G_k'$ is contained in the centre of G. Then $G_1 \equiv G/N$ is a connected simply connected nilpotent Lie group, $H \cap N$ is a syndetic subgroup of N, the groups NH/H and $N/(H \cap N)$ are isomorphic, and HN/H is a syndetic subgroup of G_1. Moreover, the length of the lower central series of G_1 is equal to $(k - 1)$. The group A/N is the commutator of the group G_1. Since the flow $(G/AH,\mathbb{R},\sigma)$ is minimal, the flow $(G/ANH,\mathbb{R},\sigma)$ is also minimal. Hence by the inductive assumption, the set G/NH is minimal under the flow induced by $(G/H,\varphi)$.

We shall prove that $(G/H,\varphi)$ is also minimal. By Corollary 2·10.8, the flow $(G/H,\varphi)$ is distal. It follows from Theorem 1·6.4 that the set $M \equiv \{H(t\varphi) | t \in \mathbb{R}\}$ is minimal under $(G/H,\varphi)$. Denote $M \cap (NH/H)$ by S. Since the group N is contained in the centre of G, every translation by an element of the group N is an automorphism of the transformation group $(G/H,\varphi)$. Hence it follows that S is a closed subsemigroup of the compact group NH/H. Since every minimal right ideal of the semigroup S contains an idempotent, namely the identity of NH/H, we conclude that S is a subgroup of NH/H. Let us prove that S coincides with NH/H and, consequently, that $(G/H,\varphi)$ is a minimal transformation group. Suppose that, to the contrary, S is a proper subgroup of NH/H. Without loss of generality we may then suppose that S is the identity subgroup. Since the flow on G/NH is minimal by the inductive hypothesis, the flow on M is mapped isomorphically onto the minimal flow $(G/NH,\varphi)$. It follows that for each element $g \in G$ there exists an element $(N\cap H)b$ of the group $N/(N\cap H)$ such that $Hg(N\cap H)b \equiv Hgb \in M$.

We shall show that $(N\cap H)b$ is uniquely defined. Suppose the contrary to be true. Then for some element $g \in G$ there are elements $b_1 \in N$ and $b_2 \in N$ such that $Hgb_1 = Hgb_2$, but $(N\cap H)b_1 \neq (N\cap H)b_2$. Set $b_0 = b_1^{-1}b_2$. Then $Hgb_2 = Hgb_1b_0 \in M\cap Mb_0$. Since the set M is minimal and N is a subset of the centre of G, the set Mb_0 is also minimal. Therefore $Hgb_2 \in M\cap Mb_0$ implies $Mb_0 = M$, and thus $Hb_0 \in M$, *i.e.*, $Hb_0 \in S$ and $Hb_0 \neq H$. This contradicts our assumption that S is the identity group.

Let us consider the map $\Phi: G/H \to (G/NH)\times(N/N\cap H)$, where $(Hg)\Phi = (NHg,(N\cap H)b)$ and let us choose the element $(N\cap H)b$ in such a way that $Hg(N\cap H)b \equiv Hgb \in M$. It follows from the above that Φ is well defined. We shall show now that Φ is bijective. If $(Hg_1)\Phi = (Hg_2)\Phi$, then $NHg_1 = NHg_2$, *i.e.*, $Hg_1 = Hg_2b$ for some $b \in N$. If the element $(N\cap H)b_1$ is such that $Hg_1b_1 \in M$, then also $Hg_2bb_1 \in M$. Therefore the second coordinate of the point $(Hg_2)\Phi$ is equal to $(N\cap H)bb_1$. From the hypothesis $(Hg_1)\Phi = (Hg_2)\Phi$ it follows that $(N\cap H)b_1 = (N\cap H)bb_1$, *i.e.*, $b \in N\cap H$. Therefore $Hg_1 = Hg_2b = Hg_2$. Thus Φ is injective.

Now we shall prove that Φ is continuous. If the net $\{Hg_i \mid i \in I\}$ converges to an element $Hg \in G/H$, then $\{NHg_i \mid i \in I\} \to NHg$. For each $i \in I$ there exists an element $(N\cap H)b_i$ such that $b_i \in N$ and $Hg_ib_i \in M$ $(i \in I)$. Let $(N\cap H)b_0$ be the second coordinate of $(Hg)\Phi$, that is, $Hgb_0 \in M$. We shall prove that $\{(N\cap H)b_i \mid i \in I\} \to (N\cap H)b_0$. Suppose the contrary holds. Since the group $N/(N\cap H)$ is compact, there exists a subnet $\{(N\cap H)b_j \mid j \in \mathcal{J}\}$ of the net $\{(N\cap H)b_i \mid i \in I\}$ such that $\{(N\cap H)b_j \mid j \in \mathcal{J}\} \to (N\cap H)b_1$, where $b_1 \in N$ and, moreover, $(N\cap H)b_1 \neq (N\cap H)b_0$. Since $Hg_ib_i \in M$, we get $Hgb_1 \in M$; hence the second coordinate of the point $(Hg)\Phi$ is equal to $(N\cap H)b_1$. This is impossible because Φ is well defined. Thus the map Φ is continuous and bijective. Since G/H is compact, Φ is a homeomorphism of G/H onto $(G/NH)\times(N/(N\cap H))$.

Since $N/(N\cap H)$ is a finite-dimensional torus, its fundamental group is commutative. By hypothesis, H is a discrete subgroup of G, hence H is the fundamental group of the space G/H (Pontrjagin [*1*]). Since the length of the lower central series of G is k and $H\cap G_i'$ is a syndetic subgroup of G_i' $(i = 1,\dots,k)$, the length of the lower central series of H is also equal to k. Similarly NH/N is the fundamental group of the space G/NH and the length of the lower central series of the group NH/N is equal to $(k - 1)$.

Thus the length of the lower central series of the fundamental group of the space G/H is k, but the length of the lower central series of the fundamental group $(NH/N) \times (N \cap H)$ of $(G/NH) \times (N/N \cap H)$ is equal to $(k - 1)$. This is impossible, because the spaces G/H and $(G/NH) \times (N/N \cap H)$ are homeomorphic, as was shown above. This contradiction shows that $(G/H,\varphi)$ is a minimal flow. ▮

2·10.12 EXAMPLE: Let G be the group of all matrices of the form

$$\begin{bmatrix} 1 & a & b \\ 0 & 1 & c \\ 0 & 0 & 1 \end{bmatrix},$$

where a,b,c are real numbers, and let H be the set of all those matrices in G which have integer entries. Clearly G is a connected simply connected nilpotent Lie group and H is a discrete syndetic subgroup of G.

Each one-parameter subgroup of G (*i.e.*, each continuous homomorphism $\varphi:\mathbb{R} \to G$) is of the form

$$t\varphi = \begin{bmatrix} 1 & \alpha t & \gamma t + \frac{1}{2}\alpha\beta t^2 \\ 0 & 1 & \beta t \\ 0 & 0 & 1 \end{bmatrix},$$

where α,β and γ are real numbers $(t \in \mathbb{R})$.

We can conclude from Theorem 2·10.11 that the flow $(G/H,\varphi)$ is minimal iff the numbers α and β are rationally independent. Indeed, $A \equiv [G,G]$ is the subgroup of G consisting of all matrices of the form

$$\begin{bmatrix} 1 & 0 & b \\ 0 & 1 & 0 \\ 0 & 0 & 1 \end{bmatrix},$$

and the flow induced by $(G/H,\varphi)$ on the adjoint two-dimensional torus $T = G/AH$ is minimal iff the numbers α and β are rationally independent. By Corollary 2·10.6, the flow $(G/H,\varphi)$ is distal for every one-parameter subgroup $\varphi:\mathbb{R} \to G$.

Thus whenever α and β are rationally independent, $(G/H,\varphi)$ is a distal minimal flow. Moreover, the flow $(G/H,\varphi)$ is not equicontinuous in this case. Indeed, the fundamental group $\pi_1(G/H)$ of the space G/H is equal to H, which is not commutative. If the minimal flow $(G/H,\varphi)$ were equicontinuous, the nilmanifold G/H would be, by Theorem 2·8.5, homeomorphic to a commutative topological group, and hence the fundamental group $\pi_1(G/H)$ would be commutative. Thus the flow $(G/H,\varphi)$ is not equicontinuous.

REMARKS AND BIBLIOGRAPHICAL NOTES

Theorems 2•10.4,5 were proved by Wu [*4*] and independently by Bronšteĭn [*11*]. Corollaries 2•10.6,8 were first obtained by Keynes [*1*]. Theorem 2•10.11 is due to Gree (*see* L. Auslander, Green and Hahn [*1*]). The proof of this theorem given here is due to Bronšteĭn [*2,5*]. Example 2•10.12 is taken from the book by L. Auslander, Green and Hahn [*1*].

§(2•11): TOPOLOGICAL PROPERTIES OF MINIMAL SETS

We shall investigate in this section various topological properties of minimal sets and we shall discuss their situation in the phase space. We shall consider several concepts of homogeneity and relations between them, as well as between homogeneity, distality and equicontinuity. We shall state some sufficient conditions for the fundamental group of a minimal set to be non-trivial. We shall also discuss the kind of spaces which can support a minimal transformation group. The exposition will be illustrated by several examples.

2•11.1 THEOREM: *Let* (X,T,π) *be a topological transformation group and let* $A \subset X$ *be a minimal set. Then the following assertions hold:*

(1): *A is open in X iff* $\mathrm{int}A \neq \emptyset$;

(2): *If* $\mathrm{int}A \neq \emptyset$, *then A is a union of components of the space X;*

(3): *If* $\mathrm{int}A \neq \emptyset$ *and the group T is connected, then A is a component of X;*

(4): *If* $\mathrm{int}A \neq \emptyset$ *and X is connected, then* $A = X$.

Proof: (1): Let $x \in \mathrm{int}A$ and $y \in A$. Since A is minimal, we have $\overline{yT} = A$. Hence there exists an element $t \in T$ with $y\pi^t \in \mathrm{int}A$. Since $\pi^t : X \to X$ is a homeomorphism, the set $B \equiv (\mathrm{int}A)\pi^{t^{-1}}$ is open in X and, moreover, $y \in B \subset A$. Thus the set A is open in X. The converse is obvious. (2): Let $\mathrm{int}A \neq \emptyset$. By (1) the set A is open in X. But A is also closed in X. Hence (2) is true. (3): Let $\mathrm{int}A \neq \emptyset$ and let T be connected. If $x \in A$, then $\overline{xT} = A$ and consequently A is connected. Therefore (2) implies that A is a component of X. (4): This statement follows at once from (2). ▮

2•11.2 Let n be a positive integer. A space X is called an *n-DIMENSIONAL MANIFOLD* if it is connected and each point of X has a neighbourhood homeomorphic to n-dimensional Euclidian space. Let X be an n-dimensional manifold. It is proved in dimension theory (*see*: Hurewicz and Wallman [*1*]) that if $A \subset X$, then

$\dim A \leq n$, where dim denotes the inductive dimension, and moreover $\dim A = n$ iff $\operatorname{int} A \neq \emptyset$.

> 2·11.3 THEOREM: *Let X be an n-dimensional manifold and let $A \subset X$ be a minimal set of (X,T,π) such that $A \neq X$. Then* $\dim A \leq n - 1$.

Proof: Suppose the contrary: $\dim A = n$. Then $\operatorname{int} A \neq \emptyset$ and since X is connected, we get $X = A$ by Theorem 2·11.1(4), but this contradicts $A \neq X$. Consequently, $\dim A \leq n - 1$.∎

2·11.4 Let n be a positive integer. A compact metrizable space X of $\dim X = n$ is called a *CANTOR MANIFOLD* if X cannot be presented as a union of two non-empty closed subsets X_1 and X_2 such that $\dim(X_1 \cap X_2) \leq n - 2$. Note that a Cantor manifold is not necessarily a manifold in the above sense.

It is not difficult to see that a Cantor manifold is connected and moreover each n-dimensional Cantor manifold is of dimension n at each of its points. By the Hurewicz-Tumarkin Theorem (*see*: Hurewicz and Wallman [*1*]), each n-dimensional compact metrizable space contains an n-dimensional Cantor manifold.

> 2·11.5 THEOREM: *Let X be a finite-dimensional metrizable compact space, T a connected group and (X,T,π) a minimal transformation group. Then X is a Cantor manifold.*

Proof: Let $n = \dim X$. Suppose to the contrary that X is not a Cantor manifold. Then there exist closed non-empty subsets A and B of X such that $X = A \cup B$ and $\dim(A \cap B) \leq n - 2$. By the Hurewicz-Tumarkin Theorem there is a subset $C \subset X$ such that $\dim C = n$ and C is a Cantor manifold.

Put $E = \{t \mid t \in T, Ct \subset A\}$ and $F = \{t \mid t \in T, Ct \subset B\}$. Since A and B are closed in X, the sets E and F are closed in T. For each $t \in T$, the set Ct is homeomorphic to C, and hence Ct is a Cantor manifold. Since

$$\dim(A \cap B) \leq n - 2, \qquad \dim Ct = n,$$

we get either $Ct \subset A$ or $Ct \subset B$. In other words, $T = E \cup F$, and $E \cap F = \emptyset$. By hypothesis, the group T is connected, hence either $T = E$ or $T = F$. Suppose, for example, that $T = E$. Then $CT \subset A$, hence $\overline{CT} \subset A$. Therefore $\overline{CT} \neq X$, but this contradicts the fact that (X,T,π) is minimal.∎

2·11.6 If the hypothesis that T is connected is omitted from Theorem 2·11.5, the above conclusion cannot be drawn. This is demonstrated by the following example.

Let A be a closed rectangle in the plane, defined by

$$A = \{(x,t) \mid a \leq x \leq a + \Delta, b \leq t \leq b + h\},$$

where $\Delta > 0$, $h > 0$. Let us denote by A^* the union of the following three rectangles:

$$A_0 = \{(x,t)|a \leqslant x \leqslant a + \frac{1}{5}\Delta, b \leqslant t \leqslant b + \frac{1}{2}h\},$$

$$A_1 = \{(x,t)|a + \frac{2}{5}\Delta \leqslant x \leqslant a + \frac{3}{5}\Delta, b \leqslant t \leqslant b + h\},$$

$$A_2 = \{(x,t)|a + \frac{4}{5}\Delta \leqslant x \leqslant a + \Delta, b + \frac{1}{2}h \leqslant t \leqslant b + h\}.$$

Define by induction a sequence of sets X_n ($n = 0,1,\ldots$). Let $X_0 = \{(x,t)|0 \leqslant x \leqslant 1, 0 \leqslant t \leqslant 1\}$, $X_1 = X^*_0$. Then X_1 is a union of three rectangles, which we shall denote by $S(1,0)$, $S(1,1)$, $S(1,2)$:

$$S(1,0) = \{(x,t)|0 \leqslant x \leqslant \frac{1}{5}, 0 \leqslant t \leqslant \frac{1}{2}\},$$

$$S(1,1) = \{(x,t)|\frac{2}{5} \leqslant x \leqslant \frac{3}{5}, 0 \leqslant t \leqslant 1\},$$

$$S(1,2) = \{(x,t)|\frac{4}{5} \leqslant x \leqslant 1, \frac{1}{2} \leqslant t \leqslant 1\}.$$

Suppose that the set X_n has already been defined and it is a union of 3^n rectangles $S(n,0), S(n,1), \ldots, S(n,3^n - 1)$. Put

$$X_{n+1} = \cup[S^*(n,i)|i = 0,1,\ldots,3^n - 1].$$

The set $S^*(n,i)$ ($i = 0,1,\ldots,3^n - 1$) consists of three rectangles, which will be denoted by $S(n + 1,i)$, $S(n + 1,i + 3^n)$, $S(n + 1,i + 2.3^n)$ corresponding to the arrangement of their projections on the x-axis from left to right, according to the positive direction of the x-axis. Then X_{n+1} is a union of 3^{n+1} closed rectangles $S(n + 1,i)$ ($i = 0,1,\ldots,3^{n+1} - 1$).

We have thus defined closed subsets $X_0, X_1, \ldots, X_n, \ldots$ with $X_{i+1} \subset X_i$ ($i = 0,1,\ldots$). Let $X = \bigcap_{i=0}^{\infty} X_i$. The set X is compact because all the sets X_i are compact. X consists of vertical line segments, some of which are degenerate (*i.e.*, consist of a single point).

For each positive integer n define a finite set $G_n = \{S(n,i)| 0 \leqslant i \leqslant 3^n - 1\}$ and a permutation $F_n: G_n \to G_n$ by

$$S(n,i)F_n = S(n,i + 1), \qquad (0 \leqslant i < 3^n - 1),$$

$$S(n,3^n - 1)F_n = S(n,0).$$

It is easy to verify that $S(n,j)F_n^3 = S(n,j)$ ($0 \leqslant j \leqslant 3^n - 1$).

Choose any numbers j_n with $0 \leqslant j_n \leqslant 3^n - 1$ and put

$$A = \bigcap_{n=1}^{\infty} S(n,j_n), \qquad B = \bigcap_{n=1}^{\infty} S(n,j_n)F_n.$$

If $S(n,j) \subset S(m,k)$, then $S(n,j)F_n \subset S(m,k)F_m$; therefore $A \neq \emptyset$ iff $B \neq \emptyset$. If $A \neq \emptyset$, then either A consists of a single point or A is a vertical segment. If A is a point, say x, then B is also a point, say y. In this case we define $xF = y$. Whenever A is a nondegenerate segment, then B is also a nondegenerate segment. We define F to be a linear map of A onto B.

Thus we have obtained a map $F:X \to X$. It is easy to see that F is continuous and bijective, hence $F:X \to X$ is a homeomorphism.

Let us show that the cascade $(X,\mathbb{Z},\pi)$, where $(x,n)\pi = xF^n$ $(x \in X,\ n \in \mathbb{Z})$, is minimal. Let $a_1 = (x_1,t_1) \in X$, $a_2 = (x_2,t_2) \in X$, and $\varepsilon > 0$. Choose n large enough so that the breadth of each rectangle $S(n,j)$ $(0 \leqslant j \leqslant 3^n - 1)$ is smaller than $\varepsilon/2$.

Let k_0 be such that $a_2 \equiv (x_2,t_2) \in S(n,k_0)$. There exist numbers m and j_0 such that $m \geqslant n$, $0 \leqslant j_0 \leqslant 3^m - 1$, $S(m,j_0) \subset S(n,k_0)$ and moreover $|t_2 - t| < \varepsilon/2$, whenever $a \equiv (x,t) \in S(m,j_0)$.

Observe that $a_1 \in S(m,\ell_0)$ for a certain number ℓ_0. It follows from the definition of the map F that there exists a number r such that $(x_3,t_3) \equiv a_1F^r \in S(m,j_0) \subset S(n,k_0)$. Then $|t_3 - t_2| < \varepsilon/2$. Since the breadth of the rectangle $S(n,k_0)$ is smaller than $\varepsilon/2$, we get $|x_3 - x_2| < \varepsilon/2$. Therefore $\rho(a_2,a_1F^r) < \varepsilon$. Since a_1 and a_2 are arbitrary points of the space X, we conclude that X is a minimal set.

It is clear that the dimension of X is equal to 0 at the point $(0,0) \in X$ and it is equal to 1 at $(\frac{1}{2},\frac{1}{2}) \in X$. Therefore X is not a Cantor manifold.

Observe that X is a locally almost periodic minimal set and that two points of X are proximal iff they belong to the same vertical segment (the proof is left to the reader).

2·11.7 A *TRANSFORMATION GROUP* (X,T,π) is called:

(1) *TOPOLOGICALLY HOMOGENEOUS* if the space X is homogeneous, *i.e.*, given two points $x,y \in X$, there exists a homeomorphism $\varphi:X \to X$ with $x\varphi = y$;

(2) *ORBITALLY HOMOGENEOUS* if for each two points $x,y \in X$ there exists a homeomorphism $\varphi:X \to X$, $x\varphi = y$, which preserves orbits (*i.e.*, $(zT)\varphi = (z\varphi)T$ for each point $z \in X$);

(3) *DYNAMICALLY HOMOGENEOUS* (or, simply, homogeneous) if for each two points $x,y \in X$ there exists an automorphism φ of the transformation group (X,T,π) sending x to y (*i.e.*, a homeomorphism $\varphi:X \to X$ such that $x\varphi = y$ and $(z\pi^t)\varphi = (z\varphi)\pi^t$ for each $z \in X$, $t \in T$);

(4) *h-HOMOGENEOUS* if for each two points $x,y \in X$ there exists an orbit-preserving homeomorphism $\varphi:X \to X$, $x\varphi = y$, which is homotopic to the identity map.

(5) *HARMONIZABLE* if there is a dynamically homogeneous trans-

formation group (X,T,π^*), which has the same orbits as (X,T,π) does.

2·11.8 Let X be a topological space, (Y,ρ) a metric space, $\{\varphi_n | n = 1,2,\ldots\}$ a sequence of continuous maps of X into Y and $x \in X$. We say that $\{\varphi_n\}$ *CONVERGES TO A MAP* $\varphi : X \to Y$ *UNIFORMLY AT THE POINT* x if for each $\varepsilon > 0$ there exists a neighbourhood U of x and a positive integer N such that $\rho(y\varphi, y\varphi_n) < \varepsilon$ for all $y \in U$, $n \geqslant N$.

2·11.9 LEMMA: *Let X be a topological space, let Y be a metric space, and let $\{\varphi_n | n = 1,2,\ldots\}$ be a sequence of continuous maps of X into Y which converges to a function $\varphi : X \to Y$ at each point $x \in X$. Let X_0 denote the set of all points $x \in X$ at which $\{\varphi_n\}$ converges to φ uniformly. Then X_0 can be represented as the intersection of a countable family of open dense subsets of X.*

Proof: Let $A(\varepsilon)$ denote the set of all points $x \in X$ such that $\rho(y\varphi, y\varphi_m) < \varepsilon$ for all points y from some neighbourhood of x and for all m, greater than some positive integer. It is clear that for each $\varepsilon > 0$ the set $A(\varepsilon)$ is open in X and $X_0 = \bigcap_{n=1}^{\infty} A(1/n)$.

Let $\varepsilon > 0$. It is enough to show that the set $B = X \setminus A(\varepsilon)$ can be represented as a union of a countable family of closed nowhere dense subsets. Set $\delta = \varepsilon/5$. For each positive integer n, let C_n denote the set of all points $x \in X$ such that $\rho(x\varphi, x\varphi_m) < \delta$ whenever $m \geqslant n$. By hypothesis, $X = \bigcup_{n=1}^{\infty} C_n$; therefore $B = \bigcup_{n=1}^{\infty} (B \cap C_n)$. Put $D_n = B \cap C_n$, then $B = \bigcup_{n=1}^{\infty} D_n$. Since the set B is closed we have $B \supset \bigcup_{n=1}^{\infty} \bar{D}_n \supset B$, and hence $B = \bigcup_{n=1}^{\infty} \bar{D}_n$. It remains to show that the sets $\bar{D}_n$, or equivalently, the D_n are nowhere dense in X. Let n be a fixed positive integer. Suppose to the contrary that $\bar{D}_n$ contains some non-empty open subset $U \subset X$. Let $x \in U$ and $p \geqslant n$. Then there exists a positive integer $q \geqslant n$ such that $\rho(x\varphi, x\varphi_q) < \delta$. There is also a point $y \in U \cap D_n \subset C_n$ with $\rho(x\varphi_p, y\varphi_p) < \delta$ and $\rho(x\varphi_q, y\varphi_q) < \delta$. Then $\rho(x\varphi, x\varphi_p) \leqslant \rho(x\varphi, x\varphi_q) + \rho(x\varphi_q, y\varphi_q) + \rho(y\varphi_q, y\varphi) + \rho(y\varphi, y\varphi_p) + \rho(y\varphi_p, x\varphi_p) < 5\delta = \varepsilon$.

Thus $\rho(x\varphi, x\varphi_p) < \varepsilon$ whenever $x \in U$ and $p \geqslant n$. By definition, $U \subset A(\varepsilon) = X \setminus B$, and hence $U \cap B = \emptyset$. On the other hand, $U \subset \bar{D} \subset \bar{B}$, hence $U \cap B \neq \emptyset$, because U is a non-empty open set. We have arrived at a contradiction. This completes the proof.▮

2·11.10 THEOREM: *A homogeneous transformation group (X,T,π) with a compact metrizable phase space is equicontinuous.*

Proof: We need only to show that the closure of the set $G = \{\pi^t | t \in T\}$ in $C(X,X)$ is compact (Section §(1·6)). Let $\{t_n | n = 1,2,\ldots\}$ be any sequence of elements $t_n \in T$. We shall prove that

$\{\pi^{t_n}\}$ contains a uniformly convergent subsequence. Let $x_0 \in X$. Without loss of generality we may suppose that $\lim_{n\to\infty} x_0\pi^{t_n}$ exists and is equal to $y_0 \in X$. Let x be an arbitrary point of the space X. There exists an automorphism $g \in A(X,T,\pi)$ with $x_0 g = x$. Then $y_0 g = \lim x_0\pi^{t_n} g = \lim x_0 g\pi^{t_n} = \lim x\pi^{t_n}$. Thus the sequence $\{\pi^{t_n} | n = 1,2,\ldots\}$ converges at each point $x \in X$. Let $h:X \to X$ be the limit function.

The space X is compact and therefore complete. By the preceding lemma and Baire's category theorem, there exists a point $x_1 \in X$ such that $\{\pi^{t_n}\}$ converges to h uniformly at x_1.

Let us show that $\{\pi^{t_n}\} \to h$ uniformly on X. Let $x \in X$ and $\varepsilon > 0$. There exists an element $g \in A(X,T,\pi)$ with $x = x_1 g$. Choose a number $\delta > 0$ such that $\rho(z_1 g, z_2 g) < \varepsilon$ for all $z_1, z_2 \in X$ with $\rho(z_1,z_2) < \delta$. Since $\{\pi^{t_n}\}$ converges uniformly to h at x_1, we can choose a neighbourhood U_1 of x_1 and an integer $N > 0$ such that $\rho(zh, z\pi^{t_n}) < \delta$ for all $z \in U_1$ and $n > N$. Denote $U_1 g$ by U. Since $x = x_1 g$ and $g:X \to X$ is a homeomorphism, U is a neighbourhood of x. If $y \in U$ and $n > N$, then $yg^{-1} \in U_1$ and therefore $\rho(yhg^{-1}, y\pi^{t_n}g^{-1}) = \rho(yg^{-1}h, yg^{-1}\pi^{t_n}) < \delta$. Then

$$\rho(yh, y\pi^{t_n}) < \varepsilon, \qquad (y \in U, n > N). \tag{11.1}$$

Thus for every point $x \in X$ there exists a neighbourhood U of x and a number N such that (11.1) holds. Since the space X is compact, it follows that the sequence $\{\pi^{t_n}\}$ converges to h uniformly on X. ▮

2·11.11 THEOREM: *If the space X is compact and the group T is commutative, then every equicontinuous minimal transformation group (X,T,π) is homogeneous.*

Proof: By Theorem 2·8.5 the transformation group (X,T,π) is isomorphic to a transformation group (S,T,ρ), where S is a compact commutative topological group, $(s,t)\rho = s(t\varphi)$ $(s \in S, t \in T)$ and $\varphi:T \to S$ is a homomorphism, satisfying the condition $\overline{T\varphi} = S$. Let $p,q \in S$. The map $\psi:S \to S$ defined by $s\psi = qp^{-1}s$ $(s \in S)$ is an automorphism taking p to q. ▮

The hypothesis that T is commutative cannot be dropped, as shown by the example of the rotation group of the two-dimensional sphere.

2·11.12 THEOREM: *Let X be a compact metric space and let T be a commutative group. A minimal transformation group (X,T,π) is homogeneous iff it is equicontinuous.*

Proof: This follows from Theorems 2·11.10,11. ▮

2·11.13 LEMMA: *Let X be a compact space, (X,T) a distal transformation group and $E = E(X,T)$ the enveloping semigroup.*

Then (E,T) *is a homogeneous minimal transformation group.*

Proof: By Theorem 1·6.3, the enveloping semigroup is in fact a group. Hence it follows from Lemma 1·4.8 that (E,T) is a minimal transformation group. Let $p,q \in E$. The map $\varphi: E \to E$ defined by $s\varphi = qp^{-1}s$ $(s \in E)$ is continuous and bijective. From this we easily deduce that φ is an automorphism of the transformation group (E,T) sending p to q.∎

2·11.14 LEMMA: *A homogeneous transformation group* (X,T) *with a compact phase space is distal.*

Proof: Since X is compact, it follows from the Theorems 1·1.7 and 1·5.4 that X contains an almost periodic point $x_0 \in X$. Since the transformation group (X,T) is homogeneous, all the points $x \in X$ are almost periodic. Moreover, the transformation group $(X \times X, T)$ is pointwise almost periodic. Therefore it follows from Theorem 1·6.3 that (X,T) is distal.∎

2·11.15 THEOREM: *A minimal transformation group* (X,T) *with a compact phase space is distal iff* $(E(X,T),T)$ *is a homogeneous transformation group.*

Proof: Let (E,T) be homogeneous. Then (E,T) is distal by Lemma 2·11.14. There is a homomorphism of (E,T) onto (X,T), hence (X,T) is distal (Corollary 1·6.7.). The second assertion follows from Lemma 2·11.13.∎

2·11.16 THEOREM: *Let* (X,T) *be a distal minimal transformation group with a compact phase space and suppose that at least one of the following conditions is satisfied:*

(a): *The group* $E(X,T)$ *is commutative;*

(b): *The space* $E(X,T)$ *is metrizable.*

Then (X,T) *is equicontinuous.*

Proof: If (a) is satisfied, then the multiplication in the compact group $E(X,T)$ is continuous in each variable separately. Hence it follows from Lemma 2·8.2 that (X,T) is equicontinuous. If condition (b) is satisfied, then the transformation group (E,T) is equicontinuous by Lemma 2·11.13 and Theorem 2·11.10. Hence it follows that (X,T) is also equicontinuous (Corollary 1·6.16).

2·11.17 LEMMA: *A minimal transformation group* (X,T,π) *with a compact phase space is homogeneous iff the map* $\varphi_x: E(X,T,\pi) \to X$*, where* $p\varphi_x = xp$ $(p \in E)$*, is injective for each* $x \in X$.

Proof: If (X,T,π) is homogeneous, then it follows from the Theorems 2·11.15 and 1·6.3 that $E(X,T,\pi)$ is a group. Let $x \in X$ and $p \in E$ be such that $p\varphi_x \equiv xp = x$. We shall show that p is the identity of the group E. Indeed, if $\{\pi^{t_i} \mid i \in I\}$ is a net that converges to p, then $\{x\pi^{t_i} \mid i \in I\}$ converges to $xp = x$. For each point $y \in X$, there exists an automorphism $\varphi \in A(X,T,\pi)$ with $x\varphi = y$.

Thus

$$\{y\pi^{t_i}\} = \{x\varphi\pi^{t_i}\} = \{x\pi^{t_i}\varphi\} \to x\varphi = y,$$

and hence $yp = y$ $(y \in X)$. But this just means that p is the identity of E. If $p\varphi_x = q\varphi_x$, then $xpq^{-1} = x$. Then, as it was shown above, $p = q$. Hence φ_x is injective.

Conversely, let φ_x be injective. The semigroup E contains at least one minimal ideal, say I. Since $xI = X$, there exists an element $p \in I$ with $p\varphi_x = x$. From the injectivity of φ_x at the point e we conclude that $e = p \in I$. Hence $I = E$; consequently E is a group (Corollary 1·4.13). By Theorem 1·6.3 the transformation group (X,T) is distal, and from Lemma 2·11.13 we deduce that (E,T) is homogeneous. Since φ_x is bijective, it is an isomorphism of (E,T) onto (X,T).

Thus the transformation group (X,T) is homogeneous.∎

2·11.18 Given a minimal transformation group (X,T,π), a point $x \in X$ is called a *CHARACTERISTIC POINT* if for every net $\{t_i\}$, $t_i \in T$, the condition $\{xt_i\} \to x$ implies $\{yt_i\} \to y$ for all $y \in X$.

2·11.19 THEOREM: *A minimal transformation group (X,T,π) with a compact phase space is homogeneous iff X has at least one characteristic point.*

Proof: If X is a homogeneous minimal set, then all of its points are characteristic. Conversely, let $x \in X$ be a characteristic point. It is easily seen from Definition 2·11.18, that if the equality $x\xi = x$ holds for some element $\xi \in E(X,T)$, then ξ is the identity map of X. Hence, the map $\varphi_x : E \to X$ is injective at the point $e \in E$. For the same reason as in the proof of the second part of Lemma 2·11.17, E is a group; hence (X,T) is distal and (E,T) is homogeneous. We shall prove that φ_x is injective. In fact, if $x\xi = x\eta$, then $x\xi\eta^{-1} = x$ and therefore $\xi\eta^{-1} = e$, that is, $\xi = \eta$. Thus $\varphi_x : (E,T) \to (X,T)$ is an isomorphism.∎

2·11.20 THEOREM: *The class of all distal minimal transformation groups with compact phase spaces and a fixed phase group T contains a universal transformation group (D_0,T) which is unique up to isomorphism.*

Proof: It is easy to verify that the class of all distal minimal transformation groups with compact phase spaces is admissible, and hence the required result is a consequence of Theorem 2·7.12.∎

2·11.21 THEOREM: *The universal distal minimal transformation group (D_0,T) is homogeneous.*

Proof: It can be seen from the proof of Theorem 2·7.12 that (D_0,T) is isomorphic to $(E(D_0,T),T)$, but the latter transformation group is homogeneous by Lemma 2·11.13.∎

2·11.22 Since equicontinuity of a transformation group is preserved under homomorphisms (Corollary 1·6.16) and there exist minimal distal flows which are not equicontinuous (Example 2·9.19-20), the transformation group $(D_0,\mathbb{R})$ is not equicontinuous. From Theorem 2·11.16 we may thus conclude that the space $E(D_0,\mathbb{R}) \equiv D_0$ is not metrizable. This shows that the metrizability condition put on the phase space in Theorem 2·11.10 cannot be dropped.

2·11.23 LEMMA: *Let X be a compact space, (X,T,π) a transformation group, $\varphi:(X,T,\pi) \to (Y,T,\rho)$ a homomorphism and let (Y,T) be minimal. If the map φ is a local homeomorphism, then (X,T,π) is pointwise almost periodic.*

Proof: Let $x \in X$, $y = x\varphi$. Since each point of Y is almost periodic, there exists an idempotent $v \in E(Y,T)$ belonging to some minimal right ideal K of $E(Y,T)$ and such that $yv = y$. By Theorem 1·4.21(9), we can choose a minimal right ideal $I \subset E(X,T)$ and an idempotent $u \in I$ so that $u\vartheta = v$, where $\vartheta:E(X,T) \to E(Y,T)$ is the homomorphism of the enveloping semigroups induced by φ. We shall prove that $xu = x$. Suppose the contrary holds. Choose a net $\{t_\alpha\}$, $t_\alpha \in T$, such that $\{\pi^{t_\alpha}\} \to u$ in $E(X,T)$. Then $\{xt_\alpha\} \to xu$, and therefore $\{yt_\alpha\} = \{x\varphi t_\alpha\} = \{xt_\alpha\varphi\} \to (xu)\varphi = x\varphi(u\vartheta) = yv = y$. On the other hand $\{xt_\alpha\} \to xu$, $\{xut_\alpha\} \to xuu = xu$ and $(xut_\alpha)\varphi = (xt_\alpha)\varphi = yt_\alpha$. Since $xut_\alpha \neq xt_\alpha$ and $\{yt_\alpha\} \to y$, this contradicts the injectivity of φ in every neighbourhood of the point xu. Thus $xu = x$ for an element $u \in I$; consequently x is an almost periodic point. ∎

2·11.24 Let Y be a connected locally pathwise connected space, $p: X \to Y$ a covering map, G a connected Lie group and (Y,G,ρ) a transformation group. Then there exists no more than one transformation group (X,G,π) covering (Y,G,ρ) (*i,e.*, $p:(X,G,\pi) \to (Y,G,\rho)$ is a homomorphism). If G is simply connected, then the covering transformation group actually exists. Let us prove this assertion. Define a map $\sigma:X \times G \to Y$ by

$$(x,g)\sigma = (xp,g)\rho, \qquad (x \in X, g \in G).$$

Let $x_0 \in X$, $y_0 = x_0p$. Since $\pi_1(G) = 0$, the image of the homomorphism $\sigma_*:\pi_1((X,x_0) \times (G,e)) \to \pi_1(Y,y_0)$ induced by σ is contained in the image of the homomorphism $p_*:\pi_1(X,x_0) \to \pi_1(Y,y_0)$. By the covering theorem, there exists a map $\pi:X \times G \to X$ such that $\pi o p = \sigma$. It is easy to verify, that (X,G,π) is the required transformation group.

2·11.25 LEMMA: *Let G be a connected Lie group, Y a compact pathwise connected space, (Y,G) a minimal transformation group, $p:X \to Y$ a covering map and let (X,G) be the covering transformation group. If X is compact, then (X,G) is a minimal transformation group. If X is not compact, then X does not contain non-empty compact invariant subsets.*

Proof: Since (Y,G) is minimal and G is connected, the space Y is also connected. The space X is connected by the very definition of a covering map.

Let X be compact. Since p is a local homeomorphism, each point $x \in X$ is almost periodic (Lemma 2·11.23). Furthermore, each fibre of $p:X \to Y$ is finite and X is connected; therefore X is a minimal set.

Now suppose that X is not compact, but contains a non-empty compact invariant set. Then X contains a compact minimal set M. If the interior of M with respect to X is non-empty, then $X = M$ by Theorem 2·11.1, but this leads to a contradiction. Thus $\mathrm{int}_X M = \emptyset$. Let $m \in M$. The point $mp \in Y$ has a connected open neighbourhood V such that each component of the preimage Vp^{-1} is open in X and if we restrict $p:X \to Y$ to Vp^{-1}, we get a homeomorphism. Let U be the component of Vp^{-1} containing the point m. Then $U \cap M$ is a relative neighbourhood of m in M. Since Y is compact and minimal, it follows that $W \equiv \mathrm{int}_Y(U \cap M)p \neq \emptyset$. In view of the fact that $p|_U:U \to V$ is a homeomorphism, $\emptyset \neq U \cap Wp^{-1} \subset U \cap M \subset M$. This contradicts $\mathrm{int}_X M = \emptyset$. ▮

2·11.26 THEOREM: *Let X be a compact connected locally pathwise connected space, Y a locally pathwise connected and semi-locally simply connected space, G a connected Lie group and let (X,G) and (Y,G) be transformation groups, where (Y,G) be minimal. Let $p:(X,G) \to (Y,G)$ be a homomorphism, $x_0 \in X$ and $y_0 = x_0 p$. Then $p_*(\pi_1(X,x_0))$ is of normal index in the group $\pi_1(Y,y_0)$ (here p_* is the homomorphism of fundamental groups induced by p). In particular, if the group $\pi_1(Y,y_0)$ is infinite, then $\pi_1(X,x_0) \neq 0$.*

Proof: Without loss of generality we may suppose that G is simply connected (it is enough to replace G by its universal covering group). According to the theorem on the existence of covering maps, we can construct a covering map $q:(Z,z_0) \to (Y,y_0)$ such that

$$q_*(\pi_1(Z,z_0)) = p(\pi_1(X,x_0)).$$

Let (Z,G) be the transformation group which covers (Y,G) by q. It follows from the covering map theorem that there exists a unique continuous map $r:(X,x_0) \to (Z,z_0)$ with $roq = p$. Since X_r is a compact invariant subset of Z, Lemma 2·11.25 implies that Z is compact. But it is clear that Z is compact iff the group $q_*(\pi_1(Z,z_0)) \equiv p_*(\pi_1(X,x_0))$ is a subgroup of normal index of the group $\pi_1(Y,y_0)$. ▮

2·11.27 Let T be a topological group, S a closed syndetic subgroup of T, K a topological space and let (K,S,ρ) be a transformation group. Define two topological transformation groups $(K \times T,S,\sigma)$ and $(K \times T,T,\tau)$ as follows:

$$(k,t)\sigma^s = (k\rho^s, s^{-1}t), \qquad (k \in K, t \in T, s \in S),$$

$$(k,t)\tau^r = (k,tr), \qquad (k \in K, t \in T, r \in T).$$

Let $(K\times T)/S$ be the quotient space of orbits of $(K\times T,S,\sigma)$. It is easy to verify that the actions σ and τ commute, and therefore τ induces a transformation group $((K\times T)/S,T,\pi)$. In this case we shall say that (K,S,ρ) is a *GLOBAL SECTION of the transformation group* $((K\times T)/S,T,\pi)$. Observe that there exists a homomorphism $q:((K\times T)/S,T,\pi)\to(T/S,T,\rho)$, where $(St,r)\rho = Str$ $(t,r\in T)$, namely, $[(k,t)S]q \equiv [\{(k\rho^s,s^{-1}t)\mid s\in S\}]q = St$ $(k\in K, t\in T)$.

Since S is a syndetic subgroup of T, the quotient space $T/S = \{St\mid t\in T\}$ is compact. The transformation group $(T/S,T,\rho)$ is, of course, minimal.

2·11.28 THEOREM: *Let (X,T) be a transformation group, S a closed syndetic subgroup of T, K a closed subset of X and assume that the following conditions are satisfied:*

(1): $KT = X$;

(2): $KS\subset K$;

(3): *If* $k\in K$ *and* $kt\in K$, *then* $t\in S$.

Then (K,S) is a global section of (X,T).

Proof: We must show that the transformation group (X,T) is isomorphic to $((K\times T)/S,T,\pi)$. By condition (1), each point $x\in X$ can be represented in the form $x = kt$, where $k\in K$, $t\in T$. This can be done in different ways, but if $x = k_1t_1$, where $k_1\in K$, $t_1\in T$, then $k_1 = ktt_1^{-1}$. Using (3) we conclude that $s_1 \equiv tt_1^{-1}\in S$. Thus $k_1 = ks_1$, $t_1 = s_1^{-1}t$. Hence, the map $f:X\to(K\times T)/S$, defined by

$$xf = (k,t)S \equiv \{(ks,s^{-1}t)\mid s\in S\},$$

for $x = kt$, $k\in K$, $t\in T$, is well defined. The inverse map f^{-1} is defined by

$$[(k,t)S]f^{-1} = kt, \qquad (k\in K, t\in T),$$

and it is clearly continuous. We shall prove that f is also continuous. Let $\{x_\alpha\}$, $x_\alpha\in X$, be a net that converges to some point $x\in X$. By (1), $x_\alpha = k_\alpha t_\alpha$, where $k_\alpha\in K$, $t_\alpha\in T$. Since the subgroup S is syndetic, there exists some compact subset N satisfying $T = SN$. Therefore, $t_\alpha = s_\alpha n_\alpha$, where $n_\alpha\in N$, $s_\alpha\in S$. Denote $\ell_\alpha = k_\alpha s_\alpha$. By condition (2), we have $\ell_\alpha\in K$. Without loss of generality we may suppose that $\{n_\alpha\}\to n\in N$. Then $\{\ell_\alpha\} = \{k_\alpha s_\alpha\} = \{x_\alpha n_\alpha^{-1}\}\to xn^{-1}\equiv\ell$ and $\ell\in K$ because K is closed. Hence $x = \ell n$ and therefore $\{x_\alpha f\} = \{(k_\alpha,t_\alpha)S\} = \{(\ell_\alpha,n_\alpha)S\}\to(\ell,n)S = xf$. ∎

2·11.29 THEOREM: *Let X be a compact space, (X,T) a minimal transformation group, H a syndetic normal subgroup of T, $x_0\in X$ and let $\overline{x_0H}\equiv K\neq X$. Then there exists a subgroup S of T such that $S\supset H$ and (K,S) is a global section of (X,T).*

Proof: Define $S = \{t\mid t\in T, Kt\subset K\}$. Then $S\supset H$ and S is a closed syndetic subsemigroup of the group T. By Lemma 2·8.17,

S is a subgroup of T. It follows from the definition of S that $KS \subset K$. Since $T = SN$ for some compact subset N of T, we have $KT = KSN \subset KN \subset X$, and hence $X = \overline{KT} \subset KN \subset X$; consequently, $X = KT$. Further, if $k \in K$ and $kt \in K$, then $\overline{kH} = \overline{ktH} = \overline{x_0 H} \equiv K$, because $K \equiv \overline{x_0 H}$ is an H-minimal set (Lemma 2·8.20), *i.e.*, $(\overline{x_0 H}, H)$ is a minimal transformation group. Thus

$$Kt = \overline{x_0 H}t = \overline{kH}t = \overline{kHt} = \overline{ktH} = K,$$

and this means that $t \in S$. We have thus proved that all the conditions of Theorem 2·11.28 are satisfied. Hence (K,S) is a global section. ▮

2·11.30 THEOREM: *Let X be a connected locally pathwise connected space, K a closed connected subset of X, T a connected Lie group, S a non-trivial discreet syndetic subgroup of T and let (X,T) be a transformation group. Suppose that moreover the conditions* (1-3) *of Theorem* 2·11.28 *are satisfied. Then the fundamental group of the space X is non-trivial:* $\pi_1(X) \neq 0$.

Proof: It follows from the restrictions put on T and S, that the canonical projection of T onto T/S is a covering map. Moreover $\pi_1(T/S) = S \neq 0$. Furthermore, $X \equiv (K \times T)/S$ is the total space of a fibre bundle with base T/S and fibre K, associated with the principal fibre bundle $T \to T/S$. Hence, we have the exact sequence of homotopy groups (Hu [*1*], p. 214):

$$\cdots \to \pi_1(X) \to \pi_1(T/S) \to \pi_0(K) \to \pi_0(X).$$

Since the fibre K is a closed connected subset of a locally arcwise connected space X, $\pi_0(K) = 0$. Therefore the map

$$\pi_1(X) \to \pi_1(T/S) \equiv S \neq 0,$$

is surjective. Thus $\pi_1(X) \neq 0$. ▮

2·11.31 COROLLARY: *Let X be a compact locally pathwise connected space and let $(X,\mathbb{R})$ be a minimal, but not totally minimal, flow. Then* $\pi_1(X) \neq 0$.

Proof: By hypothesis, there exists a subgroup $S \subset \mathbb{R}$, $S = a\mathbb{Z}$, $a \neq 0$, such that the cascade $(X, a\mathbb{Z})$ is not minimal. Let $x \in X$ and $K = \overline{xS}$; then $K \neq X$. By Theorem 2·11.29 we may without loss of generality suppose that (K,S) is a global section of the flow $(X,\mathbb{R})$. To conclude the proof we refer to Theorem 2·11.30. ▮

2·11.32 THEOREM: *Let (X,d) be a compact metric locally arcwise connected space, and $(X,\mathbb{R},\pi)$ be a minimal flow with a global section. If the flow $(X,\mathbb{R},\pi)$ is h-homogeneous, then it is dynamically homogeneous (and consequently almost periodic).*

Proof: By hypothesis, there exists a homomorphism φ of the flow $(X,\mathbb{R},\pi)$ on some periodic flow $(S^1,\mathbb{R},\rho)$. Without loss of generality we may suppose, that the period of $(S^1,\mathbb{R},\rho)$ is 1. Let $x_0 \in X$ be an arbitrary but fixed point. Let us show that x_0 is a characteristic point. To verify this, we may restrict ourselves to consider only points $y_0 \in X$ with $x_0\varphi = y_0\varphi$. Let ε be an arbitrary positive number. There exists a homeomorphism $h:X \to X$ which is homotopic to the identity map, preserves orbits and sends x_0 to y_0. The map $h\circ\varphi$ is homotopic to $\varphi:X \to S^1$, therefore the induced homomorphisms of fundamental groups coincide, *i.e.*

$$(h\circ\varphi)_* = \varphi_*. \tag{11.2}$$

Let $z_0 = x_0\varphi$. The fundamental group $\pi_1(S^1,z_0)$ is isomorphic to the group $\mathbb{Z}$.

If σ is a loop at the point x_0, then σh is a loop at $y_0 = x_0h \in X$. Let $[\sigma]$ be the class of loops homotopic to the loop σ; then $[\sigma] \in \pi_1(X,x_0)$ and $[\sigma\varphi]$, $[\sigma h\varphi] \in \pi_1(S^1,z_0)$. By (11.2)

$$[\sigma\varphi] \equiv [\sigma]\varphi_* = [\sigma](h\circ\varphi)_* \equiv [\sigma h\varphi]. \tag{11.3}$$

Since X is a compact set, given $\varepsilon > 0$ there exists a number $\Delta > 0$ such that $\Delta < 1/3$ and $d(z,z\pi^t) < \varepsilon/2$ for each $z \in X$ and $t \in (-\Delta,\Delta)$. Put $S = \{x \mid x \in X, x\varphi = z_0\}$ and $U = \{x\pi^t \mid x \in S, |t| < \Delta\}$. Since U is an open subset of X and the space X is locally pathwise connected, we can choose a small enough number $\alpha > 0$ so that $\alpha < \varepsilon$, $S(y_0,\alpha) \subset U$ and every two points from $S(y_0,\alpha)$ can be joined by a path situated in U. Choose a number $\Delta' > 0$ such that $\Delta' < \Delta/2$ and $d(z,z\pi^t) < \alpha/2$ for all $z \in X$ and $t \in (-\Delta',\Delta')$. The open subset $\{x\pi^t \mid x \in S, t \in (-\Delta',\Delta')\}$ will be denoted by U'. There exists a number $\alpha' > 0$ with $\alpha' < \alpha/2$ and $S(y_0,\alpha') \subset U'$. Since h is a homeomorphism, $S(x_0,\beta)h \subset S(y_0,\alpha')$ for some $\beta > 0$. There exists a number $\beta' > 0$ such that $\beta' < \beta$ and every two points of $S(x_0,\beta')$ may be joined by a path lying in $S(x_0,\beta)$. Further, choose a number $\Delta'' > 0$ such that $\Delta'' < \Delta/2$ and $d(z,z\pi^t) < \beta'/2$ for all $z \in X$, $t \in (-\Delta'',\Delta'')$. We get an open subset $U'' = \{x\pi^t \mid x \in S, |t| < \Delta''\}$. Finally, there exists a number $\delta > 0$ with $\delta < \beta'/2$ and $S(x_0,\delta) \subset U''$.

Suppose that $d(x_0,x_0\pi^t) < \delta$ for some $t \in \mathbb{R}$. Then $x_0\pi^t \in U''$, hence $t = m + \mu$, where m is an integer and $|\mu| < \Delta''$. Since $d(x_0, x_0\pi^t) < \delta < \beta'$, there exists a path ℓ joining x_0 to $x_0\pi^t$ and lying in $S(x_0,\beta)$. By the choice of Δ'' and δ we have that

$$d(x_0\pi^t,x_0\pi^m) = d(x_0\pi^t,x_0\pi^t\pi^{-\mu}) < \beta'/2,$$

$$d(x_0,x_0\pi^m) \leqslant d(x_0,x_0\pi^t) + d(x_0\pi^t,x_0\pi^m) < \beta'.$$

Therefore there exists a path ℓ' lying in $S(x_0,\beta)$ and joining x_0 to $x_0\pi^m$. Let us consider three loops σ, σ' and σ'' based at the point x. To traverse the loop σ, we must go from x_0 to $x_0\pi^t$ along the orbit within a time $\frac{1}{2}$ and then return to x_0 along the path inverse to ℓ with doubled velocity. To get the loop σ', we must firstly go from x_0 to $x_0\pi^m$ along the orbit curve $(x_0,[0,m])\pi$ and then return to x_0 along the path inverse to ℓ'. Finally, traversing first the path ℓ' and then passing the orbit curve from $x_0\pi^m$ to $x_0\pi^t$ and then returning along the path inverse to ℓ, we obtain the loop σ''.

Clearly, $[\sigma] = [\sigma']\cdot[\sigma'']$. Since φ is a continuous map, $\sigma'\varphi$ is a loop in the space S^1 with base point z_0 with $[\sigma'\varphi] = m$. By definition of the induced homomorphism $\varphi_*:\pi_1(X,x_0) \to \pi_1(S,z_0)$ we have that

$$[\sigma']\varphi_* = [\sigma'\varphi] = m. \tag{11.4}$$

Since the loop σ'' is situated in U, the loop $\sigma''\varphi$ lies inside the arc $(z_0,[-\Delta,\Delta])\rho$, where $\Delta < 1/3$. Therefore $[\sigma'']\varphi_* = [\sigma''\varphi] = 0$ and hence

$$[\sigma\varphi] = [\sigma]\varphi_* = [\sigma']\varphi_* + [\sigma'']\varphi_* = m.$$

Since the homeomorphism h preserves orbits, $(x_0\pi^t)h = y_0\pi^\tau$ for some $\tau \in \mathbb{R}$. It follows from the choice of β and α' that

$$y_0\pi^\tau \in S(y_0,\alpha') \subset U' = \{x\pi^t \mid x \in S, |t| < \Delta'\};$$

therefore $\tau = k + \nu$, where $k \in \mathbb{Z}$ and $|\nu| < \Delta'$. Since the path ℓ lies in $S(x_0,\beta)$, the path $\xi \equiv \ell h$ is situated in $S(y_0,\alpha')$. Denote the loop σh based at y_0 by ϑ. Then by the choice of α' and Δ' we have that

$$d(y_0,y_0\pi^k) \leqslant d(y_0,y_0\pi^\tau) + d(y_0\pi^\tau,y_0\pi^\tau\pi^{-\nu}) < \alpha.$$

It follows from the choice of α that the points y_0 and $y_0\pi^k$ can be joined by a path ξ' lying in U. Similarly as above, we get three loops $\vartheta = \sigma h$, ϑ' and ϑ'' with $[\vartheta']\varphi_* = k$ and $[\vartheta'']\varphi_* = 0$. Therefore

$$[\vartheta]\varphi_* = [\vartheta\varphi] = [\sigma h\varphi] = [\vartheta']\varphi_* + [\vartheta'']\varphi_* = k. \tag{11.5}$$

It follows from the equalities (11.3-5) that $k = m$. Hence $|t - \tau| = |\mu - \nu| \leqslant |\mu| + |\nu| < \Delta' + \Delta'' < \Delta$. Thus $d(y_0\pi^\tau,y_0\pi^t) = d(y_0\pi^\tau,y_0\pi^\tau\pi^{t-\tau}) < \varepsilon/2$ by the choice of Δ. Moreover $d(y_0,y_0\pi^\tau) < \alpha' < \alpha/2 < \varepsilon/2$. Consequently

$$d(y_0,y_0\pi^t) \leqslant d(y_0,y_0\pi^\tau) + d(y_0\pi^\tau,y_0\pi^t) < \varepsilon.$$

Thus, for every $\varepsilon > 0$ and every point $y_0 \in X$ there exists a number $\delta > 0$ such that $d(x_0,x_0\pi^t) < \delta$ implies $d(y_0,y_0\pi^t) < \varepsilon$. Therefore x_0 is a characteristic point. By Theorem 2·11.19, the flow $(X,\mathbb{R},\pi)$ is homogeneous, and consequently almost periodic (Theorem 2·11.10).▮

2·11.33 COROLLARY: *A minimal harmonizable flow with a compact metrizable locally pathwise connected phase space is either almost periodic or admits no global sections (i.e., it does not admit a homomorphism onto a periodic flow).*

Proof: By Theorems 2•11.10 and 2•8.5, a harmonizable flow is h-homogeneous.

2•11.34 COROLLARY: *The distal minimal flow on the three-dimensional torus described in* 2•9.19-20 *is not harmonizable.*

2•11.35 We will now give an example of a topological homogeneous, but orbitally non-homogeneous, minimal flow.

Let G be the group of all real matrices of the form

$$\begin{bmatrix} 1 & a & c \\ 0 & 1 & b \\ 0 & 0 & 1 \end{bmatrix}. \tag{11.6}$$

Subsequently we shall write for the sake of brevity $\langle a,b,c\rangle$ instead of (11.6). Let N denote the subgroup of G consisting of all matrices with integer elements. The homogeneous space $X_0 = G/N = \{Ng \mid g \in G\}$ is a compact three-dimensional nil-manifold and N is its fundamental group.

Each homeomorphism Φ of X_0 onto itself, for which $N\Phi = N$, induces an automorphism

$$\Phi_*:\pi_1(X_0,N) \to \pi_1(X_0,N),$$

i.e., an automorphism $\Phi_*:N \to N$.

Every homomorphism φ_0 of $\mathbb{R}$ into G is of the form

$$t\varphi_0 = \langle \alpha t, \beta t, \gamma t + \tfrac{1}{2}\alpha\beta t^2\rangle, \qquad (t \in \mathbb{R}),$$

where α, β and γ are arbitrary real parameters. The homomorphism $\varphi_0:\mathbb{R} \to G$ induces a flow $(X_0:\mathbb{R},\pi)$:

$$(Ng,t)\pi = Ng(t\varphi_0), \qquad (Ng \in X_0, t \in \mathbb{R}).$$

If α and β are rationally independent, then $(X_0,\mathbb{R},\pi)$ is a distal minimal (but not an equicontinuous) transformation group (*see*: Example 2•10.12).

We prove next several auxiliary results. The identity $\langle 0,0,0\rangle$ of the group G will be denoted by e.

2•11.36 LEMMA: *For every homeomorphism Φ of X_0 onto itself carrying the point $N \in G/N \equiv X_0$ to a point $Q = Nq$, there exists a unique homeomorphism $\varphi:G \to G$ with $e\varphi = q$ and $N(g\varphi) = (Ng)\Phi$ $(g \in G)$.*

Proof: This follows from the theorem on fibre maps (Hu [*1*], Theorem 2•16.2), if we take into account that G is the universal covering space of X_0.▮

2·11.37 LEMMA: *Let Φ and Ψ be homeomorphisms of X_0 onto itself such that $N\Phi = N\Psi = N$. If the induced homomorphisms $\Phi_*:N \to N$ and $\Psi_*:N \to N$ coincide, then there exists a bounded map $\vartheta:G \to G$ such that $g\varphi = (g\psi)(g\vartheta)$ for all $g \in G$, where the maps φ and ψ are chosen according to Lemma 2·11.36 with $e\varphi = e\psi = e$.*

Proof: If $k \in N$, then $N = (Nk)\Phi = N(k\varphi)$, hence $k\varphi \in N$. Similarly, $k\psi \in N$ for all $k \in N$. Let $k \in N$. Consider the map $\varphi_k:G \to G$, where $g\varphi_k = (k\varphi)^{-1}((kg)\varphi)$ and $(g \in G)$. Clearly, $e\varphi_k = e$. Furthermore

$$N(g\varphi_k) = N(k\varphi)^{-1}((kg)\varphi) = N((kg)\varphi) = (Nkg)\Phi = (Ng)\Phi.$$

Thus it follows from Lemma 2·11.36 that $\varphi_k = \varphi$ $(k \in N)$. Then $g\varphi = g\varphi_k = (k\varphi)^{-1}((kg)\varphi)$, and consequently $(kg)\varphi = (k\varphi)(g\varphi)$ $(g \in G, k \in N)$. A similar equality holds for ψ.

Define a map $\vartheta:G \to G$ by $g\vartheta = (g\psi)^{-1}\cdot(g\varphi)$. Let us show that $(kg)\vartheta = g\vartheta$ for all $k \in N$. In fact, $(kg)\vartheta = ((kg)\psi)^{-1}\cdot((kg)\varphi) = (g\psi)^{-1}\cdot(k\psi)^{-1}\cdot(k\varphi)\cdot(g\varphi)$. Since G is the universal covering space of X_0, we have $k\varphi = k\Phi_*$ and $k\psi = k\Psi_*$ for all $k \in N$. By hypothesis, $\Phi_* = \Psi_*$. Therefore $(kg)\vartheta = (g\psi)^{-1}(g\varphi) = g\vartheta$ and we can define a continuous map $\xi:X_0 \to G$ by $(Ng)\xi = g\vartheta$ $(g \in G)$. Since X_0 is compact, $G\vartheta = X_0\xi$ is a compact subset of G.▮

2·11.38 If φ is an automorphism of the topological group G and $N\varphi = N$, then φ induces a homeomorphism $\Phi:X_0 \to X_0$.

Let us describe all automorphisms of N onto itself. If $k \in N$, *i.e.*, $k = \langle x,y,z\rangle$ $(x,y,z \in \mathbb{Z})$, then k can be represented in the form

$$k = e_1{}^y e_2{}^x e_3{}^z,$$

where $e_1 = \langle 0,1,0\rangle$, $e_2 = \langle 1,0,0\rangle$, $e_3 = \langle 0,0,1\rangle$. Observe that e_3 belongs to the centre of the group G. If $\varphi:N \to N$ is an automorphism of N, then $e_3\varphi$ also belongs to the centre of N, and hence $e_3\varphi = \langle 0,0,p_3\rangle$, because only such elements belong to the centre of N. Let $e_1\varphi = \langle m_1,n_1,p_1\rangle$, $e_2\varphi = \langle m_2,n_2,p_2\rangle$. Since φ is an automorphism

$$k\varphi = \langle m_1,n_1,p_1\rangle^y\cdot\langle m_2,n_2,p_2\rangle^x\cdot\langle 0,0,p_3\rangle^z.$$

After reduction we get

$$\langle x,y,z\rangle\varphi = \langle m_1y + m_2x, n_1y + n_2x, p_1y + p_2x + m_1n_1\tfrac{1}{2}y(y - 1) + m_2n_2\tfrac{1}{2}x(x - 1) + m_1n_2yx + p_3z\rangle. \qquad (11.7)$$

Since φ is a bijection of N onto N, we must have

$$\Delta \equiv \begin{vmatrix} m_1 & m_2 \\ n_1 & n_2 \end{vmatrix} = \pm 1, \qquad p_3 = -\Delta. \qquad (11.8)$$

A direct calculation shows that the formulae (11.7,8) define together an automorphism of N. If (11.8) is satisfied, then the same formula (11.7), where x, y and z are assumed to be arbitrary real numbers, defines an automorphism of the group G (recall that $m_1, n_1, p_1, m_2, n_2, p_2, p_3 \in \mathbb{Z}$).

2·11.39 LEMMA: *For each homeomorphism F of X_0 onto itself there exists a uniquely defined homeomorphism $f:G \to G$, an automorphism $\varphi:G \to G$ and a continuous bounded map $\lambda:G \to G$ such that*

(1): $(Ng)F = N(gf)$ $(g \in G)$;

(2): $N\varphi = N$;

(3): $gf = (g\varphi)(g\lambda)$ $(g \in G)$.

Proof: Let $F:X_0 \to X_0$ be a homeomorphism and let $NF = Ng_0$. Define a map $R_q:X_0 \to X_0$ $(q \in G)$ by

$$(Ng)R_q = N(gq), \qquad (g \in G).$$

The homeomorphism $\tilde{\Phi} = F \circ R_{g_0^{-1}}$ of X_0 onto X_0 satisfies the condition $N\tilde{\Phi} = N$. It induces an automorphism $\tilde{\Phi}_*$ of the fundamental group $\pi_1(X_0, N) = N$. By the previous considerations, the automorphism $\tilde{\Phi}_*:N \to N$ can be extended to an automorphism φ of the group G onto itself, which induces a homeomorphism Φ of X_0 onto itself. It is easy to see that $\Phi_* = \varphi|_N = \tilde{\Phi}_*$.

By Lemmas 2·11.34,35, there exists a continuous bounded map ϑ: $G \to G$ with $g\tilde{\varphi} = (g\varphi)\cdot(g\vartheta)$ $(g \in G)$, where $\tilde{\varphi}:G \to G$ is the unique homeomorphism satisfying the conditions $e\tilde{\varphi} = e$ and $N(g\tilde{\varphi}) = (Ng)\tilde{\Phi}$ $(g \in G)$. Define the maps $f:G \to G$ and $\lambda:G \to G$ as follows:

$$gf = (g\tilde{\varphi})g_0, \qquad g\lambda = (g\vartheta)g_0, \qquad (g \in G).$$

It is easy to verify that the conditions (1), (2) and (3) are satisfied.▌

2·11.40 LEMMA: *If α is a transcendental number, $\beta = 1$ and $\gamma = 0$, then the distal minimal flow $(X_0, \mathbb{R}, \pi)$ is not orbitally homogeneous.*

Proof: Suppose that, to the contrary, for each point $Ng_0 \in X_0$ there exists a homeomorphism F of X_0 onto itself preserving orbits and carrying N to Ng_0. For this homeomorphism F we can find maps $f:G \to G$, $\varphi:G \to G$ and $\lambda:G \to G$ with the properties stated above.

Let $g \in G$ and $t \in \mathbb{R}$. There exists a number $\tau \in \mathbb{R}$ with $((Ng)\pi^t)F = ((Ng)F)\pi^\tau$. Then

$$N((g(t\varphi_0))f) = N(gf)(\tau\varphi_0).$$

Thus, for each $t \in \mathbb{R}$ there is an element $k_t \in N$ with $k_t = (gf)(\tau\varphi_0)((g(t\varphi_0))f)^{-1}$. By the continuity of the right side of this equality, k_t is a continuous function defined on the real line and taking its values in the discreet group N. Therefore k_t is a constant. Since $\tau = 0$ whenever $t = 0$, we get $k_t = k_0 = e$ for all $t \in \mathbb{R}$. Thus $(g(t\varphi_0))f = (gf)(\tau\varphi_0)$. Since $gf = (g\varphi)(g\lambda)$ $(g \in G)$, we have

$$(g(t\varphi_0))\varphi\cdot(g(t\varphi_0))\lambda = (g\varphi)(g\lambda)(\tau\varphi_0). \tag{11.9}$$

Recall that φ is an automorphism of G, *i.e.*, φ has the form (11.7) and thus (11.8) is satisfied. Let $g = \langle x,y,z\rangle$ and $g\lambda = \langle x_\lambda,y_\lambda,z_\lambda\rangle$. Since $\lambda:G \to G$ is bounded, the functions $x_\lambda,y_\lambda,z_\lambda$ are also bounded. Comparing the co-ordinates of the left and of the right sides of (11.9), we get three equalities of which we need only the following two:

$$m_1y + m_2x + m_1t + m_2\alpha t + (x + \alpha t)_\lambda = m_1y + m_2x + x_\lambda + \alpha\tau,$$

$$n_1y + n_2x + n_1t + n_2\alpha t + (y + t)_\lambda = n_1y + n_2x + y_\lambda + \tau.$$

After reduction we get

$$\begin{aligned} \alpha\tau &= m_2\alpha t + m_1t + (x + \alpha t)_\lambda - x_\lambda, \\ \tau &= n_2\alpha t + n_1t + (y + t)_\lambda - y_\lambda, \end{aligned} \tag{11.9'}$$

which holds for all $t \in \mathbb{R}$. Therefore, dividing the first equality by the second and taking the limit as $t \to \infty$, we get

$$\alpha = \frac{m_2\alpha + m_1}{n_2\alpha + n_1},$$

i.e.,

$$n_2\alpha^2 - (m_2 - n_1)\alpha - m_1 = 0. \tag{11.10}$$

Since α is a transcendental number, equality (11.10) holds only when $m_1 = n_1 = 0$, $m_2 = n_1$. Recalling the hypothesis (11.8), we find that

$$\Delta \equiv m_1n_2 - m_2n_1 = -m_2^2 = -1.$$

Consequently, either

$$m_1 = n_2 = 0, \qquad m_2 = n_1 = 1, \tag{11.11}$$

or

$$m_1 = n_2 = 0, \qquad m_2 = n_1 = -1. \tag{11.12}$$

Now we shall show that the homeomorphism $F:X_0 \to X_0$ can always be chosen in such a way that the condition (11.11) is satisfied. Indeed, define a map $\psi_0:G \to G$ by

$$\langle x,y,z\rangle\psi_0 = \langle -x,-y,z\rangle, \qquad (x,y,z \in \mathbb{R}).$$

It is easy to verify that ψ_0 is an automorphism of the group G. Since $N\psi_0 = N$, ψ_0 induces a homeomorphism $\Psi_0:X_0 \to X_0$. One can show directly that

$$((Ng)\pi^t)\Psi_0 = ((Ng)\Psi_0)\pi^{-t}, \qquad (g \in G, t \in \mathbb{R}).$$

Hence, Ψ_0 is orbit-preserving. The homeomorphism $\Psi \equiv \Psi_0 \circ F$ also preserves the orbits and it carries N to Ng_0. It is clear that $\psi \equiv \psi_0 \circ f$ satisfies the condition (11.11). Let $S^1 = \mathbb{R}/\mathbb{Z}$ and let $(\mathbb{Z} + h,t)\rho = \mathbb{Z} + h + t$. Then $(S^1,\mathbb{R},\rho)$ is a periodic flow of period 1. Define a continuous map $\varphi_0:X_0 \to S^1$ by $(N\langle x,y,z\rangle)\varphi_0 = \mathbb{Z} + y$ $(x,y,z \in \mathbb{R})$.

It is easy to verify that φ_0 is a homomorphism of $(X_0,\mathbb{R},\pi)$ onto $(S^1,\mathbb{R},\rho)$. The map φ_0 induces a homomorphism of the fundamental groups:

$$\varphi_{0*}:\pi_1(X_0,N) \to \pi_1(S^1,\mathbb{Z}),$$

where

$$\langle m,n,p\rangle\varphi_{0*} = n, \qquad (m,n,p \in \mathbb{Z}).$$

Let Ng_0 be an arbitrary point of X_0 with the property that $g_0 = \langle x,y,z\rangle$ satisfies the condition $y \in \mathbb{Z}$. By the choice of the homeomorphism Ψ, we have

$$\langle m,n,p\rangle\Psi_* = \langle m,n,p_1m + p_2n + p\rangle,$$

where p_1 and p_2 are some integers. Therefore $(\Psi \circ \varphi_0)_* = (\varphi_0)_*$ and it follows from the proof of Theorem 2·11.32 that the flow $(X_0,\mathbb{R},\pi)$ is almost periodic. But this is impossible because the fundamental group $\pi_1(X_0,N) \equiv N$ is not commutative.∎

2·11.41 REMARK: It is not hard to verify that (11.9′) implies that either $\tau = t$ or $\tau = -t$. In the second case we replace the homeomorphism F by $\Psi_0 \circ F$ and we conclude that $(X_0,\mathbb{R},\pi)$ is not orbitally homogeneous, because it is not dynamically homogeneous. Thus we obtain a new proof of Theorem 2·11.40 which does not depend on Theorem 2·11.32.

2·11.42 We give an example of an orbitally homogeneous, but not harmonizable minimal transformation group with 'two-dimensional time'.

Let X_0 be the above space G/N and let T be the Cartesian product $\mathbb{R}\times\mathbb{R}$. Define a transformation group (X_0,T,σ) as follows. If $(t,\tau)\in T$ and $g = \langle x,y,z\rangle$, we set

$$(Ng)\sigma^{(t,\tau)} = N\langle x + \alpha t, y + \beta t, z + x\beta t + \gamma t + \tfrac{1}{2}\alpha\beta t^2 + \tau\rangle .$$

In other words,

$$(Ng)\sigma^{(t,\tau)} = (Ng)(t\varphi_0)\langle 0,0,\tau\rangle .$$

If the numbers α and β are rationally independent, then the transformation group (X_0,T,σ) is minimal. Since T is commutative, whereas the group $\pi_1(X_0,N) \equiv N$ is not commutative, we conclude that (X_0,T,σ) is not harmonizable.

Let g_0 be an arbitrary element of G. Define a map $\Phi: X_0 \to X_0$ by $(Ng)\Phi = Ngg_0$ $(g\in G)$. Clearly Φ is a homeomorphism which maps N onto Ng_0. It remains to verify that Φ maps orbits of (X_0,T,σ) into orbits. Let $g_0 = \langle x_0,y_0,z_0\rangle$ and $g = \langle x,y,z\rangle$. Then

$$\begin{aligned}(Ng)\sigma^{(t,\tau)}\Phi &= (Ng(t\varphi_0)\langle 0,0,\tau\rangle)\Phi \\ &= N\langle x,y,z\rangle(t\varphi_0)\langle 0,0,\tau\rangle\langle x_0,y_0,z_0\rangle \\ &= N\langle x,y,z\rangle\langle x_0,y_0,z_0\rangle(t\varphi_0)\langle 0,0,\tau'\rangle ,\end{aligned}$$

where $\tau' = \tau + y_0\alpha t - x_0\beta t$. Hence

$$(Ng)\sigma^{(t,\tau)}\Phi = (Ng\Phi)\sigma^{(t,\tau')}.$$

2·11.43 The following interesting question arises in the study of minimal transformation groups: which topological properties characterise a space X which is a phase space of some minimal transformation group (X,T,π)? This problem is far from being solved even if $T = \mathbb{R}$ or $T = \mathbb{Z}$. In the next subsections we present some partial results obtained in this area.

2·11.44 Let M^m be a C^∞-smooth closed (*i.e.*, without boundary) connected compact manifold and let $S^1 = \{\xi \mid \xi\in\mathbb{C}, |\xi| = 1\}$. We shall show that there exists a diffeomorphism F of the manifold $M^m\times S^1$ onto itself such that the cascade $(M^m\times S^1,F)$ generated by F is minimal.

Provide M^m with a Riemannian metric of class C^∞. Let $\mathrm{Diff}^\infty(M^m\times S^1)$ denote the set of all C^∞-diffeomorphisms of the manifold $M^m\times S^1$ onto itself. For each positive integer k let $\rho_k(F_1,F_2)$ denote a metric in $\mathrm{Diff}^\infty(M^m\times S^1)$ which measures the proximity of the k-jets not only of F_1 and F_2 but also of F_1^{-1} and F_2^{-1}. Define a periodic flow $(M^m\times S^1,\mathbb{R},\sigma)$ as follows:

$$(x,\xi,t)\sigma \equiv (x,\xi)\sigma^t = (x,\xi e^{2\pi i t}), \qquad (x\in M^m, \xi\in S^1, t\in\mathbb{R}).$$

Let q be any positive integer. We introduce the following notations:

$$\delta_q = \{e^{2\pi it} | 0 \leqslant t < 1/q\}, \qquad \delta'_q = \{e^{2\pi it} | 0 < t < 1/q\},$$

$$\Delta_q = M^m \times \delta_q, \qquad \Delta'_q = M^m \times \delta'_q,$$

$$\Delta_{k,q} = \Delta_q \sigma^{k/q}, \qquad \Delta'_{k,q} = \Delta'_q \sigma^{k/q}, \qquad (k = 0,1,\ldots,q-1).$$

We shall inductively define four sequences of positive integers $\{p_n\}$, $\{q_n\}$, $\{k_n\}$, $\{\ell_n\}$ ($n = 0,1,2,\ldots$) and two sequences $\{A_n\}$ and $\{F_n\}$ of diffeomorphisms in $\mathrm{Diff}^\infty(M^m \times S^1)$ such that the following conditions will be satisfied:

(1_n): $F_n = B_n \sigma^{\alpha_n} B_n^{-1}$, where $B_n = A_1 \circ \ldots \circ A_n$, $\alpha_n = p_n/q_n$, p_n and q_n are relatively prime, $\alpha_{n+1} = \alpha_n + \beta_n$, where $\beta_n = 1/(k_n \ell_n q_n^2)$ and $\sigma^{\alpha_n} \circ A_{n+1} = A_{n+1} \circ \sigma^{\alpha_n}$. Thus:

$$p_{n+1} = k_n \ell_n q_n p_n + 1, \qquad q_{n+1} = k_n \ell_n q_n^2,$$

$$F_{n+1} = B_n \circ \sigma^{\alpha_n} \circ A_{n+1} \circ \sigma^{\beta_n} \circ A_{n+1}^{-1} \circ B_n^{-1}.$$

(2_n): $p_{n+1}(F_{n+1}^2, F_n^2) < 2^{-n-1}$ $\qquad (r = 0,1,2,\ldots,q_n - 1)$.

(3_n): Each orbit of the cascade $(M^m \times S^1, F_n)$ approximates the space $M^m \times S^1$ within 2^{-n}.

To begin with the construction, let p_0 and q_0 be any two relatively prime positive integers. Suppose that the numbers $\alpha_0,\ldots,\alpha_n$ and the diffeomorphisms $A_1,\ldots,A_n$ have already been defined so that the conditions (1_k), (2_k), (3_k) are satisfied for all $k < n$. Since the map B_n^{-1} is uniformly continuous, we can select a number $\gamma_{n+1} > 0$ such that $d(xB_n^{-1}, yB_n^{-1}) < 2^{-n-1}$, whenever $d(x,y) < \gamma_{n+1}$ (where d is a fixed metric on $M^m \times S^1$ which is invariant under σ^t ($t \in \mathbb{R}$)). Choose a family of k_n open sets $E_\ell \subset \Delta'_{0,q_n}$ ($\ell = 0,\ldots,k_n - 1$) such that

$$\mathrm{diam}\bar{E}_\ell < \gamma_{n+1}/2, \qquad \bar{E}_\ell \cap \bar{E}_m = \emptyset, \qquad (\ell \neq m),$$

and $\bigcup_\ell S(E_\ell, \gamma_{n+1}/2) \supset \Delta_{0,q_n}$.

Let h_n be an integer large enough to ensure that there exists a covering $\{G_0,\ldots,G_{h_n-1}\}$ of the space M^m by open sets G_r ($r = 0,\ldots,h_n - 1$), the closure of each of which is diffeomorphic to the disk $\mathbf{D}^m$. Write

$$G_{\ell,r} = G_r \times \delta'_{3\ell h_n+3r+1,\, 3k_n h_n q_n},$$

$$(r = 0,1,\ldots,h_n - 1; \ell = 0,1,\ldots,k_n - 1).$$

Observe that

$$G_{\ell,r} \subset G_r \times \delta'_{\ell h_n, h_n k_n q_n} \subset \Delta'_{0,q_n}.$$

Clearly, the set $\bar{G}_{\ell,r}$ is diffeomorphic to $\mathbb{D}^{m+1}$ and

$$\bar{G}_{\ell,r} \cap \bar{G}_{m,s} = \emptyset$$

whenever either $\ell \neq m$ or $r \neq s$. We choose h_n open sets $E_{\ell,r}$ $(r = 0,1,\ldots,h_n - 1)$ inside of each set E_ℓ in such a way that

(α): $\bar{E}_{\ell,r} \cap \bar{E}_{\ell,s} = \emptyset$, whenever $r \neq s$;

(β): $\bar{E}_{\ell,r}$ is diffeomorphic to $\mathbb{D}^{m+1}$ $(r = 0,1,\ldots,h_n - 1)$.

We shall now use the following proposition, which follows from results obtained by Palais [*1*] (*see*: Anosov and Katok [*1*]).

THEOREM: *Let O^m be a connected open smooth manifold of class C^∞. Let E_ℓ and G_ℓ $(\ell = 1,\ldots,n)$ be two families of open subsets of O^m whose closures are C^∞-diffeomorphic to the m-dimensional ball $\mathbb{D}^m$, and moreover:*

$$\bar{E}_r \cap \bar{E}_s = \bar{G}_r \cap \bar{G}_s = \emptyset, \qquad (r \neq s).$$

Then there exists a C^∞-diffeomorphism $S: O^m \to O^m$, which agrees with the identity map outside some compact set $N_1 \subset O^m$ and which maps E_ℓ onto G_ℓ $(\ell = 1,\ldots,n)$.

We apply this theorem to the manifold Δ'_{0,q_n} and to the families of subsets $E_{\ell,r}$ and $G_{\ell,r}$ $(\ell = 0,1,\ldots,k_n - 1; r = 0,1,\ldots,h_n - 1)$. Since the map S is the identity outside some compact $N_1 \subset \Delta'_{0,q_n}$, it can be extended to Δ_{0,q_n}, and then to $M^m \times S^1$ as follows:

$$xS = x\sigma^{-r/q_n} S \sigma^{r/q_n}$$

$$(x \in \Delta_{r,q_n}, r = 1,\ldots,q_n - 1).$$

Such an extension ensures that S and σ^{α_n} commute. Denote S by A_{n+1}.

Since F_{n+1} tends to F_n as $\beta_n \to 0$, condition (2_n) will be satisfied if ℓ_n is chosen to be a large enough multiple of h_n.

It remains to verify condition (3_n). First we observe that

$$F_{n+1}{}^{q_n} = B_n \circ A_{n+1} \circ \sigma^{\beta_n q_n} \circ A_{n+1}{}^{-1} \circ B_n{}^{-1};$$

and therefore

$$F_{n+1}^{sq_n} = B_n \circ A_{n+1} \circ \sigma^{s\beta_n q_n} \circ A_{n+1}^{-1} \circ B_n^{-1}, \qquad (s = 1,2,\ldots).$$

Let $(x,\gamma) \in M^m \times S^1$ and $(y,\xi) = (x,\gamma)B_n A_{n+1}$. There exists a number r such that $y \in G_r$. If $\ell_n > 3h_n$, then for each ℓ ($\ell = 0,1,\ldots,k_n - 1$) there is a number $s(\ell)$ with $(y,\xi)\sigma^{s(\ell)\beta_n q_n} \in G_{\ell,r}$. Hence, $(y,\xi)\sigma^{s(\ell)\beta_n q_n} A_{n+1}^{-1} \in E_{\ell,r} \subset E_\ell$. Therefore the finite set

$$\Gamma \equiv \{(x,\gamma)B_n A_{n+1}\sigma^{s(\ell)\beta_n q_n} A_{n+1}^{-1} \mid \ell = 0,1,\ldots,k_n - 1\}$$

satisfies the condition $S(\Gamma,\gamma_{n+1}) \supset \Delta_{0,q_n}$. It follows by the choice of γ_{n+1} that (3_n) holds.

By (2_n) the limit $F = \lim_{n\to\infty} F_n$ exists and $F \in \mathrm{Diff}^\infty(M^m \times S^1)$. Conditions (2_n) and (3_n) imply that $(M^m \times S^1, F)$ is a minimal cascade.

2·11.45 A slight modification of the above construction enables us to construct a diffeomorphism F such that $(M^m \times S^1, F)$ is not only minimal but also weakly mixing (*i.e.*, the direct product $(M^m \times S^1, F) \times (M^m \times S^1, F)$ has a dense orbit).

It can also be shown that if M^m is a smooth compact connected manifold without boundary and the group S^1 acts on M^m in a smooth way and without fixed points (*i.e.*, M^m is the total space of a principal smooth fibre bundle with S^1 as the structure group), then a minimal cascade with phase space M^m can be defined. In particular, there exists a minimal cascade on each sphere S^{2n+1} (Katok [*1*]). One unresolved conjecture is that it is impossible to define a minimal flow on the three-dimensional sphere. (See the survey paper by Smale [*1*]).

2·11.46 Let S^1 be the one-dimensional sphere $\{z \mid z \in \mathbb{C}, |z| = 1\}$, I the segment $[0,1]$ and $W = S^1 \times I$. Identifying every two points of W of the form $(z,0)$ and $(z^{-1},1)$ ($z \in S^1$), we obtain a non-orientable surface which is called the Klein bottle (Hu Sze-tsen [*1*]). It will be shown in the Subsection 3·18.16 that the Klein bottle supports a distal minimal cascade, and the Cartesian product of the Klein bottle and S^1 is minimal under some distal flow.

2·11.47 We present (without proof) some more results concerning topological properties of minimal sets.

(a): Jones (see Gottschalk and Hedlund [*2*], p. 151) constructed a minimal cascade whose phase space is a one-dimensional planar continuum that is locally connected at some but not all points.

(b): Hedlund [*1*] has shown that for each positive integer $p \geqslant 2$ the space X of all unit tangent vectors of a compact orientable surface of genus p is a minimal set under the horocycle flow $(X,\mathbb{R},\pi)$. Moreover,

$$\overline{P(X)} = Q(X) = X \times X.$$

(c): Chu [3] has proved that neither the universal curve of Sierpiński nor the universal curve of Menger (Aleksandrov [1]) can be the phase space for a minimal flow. Gottschalk [5] has shown that there does not exist a minimal cascade on the universal curve of Sierpiński. However, the universal curve of Menger admits a minimal cascade (Anderson [1]).

REMARKS AND BIBLIOGRAPHICAL NOTES

Theorems 2·11.1,3,5 are taken from the book by Gottschalk and Hedlund [2]. Theorem 2·11.3 is a generalization of a theorem due to Hilmy [1]. Theorem 2·11.5 was proved by Markov [1] in the case $T = \mathbb{R}$. An interesting generalization of Theorem 2·11.5 was found by Chu [3]. Example 2·11.6 is due to Floyd [1] and J. Auslander [1]. The concept of a homogeneous minimal set was introduced by Birkhoff [2]. In that paper (under the numbers 2 and 3), the following problem is posed: to prove that each minimal set is homogeneous in the sense that 'the stream lines are topologically indistinguishable from one another'. But it is clear that by embedding the cascade of Example 2·11.6 in a flow we get a topologically non-homogeneous minimal flow. Nemytskiĭ [2] posed the question, whether or not a minimal flow on a manifold is homogeneous in the sense of Birkhoff. Example 2·11.40, constructed by Bronšteĭn and Kholodenko [1], answers negatively this question. Theorem 2·11.10 is due to Gottschalk [2]. Related results were obtained by Židkov [1] and Grabar' [1]. The propositions 2·11.13-23 follow from results of J. Auslander [4], J. Auslander and Hahn [1], L. Auslander and Hahn [1], Bronšteĭn [1]. The notion of harmonizability was introduced by Markov [3]. Lemma 2·11.23 was proved by Ellis [9] (*see also* Eisenberg [2]). The material in Subsections 2·11.24-26 is taken from the paper by Kahn and Knapp [1]. The results in Subsections 2·11.28-30 are due to Ellis [8]. The fundamental group and the first cohomology group of a minimal set were considered by Chu and Geraghty [1]. The material presented in Subsections 2·11.23-43 is taken from the paper by Bronšteĭn and Kholodenko [1]. Example 2·11.44 is a particular case of a more general construction due to Anosov and Katok [1].

CHAPTER III

EXTENSIONS OF MINIMAL TRANSFORMATION GROUPS

§(3·12): THE BASIC THEORY OF EXTENSIONS

We shall introduce in this section the notion of an extension of a transformation group and we shall study some basic properties of extensions. We shall consider relations between the most important classes of extensions and we shall also investigate the problem of the openness of homomorphisms of minimal sets.

3·12.1 Let T be a fixed topological group, X and Y topological spaces, (X,T,π) and (Y,T,ρ) transformation groups and $\varphi: X \to Y$ a homomorphism of (X,T,π) onto (Y,T,ρ). In this case we say that (X,T,π) is an *EXTENSION OF* (Y,T,ρ) *BY THE HOMOMORPHISM* φ and that (Y,T,ρ) is a *FACTOR OF THE TRANSFORMATION GROUP* (X,T,π). *Often the mapping* $\varphi:(X,T,\pi) \to (Y,T,\rho)$ *itself is called an extension.* For each point $y \in Y$ the subspace $X_y = \{x \mid x \in X, x\varphi = y\}$ is called the *FIBER OF THE EXTENSION* φ *OVER* y. If each fiber X_y ($y \in Y$) consists of one point (*i.e.*, φ is injective and therefore a homeomorphism, whenever X is compact), then the extension φ is called a *TRIVIAL EXTENSION*. If the extension is not trivial, then it is called a *PROPER EXTENSION*.

An extension $\varphi:(X,T) \to (Y,T)$ is said to be a *MINIMAL EXTENSION* whenever (X,T) is a minimal transformation group.

3·12.2 Let $\varphi_i:(X_i,T) \to (Y_i,T)$ ($i = 1,2$) be extensions. A pair of homomorphisms $\psi:(X_1,T) \to (X_2,T)$, $\lambda:(Y_1,T) \to (Y_2,T)$ is called a *MORPHISM OF THE EXTENSION* $\varphi_1:(X_1,T) \to (Y_1,T)$ *INTO* $\varphi_2: (X_2,T) \to (Y_2,T)$ if $\varphi_1 \circ \lambda = \psi \circ \varphi_2$, *i.e.*, the diagram

$$\begin{array}{ccc} X_1 & \xrightarrow{\psi} & X_2 \\ {\scriptstyle \varphi_1}\downarrow & & \downarrow{\scriptstyle \varphi_2} \\ Y_1 & \xrightarrow{\lambda} & Y_2 \end{array}$$

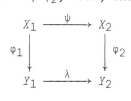

is commutative.

Let $\varphi_1:(X_1,T) \to (Y,T)$ and $\varphi_2:(X_2,T) \to (Y,T)$ be extensions of the same transformation group (Y,T). A homomorphism $\psi:(X_1,T) \to (X_2,T)$ is called a *MORPHISM OF THE EXTENSION* φ_1 *INTO THE EXTENSION* φ_2 *OVER* (Y,T) if $\varphi_1 = \psi\circ\varphi_2$. In this case we shall also say that the *extension* φ_2 *is SUBORDINATE to the extension* φ_1.

3·12.3 Let $\varphi_i:(X_i,T) \to (Y,T)$ $(i = 1,2)$ be two extensions. By $(X_1\cdot X_2,\varphi_1,\varphi_2) \equiv X_1\cdot X_2$ we *denote the subspace* $\{(x_1,x_2)|x_1 \in X_1, x_2 \in X_2, x_1\varphi_1 = x_2\varphi_2\}$ *of the Cartesian product* $X_1 \times X_2$. Clearly $(X_1\cdot X_2,\varphi_1,\varphi_2)$ is a closed invariant subset of the Cartesian product of the transformation groups (X_1,T) and (X_2,T). Hence a transformation group $((X_1\cdot X_2,\varphi_1,\varphi_2),T)$ is defined which for brevity will be denoted by $(X_1\cdot X_2,T)$, if there is no danger of ambiguity. In the particular case when (Y,T) is a trivial transformation group, $(X_1\ X_2,T)$ coincides with the Cartesian product $(X_1,T) \times (X_2,T) \equiv (X_1 \times X_2,T)$.

The map $\varphi_1\cdot\varphi_2:(X_1\cdot X_2,T) \to (Y,T)$, defined by $(x_1,x_2)(\varphi_1\cdot\varphi_2) = x_1\varphi_1 = x_2\varphi_2$ $((x_1,x_2) \in X_1\cdot X_2)$, is a homomorphism. The extension $\varphi_1\cdot\varphi_2$ is called the *WHITNEY SUM* (or the *FIBERED PRODUCT*) of the extensions φ_1 and φ_2 over (Y,T).

3·12.4 Let X be a uniform space with the uniformity $\mathfrak{U}[X]$, $\varphi: (X,T) \to (Y,T)$ an extension and $R = \{(x_1,x_2)|x_1,x_2 \in X, x_1\varphi = x_2\varphi\}$.

A point $x \in X$ is called *DISTAL RELATIVE TO* φ, or *R-DISTAL*, if it is distal from each point $x' \in X$ satisfying the conditions $x\varphi = x'\varphi$, $x \neq x'$. In other words, a point $x \in X$ is called R-distal if for each point $x' \in xR$ distinct from x there exists an index $\alpha \in \mathfrak{U}[X]$ with $(x't,xt) \notin \alpha$ for all $t \in T$.

An extension $\varphi:(X,T) \to (Y,T)$ is said to be a *DISTAL EXTENSION* if each point $x \in X$ is distal relative to φ, *i.e.*, if $P \cap R = \Delta$, where $P = P(X,T)$ is the proximal relation and $\Delta = \{(x,x)|x \in X\}$.

The extension $\varphi:(X,T) \to (Y,T)$ is said to be an *EXTENSION STABLE IN THE FIBER* X_y (and the *FIBER* X_y *is said to be STABLE WITH RESPECT TO* φ) if for each index $\alpha \in \mathfrak{U}[X]$ we can choose an index $\beta \in \mathfrak{U}[X]$ such that $(x_1t,x_2t) \in \alpha$ $(t \in T)$ whenever $(x_1,x_2) \in \beta$ and $x_1\varphi = x_2\varphi = y$. If the extension is stable in each fiber X_y $(y \in Y)$, then it is called a *STABLE EXTENSION*.

Recall that an extension $\varphi:(X,T) \to (Y,T)$ is called an *EQUICONTINUOUS EXTENSION* if for each index $\alpha \in \mathfrak{U}[X]$ there exists an index $\beta \in \mathfrak{U}[X]$ such that $(x_1,x_2) \in \beta$ and $x_1\varphi = x_2\varphi$ imply $(x_1t,x_2t) \in \alpha$ for all $t \in T$. Each equicontinuous extension is stable, but the converse fails to be true in general. If each fiber of an extension φ consists of a finite number of points, then φ is stable (but in general not equicontinuous).

An extension $\varphi:(X,T) \to (Y,T)$ is said to be an *EXTENSION PROXIMAL IN A FIBER* X_y if each two points $x_1,x_2 \in X_y$ are proximal. An extension φ is called a *PROXIMAL EXTENSION* if each two points belonging to the same fiber are proximal (*i.e.*, $R \subset P$). If for some point $y \in Y$ the fiber X_y consists of only one point, then the extension φ is called an *ALMOST AUTOMORPHIC EXTENSION*.

3·12.5 LEMMA: *Each equicontinuous extension is distal.*

Proof: Let $\varphi:(X,T) \to (Y,T)$ be an equicontinuous extension, and let $x_1,x_2 \in X$, $x_1\varphi = x_2\varphi$, but $x_1 \neq x_2$. Choose an index $\alpha \in \mathcal{U}[X]$ such that $(x_1,x_2) \notin \alpha$. Then choose an index $\beta \in \mathcal{U}[X]$ corresponding to α in the definition of an equicontinuous extension. Let us prove that $(x_1t,x_2t) \notin \beta$ for all $t \in T$. Suppose the contrary, *i.e.*, that $(x_1t_0,x_2t_0) \in \beta$ for some $t_0 \in T$. Then $(x_1,x_2) = (x_1t_0, x_2t_0)t_0^{-1} \in \alpha$, which contradicts the choice of α. ∎

3·12.6 LEMMA: *Let X be a compact space. Then an extension $\varphi:(X,T) \to (Y,T)$ is equicontinuous iff $Q(R) = \Delta$.*

Proof: If φ is equicontinuous, then for each index $\alpha \in \mathcal{U}[X]$ there is an index $\beta \in \mathcal{U}[X]$ with $(\beta \cap R)T \subset \alpha$. Hence $Q(R) \subset \overline{(\beta \cap R)T} \subset \bar{\alpha}$ for each $\alpha \in \mathcal{U}[X]$. Therefore $Q(R) = \Delta$.

Conversely, let $Q(R) = \Delta$, that is, $\bigcap_{\alpha \in \mathcal{U}[X]} \overline{(\alpha \cap R)T} = \Delta$. Given an open index $\alpha_0 \in \mathcal{U}[X]$, we shall prove that $\overline{(\beta \cap R)T} \subset \alpha_0$ for some index $\beta \in \mathcal{U}[X]$. If this is not so, then $\{\overline{(\beta \cap R)T} \cap ((X \times X) \setminus \alpha_0) \mid \beta \in \mathcal{U}[X]\}$ is a filter of closed subsets of the compact space $X \times X$. Hence $\bigcap_{\beta \in \mathcal{U}} \overline{(\beta \cap R)T} \cap (X \times X) \setminus \alpha_0) \neq \emptyset$, which contradicts $Q(R) = \Delta$. ∎

3·12.7 LEMMA: *Let X be a compact space and let $\varphi:(X,T) \to (Y,T)$ be a minimal almost automorphic extension. Then φ is proximal.*

Proof: Suppose that the fiber X_y consists of only one point x_0. Let $x_1\varphi = x_2\varphi$. For each index $\alpha \in \mathcal{U}[X]$ there exists an index $\beta \in \mathcal{U}[X]$ and a neighbourhood V of y such that $V\varphi^{-1} \subset x_0\beta$, $\beta \circ \beta \subset \alpha$, $\beta = \beta^{-1}$. Since (Y,T) is a minimal transformation group, we can find an element $t \in T$ with $x_1\varphi t = x_2\varphi t \in B$. Then $x_1t \in x_0\beta$, $x_2t \in x_0\beta$, hence $(x_1t,x_2t) \in \alpha$. Thus each two points x_1 and x_2 from the same fiber of X are proximal. ∎

3·12.8 COROLLARY: *Let X be a compact space and let $\varphi:(X,T)$ (Y,T) be a distal minimal extension. If some fiber X_y consists of a single point, then φ is trivial.*

Proof: By the preceding lemma, φ is both distal and proximal. But this can happen only in the case where φ is injective. ∎

3·12.9 Let X and Y be uniform spaces, (X,T,π) and (Y,T,ρ) transformation groups and $\varphi:(X,T,\pi) \to (Y,T,\rho)$ a homomorphism. Suppose that for each point $x \in X$ the set $\overline{xT}$ is compact. Then the *enveloping semigroup* $E = E(X,T,\pi)$ is defined an compact. Moreover, $xE = \overline{xT}$ $(x \in X)$. Let

$$E_y = \{\xi \mid \xi \in E, X_y\xi \subset X_y\}, \qquad (y \in Y).$$

3·12.10 LEMMA: *The following two conditions are equivalent:*

(1): $\xi \in E_y$;

(2): *$\xi \in E$ and there exists a point $x_0 \in X_y$ such that $x_0\xi \in X_y$.*

Proof: Obviously (1) implies (2). Let a point $x_0 \in X_y$ and an element $\xi_0 \in E$ be such that $x_0\xi_0 \in X$, *i.e.*, $x_0\xi_0\varphi = y$. If x is an arbitrary point of the fiber X_y, then $x\varphi = y$ and $x\pi^t\varphi = x\varphi\rho^t = x_0\varphi\rho^t = x_0\pi^t\rho$. Then it follows that $x\xi\varphi = x_0\xi\varphi$ for each element $\xi \in E$. But $x_0\xi_0\varphi = y$; hence $x\xi_0\varphi = y$, that is, $x\xi_0 \in X_y$ for all $x \in X_y$. Thus $\xi_0 \in E$.▌

3·12.11 LEMMA: *E_y is a compact Hausdorff subsemigroup of the semigroup E and $xE_y = \overline{xT} \cap X_y$ $(y \in Y, x \in X_y)$.*

Proof: It is easy to verify that the set E_y is closed in E; hence E_y is a compact subspace. Clearly E_y is a subsemigroup of E. Let $x_1 \in X_y \cap \overline{xT}$; then $x_1 = x\xi$ for some $\xi \in E$. By the preceding lemma, $\xi \in E_y$. Thus $X_y \cap \overline{xT} \subset xE_y$. The converse inclusion follows directly from the definition of E_y.▌

3·12.12 Let us define an equivalence relation R_y in E_y as follows: $(\xi_1,\xi_2) \in R_y$ if $x\xi_1 = x\xi_2$ for each point $x \in X_y$. The relation R_y is bilaterally stable, *i.e.*, $(\xi_1,\xi_2) \in R_y$ implies

$$(\xi\xi_1,\xi\xi_2) \in R_y, \qquad (\xi_1\xi,\xi_2\xi) \in R_y,$$

for each element $\xi \in E_y$. Indeed, if $(\xi_1,\xi_2) \in R_y$, then $x\xi_1 = x\xi_2$ $(x \in X_y)$, and therefore $x\xi_1\xi = x\xi_2\xi$ for each $\xi \in E_y$, consequently $(\xi_1\xi,\xi_2\xi) \in R_y$. If $\xi \in E_y$, then $x\xi\xi_1 = x\xi\xi_2$ for all $x \in X_y$, because $X_y\xi \subset X_y$. Thus $(\xi\xi_1,\xi\xi_2) \in R_y$.

Let $H_y^\xi = \{\lambda \mid \lambda \in E_y, (\xi,\lambda) \in R_y\}$, and let $G_y = \{H_y^\xi \mid \xi \in E_y\}$ be the quotient semigroup of the semigroup E_y with respect to the relation R_y (Liapin [*1*]). Provide G_y with the quotient topology. It is easy to show that G_y is a compact Hausdorff space. The semigroup G_y acts on X_y in the natural way:

$$xH_y^\xi = x\xi, \qquad (x \in X_y, \xi \in E_y).$$

Thus $xG_y = \overline{xT} \cap X_y$ $(x \in X_y)$.

3·12.13 Let X be a compact space, (X,T) a transformation group, φ a homomorphism of (X,T) onto (Y,T) and ψ a homomorphism of (Y,T) onto (Z,T). Let $x_0 \in X$, $x_0\varphi = y_0$, $x_0\varphi\psi = z_0$. The map $\varphi\psi \equiv \varphi\circ\psi$ is a homomorphism of (X,T) onto (Z,T).

The homomorphism $\varphi:(X,T) \to (Y,T)$ induces a canonical homomorphism $\vartheta:E(X,T) \to E(Y,T)$ of the semigroup $E(X,T)$ onto $E(Y,T)$ (Theorem 1·4.21). Let

$$E_1 = \{\xi \mid \xi \in E(X,T), x_0\xi\varphi\psi = z_0\}, \qquad E_2 = \{\eta \mid \eta \in E(Y,T), y_0\eta\psi = z_0\}.$$

Let us prove that $E_1\vartheta = E_2$. Let $\xi \in E_1$ and $\eta = \xi\vartheta$. Then $y_0\eta\psi =$

$x_0\varphi(\xi\rho)\psi = x_0\xi\varphi\psi = z_0$. Thus $\eta \in E_2$ and therefore $E_1\vartheta \subset E_2$. Let $\eta \in E_2$. Since $E(X,T)\vartheta = E(Y,T)$, there exists an element $\xi \in E(X,T)$ with $\xi\vartheta = \eta$. We shall prove that $\eta \in E_1$. Indeed, $x_0\xi\varphi\psi = x_0\varphi(\xi\vartheta)\psi = y_0\eta\psi = z_0$. Therefore $\xi \in E_1$ and hence $E_2 \subset E_1\vartheta$. Thus $E_1\vartheta = E_2$.

Let $\xi \in E_1$, $\eta \in E_2$, $X_0 = \{x \mid x \in X, x\varphi\psi = x_0\varphi\psi\}$, $Y_0 = \{y \mid y \in Y, y\psi = y_0\psi\}$, $H_1^{\xi} = \{\lambda \mid \lambda \in E_1, x\xi = x\lambda$ for each point $x \in X_0\}$, $H_2^{\xi} = \{\mu \mid \mu \in E_2$, $y\eta = y\mu$ for each point $y \in Y_0\}$, $G_1 = \{H_1^{\xi} \mid \xi \in E_1\}$, $G_2 = \{H_2^{\xi} \mid \eta \in E_2\}$.

It was shown above that E_1, E_2, G_1, G_2 are semigroups. Define a map $\tilde{\vartheta}: G_1 \to G_2$ by $H_1^{\xi}\tilde{\vartheta} = H_2^{\xi\vartheta}$ $(\xi \in E_1)$.

3·12.14 LEMMA: *$\tilde{\vartheta}$ is a semigroup homomorphism of G_1 onto G_2.*

Proof: It was proved above that $\vartheta: E_1 \to E_2$ is a homomorphism of E_1 onto E_2. If $x\xi = x\lambda$ for every point $x \in X_0$, then $x\xi\varphi = x\lambda\varphi$, and therefore $x\varphi(\xi\vartheta) = x\varphi(\lambda\vartheta)$. Since x is an arbitrary point of X_0, $y = x\varphi$ is an arbitrary point of Y_0. Therefore $y(\xi\vartheta) = y(\lambda\vartheta)$ $(y \in Y_0)$. Hence the map $\tilde{\vartheta}$ is well defined. Since ϑ is a semigroup homomorphism of E_1 onto E_2, $\tilde{\vartheta}$ is also a homomorphism of G_1 onto G_2.▮

3·12.15 LEMMA: *Let X be a compact space, (X,T,π) a minimal transformation group and φ a homomorphism of (X,T,π) onto (Y,T,ρ). If $U \subset X$ is an open and non-empty set, then* $\operatorname{int}(U\varphi) \neq \emptyset$.

Proof: Let $x_0 \in U$. There exists an open set V such that $x_0 \in V$ and $\bar{V} \subset U$. Since the space X is compact and (X,T,π) is a minimal transformation group, there exists a finite family of elements $t_i \in T$ $(i = 1,\dots,n)$ such that $\bigcup_{i=1}^{n} \bar{V}\pi^{t_i} = X$. Since $X\varphi = Y$, we get $X\varphi = \bigcup_{i=1}^{n} (\bar{V}\pi^{t_i})\varphi = \bigcup_{i=1}^{n} \bar{V}\varphi\rho^{t_i} = Y$. The set $V\varphi$ is closed in Y because the map φ is closed. Hence the set $\bar{V}\varphi\rho^{t_i}$ is closed in Y for each t_i, $i = 1,\dots,n$. Since Y is represented as an union of a finite family of closed sets, the interior of at least one member of this family is non-empty. Since all the sets $(\bar{V}\varphi)\rho^{t_i}$ $(i = 1,\dots,n)$ are homeomorphic to each other, it follows that $\operatorname{int}(\bar{V}\varphi) \neq \emptyset$, whence $\operatorname{int}(\bar{U}\varphi) \neq \emptyset$.▮

3·12.16 We recall some well known definitions. A subset A of a topological space X is said to be of the *FIRST CATEGORY* if A can be represented as a union of a countable family of nowhere dense sets: $A = \bigcup_{i=1} A_i$, $\operatorname{int}\bar{A}_i = \emptyset$, $(i = 1,\dots)$. A subset $A \subset X$ is called a *RESIDUAL SUBSET* if the complement $X \setminus A$ is of the first category. A subset $A \subset X$ is called a *BAIRE SUBSET* if A can be represented as an intersection of a countable family of open dense subsets of X:

$$A = \bigcap_{i=1}^{\infty} G_i, \qquad \mathrm{int}G_i = G_i, \qquad \bar{G}_i = X, \qquad (i = 1,2,\dots).$$

Each residual subset contains a Baire subset. In fact, if $A \subset X$ is residual, then:

$$X \setminus A = \bigcup_{i=1}^{\infty} E_i \subset \bigcup_{i=1}^{\infty} \bar{E}_i, \qquad \mathrm{int}\bar{E}_i = \emptyset, \qquad (i = 1,2,\dots),$$

and hence:

$$A \supset \bigcap_{i=1}^{\infty} (X \setminus E_i).$$

Clearly each set $X \setminus \bar{E}_i$ is open and dense in X.

3·12.17 LEMMA: *Let X be a compact space, (X,T) a minimal transformation group and $\varphi:(X,T) \to (Y,T)$ an extension. If A is a residual (Baire) subset of Y, then $A\varphi^{-1}$ is a residual (respectively, Baire) subset of X.*

Proof: It will suffice to consider the case where $A \subset Y$ is a Baire subset, *i.e.*, $A = \cap [G_i | i = 1,2,\dots]$, where the G_i are open and dense in Y. Then $A\varphi^{-1} = \cap [G_i\varphi^{-1} | i = 1,2,\dots]$. Each of the sets $G_i\varphi^{-1}$ is open because the map φ is continuous, thus it follows from Lemma 3·12.15 that the sets $G_i\varphi^{-1}$ $(i = 1,2,\dots)$ are dense in X. ∎

3·12.18 LEMMA: *Let X be a compact metric space, (X,T,π) be a minimal transformation group and $\varphi:(X,T,\pi) \to (Y,T,\rho)$ be a homomorphism. Then the set X_0 of all points $x \in X$, at which the map φ is open is a Baire subset.*

Proof: Let G denote the set of all points $x \in X$ such that $x \in U \cap (\mathrm{int}(U\varphi))\varphi^{-1}$ for some open set U whose diameter is less than $1/n$ $(n = 1,2,\dots)$. By Lemma 3·12.15, the set G_n $(n = 1,2,\dots)$ is dense in X. It is clear that each G_n is open and $X_0 = \bigcap_{n=1}^{\infty} G_n$. ∎

3·12.19 LEMMA: *Let X be a compact metric space, (X,T) a minimal transformation group and $\varphi:(X,T) \to (Y,T)$ an extension. If K is a residual subset of X, then the set*

$$A_K = \{y | y \in Y, K \cap X_y \text{ is a residual subset of } X_y\}$$

is residual in Y.

Proof: By hypothesis, there exists a sequence $F_1,F_2,\dots$ of closed nowhere dense subsets of X such that $X \setminus K \subset \bigcup_{\ell=1}^{\infty} F_\ell$. Define

$K_y = K \cap X_y$ ($y \in Y$). It is easy to verify that

$$X_y \setminus K_y \subset \bigcup_{\ell=1}^{\infty} F_\ell \cap X_y.$$

Let B be the set of all points $y \in Y$ such that for some F the interior of $F_\ell \cap X_y$ relative to X_y is non-empty. If $y \notin B$, then K_y is residual in X_y. In other words, $B \subset (Y \setminus A_K)$. Hence it will suffice to show that B is a set of the first category in Y.

Let $\{G_m | m = 1,2,\dots\}$ be a base of the topology in X. We denote by $G_{m\ell}$ the set of those points $y \in Y$ for which

$$G_m \cap X_y \neq \emptyset, \qquad G_m \cap X_y \subset F_\ell.$$

Clearly

$$B = \bigcup_{m,\ell=1}^{\infty} G_{m\ell}.$$

It remains to prove that each set $G_{m\ell}$ is nowhere dense in Y.

Set $D_{m\ell} = \bar{G}_{m\ell}$ and suppose that $\mathrm{int} D_{m\ell} \neq \emptyset$ for some m and ℓ. If $y \in G_{m\ell} \cap \mathrm{int} D_{m\ell}$, then $X_y \cap G_m \neq \emptyset$. Hence $E = (\mathrm{int} D_{m\ell})\varphi^{-1} \cap G_m$ is a non-empty open subset of X.

By Lemma 3·12.18 the set E contains a dense subset E_1 consisting of those points at which the map $\varphi: X \to Y$ is open. If $z \in E_1$, then $z\varphi \in \mathrm{int} D_{m\ell}$ and therefore there exists a sequence $\{y_n\}$, $y_n \in G_{m\ell}$ ($n = 1,2,\dots$) with $\{y_n\} \to z\varphi$. Since the map φ is open at z, we can choose a sequence $\{z_n\}$, $z_n \in X$, such that $\{z_n\} \to z$ and $z_n\varphi = y_n$ ($n = 1,2,\dots$). Since $z \in E_1 \subset E \subset G_m$, we may suppose without loss of generality that $z_n \in G_m$ ($n = 1,2,\dots$). Then $z_n \in G_m \cap X_{y_n} \subset F_\ell$, because $z_n\varphi = y_n \in G_{m\ell}$. Hence $z \in \bar{F}_\ell = F_\ell$ for every point $z \in E_1$. But since the set E_1 is dense in E, we get $E \subset F_\ell$ and this completes the proof. ▌

Observe that we did not use directly the hypothesis that (X,T) is minimal. It was essential only that the set X_0 of those points at which the mapping $\varphi: X \to Y$ is open is a dense subset of X.

3·12.20 Let X be a compact space, (X,T) a transformation group, $\varphi:(X,T) \to (Y,T)$ an extension and $R = \{(x_1,x_2) | x_1,x_2 \in X, x_1\varphi = x_2\varphi\}$. Suppose that $\overline{aT} = X$ for some point $a \in X$. Set $a\varphi = b$ and define

$$X_b = \{x | x \in X, x\varphi = b\}, \qquad E_b = \{\xi | \xi \in E(X,T), X_b\xi \subset X_b\}.$$

By Lemma 3·12.11, E_b is a compact Hausdorff subsemigroup of the semigroup E and $aE_b = X_b$.

> 3·12.1 LEMMA: *If the point a is R-distal, then $au = a$ for each idempotent $u \in E_b$.*

Proof: If $u \in E_b$ and $u^2 = u$, then $(au)u = au$ and thus the points a and au are proximal. Since $u \in E_b$, we have $au \in X_b$, *i.e.*, $au\varphi = a\varphi$. Hence $au = a$, because a is R-distal. ▮

3·12.22 LEMMA: *Let a be an R-distal point. Then for each* $\gamma \in E_b$ *there exists a point* $x = x_\gamma \in X_b$ *such that* $x\gamma \equiv x_\gamma\gamma = a$.

Proof: Let I_b be a minimal right ideal of the semigroup E_b. By Lemma 1·4.11 the set $\mathcal{J}_b$ of idempotents of I_b is non-empty. Let $\gamma \in E_b$ and let $u \in \mathcal{J}_b$. Then $u\gamma \in I_b$. By Lemma 1·4.2(6) there exists an idempotent $v \in \mathcal{J}_b$ such that $u\gamma \in I_b v$. Choose $\delta \in I_b v$ with $\delta(u\gamma) = v = (u\gamma)\delta$. This is possible by Lemma 1·4.2(4). If $x = a\delta u$, then $x\gamma = a\delta u\gamma = av$. Since $v \in \mathcal{J}_b$, it follows by Lemma 1·12.21 that $av = a$. Thus $x\gamma = a$. Since $\delta \in I_b v$ and $u \in \mathcal{J}_b$, we get $\delta u \in I_b \subset E_b$ and therefore $a\delta u \in X_b$. Consequently $x_\gamma \equiv a\delta u$ is the required point. ▮

3·12.23 LEMMA: *Let a be an R-distal point. Then for each index* $\alpha \in \mathcal{U}[X]$ *there exist points* $x_1, \ldots, x_p \in X_b$ *satisfying the following condition: Given any element* $\gamma \in E_b$, *we can choose an* x_i, $1 \leqslant i \leqslant p$, *such that* $(x_i\gamma, a) \in \alpha$.

Proof: Let $\alpha \in \mathcal{U}[X]$. Without loss of generality we may suppose that α is symmetric and open. For every element $\gamma \in E_b$ we find a point $x_\gamma \in E_b$ satisfying the assertion of the preceding lemma and we define $B_\alpha(\gamma) = \{\xi \mid \xi \in E_b, (x_\gamma\xi, a) \in \alpha\}$. Since $x_\gamma\gamma = a$, we have $\gamma \in B_\alpha(\gamma)$ for all $\gamma \in E_b$, $\alpha \in \mathcal{U}[X]$. Since the space E_b is compact, we can select a finite covering $\{B_\alpha(\gamma_1), \ldots, B_\alpha(\gamma_p)\}$ of E_b. Set $x_i = x_{\gamma_i}$ $(i = 1, \ldots, p)$. If $\gamma \in E_b$, then $\gamma \in B_\alpha(\gamma_i)$ for some γ_i, $1 \leqslant i \leqslant p$, that is, $(x_i\gamma, a) \in \alpha$. ▮

3·12.24 THEOREM: *Let X be a compact space,* (X,T) *a transformation group,* $\varphi:(X,T) \to (Y,T)$ *an extension and* $R = \{(x_1, x_2) \mid x_1, x_2 \in X, x_1\varphi = x_2\varphi\}$. *If* $a \in X$ *is an R-distal point with* $\overline{aT} = X$, *then the map* $\varphi: X \to Y$ *is open at a.*

Proof: Suppose that to the contrary there exists an index $\alpha_0 \in \mathcal{U}[X]$ such that $a\alpha_0\varphi$ is not a neighbourhood of the point $a\varphi = b$. Then there exists a net $\{y_\nu\} \to b$ with $y_\nu \notin a\alpha_0\varphi$, *i.e.*

$$X_{y_\nu} \cap a\alpha_0 = \emptyset. \tag{12.1}$$

For $\alpha_0 \in \mathcal{U}[X]$ choose points $x_1, \ldots, x_p$ as in the preceding lemma. From each fiber X_{y_ν} choose some point w_ν. Since $X = \overline{aT} = aE$, there exist elements $\gamma_\nu \in E$ with $a\gamma_\nu = w_\nu$. Let γ be a limit point of the net $\{\gamma_\nu\}$. Then, obviously, $a\gamma \in X_b$. By Lemma 3·12.10, $\gamma \in E_b$. Using Lemma 3·12.23, select x_i, $1 \leqslant i \leqslant p$, with $(x_i\gamma, a) \in \alpha_0$. There exists a γ_ν such that $(x_i\gamma_\nu, a) \in \alpha_0$. Since $x_i\varphi = a\varphi$, we have $x_i\gamma_\nu\varphi = a\gamma_\nu\varphi = w_\nu\varphi = y$, hence $x_i\gamma_\nu \in X_{y_\nu}$, and therefore $X_{y_\nu} \cap a\alpha_0 \neq \emptyset$. This contradicts (12.1). ▮

3·12.25 COROLLARY: *Let X be a compact space,* (X,T) *a minimal transformation group and* $\varphi:(X,T) \to (Y,T)$ *a distal extension. Then the map* $\varphi:X \to Y$ *is open.*

3·12.26 Let X be a compact topological space and let $\mathcal{U}[X]$ be the uniformity compatible with the topology of X. Let 2^X denote the family of all non-empty closed subsets of X. For each index $\alpha \in \mathcal{U}[X]$ we define

$$\alpha^* = \{(A,B) \mid A,B \in 2^X, A \subset B\alpha, B \subset A\alpha\}.$$

The system $\{\alpha^* \mid \alpha \in \mathcal{U}[X]\}$ is a uniformity base of some uniformity $\mathcal{U}^*$ and, moreover, the space $(2^X,\mathcal{U}^*)$ is compact. If X is metrizable, then $(2^X,\mathcal{U}^*)$ is also metrizable. The distance between two subsets A and B can be defined by (the so-called *HAUSDORFF METRIC*)

$$\alpha(A,B) = \inf\{\varepsilon \mid \varepsilon > 0, A \subset S(B,\varepsilon), B \subset S(A,\varepsilon)\}$$

Let (X,T,π) be a transformation group with a compact phase space X. Then we can define a transformation group $(2^X,T,\sigma)$ by $(A,t)\sigma = A\pi^t$ $(A \in 2^X, t \in T)$.

3·12.27 LEMMA: *Let* X *be a compact metric space,* (X,T,π) *a minimal transformation group and* $\varphi:(X,T) \to (Y,T)$ *an extension. Then there exist compact metric spaces* X^* *and* Y^**, minimal transformation groups* (X^*,T) *and* (Y^*,T)*, and homomorphisms* $p:(X^*,T) \to (X,T)$*,* $q:(Y^*,T) \to (Y,T)$ *and* $\psi:(X^*,T) \to (Y^*,T)$ *such that*

(1): *The diagram*

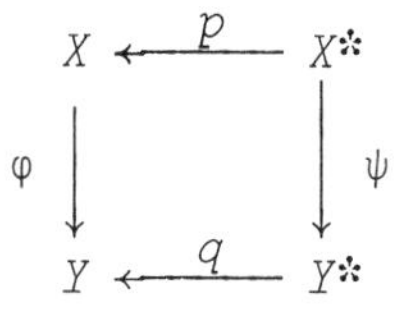

is commutative;

(2): p *and* q *are almost automorphic extensions;*

(3): ψ *is an open map.*

Proof: Let $R = \{(x_1,x_2) \mid x_1 \in X, x_2 \in X, x_1\varphi = x_2\varphi\}$. We assign to each point $y \in Y$ the compact set $X_y = \{x \mid x \in X, x\varphi = y\}$. The function $F:Y \to 2^X$, $y \to X_y$ defined in this way is upper semi-continuous because X is compact and the relation R is closed. Therefore the set $\mathcal{C}(F)$ of points of continuity of the map $F:Y \to 2^X$ is residual in Y. Since Y is a compact minimal set, each point $y \in Y$ is almost periodic. Hence if $a \in \mathcal{C}(F)$, then $X_a \equiv a\varphi^{-1}$ is an almost periodic point of the transformation group $(2^X,T)$ and since 2^X is compact, the orbit closure of the point X_a in 2^X is a compact minimal set. Denote it by Y^*. It is easy to verify that Y^* does not depend on the choice of the point $a \in \mathcal{C}(F)$. Moreover, the closure of the set $\{X_y | y \in Y\}$ in 2^X contains exactly one minimal set. Indeed, let $y_n \in Y$ and let $\lim X_{y_n} = A$ be an almost periodic point. We may suppose that $\{y_n\}$ converges to some point $y \in Y$. For each point $a \in \mathcal{C}(F)$ we can choose a sequence $\{t_k\}$, $t_k \in T$, such that $\{yt_k\} \to a$ and the limit $\lim(At_k)$ exists and equals A_0. Choose a subsequence $\{n_k\}$ such that $\{y_{n_k}t_k\} \to a$ and $A_0 \equiv \lim(At_k) = \lim(y_{n_k}\varphi^{-1}t_k) = \lim(y_{n_k}t_k\varphi^{-1})$. It follows from $a \in \mathcal{C}(F)$ that $\{y_{n_k}t_k\varphi^{-1}\} \to X_a$, and hence $\{At_k\} \to X_a$. Since A and X_a are almost periodic points, we conclude that they belong to the same minimal set.

It is easy to see that (x,X_a), where $a \in \mathcal{C}(F)$, $x \in X_a$, is an almost periodic point of the transformation group $(X \times 2^X,T) \equiv (X,T) \times (2^X,T)$. Therefore the orbit-closure of the point (x,X_a) in $X \times 2^X$ is a compact minimal set. Denote it by X^*. One can show that X^* is the only minimal set of the transformation group $(X,T) \times (Y^*,T)$.

The projection $(X \times 2^X) \to X$ determines a homomorphism $p:(X^*,T) \to (X,T)$ and the projection $X \times 2^X \to 2^X$ defines a homomorphism $\psi: (X^*,T) \to (Y^*,T)$. If $B \in Y^*$ and $x_1,x_2 \in B$, then $x_1\varphi = x_2\varphi$. This enables us to define a mapping $q:Y^* \to Y$ by $Bq = x\varphi$ $(x \in B, B \in Y^*)$. Clearly, $p\circ\varphi = \psi\circ q$. If $a \in \mathcal{C}(F)$, then aq^{-1} consists of a single point $X_a \in Y^*$. Hence the extension $q:(Y^*,T) \to (Y,T)$ is almost automorphic. Let $a \in \mathcal{C}(F)$. Let us prove that $(x,X_a) \in X^*$ for each point $x \in X_a$. Since $X^*\psi = Y^*$, there exists a point $x_0 \in X_a$ with $(x_0,X_a) \in X^*$. Choose a sequence $\{t_k\}$, $t_k \in T$ such that $\{x_0t_k\} \to x$. Then $\{x_0t_k\varphi\} \to x\varphi = a$ and since $a \in \mathcal{C}(F)$, we get $\{(x_0t_k\varphi)\varphi^{-1}\} \to a\varphi^{-1} \equiv X_a$. Thus

$$\{(x_0,X_a)t_k\} = \{(x_0t_k,(x_0t_k\varphi)\varphi^{-1})\} \to (x,X_a),$$

and this just means that $(x,X_a) \in X^*$ $(x \in X_a, a \in \mathcal{C}(F))$. Hence it follows that for $a \in \mathcal{C}(F)$ the set $X_a\psi^{-1}$ is homeomorphic to the set X_a. It is easy to show that the pre-image of each point $x \in X_a$, $a \in \mathcal{C}(F)$, consists of a single point (x,X_a). Therefore the extension $p:(X^*,T) \to (X,T)$ is almost automorphic.

Let us now verify that ψ is open. Let $A \in Y^*$ and $x \in A$. We shall prove that $(x,A) \in X^*$. There exists a sequence $\{a_n\}$, $a_n \in \mathcal{C}(F)$ such that $\{a_n\varphi^{-1}\} \to A$ in the space 2^X. Hence we can choose points $x_n \in a_n\varphi^{-1}$ such that $\{x_n\} \to x$. Since $a_n \in \mathcal{C}(F)$, we have $(x_n,a_n\varphi^{-1}) \in X^*$, as was shown above. Hence $(x,A) = \lim\{(x_n,a_n\varphi^{-1})\} \in X^*$. Thus, if $A \in Y^*$ and $x \in A$, then $(x,A) \in X^*$. It is not difficult to deduce from this that the map $\psi:X^* \to Y^*$ is open. ▮

3.12.28 LEMMA: *Let X be a compact space and let $\varphi:(X,T) \to (Y,T)$ and $\psi:(Y,T) \to (Z,T)$ be distal (proximal) extensions. Then $\varphi\circ\psi$ is also distal (respectively, proximal).*

Proof: First suppose that φ and ψ are distal extensions and $x_1\varphi\psi = x_2\varphi\psi$, $x_1 \neq x_2$. If $x_1\varphi = x_2\varphi$, then the points x_1 and x_2

are distal by hypothesis. Suppose that $x_1\varphi \neq x_2\varphi$. Since ψ is a distal extension, the points $x_1\varphi$ and $x_2\varphi$ are distal and since φ is uniformly continuous, the points x_1 and x_2 are also distal.

Let us now consider the second case. If $x_1\varphi = x_2\varphi$, then x_1 and x_2 are proximal by hypothesis. Suppose that $x_1\varphi \neq x_2\varphi$, but $x_1\varphi\psi = x_2\varphi\psi$. Let us prove that x_1 and x_2 are proximal. Let I be a minimal right ideal of the semigroup $E(X,T)$ and let $v \in I$ be an idempotent. It will suffice to prove that $x_1v = x_2v$. Since $(x_1v,x_2v) \equiv (x_1,x_2)v$ is an almost periodic point of the transformation group $(X \times X,T)$ and $x_1v\varphi\psi = x_2v\varphi\psi$, the point $(x_1v\varphi, x_2v\varphi)$ is also almost periodic and satisfies the condition $(x_1v\varphi, x_2v\varphi) \in R\psi$. But the extension ψ is proximal, hence it follows from Lemma 1·6.2 that $x_1v\varphi = x_2v\varphi$. Since φ is proximal, we get $x_1v = x_2v$. ▮

3·12.29 LEMMA: *Let* $\{(X_\lambda,T)|0 \leqslant \lambda \leqslant \vartheta\}$ *be a projective system with homomorphisms* φ_ν^λ $(0 \leqslant \nu \leqslant \lambda \leqslant \vartheta)$. *If each extension* $\varphi_\lambda^{\lambda+1}$ $(0 \leqslant \lambda \leqslant \vartheta)$ *is distal (proximal), then* $\varphi_0^\vartheta:(X_\vartheta,T) \to (X_0,T)$ *is also distal (proximal).*

Proof: We show by induction that each extension $\varphi_0^\lambda:(X_\lambda,T)$ (X_0,T) is distal (respectively, proximal) $(0 < \lambda \leqslant \vartheta)$. By hypothesis, the statement is true for $\lambda = 1$. Suppose that it is true for all λ, $\lambda < \mu$. There are two cases which we consider separately: (1) the ordinal μ has an immediate predecessor, $\mu = \lambda + 1$. Since $\varphi_\lambda^{\lambda+1} = \varphi_\lambda^{\lambda+1}\circ\varphi_0^\lambda$, the extension $\varphi_0^{\lambda+1}$ is distal (respectively, proximal) by Lemma 3·12.28; (2) μ is a limit ordinal. Suppose that each extension φ_0^λ $(0 < \lambda < \mu)$ is distal. We shall prove that φ_0^μ is also distal. Suppose that, to the contrary, there exist two distinct proximal points $a,b \in X$ such that $a\varphi_0^\mu = b\varphi_0^\mu$. Then for each λ, $\lambda < \mu$, the points $a\varphi_\lambda^\mu$ and $b\varphi_\lambda^\mu$ are also proximal, but this is possible only when $a\varphi_\lambda^\mu = b\varphi_\lambda^\mu$ $(\lambda < \mu)$. Since (X_μ,T) is the inverse limit of the projective system $\{(X_\mu,T)|\lambda < \mu\}$, we have $a = b$, which contradicts our assumption that $a \neq b$.

Now we suppose that each extension φ_0^λ $(0 < \lambda < \mu)$ is proximal and we proceed to prove the proximality of φ_0^μ. Assume the contrary. Then there exist distal points a and b in X_μ with $a\varphi_0^\mu = b\varphi_0^\mu$. By Lemma 1·6.2 we may suppose without loss of generality that (a,b) is an almost periodic point of the transformation group $(X_\mu \times X_\mu,T)$, and hence $(a\varphi_\lambda^\mu,b\varphi_\lambda^\mu)$ is almost periodic in $(X_\mu \times X_\mu,T)$. Since φ_0^λ is proximal, we get $a\varphi_\lambda^\mu = b\varphi_\lambda^\mu$. Consequently $a = b$, which contradicts the hypothesis. ▮

3·12.30 An extension $\varphi:(X,T) \to (Y,T)$ is called an *AE-EXTENSION* if φ can be represented as a composition $\lambda\circ\mu$ of an equicontinuous extension $\lambda:(X,T) \to (Y,T)$ and an almost automorphic extension $\mu:(Z,T) \to (Y,T)$. If, moreover, λ is a proper extension, then φ is called a *PROPER AE-EXTENSION*. Thus each almost automorphic extension is considered to be an (improper) *AE*-extension.

3·12.31 Let $\{(X_\lambda,T)|0 \leqslant \lambda \leqslant \vartheta\}$ be a projective system with homomorphisms φ_μ^λ $(0 \leqslant \mu \leqslant \lambda \leqslant \vartheta)$. We introduce the following definitions. The system $\{(X_\lambda,T)|0 \leqslant \lambda \leqslant \vartheta\}$ is called an *F-SYSTEM* if each extension $\varphi_\lambda^{\lambda+1}$ $(0 \leqslant \lambda < \vartheta)$ is equicontinuous. Suppose that $\varphi_\lambda^{\lambda+1}$ is an *AE*-extension for all $\lambda < \vartheta$ and, moreover, $\varphi_\lambda^{\lambda+1}$ is a proper *AE*-extension whenever $\lambda + 1 < \vartheta$. Then $\{(X_\lambda,T)|0 \leqslant \lambda \leqslant \vartheta$ is called a *V-SYSTEM*. If each extension $\varphi_\lambda^{\lambda+1}$ $(\lambda < \vartheta)$ is either proximal or distal (equicontinuous), then $\{(X_\lambda,T)|0 \leqslant \lambda \leqslant \vartheta\}$ is said to be a *PD-SYSTEM* (respectively, a *PE-SYSTEM*).

An extension $\varphi:(X,T) \to (Y,T)$ is called an *F-EXTENSION* if there exists an *F*-system $\{(X_\lambda,T)|0 \leqslant \lambda \leqslant \vartheta\}$ such that $(X_0,T) = (Y,T)$, $(X_\vartheta,T) = (X,T)$ and $\varphi_0^\vartheta = \varphi$. Similarly one defines the notions of *V-EXTENSION*, *PD-EXTENSION*, and *PE-EXTENSION*. We shall say that a transformation group belongs to the *CLASS F* (respectively, *V*, *AE*,*PE*,*PD*) if it is an extension of the type *F* (respectively, *V*, *AE*,*PE*,*PD*) of the trivial transformation group.

REMARKS AND BIBLIOGRAPHICAL NOTES

Furstenberg's theorem on the structure of distal minimal sets has stimulated the investigation of equicontinuous and distal extensions for their own sake. The notion of proximality led naturally to proximal extensions. Almost automorphic extensions arose in connection with the theory of almost automorphic functions (Bochner [*2*,*3*], Veech [*1*,*2*]). Extension of the types *AE* and *V* were introduced by Veech [*4*,*5*]. *PD*-extensions were considered by Shapiro [*2-4*]. Related problems were studied by Petersen [*2*]. The material in Subsections 3·12.19-24 is an extension of some results due to Veech [*5*] to the case of *R*-distal points (*see also* Furstenberg [*1*]). Lemma 3·12.27 is due to Veech. His method was somewhat generalised by Bronšteĭn [*16*].

§(3·13): EQUICONTINUOUS AND STABLE EXTENSIONS

We shall study in this section equicontinuous and stable minimal extensions. Primary attention will be paid to the description of the smallest closed invariant equivalence relation containing $Q(R)$. These results will be used in the following sections, in particular, in the study of distal minimal extensions.

3·13.1 LEMMA: *Let* (X,T) *be a minimal transformation group with a compact phase space,* $R \subset X \times X$ *a closed invariant equivalence relation and* $(x,y) \in Q(R)$. *Then for each index* $\alpha \in \mathfrak{U}[X]$ *and every non-empty open subset* U *of* X *there exist points* $x_1 \in x\alpha$, $y_1 \in y\alpha$ *and an element* $t_1 \in T$ *such that* $(x_1,y_1) \in R$ *and* $x_1t_1 \in U$, $y_1t_1 \in U$.

Proof: Let $x_0 \in U$ and let $\beta \in \mathfrak{U}[X]$ be an index such that $x_0\beta^2 \subset U$. Since (X,T) is a minimal transformation group, there exists a compact subset K of T with $X = (x_0\beta)K$. Then $xK^{-1} \cap x_0\beta \neq \emptyset$ for each point $x \in X$. Choose an index $\gamma \in \mathfrak{U}[X]$ such that $\gamma K^{-1} \subset \beta$.

Since $(x,y) \in Q(R)$, given an arbitrary index $\alpha \in \mathfrak{U}[X]$ there exist points $x_1 \in x\alpha$, $y_1 \in y\alpha$ and an element $\tau_1 \in T$ such that $(x_1,y_1) \in R$ and $(x_1\tau_1,y_1\tau_1) \in \gamma$. Choose an element $k_1 \in K^{-1}$ with $x_1\tau_1k_1 \in x_0\beta$. Since $(x_1\tau_1,y_1\tau_1) \in \gamma$, we have $(x_1\tau_1k_1,y_1\tau_1k_1) \in \beta$. Then $x_1\tau_1k_1 \in x_0\beta \subset U$, $y_1\tau_1k_1 \in x_1\tau_1k_1\beta \subset x_0\beta^2 \subset U$. Hence $t_1 = \tau_1k_1$ is the required element. ∎

3·13.2 THEOREM: *Let* (X,T) *be a minimal transformation group with a compact phase space and let* S *and* R *be two closed equivalence relations invariant under* (X,T) *and satisfying the condition* $S \subset R$. *Let* $\psi: X \to X/S$ *and* $\varphi: X/S \to X/R$ *be the canonical projections. Then*

$$Q(X,R)(\psi \times \psi) = Q(X/S,R/S),$$

where $R/S = R(\psi \times \psi)$ *and* $(a,b)(\psi \times \psi) = (a\psi,b\psi)$ $(a,b \in X)$.

Proof: Since the canonical map $\psi: X \to X/S$ is uniformly continuous, we get $Q(X,R)(\psi \times \psi) \subset Q(X/S,R/S)$.

The converse inclusion holds also. Assume the contrary. Then there exist two points a and b in X such that $(a\psi,b\psi) \in Q(X/S,R/S)$, but $(x,y) \notin Q(X,R)$ for any points $x \in aS$, $y \in bS$. Using the compactness of aS and bS, we deduce that there exist indices $\alpha_0 \in \mathfrak{U}[X]$ and $\delta_0 \in \mathfrak{U}[X]$ satisfying the condition

$$(aS\delta_0 \times bS\delta_0) \cap (\alpha_0 \cap R)T = \emptyset. \tag{13.1}$$

Let $\alpha_1 \in \mathfrak{U}[X]$ be a symmetric index with $\alpha_1^2 \subset \alpha_0$ and let x_0 be some point of X. By Lemma 3·12.15,

$$\operatorname{int}(x_0\alpha_1\psi) \neq \emptyset. \tag{13.2}$$

Since S is a closed equivalence relation, the set $aS\delta_0$ contains an S-saturated neighbourhood of the set aS and similarly $bS\delta_0$ contains an S-saturated neighbourhood of bS. By (13.2) the set $x_0\alpha_1S$ contains a non-empty open S-saturated set.

Since $(a\psi,b\psi) \in Q(X/S,R/S)$, it follows from Lemma 3·13.1 that there exist points x and y of X and an element $t \in T$ such that $(x,y) \in R$, $xS \subset aS\delta_0$, $yS \subset bS\delta_0$, $xSt \subset x_0\alpha_1S$, $ySt \subset x_0\alpha_1S$. Therefore $xSt \cap x_0\alpha_1 \neq \emptyset$ and $ySt \cap x_0\alpha_1 \neq \emptyset$. Then $(xS \times yS)t \cap \alpha_1^2 \neq \emptyset$ and $(xS \times yS)t \subset R$, and since $\alpha_1^2 \subset \alpha$, this contradicts (13.1). ∎

3·13.3 COROLLARY: *Suppose that the hypotheses of Theorem* 3·13.2 *are satisfied. Then the extension* $\varphi:(X/S,T) \to (X/R,T)$ *is equicontinuous iff* $Q(X,R) \subset S$.

Proof: By Lemma 3·12.6 the extension φ is equicontinuous iff $Q(X/S,R/S) = \Delta(X/S)$. This is equivalent to $Q(X,R) \subset S$, as is seen from the equality $Q(X,R)(\psi \times \psi) = Q(X/S,R/S)$. ∎

3·13.4 COROLLARY: *Let* X *be a compact space,* (X,T) *a minimal transformation group and let* $\varphi:(X,T) \to (Y,T)$ *and* $\psi:(Y,T) \to (Z,T)$ *be extensions. If* $\varphi\circ\psi$ *is an equicontinuous extension, then both* φ *and* ψ *are also equicontinuous.*

Proof: Clearly, φ is an equicontinuous extension. Let S and R be closed invariant equivalence relations corresponding to φ and $\varphi\circ\psi$ respectively, that is

$$S = \{(x_1,x_2) \mid x_1,x_2 \in X, x_1\varphi = x_2\varphi\},$$

$$R = \{(x_1,x_2) \mid x_1,x_2 \in X, x_1\varphi\psi = xx_2\varphi\psi\}.$$

It is clear that $S \subset R$. Since $\varphi\circ\psi$ is an equicontinuous extension, $Q(X,R) = \Delta(X)$ according to Lemma 3·12.6. Thus $Q(X,R) \subset S$, and therefore ψ is equicontinuous (Corollary 3·13.3). ∎

3·13.5 Let X be a compact space, (X,T) a minimal transformation group, $\varphi:(X,T) \to (Y,T)$ an extension and let $R = \{(x_1,x_2) \mid x_1,x_2 \in X, x_1\varphi = x_2\varphi\}$. By $Q^*(R) \equiv Q^*(X,R) \equiv Q^*(\varphi)$ we denote the smallest closed invariant equivalence relation containing $Q(R) \equiv Q(X,R) \equiv Q(\varphi)$. The extension $\psi:(X/Q^*(R),T) \to (X/R,T)$ is equicontinuous by Corollary 3·13.3 and is the 'greatest' equicontinuous extension of the transformation group $(Y,T) \equiv (X/R,T)$, subordinated to the extension φ in the following sense: given a commutative diagram

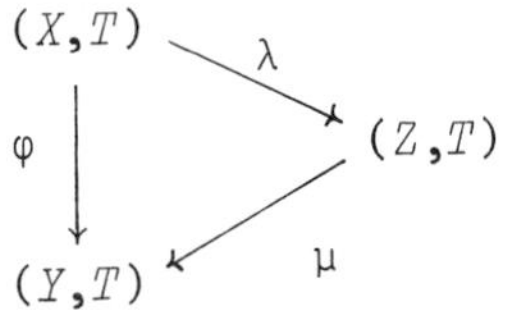

where μ is an equicontinuous extension, there exists a homomorphism $\nu:(X/Q^*(R),T) \to (Z,T)$ making the diagram

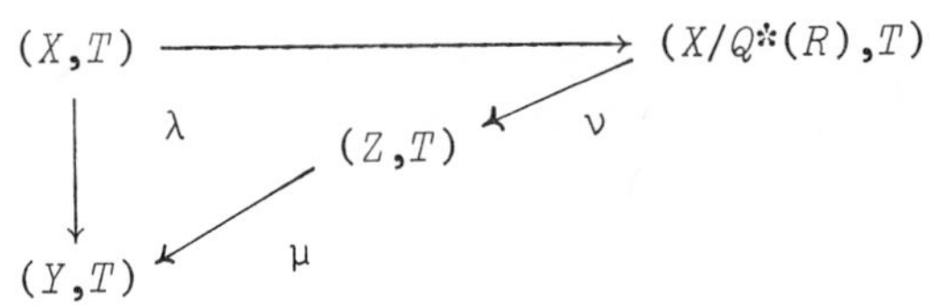

commute.

In fact, let $S = \{(x_1,x_2) \mid x_1\lambda = x_2\lambda\}$. Then $S \subset Q(R)$ by Corollary 3·13.3, and hence $S \supset Q^*(R)$.

3·13.6 Let X be a compact uniform space, (X,T) a minimal transformation group, $R \subset X \times X$ a closed invariant equivalence relation and let $\mathcal{J} = \{u \mid u \in E(X,T), u^2 = u, uE$ be a minimal right ideal of the semigroup $E\}$. Then $R\mathcal{J} = \{(xu,yu) \mid (x,y) \in R, u \in \mathcal{J}\}$ is the set of all almost periodic points in R.

In Subsection 2·9.4 we introduced the relation $Q(R)$. We recall that

$$Q(R) = \bigcap_{\alpha \in \mathfrak{U}[X]} \overline{\alpha T \cap R}.$$

This formula can be applied to any subset $R \subset X \times X$. In particular, define:

$$Q(R\mathcal{J}) = \bigcap_{\alpha \in \mathfrak{U}[X]} \overline{\alpha T \cap R\mathcal{J}}.$$

It is easy to verify that $Q(R\mathcal{J}) = Q(\overline{R\mathcal{J}})$.

Let

$$Q_1(R) = \{(x,y) \mid (x,y) \in R, (xu,yu) \in Q(R), u \in \mathcal{J}\}.$$

The relation $Q_1(R)$ is reflexive, symmetric and invariant. Moreover, $Q(R) \subset Q_1(R)$. Indeed, if $(x,y) \in Q(R) \subset R$, then $(xu,yu) \in Q(R)u \subset Q(R)$ $(u \in \mathcal{J})$, because the relation $Q(R)$ is invariant and closed.

3·12.7 LEMMA: *If $Q(R)\mathcal{J} \subset Q(R\mathcal{J})$, then $Q_1(R)$ is an equivalence relation.*

Proof: We need only to show that $Q_1(R)$ is transitive. Let $(a,b) \in Q_1(R)$, $(b,c) \in Q_1(R)$. We shall prove that $(a,c) \in Q_1(R)$. Choose any element $u \in \mathcal{J}$ and let $au = x$, $bu = y$, $cu = z$. We shall show that $(x,z) \in Q(R)$.

By hypothesis, $(x,y) \in Q(R)$ and $(y,z) \in Q(R)$. Let $\alpha \in \mathfrak{U}[X]$. Choose $\beta \in \mathfrak{U}[X]$ such that $\beta^3 \subset \alpha$. The element $u \in \mathcal{J}$ belongs to some minimal right ideal I of the semigroup $E = E(X,T)$. Set $U = \{\xi \mid \xi \in I, x\xi \in x\alpha, y\xi \in y\alpha, z\xi \in z\alpha\}$. According to the definition of the topology in $E(X,T)$, the set U is open in I. Since $u \in I$ and $xu = x$, $yu = y$, $zu = z$, we have $U \neq \emptyset$. The map $\varphi_0 : I \to X$, defined by $\xi\varphi_0 = y\xi$ $(\xi \in I)$, is a homomorphism of (I,T) onto (X,T). Since (I,T) is a minimal transformation group, it follows from Lemma 3·12.15 that $V \equiv \operatorname{int}(U\varphi_0) \equiv \operatorname{int}(yU) \neq \emptyset$.

Let y_1 be an arbitrary point of V. Then there exists an element $\xi_1 \in U$ with $y_1 = y\xi_1$. By the definition of U, $x_1 \equiv x\xi_1 \in x\alpha$. Since $Q(R)$ is a closed invariant relation, $(x,y) \in Q(R)\mathcal{J}$ implies that $(x_1,y_1) = (x\xi_1,y\xi_1) \in Q(R)\mathcal{J} \subset Q(R\mathcal{J})$. Hence there exists a point y_2 in the neighbourhood V of y_1 and there exists a point x_2 in the neighbourhood $x\alpha$ of x_1 such that $(x_2,y_2) \in \alpha T \cap R\mathcal{J}$. Thus (x_2,y_2) is an almost periodic point and $(x_2,y_2)t \in \beta$ for some element $t \in T$. Hence it follows that there is a syndetic set $A \subset T$ with

$$(x_2, y_2)A \subset \beta. \tag{13.3}$$

Hence $T = AK_0$ for some compact subset $K_0 \subset T$. It follows from (13.3) that $(x_2, y_2)T \subset \beta K_0$, therefore $(x_2, y_2)E \subset \bar{\beta}K_0 \subset \beta^2 K_0$ and $(x_2\xi, y_2\xi)T \subset \beta^2 K_0$ $(\xi \in E)$. Consequently, for each element $\xi \in E$ there exists a subset $A(\xi) \subset T$ such that

$$A(\xi)K_0 = T, \qquad (x_2\xi, y_2\xi)A(\xi) \subset \beta^2. \tag{13.4}$$

Since $y_2 \in V \subset yU$, we have $y_2 = y\xi_2$ for some $\xi_2 \in U$. Put

$$W = \{\xi \mid \xi \in I, x_2\xi \in x\alpha, y_2\xi \in V\}.$$

The set W is open in I. Since (x_2, y_2) is an almost periodic point, there exists an element $\nu \in I$ with $x_2\nu = x_2$, $y_2\nu = y_2$, hence satisfying $\nu \in W$.

The map $\varphi_1 : I \to X$, where $\xi\varphi_1 = y_2\xi$ $(\xi \in I)$, is a homomorphism of (I,T) onto (X,T). By Lemma 3·12.15

$$V_1 \equiv \operatorname{int}(W\varphi_1) \equiv \operatorname{int}(y_2 W) \neq \emptyset.$$

Let y_3 be any point of V_1. Since $y_3 \in V_1 \subset y_2 W \subset V \subset yU$, there exists an element $\xi_3 \in U$ with $y_3 = y\xi_3$. Hence $z_3 \equiv z\xi_3 \in z\alpha$. As $(y,z) \in Q(R)\mathcal{J}$ and $Q(R)$ is a closed invariant relation, we get that

$$(y_3, z_3) \equiv (y\xi_3, z\xi_3) \in Q(R)\mathcal{J} \subset Q(R\mathcal{J}).$$

Let $\delta \in \mathfrak{U}[X]$ be such that $\delta K_0^{-1} \subset \beta$. There exist a point y_4 in the neighbourhood V_1 of y_3 and a point z_4 in the neighbourhood $z\alpha$ of z_3 such that $(y_4, z_4) \in R\mathcal{J} \cap \delta T$. Since (y_4, z_4) is an almost periodic point, we can find a syndetic subset $B \subset T$ such that $(y_4, z_4)B \subset \delta$. Then

$$(y_4, z_4)BK_0^{-1} \subset \delta K_0^{-1} \subset \beta. \tag{13.5}$$

Since $y_4 \in V_1 \subset y_2 W$, we have $y_4 = y_2\xi_4$ for some element $\xi_4 \in W$. Denote $x_2\xi_4$ by x_4; then $x_4 \in x\alpha$. It follows from (13.4) that

$$(x_4, y_4)A(\xi_4) \equiv (x_2\xi_4, y_2\xi_4)A(\xi_4) \subset \beta^2. \tag{13.6}$$

Define $C = A(\xi_4) \cap BK_0^{-1}$. By Lemma 1·3.4 the set C is syndetic. From (13.5,6) we deduce that

$$(x_4, z_4)C \subset \beta^3 \subset \alpha. \tag{13.7}$$

Recall that $x_4 \in x\alpha$, $z_4 \in z\alpha$. Since α is an arbitrary index of the uniformity $\mathcal{U}[X]$, we get $(x,z) \in Q(R)$. ∎

3·13.8 LEMMA: *If* $Q(R)\mathcal{J} \subset Q(R\mathcal{J})$, *then* $Q_1(R) = Q(R)\mathcal{J}\circ(P\cap R) = Q(R)\circ Q(R)$, *where* $P = P(X,T)$ *is the proximal relation.*

Proof: According to the preceding lemma, the relation $Q_1(R)$ is transitive. Since $Q(R) \subset Q_1(R)$, we have:

$$Q(R)\mathcal{J}\circ(R\cap P) \subset Q(R)\circ Q(R) \subset Q_1(R)\circ Q_1(R) = Q_1(R).$$

Conversely, let $(x,y) \in Q_1(R)$. By Lemma 1·5.10 there exists an element $u \in \mathcal{J}$ with $xu = x$. Then $(x,yu) = (xu,yu) \in Q(R)$ in accordance with the definition of $Q_1(R)$. Since $u^2 = u$, it follows that $(yu,y) \in P$. From $(x,y) \in Q(R) \subset R$ and $(x,yu) \in Q(R) \subset R$ we deduce that $(yu,y) \in R$. Hence $(yu,y) \in P\cap R$. Therefore $(x,y) = (x,yu)\circ(yu,y) = (xu,yu)\circ(yu,y) \in Q(R)\mathcal{J}\circ(P\cap R)$. ∎

3·13.9 THEOREM: *Let* X *be a compact space,* (X,T) *a minimal transformation group and* $R \subset X\times X$ *a closed invariant equivalence relation. If* $Q(R)\mathcal{J} \subset Q(R\mathcal{J})$, *then* $Q_1(R) = Q^*(R)$.

Proof: Lemma 3·13.7 asserts that $Q_1(R)$ is an invariant equivalence relation. By Lemma 3·13.8, $Q_1(R) = Q(R)\circ Q(R)$. Since $Q(R)$ is closed, it follows that the relation $Q_1(R)$ is also closed. Further, $Q(R) \subset Q_1(R) = Q(R)\circ Q(R) \subset Q^*(R)$; hence $Q_1(R) = Q^*(R)$ in accordance with the definition of $Q^*(R)$. ∎

3·13.10 LEMMA: *If* $R = \overline{R\mathcal{J}}$, *then* $Q(R)\mathcal{J} \subset Q(R\mathcal{J})$.

Proof: In fact, if $R = \overline{R\mathcal{J}}$, then $Q(R\mathcal{J}) = Q(\overline{R\mathcal{J}}) = Q(R)$ and since $Q(R)$ is invariant and closed, we get $Q(R)\mathcal{J} \subset Q(R)$. ∎

3·13.11 LEMMA: *Let* X *be a compact metrizable space,* (X,T) *a minimal transformation group and* $R \subset X\times X$ *a closed invariant equivalence relation satisfying the condition* $Q(R)\mathcal{J} \subset Q(R\mathcal{J})$. *Then* $Q^*(R) = R$ *holds iff* $\overline{R\mathcal{J}} = \overline{P\cap R\mathcal{J}}$.

Proof: As $Q(R)\mathcal{J} \subset Q(R\mathcal{J})$, we have $Q^*(R) = Q(R)\mathcal{J}\circ(P\cap R) = Q_1(R)$. If $Q^*(R) = R$ and $(x,y) \in R \equiv Q_1(R)$, then $(xu,yu) \in Q(R)$ $(u \in \mathcal{J})$. Thus $R\mathcal{J} \subset Q(R)\mathcal{J} \subset Q(R\mathcal{J})$ and therefore $\overline{R\mathcal{J}} \subset Q(R\mathcal{J}) \subset \overline{R\mathcal{J}}$. Hence $Q(R\mathcal{J}) = \overline{R\mathcal{J}}$, that is

$$\bigcap_{n=1}^{\infty} \overline{\alpha_n T\cap\overline{R\mathcal{J}}} = \overline{R\mathcal{J}}, \tag{13.8}$$

where $\{\alpha_n | n = 1,2,\dots\}$ is a uniformity base of the compact metrizable space X. We may suppose that each index α_n is open in $X\times X$. It follows from (13.8) that $\alpha_n T\cap\overline{R\mathcal{J}}$ is an open dense subset of $\overline{R\mathcal{J}}$. By Baire's Theorem

$$\overline{P\cap\overline{R\mathcal{J}}} = \overline{\bigcap_{n=1}^{\infty} \alpha_n\, T\cap\overline{R\mathcal{J}}} = R\mathcal{J}.$$

The converse assertion is obvious.

3·13.12 THEOREM: *Let X be a compact metric space with the metric ρ and let (X,T) be a minimal transformation group. Suppose there is a point $a \in X$ which is distal from all other points $x \in X$ and assume that X contains more than one point. Then $Q^* \neq X \times X$.*

Proof: The function $f(x) = \inf_{t \in T} \rho(at,xt)$ $(x \in X)$ is upper semi-continuous, hence there is a point $b \in X$ $(b \neq a)$ of continuity of f. Since the point a is distal from b, we have $f(b) > 0$; hence there exist numbers $\alpha > 0$ and $\beta > 0$ such that $f(x) > \alpha$ for all points $x \in X$ satisfying the condition $\rho(x,b) < \beta$. Let us prove that (a,b) does not belong to $\bar{P}$. Suppose the contrary. Then there exist points $a',b' \in X$ with $\rho(a,a') < \alpha/2$, $\rho(b,b') < \beta$, $(a',b') \in P(X,T)$. Hence there is an idempotent $u \in E(X,T)$ such that $b'u = a'$ (Theorem 1·5.11). As the point a is distal from all other points $x \in X$, we have $au = a$ and $\rho(b'u,au) = \rho(a',a) < \alpha/2$. Consequently there exists an element $t \in T$ with $\rho(b't,at) < \alpha$, thus $f(b') < \alpha$, and this contradicts the hypothesis that $\rho(b,b') < \beta$. Thus $\bar{P} \neq X \times X$. Applying Lemma 3·13.11 with $R = X \times X$, we get the required result.▌

3·13.13 LEMMA: *Let X be a compact metric space, T a commutative group and let (X,T,π) be a minimal transformation group. Then the following conditions are equivalent:*

(1): $Q^* \equiv Q^*(X \times X) = X \times X$;

(2): $Q = \bar{P} = X \times X$.

Proof: (2) evidently implies (1). Let (1) be satisfied and let $x \in X$. Since the group T is commutative, the points of the form $(xt,x\tau)$ $(t,\tau \in T)$ are almost periodic under $(X \times X,T)$. This and the minimality of (X,T) imply that the relation $R \equiv X \times X$ satisfies the condition $R = \overline{R\mathcal{J}}$. By Lemma 3·13.11, condition (1) is equivalent to $\bar{P} = X \times X$. Thus (1) implies (2).▌

3·13.14 A transformation group (X,T) is called a *WEAKLY MIXING TRANSFORMATION GROUP* if $Q(X,T) = X \times X$. If X is a compact metric space, then the transformation group (X,T) is weakly mixing iff $\overline{(a,b)T} = X \times X$ for some point $(a,b) \in X \times X$. The proof of this statement is left to the reader.

3·13.15 THEOREM: *Let X be a compact space, (X,T) a minimal transformation group and let $R \subset X \times X$ be an invariant equivalence relation that is both closed and open. Further, let $R = \overline{R\mathcal{J}}$. Then $Q(R)$ is a closed invariant equivalence relation.*

Proof: It can be seen from Lemma 3·13.10, Theorem 3·13.9 and Lemma 3·13.8 that

$$Q^*(R) = Q(R)\mathcal{J}\circ(P \cap R),$$

holds. To prove the theorem, it suffices to show that $Q(R) \subset Q(R)\mathcal{J} \circ (P \cap R)$. Let $(x,y) \in Q(R)\mathcal{J}$ and $(y,z) \in P \cap R$. We shall prove that $(x,z) \in Q(R)$. Let U and V be arbitrary neighbourhoods of x and z respectively. Let α be an index in $\mathcal{U}[X]$. Choose $\beta \in \mathcal{U}[X]$ with $\beta^3 \subset \alpha$.

Since $(y,z) \in P \cap R$, we have $(yt_0, zt_0) \in \beta$ for some element $t_0 \in T$. By continuity we can choose neighbourhoods W of y and V_1 of z such that $V_1 \subset V$ and

$$(W \times V_1)t_0 \subset \beta. \tag{13.9}$$

Since $(y,z) \in R$ and since the relation R is open, the neighbourhood W can be chosen small enough so as to satisfy

$$(\forall y' \in W)(\exists z' \in V_1):(y',z') \in R. \tag{13.10}$$

Let I be a minimal right ideal of the semigroup $E = E(X,T)$. Since $(x,y) \in Q(R)\mathcal{J}$, the point (x,y) is almost periodic under $(X \times X, T)$. By Lemma 1·5.10 there exists an idempotent $u \in I$ with $(x,y) = (xu,yu)$. Hence it follows that the set $\tilde{U} = \{\xi \mid \xi \in K, x\xi \in U, y\xi \in W\}$ is non-empty. Moreover, it is open in I and $W_1 \equiv \operatorname{int}(y\tilde{U}) \neq \emptyset$ by Lemma 3·12.15.

Let $y_1 \in W_1 \subset W$. According to (13.10) there is an element $z_1 \in V_1$ with $(y_1,z_1) \in R$. As $R = \overline{R\mathcal{J}}$, we can choose points $y_2, z_2 \in X$ such that

$$(y_2,z_2) \in R\mathcal{J} \cap (W_1 \times V_1). \tag{13.11}$$

Then by (13.9)

$$(y_2,z_2)t_0 \in \beta. \tag{13.12}$$

It follows from (13.11) that (y_2,z_2) is an almost periodic point of the transformation group $(X \times X, T)$, and hence there is a compact set $K_0 \subset T$ such that $(y_2,z_2)tK_0 \cap \beta \neq \emptyset$ $(t \in T)$. Consequently, for each element $\xi \in E$

$$(y_2\xi, z_2\xi)tK_0 \cap \beta^2 \neq \emptyset, \qquad (t \in T). \tag{13.13}$$

The set $\tilde{W} = \{\xi \mid \xi \in I, y_2\xi \in W_1, z_2\xi \in V_1\}$ is non-empty and open in I. Using Lemma 3·12.15 again, we get $W_2 \equiv \operatorname{int}(y_2\tilde{W}) \neq \emptyset$.

Let $y_3 \in W_2$. Since $W_2 \subset W_1 \subset y\tilde{U}$, there exists an element $\xi_3 \in \tilde{U}$ with $y_3 = y\xi_3$. Therefore $x_3 \equiv x\xi_3 \in U$, as follows from the definition of $\tilde{U}$. Since $(x,y) \in Q(R)$ and since $Q(R)$ is a closed invariant relation, we get $(x_3,y_3) = (x\xi_3, y\xi_3) \in Q(R)$.

Choose $\delta \in \mathcal{U}[X]$ such that $\delta K_0 \subset \beta$. As $(x_3,y_3) \in Q(R)$, there exist points $x_4 \in U$ and $y_4 \in W_2$ and an element $s_0 \in T$ satisfying the conditions

$$(x_4,y_4) \in R, \qquad (x_4,y_4)s_0 \in \delta. \tag{13.14}$$

Then we have by the choice of δ that

$$(x_4,y_4)s_0K_0 \subset \beta. \tag{13.15}$$

Since $y_4 \in W_2 \equiv \text{int}(y_2W)$, we get $y_4 = y_2\xi_4$ for some element $\xi_4 \in W$. Put $z_4 = z_2\xi_4$. Then $z_4 \in V_1 \subset V$ by the definition of the set $\tilde{W}$. Further, it follows from (13.13) that

$$(y_2\xi_4,z_2\xi_4)s_0K_0 \cap \beta^2 \neq \emptyset.$$

Therefore

$$(y_4,z_4)s_0k_0 \equiv (y_2\xi_4,z_2\xi_4)s_0k_0 \in \beta^2, \tag{13.16}$$

for some element $k_0 \in K_0$. From (13.15,16) we deduce that $(x_4,z_4) \circ s_0k_0 \in \beta^3 \subset \alpha$.

Recall that $x_4 \in U$, $z_4 \in V$. It follows from (13.11) that $(y_4,z_4) = (y_2,z_2)\xi_4 \in R\xi_4 \subset R$, and hence $(x_4,z_4) = (x_4,y_4)(y_4,z_4) \in R\circ R = R$ by (13.14). ▌

3·13.16 LEMMA: *Let X be a compact topological space, (X,T) a minimal transformation group and $\varphi:(X,T) \to (Y,T)$ an extension. Further, let $b \in Y$ be a continuity point of the map $F:Y \to 2^X$ defined by $yF = Xy$ $(y \in Y)$ and assume that the fiber X_b is stable with respect to the extension φ. Then $X_b \times X_b$ consists only of almost periodic points of the transformation group $(X\times X,T) = (X,T)\times(X,T)$.*

Proof: Let I be a minimal right ideal of the semigroup $E = E(X,T)$ and let $a \in X_b$. There is an idempotent $u \in I$ with $au = a$. By Lemma 3·12.10, $u \in E$, *i.e.*, $X_bu \subset X_b$. We shall prove that $X u = X_b$. This will imply that if $x_1,x_2 \in X_b$, then there exist points $a_1,a_2 \in X_b$ satisfying the equalities $x_1 = a_1u$, $x_2 = a_2u$, and hence $(x_1,x_2) = (a_1,a_2)u$ is an almost periodic point.

Thus we need to show the inclusion $X_b \subset X_bu$. Let $\alpha \in \mathfrak{U}[X]$. Choose $\beta \in \mathfrak{U}[X]$ with $\beta^5 \subset \alpha$ and $\beta = \beta^{-1}$. For β we choose an index $\delta \in \mathfrak{U}[X]$, $\delta = \delta^{-1}$, according to the stability of the fiber X_b. Since X_b is compact, $X_b \subset \bigcup_{i=1}^{n} x_i\delta$ for some finite family $x_1,\ldots,x_n \in X_b$. By the definition of the semigroup $E(X,T)$, there exists an element $t_0 \in T$ with $(x_it_0,x_iu) \in \beta$ $(i = 1,\ldots,n)$. Since $x_iu\varphi = b$ and the map $F:Y \to 2^X$ is continuous at the point b, the element $t_0 \in T$ can be chosen in such a way that

$$X_bt_0 \subset X_b\beta, \qquad X_b \subset (X_bt_0)\beta. \tag{13.17}$$

Let $x \in X_b$. Then $(x,x_i) \in \delta$ for some x_i, $1 \leqslant i \leqslant n$. By the choice of δ, $(xt,x_it) \in \beta$ for all $t \in T$, hence $(xu,x_iu) \in \beta^2$. Then $(xu,xt_0) = (xu,x_iu)\circ(x_iu,x_it_0)\circ(x_it_0,xt_0) \in \beta^2\circ\beta\circ\beta = \beta^4$ $(x \in X_b)$. Therefore $X_bt_0 \subset (X_bu)\beta^4$ and $X_bu \subset (X_bt_0)\beta^4$. Taking (13.17) into account, we get $X_b \subset (X_b t_0)\beta \subset (X_b u)\beta^5 \subset (X_b u)\alpha$. Since the fiber X_b is compact and stable, it follows that the set $X_b u$ is also compact and consequently closed. Since α is an arbitrary index of $\mathfrak{U}[X]$, we finally obtain $X_b \subset \bigcap_{\alpha\in\mathfrak{U}} (X_b u)\alpha = \overline{X_b u} = X_b u$. ∎

3·13.17 THEOREM: *Let X be a compact space. Then a minimal extension $\varphi:(X,T) \to (Y,T)$ is equicontinuous iff the map φ is open and at least one fiber X_y is stable.*

Proof: Let $\varphi:(X,T) \to (Y,T)$ be an equicontinuous extension. By Lemma 3·12.5, φ is distal and consequently open (Corollary 3·12.25).

Conversely, suppose that φ is open and some fiber $X_0 = \{x | x \in X, x\varphi = y_0\}$ is stable. We must show that $\varphi:(X,T) \to (Y,T)$ is an equicontinuous extension.

Since φ is open and since X and Y are compact, the map $F:Y \to 2^X$ defined by $yF = X_y$ $(y \in Y)$ is equicontinuous. By Lemma 3·13.16 each pair $(a,b) \in X_0 \times X_0$ is an almost periodic point of the transformation group $(X \times X,T)$. Since (X,T) is minimal and φ is open, the set of almost periodic points of the form (at,bt) $((a,b) \in X_0 \times X_0, t \in T)$ is dense in $R = \{(x_1,x_2) | x_1\varphi = x_2\varphi\}$. Hence $R = \overline{R\mathfrak{I}}$.

Suppose that the extension φ is not equicontinuous. Then for some index $\beta_0 \in \mathfrak{U}[X]$ and any index $\alpha \in \mathfrak{U}[X]$ there exist points a_α, $b_\alpha \in X$ and an element $t_\alpha \in T$ such that

$$(a_\alpha,b_\alpha) \in \alpha, \qquad a_\alpha\varphi = b_\alpha\varphi, \qquad (a_\alpha t_\alpha, b_\alpha t_\alpha) \in \beta_0. \tag{13.18}$$

Since $R = \overline{R\mathfrak{I}}$, we may suppose without loss of generality that

$$(a_\alpha,b_\alpha) \in R\mathfrak{I}, \qquad (\alpha \in \mathfrak{U}), \tag{13.19}$$

i.e., (a_α,b_α) is an almost periodic point of the transformation group $(X \times X;T)$.

Let c be a specified point of the fiber X_0. Since X is a minimal set, there is an element $\xi_\alpha \in E(X,T)$ with $a_\alpha t_\alpha \xi_\alpha = c$ $(\alpha \in \mathfrak{U})$. There is no loss of generality in assuming that $\{b_\alpha t_\alpha \xi_\alpha | \alpha \in \mathfrak{U}\} \to d$, $d \in X$. The following equalities hold

$$y_0 = c\varphi = a_\alpha t_\alpha \xi_\alpha \varphi = b_\alpha t_\alpha \xi_\alpha \varphi. \tag{13.20}$$

Hence $d\varphi = y_0$, *i.e.*, $d \in X_0$.

Let us consider two possible cases: (1) $c = d$, *i.e.*, $\{b_\alpha t_\alpha \xi_\alpha\} \to c$: Since (a_α,b_α) is almost periodic by (13.19), there exists an element $\eta_\alpha \in E(X,T)$ with

$$(a_\alpha t_\alpha \xi_\alpha, b_\alpha t_\alpha \xi_\alpha)\eta_\alpha = (a_\alpha t_\alpha, b_\alpha t_\alpha). \tag{13.21}$$

But $\{b_\alpha t_\alpha \xi_\alpha | \alpha \in \mathfrak{U}\} \to c = a_\alpha t_\alpha \xi_\alpha$ and this, in view of (13.20,21) and (13.18) contradicts the stability of the fiber X_0. (2) $c \neq d$: By Lemma 3·13.16, (c,d) is an almost periodic point. Since (a_α,b_α) is also almost periodic, we have that

$$(a_\alpha t_\alpha \xi_\alpha, b_\alpha t_\alpha \xi_\alpha)\lambda_\alpha = (a_\alpha, b_\alpha) \tag{13.22}$$

for some element $\lambda_\alpha \in E(X,T)$. Since $\{b_\alpha t_\alpha \xi_\alpha\} \to d \in X_0$ and the fibre X_0 is stable, the points $b_\alpha t_\alpha \xi_\alpha \lambda_\alpha$ and $d\lambda_\alpha$ eventually become arbitrarily close to each other as α runs over $\mathcal{U}[X]$. Taking (13.22,18) into account, we conclude that for a suitable choice of α the pair $(c\lambda_\alpha, d\lambda_\alpha) = (a_\alpha t_\alpha \xi_\alpha \lambda_\alpha, b_\alpha t_\alpha \xi_\alpha \lambda_\alpha) \circ (b_\alpha t_\alpha \xi_\alpha \lambda_\alpha, d\lambda_\alpha) = (a_\alpha, b_\alpha) \circ (b_\alpha t_\alpha \xi_\alpha \lambda_\alpha, d\lambda_\alpha)$ comes to lie in any neighbourhood of the diagonal $\Delta = \{(x,x) \mid x \in X\}$. Since the orbit closure of the point $(c,d) \in X \times X$ is a minimal set, the preceding considerations show that $(c,d) \in \Delta$, and this contradicts the hypothesis $c \neq d$. ▮

3·13.18 THEOREM: *Let X be a compact metrizable space, (X,T) a minimal transformation group and $\varphi:(X,T) \to (Y,T)$ an extension with at least one stable fibre. Then there exist minimal transformation groups (X^*,T) and (Y^*,T) with compact metrizable phase spaces and there exist homomorphisms p: $(X^*,T) \to (X,T)$, $q:(X^*,T) \to (Y^*,T)$, $r:(Y^*,T) \to (Y,T)$ such that*

(1): *p and r are almost automorphic extensions;*

(2): *q is an equicontinuous extension;*

(3): *$p \circ \varphi = q \circ r$.*

Proof: By Lemma 3·12.27 there exist minimal transformation groups (X^*,T) and (Y^*,T) with compact metrizable phase spaces and homomorphisms $p:(X^*,T) \to (X,T)$, $q:(X^*,T) \to (Y^*,T)$, $r:(Y^*,T) \to (Y,T)$ such that p and r are almost automorphic extensions, q is open and $p \circ \varphi = q \circ r$. If at least one fiber of the extension $\varphi:(X,T) \to (Y,T)$ is stable, then it is not hard to verify that the extension $q:(X^*,T) \to (Y^*,T)$ also has a stable fibre. Hence it follows from Theorem 3·13.17 that q is an equicontinuous extension. ▮

3·13.19 COROLLARY: *If X is a compact metrizable space and $\varphi:(X,T) \to (Y,T)$ is a minimal extension which has at least one stable fibre and at least one proximal fibre, then φ is almost autormorphic.*

Proof: Applying Theorem 3·13.18 we get that q is an isomorphism. Hence φ is an almost autormorphic extension. ▮

3·13.20 Example 2·8.42 shows that the metrizability hypothesis for X in Theorem 3·13.18 and in Corollary 3·13.19 cannot be dropped.

3·13.21 If some fibre consists of finitely many points, then it is stable, and therefore Theorem 3·13.18 can be applied to extensions having at least one finite fiber. In this case q is not only an equicontinuous extension but also a finite-to-one locally homeomorphic map.

REMARKS AND BIBLIOGRAPHICAL NOTES

The results of Subsections 3·13.6-11 are due to Bronšteĭn [*15*,*18*]. Theorem 3·13.12 was proved by Veech [*4*,*5*]. Lemma 3·13.13 was obtained by Keynes and Robertson [*1*,*2*] in a different way than here. Theorem 3·13.15 was proved by Bronšteĭn [*18*]. Some results concerning the relation $Q^*(R)$ are in the papers by Ellis and Keynes [*1*,*2*]. Theorems 3·13.17,18 were proved by Bronšteĭn [*18*]. Weakly mixing transformation groups were studied by Furstenberg [*2*], Keynes and Robertson [*1*,*2*], Keynes [*5*] and others.

§(3·14): THE STRUCTURE OF DISTAL AND ALMOST DISTAL EXTENSIONS

In this section there will be investigated the distal, proximal and almost distal extensions; several theorems will be proved.

3·14.1 LEMMA: *Let X be a compact space, (X,T) a minimal transformation group and let S and R be closed invariant equivalence relations on X with $S \subset R$. Let $\psi: X \to X/S$ and $\varphi: X/S \to X/R$ be the canonical projections. Then*

$$((P(X,T) \cap R)(\psi \times \psi) = P(X/S,T) \cap R(\psi \times \psi).$$

Proof: As ψ is a uniformly continuous map, $((P(X,T) \cap R)(\psi \times \psi) \subset P(X/S,T) \cap R(\psi \times \psi)$. Let us prove the reverse inclusion. Let $(a,b) \in R$ and $(a\psi, b\psi) \in P(X/S,T)$. According to Theorem 1·5.11 there exists an idempotent $v \in E(X/S,T)$ with $a\psi = b\psi v$. Let $\psi^*: E(X,T) \to E(X/S,T)$ denote the homomorphism of the eveloping semigroups induced by ψ. Choose an idempotent $u \in E(X,T)$ such that $u\psi^* = v$ (Theorem 1·4.21(9)). Then

$$a\psi = b\psi v = (b\psi)(u\psi^*) = (bu)\psi, \qquad (bu,b)(\psi \times \psi) = (a\psi, b\psi)$$

and, moreover, $(bu,b) \in P(X,T)$. Since $a\psi = (bu)\psi$, we get $(a,bu) \in S \subset R$. By hypothesis, $(a,b) \in R$. Hence $(bu,b) \in R$. ∎

3·14.2 COROLLARY: *Let the hypothesis of Lemma 3·14.1 be fulfilled. Then the extension $\varphi: (X/S,T) \to (X/R,T)$ is distal iff $S \supset P(X,T) \cap R$.*

Proof: The distality of φ is equivalent to $P(X/S,T) \cap R(\psi \times \psi) = \Delta(X/S)$. In view of Lemma 3·14.1 this means that $P(X,T) \cap R \subset S$. ∎

3·14.3 LEMMA: *Let X be a compact space and let $\varphi: (X,T) \to (Y,T)$ be an extension where (Y,T) is a minimal transformation group. Then for every point $y \in Y$ and every minimal right ideal I of the semigroup $E = E(X,T)$ there is an idempotent $u \in I \cap E_y$.*

Proof: Let I be any minimal right ideal of E and let $a \in X_y$. Then aI is a minimal set (Lemma 1·4.10) and therefore $(aI)\varphi = Y$.

Hence there is an element $\xi \in I$ with $a\xi\varphi = y$. Lemma 3·12.10 asserts that $\xi \in E_y$. Thus $I \cap E_y$ is a non-empty subsemigroup of E. According to Lemma 1·4.11 there exists an idempotent $u \in I \cap E_y$.∎

3·14.4 LEMMA: *Let $\varphi:(X,T) \to (Y,T)$ be an extension, where X is a compact space and Y is a minimal set. If a point $a \in X$ is distal relative to the extension φ, then $\overline{aT}$ is a compact minimal set. If, moreover, $x \in \overline{aT}$ and $a\varphi = x\varphi$, then we have that (a,x) is an almost periodic point of the transformation group $(X \times X,T)$.*

Proof: According to Lemma 3·14.3 there is an idempotent $u \in I \cap E_b$, where $b \equiv a\varphi$. As a is distal from all other points of the fiber X_b, we have $a = au$ (Lemma 3·12.21), which means, in view of $u \in I$, that a is an almost periodic point, and hence $\overline{aT}$ is a minimal set. Let $x \in \overline{aT}$ and $a\varphi = x\varphi$. Choose an idempotent $u \in I$ with $xu = x$. Then $au\varphi = xu\varphi = x\varphi = a\varphi$, and hence $u \in E_b$ (Lemma 3·12.10). By Lemma 3·12.21, $au = a$. Hence $(a,x) = (a,x)u$ is an almost periodic point.∎

3·14.5 LEMMA: *Let X be a compact space, (X,T) a minimal transformation group and let $\varphi:(X,T) \to (Y,T)$ be an extension. Let $y \in Y$ be such that each two distinct points of the fiber X_y are distal. Then G_y is a group.*

Proof: According to Subsection 3·12.12, G_y is a compact Hausdorff semigroup. If $e \in E(X,T)$ is the identity map of X onto itself, then H_y^e is the identity of the semigroup G_y. We shall show that H_y^e is the only idempotent of G_y. Let $H_y^u \in G_y$ be an idempotent, *i.e.*, $H_y^u \circ H_y^u = H_y^u$. Then $xu^2 = x$ for each point $x \in X_y$. Let us prove that $xu = x$ $(x \in X_y)$. Suppose that, to the contrary, there exists a point $x_0 \in X_y$ such that $x_0u \neq x_0$. Clearly, the points x_0u and x_0 are proximal and, moreover, $x_0 \in X_y$ and $x_0u \in X_y$. But this contradicts the distality of the fiber X_y. Hence $xu = x$ $(x \in X_y)$, and therefore $H_y^u = H_y^e$. It follows that G_y does not contain a proper right ideal; therefore G_y is a group (Corollary 1·4.3).∎

3·14.6 LEMMA: *Let X be a compact space, $\varphi:(X,T) \to (Y,T)$ an extension and let (Y,T) be a minimal transformation group. If for some point $y \in Y$ the semigroup G_y is in fact a group, then for every $x_1,\dots,x_n \in X_y$ the point $(x_1,\dots,x_n)$ is an almost periodic point of the Cartesian product (X^n,T).*

Proof: By Lemma 3·14.3 there exists an idempotent $u \in I \cap E_y$. Then H_y^u is an idempotent of G_y. However, since G_y is a group, $H_y^u = H_y^e$, *i.e.*, $xu = x$ for all $x \in X_y$. In particular, $(x_1u,\dots,x_nu) = (x_1,\dots,x_n)$. Since the enveloping semigroup $E(X^n,T)$ is canonically isomorphic to $E(X,T)$, $(x_1,\dots,x_n) \equiv (x_1,\dots,x_n)u$ is an almost periodic point.∎

3·14.7 THEOREM: *Let $\varphi:(X,T) \to (Y,T)$ be a minimal extension, let X be a compact space and let $\iota \in Y$. The following statements are mutually equivalent:*

(1): *The extension* $\varphi:(X,T) \to (Y,T)$ *is distal in the fiber* X_y;

(2): G_y *is a group;*

(3): *If* $n \geqslant 2$ *and* $x_1,\ldots,x_n \in X_y$, *then* $(x_1,\ldots,x_n) \in X^n$ *is an almost periodic point of* (X^n,T);

(4): *For each two points* x_1 *and* x_2 *of* X_y *the pair* $(x_1,x_2) \in X \times X$ *is almost periodic under* $(X \times X,T)$.

Proof: By Lemma 3·14.5, (1) implies (2). Statements (3) and (4) follows from (2) in accordance with Lemma 3·14.6. By Lemma 1·6.2, (4) implies (1).▮

3·14.8 LEMMA: *Let* X *be a compact space,* $\varphi:(X,T) \to (Y,T)$ *a minimal extension and* $a \in X$, $b \in X$, $a\varphi = y$, $b\varphi = z$. *Further, let* $\lambda,\mu \in E(X,T)$ *and* $a\lambda = b$, $b\mu = a$. *Then* $\mu E_y\lambda \subset E_z$, $\lambda E_z\mu \subset E_y$.

Proof: Let $\xi \in E_y$. Then $a\xi\varphi = a\varphi = y$. Therefore $b\mu\xi\varphi = a\xi\varphi = a\varphi = b\mu\varphi$, and hence $b\mu\xi\lambda\varphi = b\mu\lambda\varphi = b\varphi$. Thus $\mu\xi\lambda \in E_z$. We have thus proved that $\mu E_y\lambda \subset E_z$. By symmetry, $\lambda E_z\mu \subset E_y$.▮

3·14.9 THEOREM: *Let* X *be a compact space,* (X,T) *a minimal transformation group,* $\varphi:(X,T) \to (Y,T)$ *a distal extension,* $a \in X$, $\lambda \in E$, $b = a\lambda$, $a\varphi = y$ *and* $b\varphi = z$. *Then* $X_y\lambda = X_z$ *and there exists an element* $\mu \in E$ *such that* $b\mu = a$ *and the map* $f:G_z \to G_y$, *defined by* $H_y^{\xi}f = H_y^{\lambda\xi\mu}$ $(\xi \in E_z)$, *is an isomorphism of the group* G_z *onto* G_y.

Proof: By Lemma 3·14.5, G is a group for every $y \in Y$. Since $\overline{bT} = bE = X$, there exists an element $\mu_0 \in E(X,T)$ with $b\mu_0 = a$. As $a\lambda\mu_0 = a$, we get $\lambda\mu_0 \in E_y$ and $H_y^{\lambda\mu_0} \in G_y$. Since G_y is a group there is an element $\nu \in E_y$ with $H_y^e = H_y^{\lambda\mu_0\nu}$, *i.e.*, $\lambda\mu_0\nu \in H_y^e$. Set $\mu = \mu_0\nu$. Then $b\mu = a\lambda\mu = a$.

Consider the map $F:E \to E$ defined by $\xi F = \lambda\xi\mu$ $(\xi \in E)$. By Lemma 3·14.8, $E_zF \subset E_y$. We shall show that $(\xi_1,\xi_2) \in R_z$ implies $(\xi_1F,\xi_2F) \in R$ (*see* Subsection 3·12.12). Indeed, let $v\xi_1 = v\xi_2$ for each point $v \in X_z$. If $u \in X_y$, then $u\lambda \in X_z$, hence $u\lambda\xi_1 = u\lambda\xi_2$, and consequently $u\lambda\xi_1\mu = u\lambda\xi_2\mu$. As u is an arbitrary point of the fiber X_y, $(\lambda\xi_1\mu,\lambda\xi_2\mu) \in R_y$, *i.e.*, $(\xi_1F,\xi_2F) \in R_y$. Therefore the map F induces a map $f:G_z \to G_y$ defined by $H_z^{\xi}f = H_y^{\lambda\xi\mu}$ $(\xi \in E_z)$.

Let us show that f is surjective. Let $\eta \in E_y$; then $\xi \equiv \mu\eta\lambda \in \mu E_y\lambda \subset E_z$. We shall prove that $H_z^{\mu\eta\lambda}f = H_y^{\eta}$. In fact,

$$H_z^{\mu\eta\lambda}f = H_y^{\lambda\mu\eta\lambda\mu} = H_y^{\lambda\mu}\cdot H_y^{\eta}\cdot H_y^{\lambda\mu} = H_y^{e}\cdot H_y^{\eta}\cdot H_y^{e} = H_y^{\eta}.$$

Now we shall show that $X_y\lambda = X_z$. Obviously $X_y\lambda \subset X_z$. Let $v \in X_z$. Then $v\mu \in X_y$. Hence it follows that $v\mu\lambda = v$. Indeed, $v\mu\lambda\mu = v\mu$, because $v\mu \in X_y$ and $\lambda\mu \in H_y^e$. Since $\varphi:(X,T) \to (Y,T)$ is a distal extension, the map $\psi:X_z \to X_y$, defined by $v\psi = v\mu$ $(v \in X_z)$, is injective. Hence $v\mu\lambda\mu = v\mu$ implies $v\mu\lambda = v$ ($v \in X_z$). Thus $X_z \subset X_y\lambda$. Finally, we get $X_y\lambda = X_z$ and $H_z^{\mu\lambda} = H_z^e$.

Let us prove that $f: G_z \to G_y$ is a group homomorphism. Let

$$H_z^{\xi_1} \in G_z, \qquad H_z^{\xi_2} \in G_z.$$

Then

$$H_z^{\xi_1} f = H_y^{\lambda\xi_1\mu}, \qquad H_z^{\xi_2} f = H_y^{\lambda\xi_2\mu}, \qquad H_y^{\lambda\xi_1\mu} \cdot H_y^{\lambda\xi_2\mu} = H_y^{\lambda\xi_1\mu\lambda\xi_2\mu}.$$

Let $u \in X_y$. Then $u\lambda \in X_z$, and hence $u\lambda\xi_1 \in X_z$ as $\xi_1 \in E_z$. As was noted above, $\mu\lambda \in H_z^e$, and therefore $u\lambda\xi_1\mu\lambda = u\lambda\xi_1$. Hence $u\lambda\xi_1\mu \cdot \lambda\xi_2\mu = u\lambda\xi_1\xi_2\mu$ for every point $u \in X_y$. It follows that

$$H_y^{\lambda\xi_1\mu\lambda\xi_2\mu} = H_y^{\lambda\xi_1\xi_2\mu} = H_z^{\xi_1\xi_2} f;$$

therefore f is a homomorphism.

We shall now prove that f is an isomorphism of G_z onto G_y. Let $\xi \in E_z$ and let $H_z^{\xi} f \equiv H_y^{\lambda\xi\mu} = H_y^e$. Then $u\lambda\xi\mu = u$ for every point $u \in X_y$, hence $u\lambda\xi\mu\lambda = u\lambda$. As u is an arbitrary point of X_y and $X_y\lambda = X_z$, we have $v\xi\mu\lambda = v$ for each point $v \in X_z$. Since $v \in X_z$ and $\xi_z \in E_z$, we have $v\xi \in X_z$. But $\mu\lambda \in H_z^e$, therefore $v\xi\mu\lambda = v\xi$. Hence $v\xi = v$ for every point $v \in X$, *i.e.*, $\xi \in H_z^e$ and $H_z^{\xi} = H_z^e$. Thus f is an isomorphism. ▮

3·14.10 LEMMA: *Let X be a compact space, (X,T) a minimal transformation group, φ a homomorphism of (X,T) onto a transformation group (Y,T) and let ψ be a homomorphism of (Y,T) onto (Z,T). If $\varphi \circ \psi: (X,T) \to (Z,T)$ is a distal extension, then*

(1): *(X,T) is a distal extension of (Y,T) by φ;*

(2): *(Y,T) is a distal extension of (Z,T) by ψ.*

Proof: The first statement is obvious. Let $x_0 \in X$, $y_0 = x_0\varphi$, $z_0 = y_0\psi$ and take G_1 and G_2 the same as in Lemma 3·12.14. Since (X,T) is a distal extension of (Z,T) by $\varphi\psi$, the semigroup G_1 is in fact a group (Theorem 3·14.7). It follows from Lemma·3 12.14 that G_2 is also a group. Using again Theorem 3·14.7, we again deduce that $\psi: (Y,T) \to (Z,T)$ is a distal extension. ▮

3·14.11 If a transformation group (X,T) is an equicontinuous extension of (Y,T) by means of a homomorphism φ, then $\varphi: (X,T) \to (Y,T)$ is a distal extension. Let $a \in X$, $\lambda \in E(X,T)$ and $b = a\lambda$. By Theorem 3·14.9 there exists an element $\mu \in E(X,T)$ such that $b\mu = a$ and the map $f: G_z \to G_y$ defined by $H_z^{\xi} f = H_y^{\lambda\xi\mu}$ ($\xi \in E_z$, $y = a\varphi$, $z = b\varphi$), is an isomorphism of the group G_z onto G_y.

3·14.12 THEOREM: *If a minimal transformation group* (X,T) *with a compact phase space is an equicontinuous extension of some transformation group* (Y,T) *by a homomorphism* φ, *then*

(1): G_y *is a topological group for each point* $y \in Y$;

(2): *The map* $f: G_z \to G_y$ *is an isomorphism of topological groups* $(z,y \in Y)$;

(3): *The map* $\pi_y: X_y \times G_y \to X_y$ *defined by* $(x,H_y^\xi)\pi_y = x\xi$ $(\xi \in E_y, x \in X_y)$ *determines a topological transformation group* (X_y,G_y,π_y) $(y \in Y)$.

Proof: Let $y \in Y$. Each element H_y of the group G_y may be regarded as a map of X_y onto itself; therefore $G_y \subset X_y{}^{X_y}$. It is easy to show that the topology induced in G_y by the space $X_y{}^{X_y}$ coincides with the initial topology of $G_y = \{H_y^\xi | \xi \in E_y\}$. As $\varphi: (X,T) \to (Y,T)$ is an equicontinuous extension, each element H_y^ξ of the group G_y determines a homeomorphism of X_y onto X_y. Therefore multiplication in the group G_y is continuous in each variable separately. By Lemma 2·8.2, G_y is a topological group, *i.e.*, (1) holds. Hence it follows from Lemma 1·6.13 that (X_y,G_y,π_y) is a topological transformation group.

To prove statement (2) it suffices to show that f is continuous. Let $\{H_z{}^i | i \in I\}$ be a net that converges in G_z to some element H_z^ξ. This means that $\{v\xi_i | i \in I\} \to v\xi$ for every point $v \in X_z$. If u is an arbitrary point of the fiber X_y, then $u \in X_z$ and therefore $\{u\lambda\xi_i | i \in I\} \to u\lambda\xi$. Since (X,T) is an equicontinuous extension of (Y,T) and $u\lambda\xi_i\varphi = u\lambda\xi\varphi$, we get $\{u\lambda\xi_i\mu\} \to u\lambda\xi\mu$. Because u is an arbitrary point of X_y, it follows that $\{H_y^{\lambda\xi_i\mu} | i \in I\} \to H_y^{\lambda\xi\mu}$. Thus f is continuous.▮

3·14.13 LEMMA: *Let* X *be a compact space,* (X,T) *a transformation group,* $\varphi: (X,T) \to (Y,T)$ *an extension and let* $R = R_\varphi$. *Then the extension* φ *is proximal iff* $R\mathcal{J} = \Delta(X)$.

Proof: Recall that $R\mathcal{J}$ is the set of all almost periodic points $(a,b) \in R$. If φ is a proximal extension, then $R\mathcal{J} = \Delta(X)$ because of Lemma 1·6.2. Conversely, suppose that $R\mathcal{J} = \Delta(X)$. Let us prove that φ is proximal. Assume that, to the contrary, there exist two distal points $a,b \in X$ such that $(a,b) \in R$. By Lemma 1·6.2 we can choose points a' and b' of X such that $(a',b') \in \overline{(a,b)T}$, $a' \neq b'$ and (a',b') is an almost periodic point. As $a\varphi = b\varphi$ and $(a',b') \in \overline{(a,b)T}$, it follows that $a'\varphi = b'\varphi$, *i.e.*, $(a',b') \in R$. But this contradicts the hypothesis $R\mathcal{J} = \Delta(X)$.▮

3·14.14 LEMMA: *Let* X *be a uniform space, let* (X,T) *be a transformation group with the property that* $\overline{xT}$ *is compact for all* $x \in X$ *and let* $\varphi: (X,T) \to (Y,T)$ *be an extension. Let* $x_0 \in X$ *be an almost periodic point. Then the point* $x_0\varphi$ *is* *also almost periodic. Conversely, if $y_0 \in Y$ is an almost periodic point, then there exists an almost periodic point $x_0 \in X$ satisfying the condition $x_0\varphi = y_0$.*

Proof: The first statement is evident. The second assertion follows from Lemma 1·5.10, Theorem 1·4.21(9), Lemma 1·4.10 and Theorem 1·5.4.∎

3·14.15 LEMMA: *Let X be a compact space, let (X,T) be a transformation group and let S and R be closed invariant equivalence relations on X with $S \subset R$. If $\psi\ X \to X/S$ is the canonical projection, then*

$$(R\mathcal{J})(\psi \times \psi) = R(\psi \times \psi)\mathcal{J}.$$

Proof: The set $R \subset X \times X$ is compact and invariant. The map $\psi \times \psi : R \to R(\psi \times \psi)$ is a homomorphism. Hence the required result follows from Lemma 3·14.14.∎

3·14.16 LEMMA: *Let X be a compact space, let (X,T) be a transformation group and let S and R be closed invariant equivalence relations on X with $S \subset R$. Then the extension $\varphi:(X/S,T) \to (X/R,T)$ is proximal iff $S \supset R\mathcal{J}$.*

Proof: This is a consequence of Lemmas 3·14.14,15.∎

3·14.17 COROLLARY: *Let (X,T) be a transformation group with a compact phase space and let $\varphi:(X,T) \to (Y,T)$ be an extension. Then there exists a greatest proximal extension subordinated to φ.*

Proof: The required extension is of the form $\psi:(X/(R\mathcal{J})^*,T) \to (X/R,T)$, where $R = \{(x_1,x_2)|x_1,x_2 \in X, x_1\varphi = x_2\varphi\}$ and $(R\mathcal{J})^*$ denotes, as usual, the smallest closed invariant equivalence relation containing $R\mathcal{J}$.∎

3·14.18 LEMMA: *Let X be a compact space. If $\lambda:(X,T) \to (Z,T)$ is a distal extension, $\mu:(Z,T) \to (Y,T)$ is a proximal extension and $\varphi = \lambda \circ \mu$, then μ is the greatest proximal extension subordinate to φ.*

Proof: Let

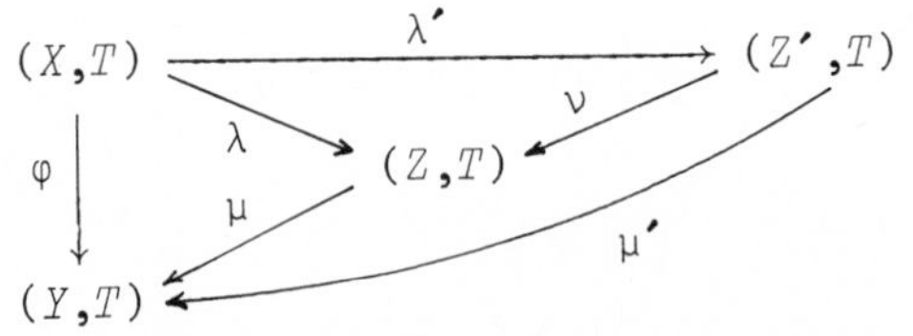

be a commutative diagram in which μ' is a proximal extension. Then ν is also proximal. On the other hand, since λ is a distal extension and $\lambda = \lambda' \circ \nu$, it follows from Lemma 3·14.10 that ν is a distal extension. This is possible only when ν is bijective.∎

3·14.19 Let X be a compact space, (X,T) a minimal transformation group, $\varphi:(X,T) \to (Y,T)$ an extension and let $R = \{(x_1,x_2)|$

$x_1, x_2 \in X, x_1\varphi = x_2\varphi\}$. For each ordinal λ we shall define inductively a relation $Q_\lambda = Q_\lambda(R) \subset X \times X$. If $\lambda = 0$, we put $Q_0 = R$. Suppose that the relations Q_λ have been defined for all λ with $\lambda < \mu$. If the ordinal μ has an immediate predecessor λ, that is, $\mu = \lambda + 1$, then we define

$$Q_\mu \equiv Q_{\lambda+1} = Q(X, Q_\lambda) = \bigcap_{\alpha \in \mathcal{U}[X]} \overline{(\alpha \cap Q_\lambda)T}.$$

If μ is a limit ordinal, then we put $Q_\mu = \bigcap_{\lambda<\mu} Q_\lambda$. Observe that

$$Q_0 \supset Q_1 \supset \cdots \supset Q_\lambda \supset Q_{\lambda+1} \supset \cdots .$$

3·14.20 LEMMA: *If $\varphi:(X,T) \to (Y,T)$ is a distal extension, then for each ordinal λ the relation $Q_\lambda = Q_\lambda(R)$ is an invariant equivalence relation which is both closed and open.*

Proof: Let $\lambda = 0$. Then $Q_0 = R$ is evidently a closed invariant equivalence relation. It is open by Corollary 3·12.25.

Suppose that the statement of the lemma is true for all λ satisfying $\lambda < \mu$. Let us prove that it holds also for μ.

Consider first the case where $\mu = \lambda + 1$. Let $\psi_\lambda: X \to X/Q_\lambda$ denote the canonical projection. Since $Q_\lambda \subset R$, the extension $\psi_\lambda: (X,T) \to (X/Q_\lambda, T)$is distal. In view of Theorem 3·14.7, if $(x_1,x_2) \in Q_\lambda$, then (x_1,x_2) is an almost periodic point of the transformation group $(X \times X, T)$. In other words, $Q_\lambda = Q_\lambda\mathcal{J}$. By the inductive hypothesis, Q_λ is both a closed and open invariant equivalence relation. Hence it follows from Theorem 3·13.15 that $Q_{\lambda+1} \equiv Q(Q_\lambda)$ is a closed invariant equivalence relation. Let $\psi_{\lambda+1}: X \to X/Q_{\lambda+1}$ be the canoncial map. By Lemma 3·14.10, $\psi_{\lambda+1}$ is a distal extension. We conclude from Corollary 3·12.25 that the map $\psi_{\lambda+1}$ is open, *i.e.*, $Q_{\lambda+1}$ is an open relation.

Now let μ be a limit ordinal. Then $Q_\mu \equiv \bigcap_{\lambda<\mu} Q_\lambda$ is a closed invariant equivalence relation. By Lemma 3·14.10 and Corollary 3·12.25, Q_μ is open. ▌

3·14.21 LEMMA: *Let X be a compact space, let (X,T) be a minimal transformation and let $R \subset X \times X$ be a closed invariant equivalence relation such that $\varphi:(X,T) \to (X/R,T)$ is a distal extension. If $R \neq \Delta$ and for each two distinct points $a_1, a_2 \in X$ satisfying the condition $(a_1,a_2) \in R$ there exists a homomorphism ψ of (X,T) onto some transformation group (Z,T) with a metrizable phase space such that $a_1\psi \neq a_2\psi$, then $Q(R) \neq R$.*

Proof: Suppose that, to the contrary, $Q(R) = R$. Let $(a_1,a_2) \in R \setminus \Delta(X)$. There exists a homomorphism $\psi:(X,T) \to (Z,T)$ such that Z is metrizable and $a_1\psi \neq a_2\psi$. Hence $R(\psi \times \psi) \neq \Delta(Z)$. Since φ is distal, each pair $(x_1,x_2) \in R$ is an almost periodic point of the transformation group $(X \times X, T)$. Because $(\psi \times \psi):(X \times X,T) \to (Z \times Z,T)$

is a homomorphism, all the points $(z_1,z_2) \in R(\psi \times \psi)$ are also almost periodic. By hypothesis

$$Q(R) \equiv \bigcap_{\alpha \in \mathcal{U}[X]} \overline{\alpha T \cap R} = R.$$

Hence $\overline{\alpha T \cap R} = R$ for every index $\alpha \in \mathcal{U}[X]$. Therefore

$$R(\psi \times \psi) = \overline{\alpha T \cap R}(\psi \times \psi) \subset \overline{\alpha(\psi \times \psi) T \cap R(\psi \times \psi)} \subset R(\psi \times \psi)$$

and consequently

$$R(\psi \times \psi) = \overline{\alpha(\psi \times \psi) T \cap R(\psi \times \psi)}, \qquad (\alpha \in \mathcal{U}[X]).$$

Since the map ψ is uniformly continuous, for each index $\beta \in \mathcal{U}[Z]$ there is an index $\alpha \in \mathcal{U}[X]$ with $\alpha(\psi \times \psi) \subset \beta$, and hence $\overline{\beta T \cap R(\psi \times \psi)} = R(\psi \times \psi)$ $(\beta \in \mathcal{U}[Z])$.

Let $\{\beta_i \mid i = 1,2,\dots\}$ be an open uniformity base of the compact metrizable space Z. Then

$$\overline{\beta_i T \cap R(\psi \times \psi)} = R(\psi \times \psi), \qquad (i = 1,2,\dots).$$

As $R(\psi \times \psi)$ is a compact metrizable space, it follows from Baire's Theorem that

$$\overline{\bigcap_{i=1}^{\infty} \beta_i T \cap R(\psi \times \psi)} = R(\psi \times \psi), \qquad (i = 1,2,\dots).$$

But

$$\bigcap_{i=1}^{\infty} \beta_i T = P(Z,T),$$

and $R(\psi \times \psi)$ consists of almost periodic points of the transformation group $(Z \times Z,T)$. Hence we get, in view of Lemma 1·6.2, that

$$\bigcap_{i=1}^{\infty} \beta_i T \cap R(\psi \times \psi) = \Delta(Z).$$

Hence $R(\psi \times \psi) = \Delta(Z)$ in spite of $(a_1,a_2) \in R$ and $a_1\psi \neq a_2\psi$. ∎

3·14.22 LEMMA: *Let X be a compact topological space, let (X,T) be a minimal transformation group and let $R \subset X \times X$ be a closed invariant equivalence relation such that φ: $(X,T) \to (X/R,T)$ is a distal extension. Suppose that at least one of the following conditions is satisfied:*

(a): *The space X is metrizable (i.e., the topology of X has a countable base);*

(b): *The group T is σ-compact.*

Then $\varphi:(X,T) \to (Y,T)$ *is an F-extension.*

Proof: Set $R = \{(x_1,x_2) | x_1,x_2 \in X, x_1\varphi = x_2\varphi\}$ and define the relations $Q_\lambda = Q_\lambda(R)$ as before (*see* Subsection 3 14.19). According to Lemma 3·14.22 there is an ordinal ϑ with $Q_\vartheta = \Delta(X)$. Thus

$$R \equiv Q_0 \supset Q_1 \supset \cdots \supset Q_\lambda \supset Q_{\lambda+1} \supset \cdots \supset Q_\vartheta \equiv \Delta(X).$$

Each Q_λ $(0 \leqslant \lambda \leqslant \vartheta)$ is a closed invariant equivalence relation (Lemma 3·14.20). Let $(X_\lambda,T) \equiv (X/Q_\lambda,T)$ and let $\varphi_\mu^\lambda:(X_\lambda,T) \to (X_\mu,T)$ $(0 \leqslant \mu < \lambda \leqslant \vartheta)$ be the canonical homomorphism. As $Q_{\lambda+1} = Q(Q_\lambda)$, it follows from Subsection 3·13.5 that $\varphi_\lambda^{\lambda+1}:(X_{\lambda+1},T) \to (X_\lambda,T)$ is the greatest equicontinuous extension subordinate to $\varphi_\lambda^\vartheta:(X,T) \to (X_\lambda,T)$. If μ is a limit ordinal, then (X_μ,T) is the projective limit of the transfinite sequence $\{(X_\lambda,T) | 0 \leqslant \lambda < \mu\}$ because $Q_\mu = \bigcap_{\lambda<\mu} Q_\lambda$. Finally, $(X_0,T) = (X/R,T) = (Y,T)$, $(X_\vartheta,T) = (X,T)$. ∎

3·14.24 REMARK: Conditions (a) and (b) in Theorem 3·14.23 as well as in Lemma 3·14.21 may be replaced by the following condition:

(c): *For each two distinct points* $a_1,a_2 \in X$, $(a_1,a_2) \in R$, *there exists a homomorphism* ψ *of* (X,T) *onto some transformation group* (Z,T) *with a metrizable phase space such that* $a_1\psi \neq a_2\psi$.

3·14.25 LEMMA: *Let* X *be a compact space, let* (X,T) *be a transformation group and let* $R \subset X \times X$ *be a closed invariant equivalence relation. Then the set of all* R*-distal points is invariant.*

Proof: Let $a \in X$ be an R-distal point and $t \in T$. We need to show that the point at is distal from all points x satisfying the conditions $(at,x) \in R$, $at \neq x$. Suppose the contrary is true. Then there exist a point $x_0 \in X$ and an element $t_0 \in T$ such that $at_0 \neq x_0$ and $(at_0,x_0) \in R \cap P(X,T)$. Then $(a,x_0t_0^{-1}) \in R \cap P(X,T)$. But this is impossible because $a \neq x_0t_0^{-1}$. ∎

3·14.26 An extension $\varphi:(X,T) \to (Y,T)$ is called an *ALMOST DISTAL EXTENSION* if the set of all points $x \in X$ which are distal relative to the extension φ is residual.

3·14.27 LEMMA: *Let* X *be a compact metric space,* (X,T) *a minimal transformation group and* $\varphi:(X,T) \to (Y,T)$ *an extension. Consider the commutative diagram*

$$\begin{array}{ccc} (X,T) & \xleftarrow{\;p\;} & (X^*,T) \\ \varphi \downarrow & & \downarrow \psi \\ (Y,T) & \xleftarrow{\;q\;} & (Y^*,T) \end{array}$$

constructed in Lemma 3·12.27. *Let* $R = R_\varphi$ *and* $R_1 = R_\psi$. *If* $a \in X$ *is an R-distal point, then the set* ap^{-1} *consists of a single* R_1*-distal point. If* φ *is an almost distal extension, then* ψ *is also almost distal.*

Proof: Let $a \in X$ be R-distal. It follows from Theorem 3·12.14 that ap^{-1} consists of the single point $a_1 = (a, X_y)$, where $y = a\varphi$. Let us prove that a_1 is R_1-distal. Let $(a_1,b_1) \in R_1$, $b_1 \neq a_1$. According to the construction of the space X^*, the points a_1 and b_1 are of the form $a_1 = (a, a_1\psi)$, $b_1 = (b, b_1\psi)$, where $b = b_1 p$. The condition $(a_1,b_1) \in R$ means that $a_1\psi = b_1\psi$. If the points a_1 and b_1 are assumed to be proximal, then the points a and b are obviously also proximal. But $a_1\psi = b_1\psi$ implies $(a,b) \in R$ and since a is an R-distal point, this can happen only when $a = b$. But this contradicts the condition $a_1 \neq b_1$.

Let $X_0 \subset X$ denote the set of all R-distal points and let $X_0^* \subset X^*$ denote the set of all R_1-distal points. It has been proved above that $X_0 p^{-1} \subset X_0^*$. Because of Lemma 3·12.17, the set $X_0 p^{-1}$ (and, moreover, the set X_0^*) is residual in X^*. ▌

3·14.28 LEMMA: *Let X be a compact space,* (X,T) *a minimal transformation group and* $R \subset X \times X$ *a both open and closed invariant equivalence relation. If there exists at least one R-distal point* $a_0 \in X$*, then* $R = \overline{R\mathfrak{J}}$ *(in other words, the set of almost periodic points of the transformation group* $(X \times X, T)$ *is dense in R).*

Proof: Let $(x,y) \in R$ and let $\alpha \in \mathfrak{U}[X]$. Since R is an open relation, there exists a neighbourhood $U \subset x\alpha$ of x such that for every point $x_1 \in U$ we can choose a point $y_1 \in y\alpha$ with $(x_1,y_1) \in R$. As (X,T) is a minimal transformation group, the point $x_1 \equiv a_0 t_1$ is in U for some element $t_1 \in T$. By Lemma 3·14.25, the point x_1 is R-distal. Choose a point $y_1 \in y\alpha$ such that $(x_1,y_1) \in R$. Using Lemma 3·14.4 we conclude that (x_1,y_1) is an almost periodic point. ▌

3·14.29 Let X be a compact metric space, let 2^X be the set of all closed subsets $A \subset X$ provided with the Hausdorff metric (*see* 3·12.26) and let D be a closed subset of the space $X \times X$. To each point $x \in X$ we assign the set $xD = \{x' \mid x' \in X, (x,x') \in D\}$ and we denote the map so obtained by $F: X \to 2^X$. Since D is closed, the function F is upper semi-continuous and therefore the set $C = C(F)$ of its continuity points is residual (Kuratowski [*1*]). Put $D_0 = (C \times X) \cap D$.

If, moreover, D is an open equivalence relation, then F is continuous, *i.e.*, $C = X$. In this case $D_0 = D$.

3·14.30 LEMMA: *Suppose that* $\bar{D}_0 = D$. *If W is a residual subset of D, then there exists a residual subset E of X such that* $xW = \{x' \mid x' \in X, (x,x') \in W\}$ *is a residual subset of* xD *for all* $x \in E$.

Proof: By hypothesis, $D \setminus W \subset \bigcup_{n=1}^{\infty} Z_n$, where the Z_n are closed nowhere dense subsets of D. It will suffice to prove that there exists a set Q_n of the first category in X such that xZ_n is nowhere dense in xD for all $x \in X \setminus Q_n$.

Now let Z be any closed nowhere dense subset of D. We shall prove that there exists a subset Q of the first category in X such that xZ is nowhere dense in xD for all $x \in X \setminus Q$.

We shall first prove that the set $Z \cap D_0$ is nowhere dense in D_0. Since $Z \cap D_0$ is closed in D_0, it will suffice to show that the complement $D_0 \setminus (Z \cap D_0)$ is dense in D_0. Taking into account that $D \setminus Z$ is open in D and $\bar{D}_0 = D$, we get

$$\overline{D_0 \setminus (Z \cap D_0)} = \overline{D_0 \cap (D \setminus Z)} = \overline{\bar{D}_0 \cap (D \setminus Z)}$$

$$= \overline{D \cap (D \setminus Z)} = \overline{D \setminus Z} = D \supset D_0.$$

Thus the set $Z \cap D_0$ is closed and nowhere dense in D_0.

Let $\{R_n | n = 1,2,\ldots\}$ be a base for the topology in X. Set $E_n = \{x | x \in X, \emptyset \neq R_n \cap xD \subset xZ\}$. It is easy to see that $E_n = \{x | x \in X, \emptyset \neq (x \times R_n) \cap D \subset Z\}$; therefore

$$(E_n \times R_n) \cap D \subset Z, \qquad (n = 1,2,\ldots). \tag{14.1}$$

Let us prove that

$$(\bar{E}_n \times R_n) \cap D_0 \subset Z, \qquad (n = 1,2,\ldots) \tag{14.2}$$

Let $(y,z) \in (E_n \times R_n) \cap D_0$. There exist points $y_k \in E_n$ $(k = 1,2,\ldots)$ such that $\lim\{y_k\} = y$. Since $(y,z) \in D_0$ we can choose points $z \in X$ with $\lim\{z_k\} = z$ and $(y_k, z_k) \in D$ $(k = 1,2,\ldots)$. We shall prove that $(y,z) \in Z$. Suppose the contrary holds. As Z is closed in $X \times X$, we can choose neighbourhoods U and V of y and z respectively such that

$$(U \times V) \cap Z = \emptyset. \tag{14.3}$$

The set R_n is open in X and $z \in R_n$, hence $z_k \in R_n$ for all sufficiently large subscripts k, and because $y_k \in E_k$ $(k = 1,2,\ldots)$ we deduce from (14.1) that

$$(y_k, z_k) \in (E_n \times R_n) \cap D \subset Z,$$

for all sufficiently large subscripts k. This contradicts (14.3).

Thus the formula (14.2) is proved. Now we shall show that the set $E_n \cap C$ is nowhere dense in X $(n = 1,2,\ldots)$. Suppose the contrary, *i.e.*, that for some E_n there exists an open subset G of X such that

$$\emptyset \neq G \subset \overline{E_n \cap C}. \tag{14.4}$$

By virtue of (14.2),

$$(G \propto R_n) \cap D_0 \subset (\bar{E}_n \times R_n) \cap D_0 \subset Z \cap D_0.$$

Let us prove that $(G \times R_n) \cap D_0 \neq \emptyset$. It follows from (14.4) that $G \cap E_n \cap C \neq \emptyset$. Let $x \in G \cap E_n \cap C$. Then $xD = xD_0$ and $R_n \cap xD \neq \emptyset$, hence $R_n \cap xD_0 \neq \emptyset$. Choose $y \in R_n \cap xD_0$. Then $(x,y) \in (G \times R_n) \cap D_0$. Thus $Z \cap D_0$ contains a non-empty open subset $(G \times R_n) \cap D_0$ of D_0; but this contradicts the fact that $Z \cap D_0$ is nowhere dense in D_0. Hence the set $E_n \cap C$ is nowhere dense in X $(n = 1,2,\dots)$.

Set

$$Q = \bigcup_{n=1}^{\infty} (E_n \cap C) \cup (X \setminus C).$$

It follows from the preceding considerations that Q is a set of the first category in X.

Let $x \in X \setminus Q = C \setminus \bigcup_{n=1}^{\infty} E_n$. Let us prove that xZ is nowhere dense in xD. Suppose the contrary to be true. Since $\{R_n \mid n = 1,2,\dots\}$ is a base of open subsets of X, there exists an R_m such that $\emptyset \neq (R_m \cap xD) \subset xZ$, *i.e.*, $x \in E_m$. By hypothesis, $x \in C$, and hence $x \in E_m \cap C \subset Q$, in spite of the choice of x. ▮

3·14.31 LEMMA: *Let X be a compact metric space, (X,T) a minimal transformation group and let $R \subset X \times X$ be both an open and a closed invariant equivalence relation. If $R \neq \Delta(X)$ and the extension $\varphi:(X,T) \to (X/R,T)$ is almost distal, then $Q^*(R)$ is a proper subset of R. In other words, there exists a proper equicontinuous extension $\psi:(Z,T) \to (X/R,T)$ subordinate to the extension φ.*

Proof: Since φ is open and almost distal, $R = \overline{R\mathcal{J}}$ (Lemma 3·14.28). Suppose $Q^*(R) = R$. Then $R = \overline{P \cap R}$ by Lemma 3·13.11, *i.e.*

$$R = \overline{\bigcap_{n=1}^{\infty} \alpha_n T \cap R}, \tag{14.5}$$

where $\{\alpha_n \mid n = 1,2,\dots\}$ is a uniformity base of the metric space X. The equality (14.5) means that $P \cap R$ is a Baire subset of R. By Lemma 3·14.30, $x(P \cap R)$ is a residual subset of xR for all x belonging to some residual subset $X_0 \subset X$. As φ is an almost distal extension, there exists a residual subset $X_1 \subset X$ such that $x(P \cap R) = x$ for all $x \in X_1$. Let $x_0 \in X_0 \cap X_1$; then by Baire's Theorem $x_0R = \overline{x_0(P \cap R)} = x_0$, but since the relation R is open and (X,T) is a minimal transformation group, $xR = x$ for all $x \in X$, *i.e.*, $R = \Delta(X)$ Thus if $R \neq \Delta(X)$, then $Q^*(R) \neq R$. ▮

3·14.32 LEMMA: *Let W be a compact metric space and $\varphi:(W,T) \to (Z,T)$ be a minimal almost automorphic extension. Then*

$$H = \{z \mid z \in Z, z\varphi^{-1} \text{ consists of a single point}\}$$

is a Baire subset of Z and $H\varphi^{-1}$ is a Baire subset of W.

Proof: Write

$$G_n = \{z \mid z \in Z, \operatorname{diam} W_z < 1/n\}, \qquad (n = 1,2,\dots).$$

Clearly, each set G_n is open and $H = \bigcap [G_n \mid n = 1,2,\dots]$. Since H is non-empty and invariant, we get $\bar{H} = Z$ by the minimality of Z, and hence H is a Baire subset of Z. The assertion concerning $H\varphi^{-1}$ follows from Lemma 3·12.17. ▮

3·14.33 COROLLARY: *In the case of compact metric phase spaces each almost automorphic minimal extension is almost distal.*

3 14.34 LEMMA: *Let* $\varphi:(X,T) \to (Y,T)$ *and let* $\psi:(Y,T) \to (Z,T)$ *be almost automorphic extensions and, moreover, let* (X,T) *be a minimal transformation group with a compact metrizable phase space. Then* $\varphi\circ\psi$ *is also an almost automorphic extension.*

Proof: Let

$$H_1 = \{z \mid z \in Z, z\psi^{-1} \text{ consists of a single point}\},$$

$$H_2 = \{y \mid y \in Y, y\varphi^{-1} \text{ consists of a single point}\}.$$

By Lemma 3·14.23, H_1 is residual in Z, H_2 is residual in Y and $H_1\psi^{-1}$ is residual in Y. Choose some point $y_0 \in H_2 \cap H_1\psi^{-1}$ and let $z_0 = y_0\psi$. Then $z_0(\varphi\circ\psi)^{-1} \equiv (z_0\psi^{-1})\varphi^{-1}$ consists of a single point. ▮

3·14.35 LEMMA: *Let* ϑ *be a countable ordinal and let* $\{(X_\lambda,T) \mid 0 \leqslant \lambda \leqslant \vartheta\}$ *be a projective system of minimal transformation groups with compact metrizable phase spaces. If each extension* $\varphi_\lambda^{\lambda+1}$ $(X_{\lambda+1},T) \to (X_\lambda,T)$ *is almost autormorphic, then* $\varphi_0^\vartheta:(X_\vartheta,T) \to (X_0,T)$ *is also almost automorphic.*

Proof: We shall prove by transfinite induction that each extension $\varphi_0^\lambda:(X_\lambda,T) \to (X_0,T)$ $(0 < \lambda \leqslant \vartheta)$ is almost automorphic. By hypothesis this statement is true for $\lambda = 1$. Let us assume that it holds for all λ, $\lambda < \mu \leqslant \vartheta$ and we prove that φ_0 is an almost automorphic extension.

If $\mu = \lambda + 1$, then $\varphi_0^{\lambda+1} \equiv \varphi_\lambda^{\lambda+1}\circ\varphi_0^\lambda$ is almost automorphic by Lemma 3·14.34. If μ is a limit ordinal, define:

$$H_\lambda = \{x \mid x \in X_0, x(\varphi_0^\lambda)^{-1} \text{ consists of a single point}\}, \quad (\lambda \leqslant \mu).$$

Since φ_0^λ is almost automorphic by the induction assumption, H_λ is a residual subset of X_0 for all $\lambda < \mu$. As μ is a countable ordinal, the set $\bigcap [H_\lambda \mid \lambda < \mu]$ is also residual. The required result follows now from the obvious equality $\bigcap [H_\lambda \mid \lambda < \mu] = H_\mu$. ▮

3·14.36 THEOREM: *Let X be a compact metrizable space and let $\varphi:(X,T) \to (Y,T)$ be a minimal almost distal extension. Then there exists a minimal transformation group (W,T) with a compact metrizable phase space and there exists homomorphisms $p:(W,T) \to (X,T)$ and $q:(W,T) \to (Y,T)$ such that*

(1): *p is an almost periodic extension;*

(2): *q is a V-extension;*

(3): *$p\circ\varphi = q$.*

Proof: Consider the commutative diagram

$$\begin{array}{ccc} (X,T) & \xleftarrow{\ \lambda\ } & (X^*,T) \\ {\scriptstyle\varphi}\downarrow & & \downarrow{\scriptstyle\psi} \\ (Y,T) & \xleftarrow{\ \mu\ } & (Y^*,T) \end{array}$$

obtained by using the construction of Lemma 3·12.27. Recall that X^* and Y^* are compact metrizable spaces, (X^*,T) and (Y^*,T) are minimal transformation groups, λ and μ are almost automorphic extensions, and ψ is an open mapping. By Lemma 3·14.27 the extension $\psi:(X^*,T) \to (Y^*,T)$ is almost distal. Denote R_φ by R and R_ψ by R_1. It is easy to verify that the extension ψ is trivial (*i.e.*, $R_1 = \Delta(X^*)$) iff φ is an almost automorphic extension.

Suppose that $R_1 \neq \Delta(X^*)$. By Lemma 3·14.31, $Q^*(R_1)$ is a proper subset of R_1. Set $(Y_1^*,T) = (X^*/Q^*(R_1),T)$ and let $\lambda_1:X^* \to Y_1^*$, $\mu_1:Y_1^* \to Y^*$ be the canonical maps. By Corollary 3·13.3 the extension μ_1 is equicontinuous. Since the extension ψ is almost distal and $R_{\lambda_1} = Q^*(R_1) \subset R_1 \equiv R_\psi$, the extension λ_1 is also almost distal. We now have the commutative diagram

$$\begin{array}{ccccc} (X,T) & \xleftarrow{\ \lambda\ } & (X^*,T) & \searrow^{\lambda_1} & \\ {\scriptstyle\varphi}\downarrow & & \downarrow{\scriptstyle\psi} & & (Y_1^*,T), \\ (Y,T) & \xleftarrow{\ \mu\ } & (Y^*,T) & \swarrow_{\mu_1} & \end{array}$$

where λ and μ are almost automorphic extensions, μ_1 is an equicontinuous extension and λ_1 is an almost distal extension. Now we can repeat the above considerations taking λ_1 as the initial extension. This leads to the following inductive construction.

Let Ω be the first non-countable ordinal. For every two ordinals α and β satisfying the condition $0 \leqslant \beta < \alpha < \Omega$ we shall define minimal transformation groups (W_α,T), (W_β,T), (Z_α,T), (Z_β,T) with compact metrizable phase spaces and homomorphisms φ_α, φ_β, p_β^α, q_β^α, s_α and s_β so that

(1): The diagram

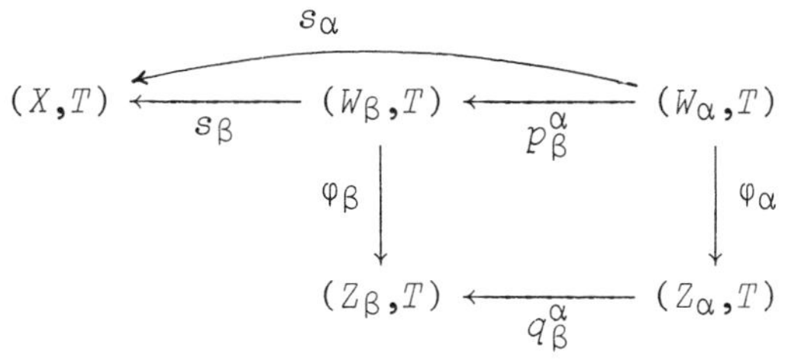

is commutative;

(2): Either the extension φ_β is almost autormorphic or $q_\beta^{\beta+1}$ is a proper *AE*-extension;

(3): s_β is an almost automorphic extension;

(4): φ_β is an almost distal extension;

(5): If α is a limit ordinal, then (W_α,T) and (Z_α,T) are projective limits of the projective systems $\{(W_\beta,T)| \beta < \alpha\}$ and $\{(Z_\beta,T)|\beta < \alpha\}$ respectively.

We proceed by induction. Set $(W_0,T) = (X,T)$, $(Z_0,T) = (Y,T)$, $\varphi_0 = \varphi$; let s_0 be the identity map. Suppose that the construction has been accomplished for all α and β satisfying the condition $0 \leqslant \beta < \alpha < \gamma < \Omega$. There are two cases to consider:

(1): γ has an immediate predecessor, say, $\gamma = \alpha + 1$. Applying the construction of Lemma 3·12.27 to the extension $\varphi_\alpha:(W_\alpha,T) \to (Z_\alpha,T)$ and then using Lemma 3·14.31, we obtain the commutative diagram

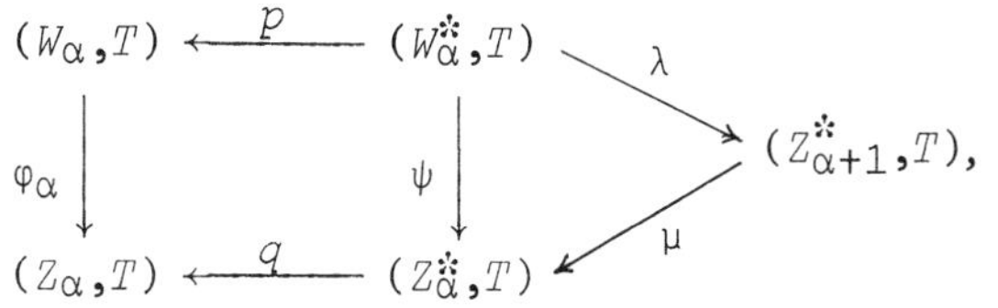

where p and q are almost automorphic extensions, μ is an equicontinuous extension and λ is an almost distal extension. If φ_α is not almost automorphic, then ψ is a proper extension; hence it follows from Lemma 3·14.31 that μ is also a proper extension.

Define $(W_{\alpha+1},T) = (W_\alpha^*,T)$, $p_\alpha^{\alpha+1} = p$, $(Z_{\alpha+1},T) = (Z_{\alpha+1}^*,T)$, $\varphi_{\alpha+1} = \lambda$, $q_\alpha^{\alpha+1} = \mu o q$, $s_{\alpha+1} = p_\alpha^{\alpha+1}\ s_\alpha$. Since s_α and $p_\alpha^{\alpha+1}$ are almost automorphic extensions, $s_{\alpha+1}$ is also almost automorphic (Lemma 3·14. 34).

(2): γ is a limit ordinal. Then transformation groups (W_γ,T) and (Z_γ,T) are uniquely defined by condition (5). It follows from Lemma 3·14.35 that s_γ is an almost autormorphic extension. Since γ is a countable ordinal, W_γ and Z_γ are compact metrizable spaces.

We shall prove that if φ_α is not an almost autormophic extension and $R_\alpha = \{(a,b)|a,b \in W_\alpha, a\varphi_\alpha = b\varphi_\alpha\}$ $(0 \leqslant \alpha < \Omega)$, then $R_{\alpha+1}(s_{\alpha+1}$

$\times s_{\alpha+1})$ is a proper subset of $R_\alpha(s_\alpha \times s_\alpha)$. Since s_α is an almost automorphic extension, it follows from Lemma 3·14.32 that there exists a residual subset $K_\alpha \subset W_\alpha$ such that the mapping s_α is injective at all points $w \in K_\alpha$. Using Lemma 3·12.19, choose a residual subset $H_\alpha \subset Z_\alpha$ in such a way that $K_\alpha \cap z\varphi_\alpha^{-1}$ is a dense subset of the fiber $z\varphi_\alpha^{-1}$ for all $z \in H_\alpha$. We see from the proof of Lemma 3·12.27 that there exists a residual set $M_\alpha \subset Z_\alpha^*$ such that for every $z \in M_\alpha$ the homomorphism $p = p_\alpha^{\alpha+1}$ maps the fiber $z\psi^{-1}$ homeomorphically onto the fiber $(zq)\varphi_\alpha^{-1}$ and, moreover, $(zq)q^{-1} = \{z\}$, *i.e.*

$$(zq)\varphi_\alpha^{-1}p^{-1} = z\psi^{-1}, \qquad (z \in M_\alpha). \tag{14.6}$$

If φ_α is not an almost automorphic extension, then $R_\psi \neq \Delta(W_\alpha^*) \equiv \Delta(W_{\alpha+1})$; therefore $Q^*(R_\psi)$ is a proper subset of R_ψ (Lemma 3·14.31). But $Q^*(R_\psi) = R_{\alpha+1}$, and since $\mu:(Z_{\alpha+1}^*,T) \to (Z_\alpha^*,T)$ is a proper equicontinuous extension, for each point $w \in W_\alpha^* \equiv W_{\alpha+1}$ the set $wR_{\alpha+1} \equiv wR_\lambda \equiv wQ^*(R_\psi)$ is a proper subset of wR_ψ. As the set H_α is residual in Z_α and q is an almost autormorphic extension, $H_\alpha q^{-1}$ is a residual subset of the space Z_α^* (Lemma 3·12.27). Let $w \in (M_\alpha \cap H_\alpha q^{-1})\psi^{-1}$, then it follows from the above that $(wR_{\alpha+1})p_\alpha^{\alpha+1} \equiv (wR_\lambda)p$ is a proper subset of the set $(wR_\psi)p = (wp)R_\alpha \equiv (wp_\alpha^{\alpha+1})R_\alpha$ Further, if $w \in (M_\alpha \cap H_\alpha q^{-1})\psi^{-1}$, then $wp\varphi_\alpha = w\psi q \in H_\alpha$, therefore the set of points of K_α at which the map s_α is injective are dense in the fiber $(wp)R_\alpha$. Hence it follows that $(wR_{\alpha+1})s_{\alpha+1} \equiv (wR_{\alpha+1})\circ p_\alpha^{\alpha+1}$ is a proper subset of $((wp_\alpha^{\alpha+1})R_\alpha)s_\alpha$, whenever $w \in (M_\alpha \cap H_\alpha q^{-1})\circ \psi^{-1}$. On account of (14.6) this completes the proof of our assertion.

Since X is a compact metric space, and

$$R_{\alpha+1}(s_{\alpha+1} \times s_{\alpha+1}) \subset R_\alpha(s_\alpha \times s_\alpha) \subset X, \qquad (\alpha < \Omega),$$

there exists an ordinal ϑ, $\vartheta < \Omega$, with $R_{\vartheta+1}(s_{\vartheta+1} \times s_{\vartheta+1}) = R_\vartheta(s_\vartheta \times s_\vartheta)$. As was already mentioned, this is possible only when is an almost automorphic extension.

We thus get the commutative diagram

$$\begin{array}{ccccccccccc}
(X,T) & \xleftarrow{p_0^1} & (W_1,T) & \leftarrow \cdots \leftarrow & (W_\alpha,T) & \xleftarrow{p_\alpha^{\alpha+1}} & (W_{\alpha+1},T) & \leftarrow \cdots \leftarrow & (W_\vartheta,T) \\
\downarrow{\scriptstyle\varphi} & & \downarrow{\scriptstyle\varphi_1} & & \downarrow{\scriptstyle\varphi_\alpha} & & \downarrow{\scriptstyle\varphi_{\alpha+1}} & & \downarrow{\scriptstyle\varphi_\vartheta} \\
(Y,T) & \xleftarrow{q_0^1} & (Z_1,T) & \leftarrow \cdots \leftarrow & (Z_\alpha,T) & \xleftarrow{q_\alpha^{\alpha+1}} & (Z_{\alpha+1},T) & \leftarrow \cdots \leftarrow & (Z_\vartheta,T)
\end{array},$$

in which all the extensions $p_\alpha^{\alpha+1}$ are almost automorphic, the extensions of the form $q_\alpha^{\alpha+1}$ are proper AE-extensions and φ_ϑ is an almost autormorphic extension. This means that $s_\vartheta:(W_\vartheta,T) \to (X,T)$ is an almost automorphic extension and $q_0^\vartheta:(Z_\vartheta,T) \to (Y,T)$ is a V-extension.

3·14.37 We now give an example of a minimal cascade which is an almost automorphic extension of some almost periodic cascade.

Let $\mathbb{K} = \{\xi | \xi \in \mathbb{C}, |\xi| = 1\}$ and let $\tau: \mathbb{K} \to \mathbb{K}$ be the homeomorphism defined by $\xi\tau = \xi e^i$ $(\xi \in \mathbb{K})$, where i is the imaginary number $\sqrt{-1}$. Let S^1 denote the circle obtained from the segment $\{t | 0 \leqslant t \leqslant 2\pi\}$ by identifying the end points. Let X be a perfect nowhere dense subset of S^1 and let $\{(\alpha_n,\beta_n) | n = 1,2,\dots\}$ be the system of its complementary intervals, where it is assumed that α_n precedes β_n in the cyclic order corresponding to the increase of t, $0 \leqslant t < 2\pi$.

We shall establish a one-to-one cyclic order-preserving correspondence between the set of intervals $\{(\alpha_n,\beta_n) | n = 1,2,\dots\}$ and the set $O \equiv \{e^{in} | n \in \mathbb{Z}\}$ of points of the circle $\mathbb{K}$ (with a specified orientation). Enumerate the points of O in the following manner: $\xi_1 = 1$, $\xi_2 = e^i$, $\xi_3 = e^{-i}, \dots$, $\xi_{2n} = e^{in}$, $\xi_{2n+1} = e^{-in}$. To the point ξ_1 we assign the interval $(\alpha_1,\beta_1) \equiv (\alpha^{(0)},\beta^{(0)})$, to the point ξ_2 we assign the interval $(\alpha_2,\beta_2) \equiv (\alpha^{(1)},\beta^{(1)})$. Further, to the point ξ_3 we assign the interval $(\alpha^{(-1)},\beta^{(-1)})$ chosen to be the interval (α_n,β_n) with the least subscript n satisfying the following condition: (α_n,β_n) is situated on that one of the two arcs between the already taken intervals for which $(\alpha^{(0)},\beta^{(0)})$, $(\alpha^{(1)},\beta^{(1)})$, $(\alpha^{(-1)},\beta^{(-1)})$ have the same cyclic order on S^1, as 1, e^i, e^{-i} do.

Suppose that to each point $\xi_1,\xi_2,\dots,\xi_N$ we have already assigned an interval in the system $\{(\alpha_n,\beta_n) | n = 1,2,\dots\}$ and let ξ_{N+1} be situated between the points ξ_k and ξ_ℓ, where $k,\ell \leqslant N$. We assign to the point ξ_{N+1} that not yet used interval with the least index n, which is situated between $(\alpha^{(k)},\beta^{(k)})$ and $(\alpha^{(\ell)},\beta^{(\ell)})$ according to the cyclic order on S^1. Thus the required correspondence is defined by induction.

Now we shall define a map $\Phi: S^1 \to \mathbb{K}$. If a point belongs to a segment $[\alpha^{(k)},\beta^{(k)}]$, then we put $x\Phi = e^{ik}$. If ϑ_0 belongs to X and ϑ_0 does not coincide with an end point of a complementary interval, then it determines a section of the interval $(\alpha^{(0)},\beta^{(0)})$ on S^1. Because the cyclic order is preserved, there is a corresponding section of the set $\{(\alpha^{(k)},\beta^{(k)}) | k \in \mathbb{Z}, k \neq 0\}$, which determines a certain point ξ_0. Then put $\vartheta_0\Phi = \xi_0$. It is easy to see that Φ is continuous.

Let p denote the restriction of Φ to X. Then ξp^{-1} consists of a single point whenever $\xi \neq e^{in}$ $(n \in \mathbb{Z})$. On the other hand, if $\xi = e^{ik}$ for some $k \in \mathbb{Z}$, then ξp^{-1} consists of the two points $\alpha^{(k)}$ and $\beta^{(k)}$.

The homeomorphism $\tau: \mathbb{K} \to \mathbb{K}$ can be lifted to S^1. Indeed, set

$$\alpha^{(k)}\sigma = \alpha^{(k+1)}, \qquad \beta^{(k)}\sigma = \beta^{(k+1)}, \qquad (k \in \mathbb{Z}).$$

Since σ preserves the cyclic order of intervals, it can be extended to X and then to S^1 so that $\sigma: S^1 \to S^1$ becomes a homeomorphism.

It is easy to see that X is a minimal set of the cascade (S^1,σ) and that $p:(X,\sigma) \to (\mathbb{K},\tau)$ is an almost automorphic extension. Moreover, any two points belonging to the same fiber are asymptotic in both directions.

3·14.38 The cascade (S^1,σ) can be embedded in a flow $(S^1 \times S^1, \mathbb{R})$ on a two-dimensional torus as a global section (*see:* Subsection 2·11.27). The orbits $\vartheta = f(\varphi,\vartheta_0)$ of this flow satisfy a differential equation of the form

$$\frac{d\vartheta}{d\varphi} = A(\varphi,\vartheta), \tag{14.7}$$

where (φ,ϑ) are the cyclic coordinates on the torus $S^1 \times S^1$, $S^1 \approx \mathbb{R}/2\pi\mathbb{Z}$; ϑ denotes the coordinate on the section and the function A is continuous on $\mathbb{R} \times \mathbb{R}$ and periodic with period 2π in each variable. The winding number α of the flow is equal to $1/2\pi$. If ϑ_1 is an end point of some interval complementary to the set X, then there exist two sequences of integers $\{p_n\}$ and $\{q_n\}$ such that

$$\{f(\varphi + 2\pi p_n,\vartheta_0) + 2\pi q_n\} \to f(\varphi,\vartheta_1),$$

uniformly on each bounded segment. If ϑ_1 and ϑ_2 are end points of the same complementary interval, then

$$\lim_{\varphi\to\infty} [f(\varphi,\vartheta_2) - f(\varphi,\vartheta_1)] = 0, \tag{14.8}$$

because the points ϑ_1 and ϑ_2 are asymptotic under (S^1,σ).

REMARKS AND BIBLIOGRAPHICAL NOTES

The results in Subsections 3·14.5-12,19-23 are due to Bronšteĭn [*11,13*]. The results in Subsections 3·14.13,15-17 are probably new. Theorem 3·14.23 on the structure of distal minimal extensions was independently proved by Ellis within the framework of his algebraic theory of minimal sets (Ellis [*6,10-13*], Horelick [*1,2*]). Lemma 3·14.30 is a slight generalization of a proposition presented in the book by Kuratowski [*1*] (Vol. I, p. 255). Almost distal extensions were first considered by Veech [*5*]. Lemma 3·14.31 was proved in the note by Bronšteĭn [*15*]. Theorem 3·14.36 on the structure of almost distal extensions was obtained by Bronšteĭn [*15*] as a generalization of the main result of the paper by Veech [*5*] which would correspond to the case of a trivial transformation group (Y,T). The proof presented above is simpler than the one given by Veech. The simplification is attained thanks to Lemma 3·14.31, which is in turn based on the description of $Q^*(R)$ given in Section §(3·13). Example 3·14.37 is taken from the book by Nemytskiĭ and Stepanov [*1*].

§(3·15): GROUPS ASSOCIATED WITH MINIMAL EXTENSIONS

We shall introduce and study here some groups associated with a given minimal extension $\varphi:(X,T) \to (Y,T)$. These groups arise in the following way. Given an extension φ, we can produce (in general, a non-unique) minimal transformation group (M,T) and a homomorphism $\lambda:(M,T) \to (X,T)$ so that if (m_1,m_2) is an almost periodic point of the transformation group $(M \times M,T)$ with $m_1\lambda\varphi = m_2\lambda\varphi$, then there exists an automorphism $g:(M,T) \to (M,T)$ sending m_1 to m_2. The totality of such automorphims forms a group which we denote by G. In the case when M is the smallest transformation group satisfying the above condition, G is called the *Galois group of the extension* φ. Further, a topology σ on the group G is introduced and the set

$$H(G) = \cap [\mathrm{cls}_\sigma V \mid V \in N_\sigma],$$

is considered, where N_σ is the neighbourhood base at the identity $e \in G$ with respect to the topology σ and $\mathrm{cls}_\sigma V$ denotes the closure of V with respect to G. It is proved that $H(G)$ is a subgroup of G.

The groups G and $H(G)$ are used in this section to develop an algebraic theory of distal minimal extensions similar to the well known Galois theory and to investigate the structure of minimal extensions belonging to a much wider class than almost distal extensions. As was shown by Ellis, these groups also play an important rôle in some other questions.

3·15.1 Let X and Y be compact spaces, let $\varphi:(X,T) \to (Y,T)$ be a minimal extension and let $R = R_\varphi$. Let $R\mathcal{J}$ denote, as usual, the set of all pairs $(x_1,x_2) \in R$ such that (x_1,x_2) is an almost periodic point of the transformation group $(X \times X,T) = (X,T) \times (X,T)$.

The extension $\varphi:X \to Y$ is said to be a *REGULAR EXTENSION* if for each pair $(x_1,x_2) \in R\mathcal{J}$ there exists an automorphism $g:(X,T) \to (X,T)$ with $x_2 = x_1 g$.

3·15.2 LEMMA: *The following statements are mutually equivalent:*

(1): *The extension φ is regular;*

(2): *If $(x_1,x_2) \in \cup [(R\mathcal{J})\ \ n = 1,2,\dots\ ]$, then $x_2 = x_1 g$ for some automorphism $g:(X,T) \to (X,T)$;*

(3): *There is a point $x_0 \in X$ such that for each $x \in x_0 R\mathcal{J}$ there exists an automorphism g with $x = x_0 g$.*

Proof: Clearly, (1) and (2) are equivalent, and (1) implies (3). Let (3) be true and let $(x_1,x_2) \in R\mathcal{J}$. Then $\overline{(x_1,x_2)T}$ is a minimal set. We can choose a point $x_3 \in X$ with $(x_3,x_0) \in \overline{(x_1,x_2)T} \subset R\mathcal{J}$. According to (3), there exists an automorphism g such that $x_0 g = x_3$. Then $x_3 t = x_0 g t = x_0 t g$ $(t \in T)$, hence for each pair $(a,b) \in \overline{(x_3,x_0)T}$ we get $a = bg$. Since $(x_1,x_2) \in \overline{(x_3,x_0)T}$, we finally get $x_1 = x_2 g$. ▌

3·15.3 A commutative diagram

$$\begin{array}{ccc} M & \xrightarrow{\lambda} & X \\ & \searrow^{\mu} & \downarrow \varphi \\ & & Y \end{array} \qquad (15.1)$$

in which M, X and Y are compact minimal sets is said to be a *REGULAR COMMUTATIVE DIAGRAM* if for any $(x_1,x_2) \in R\mathcal{J}$ there exist elements $m_1,m_2 \in M$ and an automorphism $g \in A(M,T)$ such that $m_2 = m_1 g$, $m_1\lambda = x_1$, $m_2\lambda = x_2$. Since $(R_\mu\mathcal{J})(\lambda \times \lambda) = R_\varphi\mathcal{J} \equiv R\mathcal{J}$ (Lemma 3·14.15), this definition means that $\mu:M \to Y$ is a regular extension.

Let I be a minimal right ideal of the semigroup $E(X,T)$, let $x_0 \in X$, let $\lambda:(I,T) \to (X,T)$ be the homomorphism defined by $\xi\lambda = x_0\xi$ $(\xi \in I)$ and let $\mu = \lambda\circ\varphi$. It follows from Theorem 2·7.9 and Lemma 2·7.6 that the diagram

$$\begin{array}{ccc} I & \xrightarrow{\lambda} & X \\ & \searrow^{\mu} & \downarrow \varphi \\ & & Y \end{array} \qquad (15.2)$$

is regular.

Let us set

$$D = \bigcup_{n=1}^{\infty} (R\mathcal{J})^n, \qquad S = \{(\xi,\eta) \mid \xi,\eta \in I, a\xi = a\eta, a \in x_0 D\}.$$

It is clear that S is a closed invariant equivalence relation. Let $(M,T) = (M(X,Y,x_0),T)$ denote the transformation group $(I/S,T)$. The elements of the space I/S may be viewed as restrictions $\xi|_{x_0D}$ of the map $\xi:X \to X$ $(\xi \in I)$ to x_0D.

Let $x_1 \in x_0R\mathcal{J}$. There exist elements $u \in \mathcal{J}(I) = \{v \mid v \in I, v^2 = v\}$ and $\xi \in Iu$ with $x_0 = x_0u$, $x_1 = x_0\xi = x_0\xi u$. Then

$$(x_0(R\mathcal{J})^n)\xi \subset (x_0\xi)(R\mathcal{J})^n \subset x_0(R\mathcal{J})^{n+1},$$

therefore $\xi \equiv \xi u$ maps the set x_0D into itself. The map $h:M \to M$ defined by

$$(\lambda|_{x_0D})h = \xi|_{x_0D}\circ\lambda|_{x_0D}, \qquad (\lambda \in I),$$

is an automorphism $u|_{x_0D}$ to $\xi u|_{x_0D}$ (that is, $(u|_{x_0D})h = \xi u|_{x_0D}$). Thus we get the regular diagram next shown:

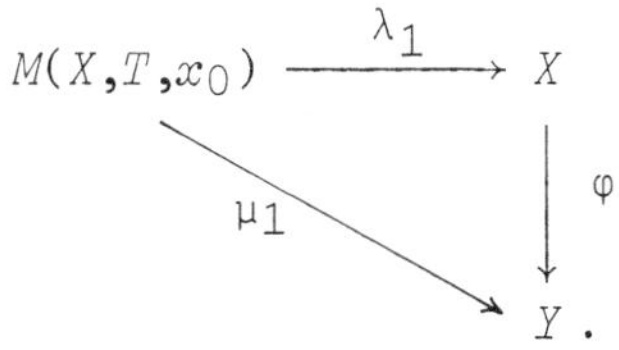

3·15.4 Let

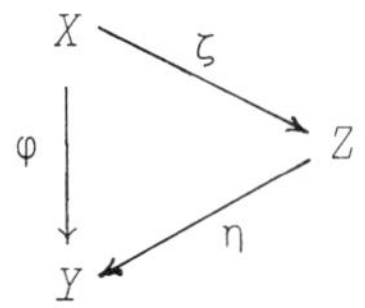

(15.3)

be a commutative diagram, $x_0 \in X$, $z_0 = x_0\zeta$ and let $\zeta^*:E(X,T) \to E(Z,T)$ be the homomorphism of enveloping semigroups induced by ζ. The set $K = I\zeta^*$ is a minimal right ideal of the semigroup $E(Z,T)$. Put:

$$S_1 = \{(\xi_1,\xi_2)\,|\,\xi_1,\xi_2 \in K, a\xi_1 = a\xi_2, a \in z_0 \bigcup_{n=1}^{\infty} (R_\eta \mathcal{J})^n\}.$$

Then $M(Z,Y,z_0) = K/S_1$. The map $\zeta^*:I \to K$ induces a homomorphism $\bar{\zeta}:M(X,Y,x_0) \to M(Z,Y,z_0)$ so that the diagram

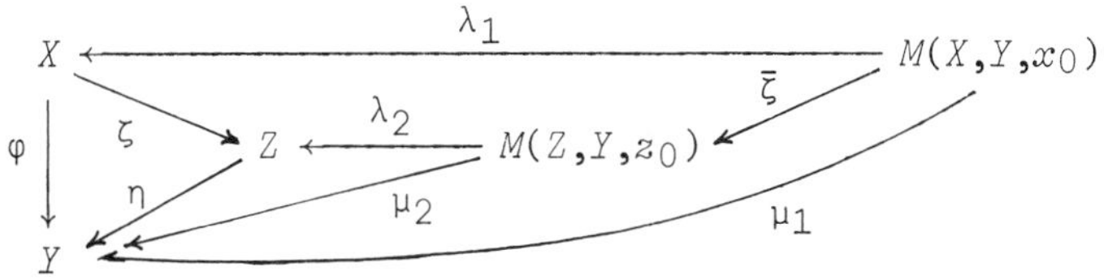

is commutative. If $\varphi:X \to Y$ is a regular extension, then $M(X,Y,x_0) = X$. In fact, if $x_0\xi_1 = x_0\xi_2$, then $x_0h\xi_1 = x_0h\xi_2$ for every automorphism $h \in A(X,T)$. By the regularity, $a\xi_1 = a\xi_2$ $(a \in x_0D)$, that is, $(\xi_1,\xi_2) \in X$.

It follows from the above considerations that, given a regular diagram (15.1), there exists a homomorphism $\psi:(M,T) \to (M(X,Y,x_0),T)$ such that the diagram

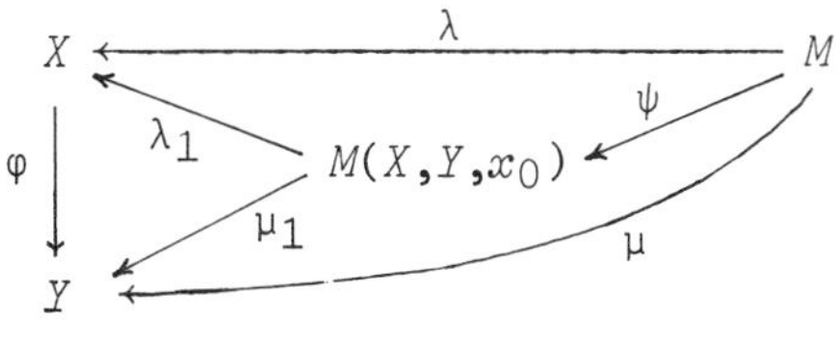

is commutative.

The family of all automorphisms h of the transformation group $(M(X,Y,x_0),T)$ satisfying the condition $m\mu_1 = mh\mu_1$ $(m \in M(X,Y,x_0))$ forms a group, which will be denoted by $G = G(X,Y) = G(X,Y,x_0)$. It is easy to verify that $G(X,Y,x_0)$ does not in fact depend on the choice of the point $x_0 \in X$. The group $G = G(X,Y)$ is called the *GALOIS GROUP OF THE EXTENSION* $\varphi:(X,T) \to (Y,T)$.

3·15.5 An extension φ is called a *HOMOGENEOUS EXTENSION* if for any two points $x_1,x_2 \in X$, $(x_1,x_2) \in R_\varphi$, there exists an automorphism $h \in A(X,T)$ with $x_1h = x_2$. Clearly, each homogeneous extension is regular, but the converse is not true in general.

3·15.6 LEMMA: *If* $\varphi:(X,T) \to (Y,T)$ *is a distal extension, then* $\mu_1:M(X,Y,x_0) \to Y$ *and* $\lambda_1:M(X,Y,x_0) \to X$ *are homogeneous extensions.*

Proof: By Lemma 3·14.28, if φ is distal, then $R_\varphi \equiv R = R\mathcal{J} = D$. Hence

$$S \equiv \{(\xi,\eta) \mid \xi,\eta \in I, a\xi = a\eta, a \in x_0D\}$$

$$= \{(\xi,\eta) \mid \xi,\eta \in I, a\xi = a\eta, a \in x_0R\}.$$

Set

$$G_0 = \{\xi|_{x_0R} \mid \xi \in I, x_0R\xi \subset x_0R\}.$$

As φ is distal, G_0 is a transitive group of transformations of the fiber x_0R onto itself (Lemmas 3·12.11,12 and Theorem 3·14.9). It was shown above that each map $\xi|_{x_0D} \equiv \xi|_{x_0R}$ satisfying the condition $x_0R\xi \subset x_0R$ is an automorphism of the transformation group $(I/S,T) \equiv (M(X,Y,x_0),T)$. Hence it follows that μ_1 is homogeneous. Consequently λ_1 is also homogeneous.∎

3·15.7 If the extension φ is equicontinuous, then Lemma 3·15.6 can be improved. As can be seen from the proof of this Lemma, the group $G(X,Y)$ may be identified with G_0. By Lemma 3·14.12, G_0 is a compact topological group. We can show that the action of $G_0 = G(X,Y)$ on $M(X,Y,x_0)$ determines a topological transformation group.

3·15.8 LEMMA: *If* X *is a compact space,* (Y,T) *is a minimal transformation group and* $\varphi:(X,T) \to (Y,T)$ *is a homogeneous extension, then* φ *is distal.*

Proof: Let $y \in Y$. There is at least one almost periodic point $x_0 \in X_y$. For any two points $x_1,x_2 \in X_y$ choose automorphisms g_1 and g_2 from $A(X,T)$ so that $x_1 = x_0g_1$, $x_2 = x_0g_2$. Clearly (x_1,x_2) is an almost periodic point of the transformation group $(X \times X,T)$. By Theorem 3·14.7 the extension φ is distal.∎

3·15.9 Let (15.1) be a given regular diagram and let G be the group of all automorphisms $g:(M,T) \to (M,T)$ preserving the fibers of the map $\mu : M \to Y$.

The group $G \times G$ acts on $M \times M$ in a natural way:

$$(m_1,m_2)(g_1,g_2) = (m_1g_1,m_2g_2), \qquad (m_1,m_2 \in M; g_1,g_2 \in G).$$

Our next goal is to introduce and to study two topologies on the group G.

3·15.10 LEMMA: *Let $\emptyset \neq A \subset G$. For every two points $m, m_0 \in M$ the equalities*

$$\overline{(m_0A,m_0)T} = \overline{(mA,m)T} = \overline{\Delta(M)(A \times e)},$$

hold, where $\Delta(M) = \{(m,m) \mid m \in M\}$ and e is the identity of G.

Proof: There exists a net $\{t_n\}$, $t_n \in T$, such that $m_0t_n \to m$. Then $m_0gt_n = m_0t_ng \to mg$, whence $(mg,m) \in \overline{(m_0A,m_0)T}$ $(g \in A)$, and therefore $\overline{(mA,m)T} \subset \overline{(m_0A,m_0)T}$. By symmetry, $\overline{(m_0A,m_0)T} = \overline{(mA,m)T}$ $(m_0,m \in M)$; hence $\overline{(mA,m)T} = \overline{\Delta(M)(A \times e)}$ $(m \in M)$. ∎

3·15.11 Let $\emptyset \neq A \subset G$. Define:

$$\mathrm{cls}_\sigma A = \{g \mid g \in G, (mg\lambda,m\lambda) \in \overline{(mA,m)T}(\lambda \times \lambda), m \in M\}.$$

By Lemma 3·15.10 the following equalities hold:

$$\mathrm{cls}_\sigma A = \{g \mid g \in G, \exists m_0 \in M : (m_0g\lambda,m_0\lambda) \in \overline{(m_0A,m_0)T}(\lambda \times \lambda)\}$$

$$= \{g \mid g \in G, \exists m_0 \in M : (m_0g\lambda,m_0\lambda) \in \overline{\Delta(M)(A \times e)}(\lambda \times \lambda)\}.$$

Whenever $A = \emptyset$, set $\mathrm{cls}_\sigma A = \emptyset$.

3·15.12 LEMMA: *The mapping $A \mapsto \mathrm{cls}_\sigma A$ $(A \subset G)$ satisfies the following conditions:*

(a): $A \subset \mathrm{cls}_\sigma A \subset \bar{A}$;

(b): *If* $A \subset B$, *then* $\mathrm{cls}_\sigma A \subset \mathrm{cls}_\sigma B$;

(c): $\mathrm{cls}_\sigma(\mathrm{cls}_\sigma A) = \mathrm{cls}_\sigma A$;

(d): $\mathrm{cls}_\sigma(A \cup B) = \mathrm{cls}_\sigma A \cup \mathrm{cls}_\sigma B$.

In other words, the map $A \mapsto \mathrm{cls}_\sigma A$ is a closure operator for some topology σ on G.

Proof: The statements (a) and (b) follow immediately from the definitions. It is also clear that

$$\overline{(mA,m)T}(\lambda \times \lambda) = \overline{(m\,\mathrm{cls}_\sigma A,m)T}(\lambda \times \lambda),$$

and hence (c) holds. The inclusion $\mathrm{cls}_\sigma A \cup \mathrm{cls}_\sigma B \subset \mathrm{cls}_\sigma(A \cup B)$ is a consequence of (b). Let $g \in \mathrm{cls}_\sigma(A \cup B)$; then

$$(mg\lambda,m\lambda) \in \overline{(m(A \cup B),m)T}(\lambda \times \lambda) = \overline{(mA,m)T \cup (mB,m)T}(\lambda \times \lambda) = \quad \text{(Contd)}$$

(Contd) $= \overline{(mA,m)T}(\lambda\times\lambda) \cup \overline{(mB,m)T}(\lambda\times\lambda)$;

therefore $g \in \mathrm{cls}_\sigma A \cup \mathrm{cls}_\sigma B$. ▮

3·15.13 LEMMA: *If* $\emptyset \neq A \subset G$, *then:*

$$\mathrm{cls}_\sigma A = \{g \mid g \in G, (\forall m \in M)(\forall \alpha \in \mathfrak{U}[X])(\exists\, h \in A) : (mg\lambda, mh\lambda) \in \alpha T\}.$$

Proof: Let $m \in M$. If $g \in \mathrm{cls}\, A$, then $(mg\lambda, mh\lambda) \in \overline{(mA,m)T}(\lambda\times\lambda)$ Hence there exists a pair $(m_1,m_2) \in \overline{(mA,m)T}$ with $m_1\lambda = mg\lambda$, $m_2\lambda = m\lambda$. Since the set $\overline{(mA,m)T}$ is closed and invariant, we may assume that $m_2 = m$ and (m_1,m) is an almost periodic point of the transformation group $(M\times M,T)$. Hence there exists an element $k \in G$ with $m_1 = mk$. Thus $(mg\lambda,m\lambda) = (mk\lambda,m\lambda)$ and $(mk,m) \in \overline{(mA,m)T}$. There exist nets $\{t_n\}$, $t_n \in T$ and $\{h_n\}$, $h_n \in A$, such that $mt_n \to m$, $mh_nt_n \to mk$. Thus $mkt_n = mt_nk \to mk$ and hence $mgt_n\lambda = mg\lambda t_n = mk\lambda t_n = mkt_n\lambda \to mk\lambda = mg\lambda$, $mh_nt_n\lambda \to mk\lambda = mg\lambda$. Therefore for any index $\alpha \in \mathfrak{U}[X]$ there exists a t with $(mgt_n\lambda, mh_nt_n\lambda) \in \alpha$. Consequently $(mg\lambda, mh_n\lambda) \in \alpha T$.

Conversely, let $g \in G$ be an element such that for every $m \in M$ and every $\alpha \in \mathfrak{U}[X]$ there is an element $h \in A$ satisfying the condition $(mg\lambda, mh\lambda) \in \alpha T$. By the minimality of the transformation group (M,T) and the uniform integral continuity (Lemma 1·1.5) we can find nets $\{t_n\}$, $t_n \in T$, and $\{h_n\}$, $h_n \in A$, such that $mgt_n\lambda \to mg\lambda$, $mh_nt_n\lambda \to mg\lambda$, $mt_n \to m$. Then $mt_n\lambda \to m\lambda$ and therefore $(mg\lambda, m\lambda) \in \overline{(mA,m)T}(\lambda\times\lambda)$. Thus $g \in \mathrm{cls}\, A$. ▮

3·15.14 LEMMA: *Let* $m_0, m_1 \in M$ *be such that* $(m_0\lambda, m_1\lambda) \in P = P(X,T)$. *Then the family* $\{V_\alpha(m_0,m_1) \mid \alpha \in \mathfrak{U}[X]\}$, *where*

$$V_\alpha(m_0,m_1) = \{g \mid g \in G, (m_0g\lambda, m_0\lambda) \in \alpha T, (m_0g\lambda, m_1\lambda) \in \alpha T\},$$

is a base of σ*-neighbourhoods of the identity of* G.

Proof: Let $\alpha \in \mathfrak{U}[X]$ be an index. Denote $V_\alpha = V_\alpha(m_0,m_0)$. Then $V_\alpha(m_0,m_1) \subset V_\alpha$. Choose $\beta \in \mathfrak{U}[X]$ such that $\beta^2 \subset \alpha$, $\beta = \beta^{-1}$. We shall show that $V_\beta \subset V_\alpha(m_0,m_1)$. Let $g \in V_\alpha$, *i.e.*, $(m_0g\lambda, m_0\lambda) \in \beta T$. Since $(m_0g\lambda, m_0\lambda)$ is an almost periodic point of the transformation group $(X\times X,T)$, there exist subsets $A \subset T$, $K \subset T$ such that K is compact, $AK = T$ and $(m_0g\lambda, m_0\lambda)A \subset \beta$. Hence for any $t \in T$ we can choose an element $k_t \in K$ with $(m_0g\lambda, m_0\lambda)tk_t{}^{-1} \in \beta$. As X is compact and $(m_0\lambda, m_1\lambda) \in P(X,T)$, we have $(m_0\lambda, m_1\lambda)t_0K^{-1} \subset \beta$ for some $t_0 \in T$. Then $(m_0g\lambda, m_1\lambda)t_0k_t{}^{-1} \in \beta^2 \subset \alpha$, and hence $(m_0g\lambda, m_1\lambda) \in \alpha T$. Therefore $g \in V_\alpha(m_0,m_1)$.

Now we shall show that if α is open in $X\times X$, then the set V_α is σ-open. Let $h \in \mathrm{cls}_\sigma(G\setminus V_\alpha)$, that is:

$$(m_0h\lambda, m_0\lambda) \in \overline{(m_0(G\setminus V_\alpha), m_0)T}(\lambda\times\lambda).$$

If $g \in G\setminus V_\alpha$, then $(m_0g\lambda, m_0\lambda) \in X\times X\setminus \alpha T$; hence $(m_0h\lambda, m_0\lambda) \in X\times X\setminus\alpha T$, and consequently $h \in G\setminus V_\alpha$. Thus $\mathrm{cls}_\sigma(G\setminus V_\alpha) = G\setminus V_\alpha$.

Finally, let U be any open σ-neighbourhood of the identity $e \in G$. The set $F \equiv G \setminus U$ is σ-closed. We shall prove that

$$(m_0 F, m_0)T(\lambda \times \lambda) \cap \alpha_0 = \emptyset,$$

for some $\alpha_0 \in \mathfrak{U}[X]$. Suppose the contrary holds. Then for each $\alpha \in \mathfrak{U}[X]$ there exist elements $t_\alpha \in T$ and $g_\alpha \in F$ such that $(m_0 g_\alpha \lambda, m_0 \lambda)t_\alpha \in \alpha$. Without loss of generality we may suppose that the net $\{m_0 t_\alpha\}$ converges to some point $n \in M$. Then $\{m_0 t_\alpha \lambda\} \to n\lambda$ and $\{m_0 g_\alpha t_\alpha \lambda\} \to n\lambda = ne\lambda$. This means that $e \in \mathrm{cls}_\sigma F = F$, contradicting the fact $e \in U = G \setminus F$. Thus there exists an index $\alpha_0 \in \mathfrak{U}[X]$ with $(m_0 F, m_0)(\lambda \times \lambda) \cap \alpha_0 T = \emptyset$, and hence $V_{\alpha_0}(m_0, m_1) \subset V_{\alpha_0} \subset U$. ▮

3·15.15 LEMMA: *The map* $L_h : (G,\sigma) \to (G,\sigma)$ *defined by* $gL_h = hg$ *(*$g \in G$*) is continuous for every* $h \in G$.

Proof: Let $\emptyset = A \subset G$, $\mathrm{cls}_\sigma A = A$ and $h \in G$. It will suffice to show that $\mathrm{cls}_\sigma(hA) = hA$. Let $g \in \mathrm{cls}_\sigma(hA)$ and $m \in M$. By Lemma 3·15.13, for the point $mh^{-1} \in M$ and the index $\alpha \in \mathfrak{U}[X]$ we can find an element $a \in A$ such that $(mh^{-1}g\lambda, mh^{-1}ha\lambda) \in \alpha T$, *i.e.*, $(mh^{-1}g\lambda, ma\lambda) \in \alpha T$. By Lemma 3·15.13, $h^{-1}g \in \mathrm{cls}_\sigma A = A$. Hence $g \in hA$. ▮

3·15.16 Let us introduce another topology on the group G. If $\emptyset = A \subset G$, then we define (taking Lemma 3·15.10 into account):

$$\mathrm{cls}\, A = \{g \mid g \in G, (mg, m) \in \overline{(mA, m)T}, m \in M\}$$

$$= \{g \mid g \in G, \exists m_0 \in M : (m_0 g, m_0) \in \overline{(m_0 A, m_0)T}\}$$

$$= \{g \mid g \in G, \exists m_0 \in M : (m_0 g, m_0) \in \overline{\Delta(M)(A \times e)}\} \subset \mathrm{cls}_\sigma A.$$

If $A = \emptyset$, then we set $\mathrm{cls}\, A = \emptyset$.

3·15.17 LEMMA: *The map* $A \mapsto \mathrm{cls}_\sigma A$ *(*$A \subset G$*) is a closure operator for some topology* τ *on* G *with* $\tau \supset \sigma$. *If* $\emptyset \neq A \subset G$, *then*

$$\mathrm{cls}_\sigma A = \{g \mid g \in G, (\forall m \in M)(\forall \alpha \in \mathfrak{U}[M])(\exists h \in A) : (mg, m) \in \alpha T\}. \quad (15.4)$$

Let, further, $m_0, m_1 \in M$ *and* $(m_0, m_1) \in P(M,T)$. *Then the system* $\{W_\alpha(m_0, m_1) \mid \alpha \in \mathfrak{U}[M]\}$, *where*

$$W_\alpha(m_0, m_1) = \{g \mid g \in G, (m_0 g, m_0) \in \alpha T, (m_0 g, m_1) \in \alpha T\},$$

is a base for the τ*-neighbourhoods of the identity* $e \in G$. *Moreover, for each* $h \in G$ *the map* $L_h : (G,\tau) \to (G,\tau)$ *is continuous.*

Proof: The operator $A \mapsto \mathrm{cls}_\tau A$ is a particular case of the operator $A \mapsto \mathrm{cls}_\sigma A$, when $X = M$ and $\lambda : M \to M$ is the identity map. Hence this lemma is a corollary of Lemmas 3·15.12-15. ▮

3·15.18 LEMMA: *(G,τ) is compact and is a T_1 space.*

Proof: Let $\{A_\alpha\}$ be a filter of τ-closed subsets $A_\alpha \subset G$ and let $m_0 \in M$. Then $\{\overline{m_0A_\alpha}\}$ is a filter of closed subsets of the compact space M; therefore $\cap\overline{m_0A_\alpha} \neq \emptyset$. Let $m_1 \in \cap\overline{m_0A_\alpha}$; then

$$\overline{(m_1,m_0)T} \subset \bigcap_\alpha \overline{(\overline{m_0A_\alpha},m_0)T} = \bigcap_\alpha \overline{(m_0A_\alpha,m_0)T}.$$

Let (m_2,m_0) be an almost periodic point of the transformation group $(M \times M,T)$ belonging to the set $\overline{(m_1,m_0)T} \subset R_\mu$. In view of the regularity of the extension $\mu:M \to Y$, there exists an element $g_0 \in G$ with $m_2 = m_0g_0$. Then

$$(m_0g_0,m_0) \in \bigcap_\alpha \overline{(m_0A_\alpha,m_0)T};$$

hence $g_0 \in \cap \mathrm{cls}_\tau A_\alpha = \cap A_\alpha$. Thus the space (G,τ) is compact.

Let $A = \{h\}$ and $m \in M$. If $(mg,m) \in \overline{(mh,m)T}$, then there exists a net $\{t_n\}$, $t_n \in T$ such that $mt_n \to m$ and $mht_n \to mg$. But $mht_n = mt_nh \to mh$, hence $mh = mg$ $(m \in M)$, *i.e.*, $h = g$. Thus $\mathrm{cls}_\sigma\{h\} = \{h\}$ $(h \in G)$. ▮

3·15.19 LEMMA: $\overline{\Delta(M)(A \times e)} = \overline{\Delta(M)(e \times A^{-1})}$ *$(\emptyset \neq A \subset G)$.*

Proof: Let $m \in M$, $g \in A$ and $n = mg$. Then $(mg,m) = (n,ng^{-1}) \in \overline{\Delta(M)(e \times A^{-1})}$, and hence $\overline{\Delta(M)(A \times e)} \subset \overline{\Delta(M)(e \times A^{-1})}$. The proof of the reverse inclusion is similar. ▮

3·15.20 LEMMA: *The maps $R_h:G \to G$ $(h \in G)$, where $gR_h = gh$ $(g \in G)$, and the map $g \mapsto g^{-1}$ $(g \in G)$ are continuous in the τ-topology on G. (Hence they are τ-homeomorphisms).*

Proof: Let $\mathrm{cls}_\tau A = A$ and $h \in G$. Let us prove that $\mathrm{cls}_\tau(Ah) = Ah$. Let $g \in \mathrm{cls}\,(Ah)$, *i.e.*, $(mg,m) \in \overline{(mAh,m)T}$. Then $(mgh^{-1},m) = (mg,m)\circ(h^{-1},e) \in \overline{(mAh,m)(h^{-1},e)T} = \overline{(mA,m)T}$. Thus, $gh^{-1} \in \mathrm{cls}_\tau A = A$; hence $g \in Ah$. It follows that the map $R_h:(G,\tau) \to (G,\tau)$ is continuous.

Let $\mathrm{cls}_\tau A = A$. We shall prove that $\mathrm{cls}_\tau(A^{-1}) = A^{-1}$. If $g \in \mathrm{cls}_\tau(A^{-1})$, then $(mg,m) \in \overline{\Delta(M)(A^{-1} \times e)} = \overline{\Delta(M)(e \times A)}$ by Lemma 3·15.19. Put $n = mg$; then $(n,ng^{-1}) \in \overline{\Delta(M)(e \times A)}$, and hence $(ng^{-1},n) \in \overline{\Delta(M)(A \times e)}$. Thus $g^{-1} \in \mathrm{cls}_\tau A = A$, and therefore $g \in A^{-1}$. ▮

3·15.21 Consider the commutative diagram

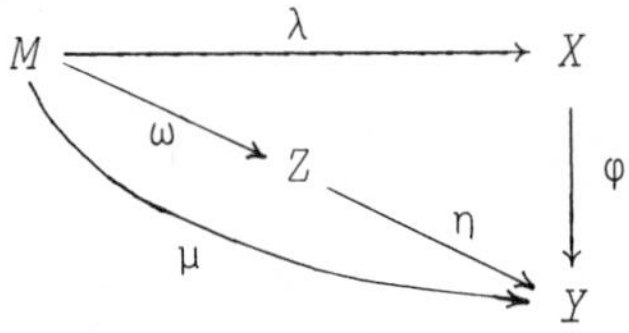

where $\mu:(M,T) \to (Y,T)$ is a regular extension. Let $G(Z)$ denote the set of all automorphisms $g:(M,T) \to (M,T)$ which preserve the fibers of the homomorphism $\omega:M \to Z$. In other words,

$$G(Z) = \{g | g \in G, mg\omega = m\omega, m \in M\}$$

$$= \{g | g \in G, \exists m_0 \in M: m_0 g\omega = m_0\omega\}.$$

3·15.22 LEMMA: *$G(Z)$ is a τ-closed subgroup of the group $G = G(Y)$ and, moreover, $G(X) \subset G(Z)$.*

Proof: Clearly $G(Z)$ is a subgroup of G and $G(Z) \supset G(X)$. Let $h \in \mathrm{cls}_\tau G(Z)$. Then $(mh,m) \in \overline{(mG(Z),m)T} \subset R_\omega$ $(m \in M)$, and hence $mh\omega = m\omega$. Thus $h \in G(Z)$ by the definition of $G(Z)$. ▌

3·15.23 Let K be a τ-closed subgroup of G satisfying the condition $K \supset G(X)$. The question arises whether there does exist a transformation group (Z,T) satisfying the commutative diagram (15.5) and the condition $G(Z) = K$. In general, the answer seems to be in the negative. We shall discuss this question under the additional hypothesis that φ is distal.

3·15.24 LEMMA: *Let $\varphi:(X,T) \to (Y,T)$ be a distal extension and let K be a τ-closed subgroup of $G = G(X,Y)$ such that $K \supset G(X)$. Then*

$$S(K) = \{(mg,m) | m \in M(X,Y), g \in K\},$$

is a closed invariant equivalence relation and, moreover, $R(\lambda_1) \subset S(K) \subset R(\mu_1)$.

Proof: Clearly $S(K)$ is an invariant equivalence relation. Let $(m_1,m_2) \in \overline{S(K)} \equiv \overline{\Delta(M)(K \times e)}$. Since $S(K) \subset R(\mu_1)$, there exists an automorphism $h \in G$ with $m_1 = m_2 h$ (Lemma 3·15.6). Thus $(m_2 h, m_2) \in \overline{\Delta(M)(K \times e)}$, hence $h \in \mathrm{cls}_\tau K = K$. This means that $(m_1,m_2) = (m_2 h, m_2) \in S(K)$. Thus the relation $S(K)$ is closed. It follows from $K \supset G(X)$ and Lemma 3·15.6 that $S(K) \supset R(\mu_1)$. ▌

Observe that $M(X,Y)/S(K)$ coincides with the quotient space $M(X,Y)/K$ of K-orbits $\{mK | m \in M(X,T)\}$.

3·15.25 THEOREM: *Let $\varphi:(X,T) \to (Y,T)$ be a distal minimal extension, let $M = M(X,Y)$ and $G = G(X,Y)$. Let $[X,Y]$ denote the set of all transformation groups (Z,T) satisfying a commutative diagram of the form* (15.3). *Let $H = G(X)$ and $[H,G] = \{K | K$ is a τ-closed subgroup of $G, K \supset H\}$. Define a map Φ: $[X,Y] \to [H,G]$ by $Z\Phi = G(Z)$ and a map $\Psi:[H,G] \to [X,Y]$ by $K\Psi = (M(X,Y)/K,T)$. Then $\Phi\circ\Psi$ is the identity transformation of $[X,Y]$ and $\Psi\circ\Phi$ is the identity transformation of $[H,G]$. Thus Φ,Ψ establish a Galois connection between $[X,Y]$ and $[H,G]$.*

Proof: Let $(Z,T) \in [X,Y]$ and $K = G(Z)$. By Lemma 3·15.22, $K \in [H,G]$. Lemma 3·15.23 asserts that $M(X,Y)/K \equiv Z_1 \in [X,Y]$. Consider the commutative diagram (15.5), where $M = M(X,Y)$. Since φ is a distal extension, $\mu:M \to Y$ is a homogeneous extension (Lemma 3·15.6). Therefore if $m_1\omega = m_2\omega$, then there exists an automorphism

$g \in A(M,T)$ such that $m_1 = m_2 g$. Then $g \in G(Z)$, according to the definition of the group $G(Z)$. Thus $R_\omega = \{(mg,m) | m \in M, g \in K = G(Z)\}$; hence $Z \equiv M(X,Y)/R_\omega = M(X,Y)/K \equiv Z_1$. This means that $\Phi \circ \Psi : [X,Y] \to [X,Y]$ is the identity map.

Let $K \in [H,G]$. By Lemma 3·15.24, $S(K) = \{(mg,m) | m \in M(X,Y), g \in K\}$ is a closed invariant equivalence relation and $M(X,Y)/S(K) = M(X,Y)/K$. Hence it follows that $G(M(X,Y)/K) \supset K$. On the other hand $G(M(X,Y)/K) \subset \mathrm{cls}_\tau K = K$. Thus $G(M(X,Y)/K) = K$. This proves that $\Psi \circ \Phi$ is the identity map of $[H,G]$.∎

3·15.26 In the remainder of this section we shall introduce and study a certain subgroup $H(G)$ of the group G, and then apply the results obtained in the investigation of extensions $\varphi:(X,T) \to (Y,T)$ satisfying the condition $Q(\overline{R_\varphi \mathcal{J}}) = \overline{R_\varphi \mathcal{J}}$. Finally, we shall prove a structure theorem which generalizes Theorem 3·14.36 on the structure of almost distal extensions.

The next lemma is a generalization of Lemma 1·4.11 to the case of T_1 spaces.

3·15.27 LEMMA: *Let E be a compact T_1 space provided with a semigroup structure in such a way that for each $x \in E$, the map $R_x : E \to E$, defined by $yR_x = yx$, is continuous and closed. Then there exists an idempotent $u \in E$.*

Proof: Let $\mathcal{A}$ be the system of all non-empty closed subsets $S \subset E$ with $S^2 \subset S$. Note that $E \in \mathcal{A}$. Order $\mathcal{A}$ by inclusion. It is easy to verify that Zorn's Lemma is applicable to the system $\mathcal{A}$, and therefore $\mathcal{A}$ contains a minimal element $M \in \mathcal{A}$.

Let $u \in M$; then $Mu \subset MR_u$ is a non-empty closed subset of M. Since $MuMu \subset MMMu \subset Mu$, we have $Mu \in \mathcal{A}$. Hence $Mu = M$, by the minimality of M. Thus there exists an element $x \in M$ with $u = xu$. The set $W = \{y | y \in M, yu = u\} \equiv uR_u^{-1}$ is non-empty and closed in M. Moreover, $W^2 \subset W$. Thus $W = M$, according to the definition of M. Hence $u^2 = uu = u$.∎

3·15.28 COROLLARY: *Let G be a compact T_1 space provided with a group structure in such a way that all the maps $R_x : G \to G$ $(x \in G)$ are continuous. If M is a non-empty closed subset of G and $M^2 \subset M$, then M is a subgroup of G.*

Proof: Since G is a group, the maps R_x $(x \in G)$ are homeomorphisms, hence they are closed. Let $x \in M$. The set $E \equiv Mx$ satisfies all the hypotheses of Lemma 3·15.27. Therefore Mx contains an idempotent. But the only idempotent in the group G is the identity e. Hence $e \in Mx$ and therefore $x^{-1} \in M$. Thus M is a subgroup of the group G.∎

3·15.29 LEMMA: *Let G be a group provided with two topologies τ and σ, satisfying the conditions:*

(1): $\sigma \subset \tau$;

(2): *(G,τ) is a compact T_1 space;*

(3): *The maps* $R_a:(G,\tau) \to (G,\tau)$ *and* $L_a:(G,\tau) \to (G,\tau)$ *are continuous for all* $a \in G$;

(4): *The map* $L_a:(G,\sigma) \to (G,\sigma)$ *is continuous for every* $a \in G$.

Let $\mathcal{N}_\sigma$ *and* $\mathcal{N}_\tau$ *denote neighbourhood bases at the identity* $e \in G$ *in the topologies* σ *and* τ *respectively, and let*

$$H = \bigcap [\mathrm{cls}_\sigma V \mid V \in \mathcal{N}_\sigma], \qquad L = \bigcap [\mathrm{cls}_\tau V \mid V \in \mathcal{N}_\sigma].$$

Then L *is a subgroup of* G *and* $H = L$.

Proof: Let $x,y \in L$. We shall prove that $xy \in L$. Let $N \in \mathcal{N}_\sigma$, and $R \in \mathcal{N}_\tau$. Then $x \in \mathrm{cls}_\tau N$. By (3), Rx is a τ-neighbourhood of x; hence $Rx \cap N \neq \emptyset$. Thus $rx \in N$ for some element $r \in R$. In view of (4) we can choose $S \in \mathcal{N}_\sigma$ so that $rxS \subset N$. Hence, by (3),

$$rx(\mathrm{cls}_\tau S) \subset \mathrm{cls}_\tau N;$$

therefore $rxy \in \mathrm{cls}_\tau N$. Thus $Rxy \cap \mathrm{cls}_\tau N \neq \emptyset$. But Rxy is a τ-neighbourhood of xy by (3). Therefore $Rxy \cap N \neq \emptyset$. Since $R \in \mathcal{N}_\tau$ and $N \in \mathcal{N}_\sigma$ are arbitrary, we get $xy \in L$. By (2) and (3), the group G and the set $M = L$ satisfy all the hypotheses of Corollary 3·15.28; hence L is a subgroup of G.

Now we shall prove that $L = H$. Since $\sigma \subset \tau$, we have $\mathrm{cls}_\tau N \subset \mathrm{cls}_\sigma N$ $(N \in \mathcal{N}_\sigma)$; therefore $L \subset H$. Let $h \in H$ and $N,R \in \mathcal{N}_\sigma$. By hypothesis (4), hN is a σ-neighbourhood of h; hence $hN \cap R \neq \emptyset$. Consequently $h(\mathrm{cls}_\tau N) \cap \mathrm{cls}_\tau R \neq \emptyset$ by condition (3). Hence it follows that

$$\{h(\mathrm{cls}_\tau N) \cap \mathrm{cls}_\tau R \mid R \in \mathcal{N}_\sigma\}$$

is a filter of τ-closed subsets. Taking condition (2) into account, we conclude that $h(\mathrm{cls}_\tau N) \cap L \neq \emptyset$. Since N is here an arbitrary σ-neighbourhood of the identity $e \in G$, we conclude from conditions (2) and (3) that $hL \cap L \neq \emptyset$. Since L is a subgroup of G, we get $h \in L$, and hence $H \subset L$. Therefore $H = L$. ▮

3·15.30 Let

$$\begin{array}{ccc} M & \xrightarrow{\ \lambda\ } & X \\ & \searrow^{\mu} & \downarrow \varphi \\ & & Y \end{array} \qquad (15.6)$$

be a commutative regular diagram, let G be the group of all automorphisms $g:(M,T) \to (M,T)$ preserving the fibers of the map $\mu:M \to Y$ and let σ and τ be the topologies on G defined in Subsections 3·15.11,12 and Subsections 3·15.16,17 respectively. Let $x_0 \in X$, $y_0 = x_0\varphi$. The set $\{m \mid m \in M, m\mu = y_0\}$ is compact and invariant under the action of the group G on M; hence it contains a G-minimal set M_0. If m_0 is a specified point of M_0, then $\overline{m_0G} = M_0$.

3·15.31 LEMMA: *Let* $g \in G$ *and let* V *be a neighbourhood of the point* $m_0 g$ *in* M. *Denote* $\{h \mid h \in G, m_0 h \in V\}$ *by* $V(G)$. *Then*

$$\operatorname{int}_\tau \operatorname{cls}_\tau V(G) \neq \emptyset.$$

Proof: Since $\overline{m_0 G}$ is a G-minimal set, $\overline{m_0 G} \subset VG$. There exists a finite family $F \subset G$ with $\overline{m_0 G} \subset VF$. Therefore $G = V(G)F$ and, moreover, $G = \operatorname{cls}_\tau V(G) \cdot F$. Sets of the form $\operatorname{cls}_\tau V(G) \cdot \ell$ ($\ell \in F$) are τ-closed and mutually τ-homeomorphic by Lemma 3·15.19. This and the equality $G = \operatorname{cls}_\tau V(G) \cdot F$ imply the required assertion because F is finite.▌

3·15.32 LEMMA: *Let* $E \subset G$ *and let* $\operatorname{cls}_\tau \operatorname{int}_\tau E = G$. *Then* $\overline{m_0 E} = \overline{m_0 G}$.

Proof: Let $g \in G$ and let V be an arbitrary neighbourhood of $m_0 g$ in M. By Lemma 3·15.31,

$$\operatorname{int}_\tau \operatorname{cls}_\tau V(G) \neq \emptyset.$$

Thus

$$\operatorname{int}_\tau \operatorname{cls}_\tau V(G) \cap \operatorname{int}_\tau E \neq \emptyset,$$

and, moreover, $\operatorname{cls}_\tau V(G) \cap \operatorname{int}_\tau E \neq \emptyset$. Hence $V(G) \cap \operatorname{int}_\tau E \neq \emptyset$, and therefore $V(G) \cap E \neq \emptyset$. It follows from the very definition of $V(G)$ that $m_0 E \cap E \neq \emptyset$. Since V is an arbitrary neighbourhood of $m_0 g$, we have $m_0 g \in \overline{m_0 E}$. Thus $\overline{m_0 G} \subset \overline{m_0 E} \subset \overline{m_0 G}$, whence $\overline{m_0 E} = \overline{m_0 G}$.▌

3·15.33 LEMMA: *Let* σ *and* τ *be the topologies on* G *defined in Subsections* 3·15.11,12 *and Subsections* 3·15,17. *Define* H *and* L *as in Lemma* 3·15.29 *and suppose that* $H = G$ *and* $V \in \mathcal{N}_\sigma$. *Then* $\overline{m_0 V} = \overline{m_0 G}$.

Proof: By Lemma 3·15.29, $L = H = G$. Denote $\operatorname{int}_\tau V$ by E. Since $E \in \operatorname{int}_\sigma V \subset \operatorname{int}_\tau V \equiv E$, E is a τ-neighbourhood of the identity $e \in G$, and since $L = G$, we get $G = \operatorname{cls}_\tau E \equiv \operatorname{cls}_\tau \operatorname{int}_\tau V$. By Lemma 3·15.32, $\overline{m_0 E} = \overline{m_0 G}$. Thus $\overline{m_0 V} = \overline{m_0 G}$.▌

3·15.34 LEMMA: *Let* $H = G$ *and suppose that there exist indices* $\alpha_i \in \mathcal{U}[X]$ *(*$i = 1, 2, \ldots$ *) such that*

$$P \cap R_\varphi = \cap [\alpha_i T \cap R_\varphi \mid i = 1, 2, \ldots].$$

If $\overline{m_0 G}$ *is a minimal set of the transformation group* (M, G) *and* $(m_0 \lambda, m_1 \lambda) \in P \cap R_\varphi$, *then the set*

$$\{m \mid m \in \overline{m_0 G}, (m\lambda, m_0 \lambda) \in P \cap R_\varphi, (m\lambda, m_1 \lambda) \in P \cap R_\varphi\},$$

is residual in $\overline{m_0 G}$.

Proof: Put

$$V_i = \{g \mid g \in G, (m_0 g \lambda, m_0 \lambda) \in \alpha_i T, (m_0 g \lambda, m_0 \lambda) \in \alpha_i T\}.$$

By Lemma 3·15.14, $V_i \in \mathcal{N}_\sigma$. Thus $\overline{m_0 V_i} = \overline{m\ G}$ ($i = 1,2,\ldots$) (Lemma 3·15.33). It is clear that

$$m_0 V_i = \{m \mid m \in m_0 G, (m\lambda, m_0\lambda) \in \alpha_i T, (m\lambda, m_0\lambda) \in \alpha_i T\}.$$

Set

$$N_i = \{m \mid m \in \overline{m_0 G}, (m\lambda, m_0\lambda) \in \alpha_i T, (m\lambda, m_0\lambda) \in \alpha_i T\}.$$

Without loss of generality we may suppose that the α_i are open in $X \times X$. Then each set N_i is open in $\overline{m_0 G}$. Since $m_0 V_i \subset N_i$, we have $\overline{N_i} = \overline{m_0 G}$. Thus the sets N_i ($i = 1,2,\ldots$) are open and dense in $\overline{m_0 G}$. Therefore the set $E = \bigcap [N_i \mid i = 1,2,\ldots]$ is residual in $\overline{m_0 G}$. If $m \in E$, then

$$(m\lambda, m_j\lambda) \in \bigcap_{i=1}^{\infty} \alpha_i T \cap R_\varphi = P \cap R_\varphi, \qquad (j = 1,2).$$

Note that $m_0\lambda\varphi = m_1\lambda\varphi = m\lambda\varphi$ whenever $m \in E \subset \overline{m_0 G}$. ∎

REMARK: If the space X is metrizable, then the requirement imposed on $P \cap R_\varphi$ in the preceding theorem is satisfied.

3·15.35 LEMMA: $\overline{R_\varphi \mathcal{J}} = \overline{\{(m_0 g t\lambda, m_0 t\lambda) \mid t \in T, g \in G\}}$ *($m_0 \in M$).*

Proof: Let $(x,y) \in R_\varphi \mathcal{J}$. Since the diagram (15.6) is regular, there exist $m \in M$ and $g_0 \in G$ such that $m\lambda = y$ and $m g_0 \lambda = x$. Therefore

$$(x,y) \in \overline{(m g_0, m)T}(\lambda \times \lambda) = \overline{(m_0 g_0, m_0)T}(\lambda \times \lambda)$$
$$\subset \overline{\{(m_0 g t\lambda, m_0 t\lambda) \mid t \in T, g \in G\}}.$$

Thus

$$\overline{R_\varphi \mathcal{J}} \subset \overline{\{(m_0 g t\lambda, m_0 t\lambda)\ t \in T, g \in G\}}, \qquad (m_0 \in M).$$

The reverse inclusion follows from the definition of G. ∎

3·15.36 LEMMA: *If* $Q(\overline{R_\varphi \mathcal{J}}) = \overline{R_\varphi \mathcal{J}}$, *then* $H = L = G$.

Proof: In view of Lemma 3·15.29, it suffices to show that $G \subset H$. Let $g_0 \in G$ and $m_0 \in M$. Then we have by the preceding lemma,

$$(m_0 g_0, m_0) \in R_\varphi \mathcal{J} \subset Q(\overline{R_\varphi \mathcal{J}}) = Q(\overline{\{(m_0 g t\lambda, m_0 t\lambda) \mid t \in T, g \in G\}})$$
$$= Q(\{(m_0 g t\lambda, m_0 t\lambda) \mid t \in T, g \in G\})$$
$$= \bigcap_{\alpha \in \mathcal{U}[X]} \overline{\{(m_0 g t\lambda, m_0 t\lambda) \mid t \in T, g \in G\} \cap \alpha T}.$$

It follows from Lemma 3·15.14 that the family $\{V_\alpha \mid \alpha \in \mathcal{U}[X]\}$, where

$V_\alpha = \{g|g \in G, (m_0 g\lambda, m_0\lambda) \in \alpha T\}$ is a base for the σ-neighbourhoods of the identity $e \in G$. The equalities (15.7) imply that

$$g_0 \in \{g|g \in G, (m_0 g\lambda, m_0\lambda) \in \overline{(m_0 V_\alpha, m_0)T}(\lambda \times \lambda)\} \equiv \mathrm{cls}_\sigma V_\alpha,$$

for each index $\alpha \in \mathfrak{U}[X]$. Hence $g_0 \in \bigcap[\mathrm{cls}_\sigma V_\alpha | \alpha \in \mathfrak{U}[X]] \equiv H$. Thus $G \subset H$. ▌

3·15.37 Let I be a minimal right ideal of the semigroup $E(X,T)$, let $x_0 \in X$, and let $\lambda:(I,T) \to (X,T)$ be the homomorphism defined by $\xi\lambda = x_0\xi$ $(\xi \in I)$ and $\mu = \lambda\circ\varphi$. As was shown in Subsection 3·15.3, the diagram

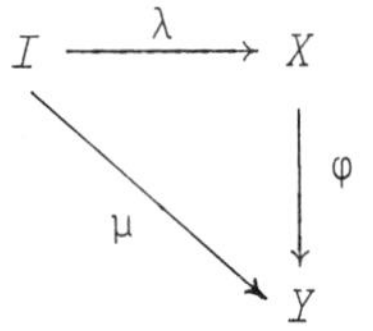

is regular. In the sequel, when using the results of Lemmas 3·15.33-36 we shall be assuming that $M = I$.

The set $I_0 = \{\xi|\xi \in I, x_0\xi\varphi = x_0\xi\}$ is a closed subsemigroup of the semigroup I; hence there exists an idempotent $u \in I_0$. Thus $G_0 \equiv I_0 u$ is a subgroup of the group Iu with identity u.

3·15.38 LEMMA: *The group G of all automorphisms $g:(I,T) \to (I,T)$ that preserve the fibers of $\mu:I \to Y$ coincides with the group $\{L_s:I \to I | s \in G_0\}$.*

Proof: Clearly, $L_s \in G$ $(s \in G_0)$. If $\xi,\eta \in I_0$ and (ξ,η) is an almost periodic point of the transformation group $(I \times I,T)$, then there exists an idempotent $v \in I_0$ with $\xi v = \xi$, $\eta v = \eta$. Set q $(\xi v)(\eta v)^{-1}$, where $(\eta v)^{-1}$ denotes the inverse element in the group $I_0 v$ with identity v. Set $s = qu$. Then $\xi = q\eta = qu\eta = s\eta$ in accordance with Lemma 1·4.2, and, moreover, $s \in I_0 vu = I_0 u = G$. The map L_s is an autormorphism sending η to ϑ. ▌

3·15.39 LEMMA: *If $p \in \bar{G}_0$ and $p = pw$ for some idempotent $w \in I$, then $p \in G_0 w$.*

Proof: If $p \in \bar{G}_0 \subset I_0$, then $p = pw \in I_0 w = I_0 uw = G_0 w$, because $uw = w$ by Lemma 1·4.2. ▌

3·15.40 LEMMA: *There exists an idempotent $w \in \bar{G}_0$ such that $\overline{G_0 w} \equiv \overline{wG}$ is a minimal set under the action of $G = \{L_s | s \in G_0\}$ on I.*

Proof: The set $\bar{G}_0$ is closed in I and invariant under the action of the group G. Hence there exists a point $p \in \bar{G}_0$ such that $\overline{G_0 p} \equiv \overline{pG}$ is a minimal set of the transformation group (I,G). Choose an idempotent $w \in I$ so that $pw = p$. By the preceding lemma there is an element $g_0 \in G_0$ with $p = g_0 w$. Then $g_0{}^{-1}p = g_0{}^{-1}g_0 w = uw = w$, and hence $w \in g_0{}^{-1}\bar{G}_0 = \overline{g_0{}^{-1}G_0} = \bar{G}_0$. Since $\overline{G_0 p} \equiv \overline{pG}$ is a minimal set and $w = g_0^{-1}p \in \overline{G_0 p}$, we get $\overline{G_0 w} = \overline{G_0 p}$. Thus $\overline{G_0 w}$ is a minimal set of (I,G). ▮

3·15.41 LEMMA: *Let X be a compact space, (X,T) a minimal transformation group, $R \subset X \times X$ a both closed and open invariant equivalence relation and let $\varphi: X \to X/R \equiv Y$ denote the canonical projection. Suppose that*

(1); $R = \overline{R\mathfrak{I}}$;

(2): $Q^*(R) = R$;

(3): $\exists \alpha_i \in \mathfrak{U}[X]: P \cap R = \cap[\alpha_i T \cap R \mid i = 1,2,\ldots]$;

and suppose that at least one of the following conditions is satisfied:

(4): *There is a point $a \in X$ such that the set $\{x \mid x \in X, (x,a) \in P \cap R\}$ is finite or countable;*

(5): *$P \cap R$ is an equivalence relation.*

Then $R = \Delta$.

Proof: According to Theorem 3·13.15, $Q^*(R) = Q(R)$, hence it follows from (1) and (2) by Lemma 3·15.36 that $H = L = G$.

Suppose first that (4) is satisfied. Consider the commutative diagram in Subsection 3·15.37, where it is assumed that $M \equiv I$ is a specified minimal right ideal of $E = E(X,T)$ and the map $\lambda: I \to X$ is defined by $\xi\lambda = a\xi$ $(\xi \in I)$. Choose an idempotent $u \in I$ with $au = a$. Denote $I_0 = \{\xi \mid \xi \in I, a\xi\varphi = a\varphi\}$ by I_0. Then $u \in I_0$ and $G_0 \equiv I_0 u$ is a group with identity u. By Lemma 3·15.40 there exists an idempotent $w \in \overline{G_0}$ such that $\overline{G_0 w} \equiv \overline{wG}$ is a minimal set under the action of the group $G \equiv \{L_s \mid s \in G_0\}$ on I. Then $(a,aw) \in P \cap R$. Without loss of generality we may suppose that the indices α_i are open in $X \times X$. Put

$$N_i = \{m \mid m \in \overline{wG}, (m\lambda, u\lambda) \in \alpha_i T, (m\lambda, w\lambda) \in \alpha_i T\}$$

$$= \{m \mid m \in \overline{G_0 w}, (am, a) \in \alpha_i T, (am, aw) \in \alpha_i T\},$$

$$W_i = \{x \mid x \in \overline{aG_0 w}, (x,a) \in \alpha_i T, (x, aw) \in \alpha_i T\}.$$

It can be seen from the proof of Lemma 3·15.34 (by setting $m_1 = u$, $m_0 = w$) that the sets N_i $(i = 1,2,\ldots)$ are open in $\overline{wG}$ and $\overline{N_i} = \overline{wG}$. Clearly $W_i = N_i\lambda$; hence

$$\overline{W_i} = \overline{aG_0 w}, \qquad (i = 1,2,\ldots).$$

Since the sets W_i are open in $\overline{aG_0 w}$, the set $E_1 \equiv \cap[W_i \mid i = 1, 2,\ldots]$ is residual in $\overline{aG_0 w}$. By (3), $E_1 \equiv \{x \mid x \in \overline{aG_0 w}, (x,a) \in P \cap R$ $(x,aw) \in P \cap R\}$. Because of (4), the set E_1 is finite or countable. Set

$$E_2 = \{x \mid x \in E_1, \{x\} \text{ is open in } \overline{aG_0 w}\}.$$

Let us show that $\overline{E_2} = \overline{aG_0w}$. Suppose that, to the contrary, $E_1 \setminus E_2$ is a finite or countable residual subset of the non-empty open set $\overline{aG_0w} \setminus \overline{E_2}$. This contradicts the definition of E_2. Thus $\overline{E_2} = \overline{aG_0w}$. Since aG_0w is dense in $\overline{aG_0w}$ and each point $x \in E_2$ is open in $\overline{aG_0w}$, it follows that $E_2 \subset aG_0w$. Let $x \in E_2$. Then $x = agw$ for some $g \in G_0$. Since $x \in E_2 \subset E_1$, we get $(agw, aw) \in P \cap R$; but (gw, w) is an almost periodic point of $I \times I$, and hence $agw = aw$. Consequently $E_2 = \{aw\}$, and thus

$$\overline{aG_0w} = \overline{E_2} = \{aw\}. \tag{15.8}$$

Let $(a,x) \in R$. Then $x = a\xi$ for some $\xi \in I_0$. Since $\xi w \in I_0w = I_0uw = G_0w$, it follows from (15.8) that $xw = a\xi w = aw$. Hence $(x,a) \in P \cap R$. As X is a minimal set, we get $R\mathfrak{J} = \Delta$. Therefore the hypothesis (1) implies $R = \Delta$. This completes the proof in the first case.

Now suppose that (5) is satisfied. Take any point $a \in X$ and define M, λ, u, G_0, w as above. It will suffice to show that $\overline{aG_0w} = \{aw\}$. Define:

$$E = \{m \mid m \in \overline{wG}, (m\lambda, u\lambda) \in P \cap R\}$$

$$\equiv \{m \mid m \in \overline{G_0w}, (am, a) \in P \cap R\}.$$

By Lemma 3·15.34 the set E is residual in $\overline{G_0w}$. Let $g \in G_0$. Since $\xi \mapsto g\xi$ $(\xi \in M = I)$ is a homeomorphism of $\overline{G_0w}$ onto itself, we have $gE \cap E \neq \emptyset$. Hence there exists an element $s \in E$ with $gs \in E$. Thus $(as, a) \in P \cap R$, $(ags, a) \in P \cap R$. Hence $(ags, as) \in P \cap R$, because $P \cap R$ is an equivalence relation. Consequently $(ag, a) \in P \cap R$, but this is possible only when $ag = g$. In fact, $g \in G_0 = I_0u$; hence $g = gu$, whence (g,u) is an almost periodic point of $(I \times I, T)$. Thus $ag = a$ for each $g \in G_0$, and therefore $agw = aw$ $(g \in G_0)$, *i.e.*, $\overline{aG_0w} = aw$. ∎

3·15.42 THEOREM: *Let X be a compact metrix space, let (X,T) be a minimal transformation group and let $\varphi:(X,T) \to (Y,T)$ be an extension. Suppose that at least one of the following conditions is satisfied:*

(a): *There is a point $a \in X$ such that the set $\{x \mid x \in X, (x,a) \in P \cap R_\varphi\}$ is finite or countable;*

(b): *$P \cap R_\varphi$ is an equivalence relation (here P denotes the proximal relation in (X,T)).*

Then there exist a minimal transformation group (W,T) with a compact metrizable phase space and homomorphisms $p:(W,T) \to (X,T)$ and $q:(W,T) \to (Y,T)$ such that

(1): *p is a proximal extension;*

(2): *q is a PE-extension;*

(3): *$p \circ \varphi = q$.*

In particular, if there is at least one R-distal point, then p is almost automorphic and q is a V-extension.

Proof: The major part of the proof consists in showing the following proposition.

PROPOSITION: *Let X be a compact metric space and let $\varphi:(X,T) \to (Y,T)$ be a minimal extension. Then there exist compact metric spaces X^* and Y^*, minimal transformation groups (X^*,T), (Y^*,T) and homomorphisms η, ξ, φ^* such that*

(1): *The diagram*

$$\begin{array}{ccc} X & \xleftarrow{\ \eta\ } & X^* \\ {\scriptstyle\varphi}\downarrow & & \downarrow{\scriptstyle\varphi^*} \\ Y & \xleftarrow{\ \xi\ } & Y^* \end{array}$$

is commutative;

(2): *ξ and η are proximal extensions;*

(3): *The restriction of η to each fiber of the extension $\varphi^*:(X^*,T) \to (Y^*,T)$ is injective;*

(4): *φ^* is an open map and $\overline{R_{\varphi^*}\mathcal{J}} = R_{\varphi^*}$.*

To prove this proposition, let us consider a regular diagram of the form (15.1). Define

$$H = G(X) = \{g \mid g \in G \equiv G(Y), mg\lambda = m\lambda, m \in M\}.$$

By Lemma 3·15.22 $G(X)$ is a τ-closed subgroup of G. Define $B = \overline{(mH,m)T}$. Clearly B is a closed invariant relation on M. Consider now the transformation group $(2^M,T)$ induced by (M,T). If $A_0 \in 2^M$ and A_1,A_2 belong to the orbit closure of A_0 in 2^M, then $A_1 \cup A_2 \subset A_0$ implies that the points A_1 and A_2 are proximal in $(2^M,T)$. In fact, let

$$\mathcal{A} = \{A \mid A \in 2^M, A \in \Sigma(A_0), A \subset A_0\},$$

where $\Sigma(A_0)$ denotes the orbit-closure of A_0 in $(2^M,T)$. Note that $A_0 \in \mathcal{A}$. If $N_1 \supset N_2 \supset \cdots \supset N_\alpha \supset N_{\alpha+1} \supset \cdots$ $(N_\alpha \in \mathcal{A})$ and $N = \bigcap_\alpha N_\alpha$, then $N = \lim\{N_\alpha\}$ in 2^M. Since $N_\alpha \in \Sigma(A_0)$, $N = \lim\{N_\alpha\} \in \Sigma(A_0)$. Moreover, $N_\alpha \subset A_0$ implies $N \subset A_0$. Hence it follows by Zorn's lemma that $\mathcal{A}$ has a minimal element, say L.

As $L \in \Sigma(A_0)$, there exists a net $\{t_n\}$, $t_n \in T$, such that $\{A_0 t_n\} \to L$ in 2^M. Without loss of generality we may suppose that $\{A_1 t_n\} \to S_1$ and $\{A_2 t_n\} \to S_1$ in 2^M. As $A_1 \cup A_2 \subset A_0$, it follows that $S_1 \cup S_2 \subset L \subset A_0$. It is also clear that $S_1, S_2 \in \Sigma(A_0)$. Hence $S_1, S_2 \in \mathcal{A}$, and $S_1 \cup S_2 \subset L$; therefore $S_1 = S_2 = L$ by the minimality of L. Thus $\{A_1 t_n\} \to L$, and $\{A_2 t_n\} \to L$. Hence A_1 and A_2 are proximal.

Fix some point $m_0 \in M$ and, as usual, set

$$m_0 B = \{m \mid m \in M, (m_0,m) \in B\}.$$

It follows easily from the above considerations that the orbit closure of (m_0,m_0B) in $(M\times 2^M,T)$ contains a unique minimal set, say M_1. Let Z denote the image of M_1 under the projection $M\times 2^M \to 2^M$. We get the commutative diagram

$$\begin{array}{ccc} M & \xleftarrow{\;q_1\;} & M_1 \\ \mu\downarrow & & \downarrow\mu_1 \\ Y & \xleftarrow{\;p_1\;} & Z \end{array}$$

where q_1 and p_1 are proximal extensions.

Let us show that $m_0B = (m_0h)B = (m_0B)h$ $(h\in H)$. Indeed, if $m\in m_0hB$, then $(m_0h,m)\in B$, hence $(m_0,m) = (m_0h,m)(h^{-1},e)\in \overline{\Delta(M)\circ(H\times e)}(h^{-1},e) = \overline{\Delta(M)(H\times e)} = B$, by Lemma 3·15.10. The proof of the remaining inclusion is similar.

Next we prove that if $(m_0,A)\in M_1$, then $(m_0h,A)\in M_1$ for each $h\in H$. Let $h\in H$, then there exists a net $\{t_n\}$, $t_n\in T$ such that $\{m_0t_n\}\to m_0h$. Without loss of generality we may assume that $\{m_0Bt_n\}\to A_1\in 2^M$. Then $(m_0h,A_1)\in\Sigma(m_0,m_0B)$, the orbit closure of $(m_0,m_0B)\in M\times 2^M$. As $(m_0,A)\in M_1\subset\Sigma(m_0,m_0B)$, it follows that $A\subset m_0B$. Similarly, $A_1\subset m_0hB = m_0B$. Since $A,A_1\in\Sigma(m_0B)$, the points A and A_1 are proximal. Hence there exists a net $\{\tau_n\}$, $\tau_n \in T$, with $\{m_0\tau_n\}\to m_0$, $\{A\tau_n\}\to A$, and $\{A_1\tau_n\}\to A$. Thus

$$\{(m_0h,A_1)\tau_n\} = \{(m_0\tau_nh,A_1\tau_n)\}\to(m_0h,A);$$

and therefore $(m_0h,A)\in\Sigma(m_0h,A_1)\subset\Sigma(m_0,m_0B)$. Clearly (m_0h,A) is an almost periodic point. Since M_1 is the only minimal set contained in $\Sigma(m_0,m_0B)$, we have $(m_0h,A)\in M_1$.

We claim that if $g\in H$ and the points (m_0,A_1), $(m_0g,A_2)\in\Sigma(m_0, m_0B)$ are such that $((m_0,A_1),(m_0g,A_2))$ is an almost periodic point of $(M\times 2^M\times M\times 2^M,T)$, then $A_1 = A_2$. Indeed, $A_1\subset m_0B$, $A_2\subset m_0gB = m_0B$ and $A_1,A_2\in\Sigma(m_0B)$, hence it follows from the above considerations that A_1 and A_2 are proximal. Thus (m_0,A_1), (m_0,A_2) are also proximal. By hypothesis, $((m_0,A_1),(m_0g,A_2))$ is an almost periodic point, and hence $((m_0,A_1),(m_0,A_2))$ is also almost periodic. This is possible only when $A_1 = A_2$.

As a consequence of what has just been proved, we get that p_1 is a proximal extension.

The homomorphism $\lambda:(M,T)\to(X,T)$ induces homomorphisms $\lambda_1:(2^M, T)\to(2^X,T)$ and $\lambda_2\equiv\lambda\times\lambda_1:M\times 2^M\to X\times 2^X$. Set $Y_1 = Z\lambda_1$ and $X_1 = M_1\lambda_2$. We get the commutative diagram on the next page.

As p_1 is proximal, the extension p is also proximal. It is easy to see that q is also proximal. It follows directly from the construction that the restriction of q to each fiber of φ_1 is injective. If φ is open and $\overline{R_\varphi\mathcal{J}} = R_\varphi$, then p and q are isomorphisms. If either $\overline{R_\varphi\mathcal{J}}\neq R_\varphi$ or φ is not open, then p is not a homeomorphism, as is seen from the construction itself. In this case

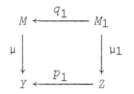

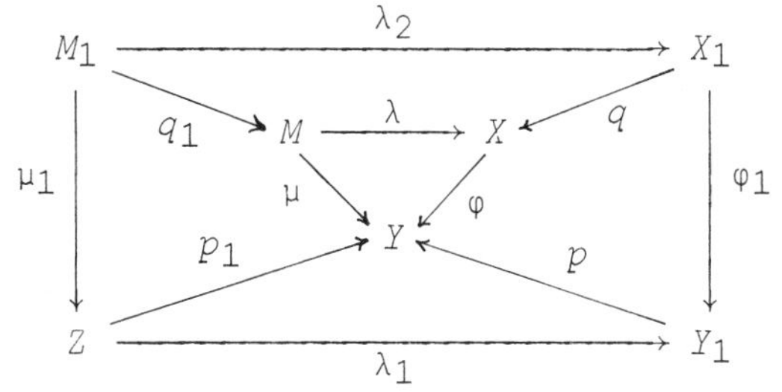

we repeat the above considerations, taking φ_1 instead of φ. Proceeding by transfinite induction, we finally obtain minimal sets X^* and Y^* with the required properties. (As a matter of fact, $X^* = X_1$, and $Y^* = Y_1$, but the proof of this will be omitted here). Thus the proof of the proposition is complete.

The remainder of the proof is very similar to the proof of Theorem 3·14.36, however, we must use Lemma 3·15.41 instead of Lemma 3·14.31. The details are left to the reader. ∎

3·15.43 REMARK: If X is a compact metrizable space and $\varphi:(X,T) \to (Y,T)$ is a minimal extension such that there is at least one R_φ-distal point, then it is not hard to deduce from Theorem 3·15.43 that the set of all R_φ-distal points is residual. Thus if X is a metrizable space and the set of all R_φ-distal points is non-empty, then it is residual.

REMARKS AND BIBLIOGRAPHICAL NOTES

The material presented in this section is essentially due to Ellis [*13,14*] (*see also* Ellis, Glasner and Shapiro *[*1*], Glasner *[*4*]), but our approach differs essentially from the original approach suggested by Ellis. We are not in a position to compare these two approaches here, since in this book we have not introduced the concepts from the algebraic theory of minimal sets that were developed by Ellis. We only note that our presentation is shorter and simpler than the one by Ellis. However, this remark is subjective and probably reflects only the preferences of the author. It should be noted that the algebraic method made possible different proofs of the results in Subsections 3·13.7-9 (*see* Ellis [*14*]) and to prove some other important statements (*see* for example, Ellis and Keynes [*1,2*]). The notion of the Galois group (in the case of a distal minimal extension) was introduced by Bronšteĭn [*17,20**].

§(3·16): GROUP EXTENSIONS

In this section we shall introduce and study group extensions. The totality of group extensions with a given structure group will be characterized in terms of homological algebra. A more complete description will be obtained under some additional conditions by using covering transformation groups.

3·16.1 Let X be a topological space and T and G be topological groups. We shall say that (X,T,π,G,σ) is a *GROUP EXTENSION* if (X,T,π) and (X,G,σ) are topological transformation groups and, moreover, for each point $x \in X$ the set $xG \equiv \{x\sigma^g | g \in G\}$ is compact and

$$xG\pi^t = x\pi^t G, \qquad (x \in X, t \in T). \tag{16.1}$$

It is evident from (16.1) that the partition $\{xG | x \in X\}$ is invariant under (X,T,π). If the set U is open in X, then UG is also open in X. Thus $\{xG | x \in X\}$ is a T-invariant star-open partition. By Subsection 1·3.2, the transformation group (X,T,π) induces a topological transformation group $(X/G,T,\rho)$ on the quotient space $X/G = \{xG | x \in X\}$. Suppose in addition that the space X/G is Hausdorff. This holds, in particular, when G is a compact group. Indeed, it follows from Lemma 1·1.3 that in this case the partition $\{xG | x \in X\}$ is star-closed.

The transformation group (X,T,π) is an extension of the transformation group $(X/G,T,\rho)$ by the canonical projection $\varphi: X \to X/G$. This justifies the above terminology.

Let (X,T,π) be an extension of (Y,T,ρ). We shall for brevity say that (X,T,π) is a group extension of (Y,T,ρ), whenever there exist a topological group G and a map $\sigma: X \times G \to X$ such that (X,T,π,G,σ) is a group extension with $(X/G,T,\rho) = (Y,T,\rho)$. It follows from (16.1) that for any given $x \in X$, $t \in T$ and $g \in G$, there exists an element $g' \in G$ satisfying the condition

$$x\sigma^g\pi^t = x\pi^t\sigma^{g'} . \tag{16.2}$$

A group extension (X,T,π,G,σ) is called a *FREE GROUP EXTENSION* if the group G acts freely on X, *i.e.*, the equality $xg \equiv x\sigma^g = x$ holds for a point $x \in X$ iff g is the identity of G. Let (X,T,π,G,σ) be a group extension. If for any two elements $t \in T$ and $g \in G$ there exists an element $g' \in G$ such that (16.2) holds for every point $x \in X$, then the group extension is called a *NORMAL GROUP EXTENSION*.

Let G be a compact group and let Aut$G$ denote the group of all automorphisms of the group G with composition as the group operation. Provide Aut$G$ with the topology of uniform convergence, *i.e.*, with the topology of a subspace of $C(G,G)$. Then Aut$G$ is a topological group.

If (X,T,π,G,σ) is a free normal group extension and the group G is compact, then (16.2) determines a map $h: T \to \text{Aut}G$, namely, for each element $t \in T$ there is a corresponding automorphism h^t such that $gh^t = g'$. We can easily verify that h is a continuous homomorphism of the group T into the group Aut$G$.

The *homomorphism* $h: T \to \mathrm{Aut}G$ is said to be *INDUCED by the free normal group extension* (X,T,π,G,σ) and the group G will be called the *STRUCTURE GROUP of this extension.*

A group extension (X,T,π,G,σ) is called a *CENTRAL GROUP EXTENSION* if

$$x\sigma^g\pi^t = x\pi^t\sigma^g, \tag{16.3}$$

for all $x \in X$, $t \in T$, $g \in G$.

Let (X,T,π,G,σ) be a free group extension with a compact structure group. This extension is said to be a *TAME EXTENSION* if (16.2) implies that

$$xkg\pi^t = xk\pi^t g' \tag{16.4}$$

for all $k \in G$, *i.e.*, for any fixed g and t the element g' in (16.2) depends only on $x\varphi \in X/G$, thus being constant on each fibre $(x\varphi)\varphi^{-1} \subset X$. Thus, if (X,T,π,G,σ) is a tame extension, then we can define a map $h: T \times Y \to \mathrm{Aut}G$ by

$$xg\pi^t = x\pi^t(gh_y^t), \qquad (x \in X, t \in T), \tag{16.5}$$

where $y = x\varphi$. If Y is compact, then $h: T \times Y \to \mathrm{Aut}G$ is continuous (the proof is left to the reader).

Each central group extension (X,T,π,G,σ) is homogeneous because each map σ^g $(g \in G)$ is an automorphism of the transformation group (X,T,π). Hence if X is a compact space, (Y,T) is a minimal transformation group and (X,T,π,G,σ) is a central group extension, then $\varphi: (X,T) \to (Y,T)$ is a distal extension.

> 3·16.2 THEOREM: *Let X be a compact space, (X,T,π) a minimal transformation group, G a compact topological group and let (X,T,π,G,σ) be a central group extension. Then the canonical projection $\varphi: (X,T) \to (X/G,T)$ is an equicontinuous extension.*

Proof: Suppose the contrary is true. Then for any index $\alpha \in \mathfrak{U}[X]$ we can choose points $a_\alpha \in X$ and $b_\alpha \in X$ and an element $t_\alpha \in T$ so that

$$(a_\alpha, b_\alpha) \in \alpha \cap R_\varphi, \qquad (a_\alpha t_\alpha, b_\alpha t_\alpha) \notin \alpha_0, \tag{16.6}$$

where α_0 is a certain fixed index.

Choose elements $g_\alpha \in G$ $(\alpha \in \mathfrak{U}[X])$ such that $b_\alpha = a_\alpha\sigma^{g_\alpha} \equiv a_\alpha g_\alpha$. Without loss of generality we may suppose that

$$\{a_\alpha\} \to a \in X, \qquad \{a_\alpha t_\alpha\} \to c \in X, \qquad \{g_\alpha\} \to g \in G. \tag{16.7}$$

Then it follows from (16.6) that $\{a_\alpha g_\alpha\} = \{b_\alpha\} \to a$, and hence $ag = a$. Since (X,T) is a minimal transformation group and the actions of T and G commute, $xg = x$ for all $x \in X$. Further, $b_\alpha t_\alpha = a_\alpha g_\alpha t_\alpha = a_\alpha t_\alpha g_\alpha$. Therefore $\{b_\alpha t_\alpha\} \to cg$ and (16.6,7) imply $(c,cg) \notin \alpha_0$, contradicting the equality $xg = x$ $(x \in X)$. ∎

3·16.3 Note that not every equicontinuous extension is a central group extension. However, it will be shown in subsection 3·17 that the investigation of minimal equicontinuous extensions can be, in some sense, reduced to the study of group extensions.

3·16.4 Two group extensions (X,T,π,G,σ) and (X_1,T,π_1,G,σ_1) are said to be *CONGRUENT GROUP EXTENSIONS*, if there exists a homeomorphism Φ of X onto X_1 such that

$$x\pi^t\Phi = x\Phi\pi_1{}^t, \qquad x\sigma^g\Phi = x\Phi\sigma_1{}^g, \qquad (x \in X, t \in T, g \in G).$$

It is easily seen that congruent normal extensions induce the same homomorphisms $h:T \to \mathrm{Aut}G$.

By $\mathrm{Ext}[(Y,T,\rho),G,h]$ we shall *denote the family of all classes of free normal group extensions* (X,T,π,G,σ) such that $(X/G,T,\rho) = (Y,T,\rho)$ and the induced homomorphism coincides with $h:T \to \mathrm{Aut}G$. Our next aim is to describe $\mathrm{Ext}[(Y,T,\rho),G,h]$.

3·16.5 A *SECTION of* (X,T,π,G,σ) is defined to be a map $u:Y \to X$ such that $u\circ\varphi$ is the identity transformation of Y. *The image of a point $y \in Y$ under the map $u:Y \to X$ will be denoted by* $u(y)$ (thus *infringing* upon our convention of *writing the function sign on the right*).

A function $f:Y \times T \to G$, $(f:Y \times T \times T \to G)$ is called a *two-dimensional* (respectively *three-dimensional*) *COCHAIN OF THE TRANSFORMATION GROUP* (Y,T) over the group G if the following equalities are satisfied

$$f(y,e) = e, \qquad (f(y,e,s) = f(y,s,e) = e), \qquad (y \in Y, s \in T).$$

3·16.6 Let $y \in Y$, $t \in T$. As,

$$u(y)\pi^t\varphi = u(y)\varphi\rho^t = y\rho^t = u(y\rho^t)\varphi, \tag{16.8}$$

there exists a uniquely defined element $f(y,t) \in G$ with

$$u(y)\pi^t = u(y\rho^t)f(y,t), \qquad (y \in Y, t \in T). \tag{16.9}$$

Then

$$u(y)\pi^{ts} = u(y\rho^{ts})f(y,ts), \qquad (y \in Y; t,s \in T). \tag{16.10}$$

On the other hand,

$$u(y)\pi^{ts} = [u(y)\pi^t]\pi^s = [u(y\rho^t)f(y,t)]\pi^s.$$

Taking (16.2,9) into account, we get

$$u(y)\pi^{ts} = u(y\rho^t)\pi^s[f(y,t)]h^s$$

$$= u(y\rho^{ts})f(y\rho^t,s)[f(y,t)]h^s. \qquad (16.11)$$

Since G acts freely on X, comparing (16.10) with (16.11) we obtain

$$f(y,ts) = f(y\rho^t,s)[f(y,t)]h^s. \qquad (16.12)$$

A function $f:Y\times T \to G$ which satisfies (16.12) will be called a *TWO-DIMENSIONAL COCYCLE*.

Let $u':Y \to X$ be some other section. Since $u'(y)\varphi = u(y)\varphi$, there exists a unique element $\lambda = \lambda(y) \in G$ with $u'(y) = u(y)\sigma^{\lambda(y)} \equiv u(y)\lambda(y)$. Let $f':Y\times T \to G$ be the cocycle corresponding to the section $u':Y \to X$, that is, $u'(y)\pi^t = u'(y\rho^t)f'(y,t)$. As $u'(y\rho^t) = u(y\rho^t)\lambda(y\rho^t)$, we have that

$$u'(y)\pi^t = u(y\rho^t)\lambda(y\rho^t)f'(y,t). \qquad (16.13)$$

On the other hand, the condition $u'(y) = u(y)\lambda(y)$ implies:

$$u'(y)\pi^t = u(y)\lambda(y)\pi^t = u(y)\pi^t[\lambda(y)]h^t$$

$$= u(y\rho^t)f(y,t)[\lambda(y)]h^t. \qquad (16.14)$$

Comparing (16.13) with (16.14) we obtain

$$f'(y,t) = [\lambda(y\rho^t)]^{-1}f(y,t)[\lambda(y)]h^t. \qquad (16.15)$$

It is easy to see that (16.15) determines an equivalence relation on the set of all cocycles, namely, $f' \sim f$ iff (16.15) is satisfied. The *set of all classes of equivalent cocycles* will be *denoted by* $C^2[(Y,T,\rho),G,h]$.

Thus, given a class of congruent extensions, *i.e.*, an element of the set $\mathrm{Ext}[(Y,T,\rho),G,h]$, we have defined a class of equivalent cocycles, *i.e.*, an element of $C^2[(Y,T,\rho),G,h]$. *In the sequel this map will be denoted by* ω.

3·16.7 Since $(x\varphi)\varphi^{-1} = xG$ $(x \in X)$ and the group G acts freely on X, each point $x \in X$ can be written uniquely in the form $x = u(x\varphi)g_x$, where $g_x \in G$. Then

$$x\pi^t = u(x\varphi)g_x\pi^t = u(x\varphi)\pi^t(g_xh^t) = u(x\pi^t\varphi)f(x\varphi,t)(g_xh^t).$$

Therefore $g_{x\pi^t} = f(x\varphi,t)(g_xh^t)$. Let $\Psi:X \to (Y\times G)$ denote the map defined by $x\Psi = (x\varphi,g_x)$ $(x \in X)$. Then $x\pi^t\Psi = (x\varphi\rho^t,f(x\varphi,t)(g_xh^t))$.

Suppose now that f is a two-dimensional cocycle, *i.e.*, $f:Y\times T \to G$ satisfies condition (16.12). Define a map $\mu:(Y\times G)\times T \to (Y\times G)$

by

$$(y,g)\mu^t = (y\rho^t, f(y,t)\cdot(gh^t)), \qquad (y \in Y, g \in G, t \in T), \quad (16.16)$$

a map $\sigma:(Y \times G) \times G \to (Y \times G)$ by

$$(y,g)\sigma^a \equiv (y,g)a = (y,ga), \qquad (y \in Y, g \in G, a \in G),$$

and a map $\eta: G \times G \to G$ by

$$(a,g)\eta = ag, \qquad (a,g \in G).$$

If the group G is compact and the function $f:Y \times T \to G$ is continuous, then $(Y \times G, T, \mu, G, \sigma)$ is a free normal group extension which will be called the *SKEW PRODUCT of the transformation groups* (Y,T,ρ) *and* (G,G,η) by means of the cocycle f. The induced homomorphism coincides with the given homomorphism $h:T \to \mathrm{Aut}G$. If the section $u:Y \to (Y \times G)$ is defined by $u(y) = (y,e)$, where e is the identity of G, then the corresponding two-dimensional cocycle is just $f:Y \times T \to G$.

Concerning the case when f is not continuous, we do not know whether or not there exists a free group extension corresponding to the cocycle f.

3·16.8 Now we shall consider the particular case when the group G is commutative. The group operation in G will be written additively.

Given a function $\lambda:Y \to G$, let $\delta\lambda:Y \times T \to G$ denote the function defined by

$$(\delta\lambda)(y,t) = -\lambda(y\rho^t) + [\lambda(y)]h^t. \quad (16.17)$$

Condition (16.12) can be rewritten in the form

$$f(y,ts) = f(y\rho^t,s) + [f(y,t)]h^s, \quad (16.18)$$

and condition (16.15) in the form

$$f'(y,t) = f(y,t) + \delta\lambda(y,t). \quad (16.19)$$

Define a group operation on the set of all two-dimensional cocycles as follows:

$$(f_1 + f_2)(y,t) = f_1(y,t) + f_2(y,t).$$

The totality of two-dimensional cocycles endowed with this operation forms a commutative group which we shall denote by $Z^2 = Z^2[(Y,T,\rho),G,h]$.

A function $\delta\lambda:Y \times T \to G$ of the form (16.17) will be called a *COBOUNDARY*. It is evident that each coboundary is a cocycle. The set $B^2 = B^2[(Y,T,\rho),G,h]$ of all coboundaries is a subgroup of the group Z^2.

Denote Z^2/B^2 by H^2. The commutative group $H^2 = H^2[(Y,T,\rho),G,h]$ is called the *SECOND COHOMOLOGY GROUP of the transformation group* (Y,T,ρ) *over the group* G *with the homomorphism* $h:T \to \mathrm{Aut}G$. Thus, if the group G is commutative, then the set C^2 can be provided with a structure of a commutative group H^2. The group $H^2[(Y,T,\rho),G,h]$ may be regarded as the second cohomology group of some complex of right T-modules (MacLane [*1*]).

3·16.9 Let (X,T,π,G,σ) be a normal group extension of the transformation group $(X/G,T,\rho) \equiv (Y,T,\rho)$. This extension is said to be *AN EXTENSION LOCALLY TRIVIAL AT A POINT* $y \in Y$ if there exist a section $u:Y \to X$ and a neighbourhood U of y in Y such that the restriction $u|_U$ of the map $u:Y \to X$ to U is a homeomorphism.

An extension is called a *LOCALLY TRIVIAL EXTENSION* if it is locally trivial at each point $y \in Y$. An extension (X,T,π,G,σ) is said to be a *TRIVIAL EXTENSION* if there exists a continuous (and, consequently, a homeomorphic) section $u:Y \to X$.

> 3·16.10 LEMMA: *Let (Y,T,ρ) be a minimal transformation group. If the extension (X,T,π,G,σ) is locally trivial at some point $y \in Y$, then it is locally trivial.*

Proof: Let $z \in Y$. By hypothesis there exist a section $u:Y \to X$ and a neighbourhood U of $y \in Y$ such that the restriction $u|_U$ is a homeomorphism. Choose an element $t \in T$ and a neighbourhood V of $z \in Y$ so that $V\rho t^{-1} \subset U$. The restriction to V of the section u': $Y \to X$ defined $u'(y) = u(y\rho t^{-1})\pi t$ is a homeomorphism. Thus the extension is locally trivial at every point $z \in Y$. ▌

> 3·16.11 LEMMA: *If (X,T,π,G,σ) is a free normal group extension which is locally trivial at some point $y \in Y \equiv X/G$ and the spaces Y and G are compact, then there exist a neighbourhood U of y and a homeomorphism $\psi_U:U \times G \to U\varphi^{-1}$ such that $(z,g)\psi_U\varphi = z$ $(z \in U, g \in G)$ (φ denotes, as usual, the canonical projection).*

Proof: By hypothesis, there exist a section $u:Y \to X$ and a neighbourhood U of $y \in Y$ such that the restriction $u|_U$ is a homeomorphism. Let $x \in U\varphi^{-1}$. As the group G acts transitively and freely on $x\varphi\varphi^{-1}$ there exists a unique element $g_x \in G$ with $x = u(x\varphi)g_x$. Define a map $\lambda_U:U\varphi^{-1} \to U \times G$ by $x\lambda_U = (x\varphi, g_x)$, then λ_U is an injective map. The inverse map $\psi_U = \lambda_U^{-1}$ is defined by $(z,g)\psi_U = u(z)g$ $(z \in U, g \in G)$ and therefore it is continuous.

Without loss of generality we may suppose that the section u: $Y \to X$ is a homeomorphism onto the closure $\bar{U}$ of U. Since $\bar{U} \times G$ is a compact set, it follows from the preceding considerations that ψ_U is a homeomorphism. If $z \in U$, $g \in G$, then $(z,g)\psi_U\varphi = u(z)g\varphi = z$. ▌

3·16.12 Let $\mathrm{Ext}_\ell[(Y,T,\rho),G,h]$ denote the set of all classes of congruent locally trivial free normal group extensions

(X,T,π,G,σ) which satisfy the condition $(X/G,T,\rho) = (Y,T,\rho)$ and induce the given homomorphism $h:T \to \mathrm{Aut}G$. Let the group G be compact.

Let (X,T,π,G,σ) belong to some class in $\mathrm{Ext}_\ell[(Y,T,\rho),G,h]$ and let $y_0 \in Y$. There exist a section $u:Y \to X$ and a neighbourhood V of y_0 such that $u|_V$ is a homeomorphism. If $f:Y\times T \to G$ is the cocycle corresponding to the section $u:Y \to X$, *i.e.*

$$u(y)\pi^t = u(y\rho^t)f(y,t), \qquad (y \in Y, t \in T),$$

then the function $f:Y\times T \to G$ is continuous on the set

$$V^* = \{(y,t) | y \in V, y\rho^t \in V\}.$$

Let $Z^2_\ell[(Y,T,\rho),G,h,y_0]$ denote the set of all those cocycles $f:Y\times T \to G$ in $Z^2[(Y,T,\rho),G,h]$ for which there exists a neighbourhood V of y_0 such that $f|_{V^*}$ is continuous.

Let $C^2_\ell[(Y,T,\rho),G,h,y_0]$ denote the set of all classes of equivalent (in the sense of (16.15)) cocycles in $Z^2_\ell[(Y,T,\rho),G,h,y_0]$. If (Y,T,ρ) is a minimal transformation group, then the set C^2_ℓ $[(Y,T,\rho),G,h,y_0]$ does not depend on $y_0 \in Y$. The proof is similar to that of Lemma 3·16.10.

Let ω_ℓ be the natural map which assigns to each class of congruent extensions in $\mathrm{Ext}_\ell[(Y,T,\rho),G,h]$ an element of $C^2_\ell[(Y,T,\rho), G,h,y_0]$, namely, the class of cocycles equivalent to the cocycle f, which corresponds to that section $u:Y \to X$ whose restriction $u|_U$ to some neighbourhood U of $y_0 \in Y$ is continuous.

3·16.13 THEOREM: *If Y is a compact space and (Y,T,ρ) is a minimal transformation group, when ω_ℓ is a bijective map of* $\mathrm{Ext}_\ell[(Y,T,\rho),G,h]$ *onto* $C^2_\ell[(Y,T,\rho),G,h,y_0]$.

Proof: Let $f:Y\times T \to G$ be a cocycle and suppose that there exists a neighbourhood U of $y_0 \in Y$ such that the restriction of f to $U^* = \{(y,t)|y \in U, y\rho^t \in U\}$ is continuous. We shall construct an extension (X,T,π,G,σ) which belongs to some class in Ext_ℓ $[(Y,T,\rho),G,h]$ and which has the given cocycle f (for a suitable section).

Let X denote the Cartesian product of the sets Y and G, $X = Y\times G$. We shall provide X with a topology which will be defined by assigning a neighbourhood base $\mathcal{B}(y,g)$ to each point $(y,g) \in Y\times G$.

Since (Y,T,ρ) is a minimal transformation group, there exists an element $t \in T$ with $yt^{-1} \equiv y\rho^{t^{-1}} \in U$. Let $V = V(y)$ be any neighbourhood of y satisfying the condition $Vt^{-1} \subset U$ and let $W(e)$ be an arbitrary neighbourhood of the identity e in the topological group G. Then $\mathcal{B}(y,g)$ is defined to be the family of all subsets of $Y\times G$ of the form

$$A(V,W,t) = \{(z,f(zt^{-1},t)\cdot[f(y,t^{-1})]h^{t}\cdot bg)\,|\,z \in V(y), b \in W(e)\}.$$

Let us show that the family $\mathcal{B}(y,g)$ in fact does not depend on the choice of the element $t \in T$ with $yt^{-1} \in U$. Let $s \in T$ and $ys^{-1} \in U$. Choose a neighbourhood $W_0(e)$ so that $W_0{}^2 \subset W$. Further choose $W_1(e)$ so that $W_1 \subset W_0$ and

$$[f(y,s^{-1})]h^{s}\cdot W_0 \supset W_1\cdot[f(y,s^{-1})]h^{s}.$$

As the restriction of $f: Y \times T \to G$ to U^* is continuous, there exists a neighbourhood $V_1 \equiv V_1(y) \subset V$ such that $V_1s^{-1} \subset U$ and

$$[f(zs^{-1},st^{-1})]h^{t} \in [f(ys^{-1},st^{-1}]h^{t}\cdot W_1$$

for all $z \in V_1(y)$. We show that $A(V_1,W_1,s) \subset A(V,W,t)$. Indeed, let $(z,b) \in A(V_1,W_1,s)$. Then $z \in V_1 \subset V$ and $b = f(zs^{-1},s)\cdot[f(y,s^{-1})]h^{s}\cdot b_1g$, where $b_1 \in W_1 \subset W_0$.

After some reductions we obtain:

$$\begin{aligned} b &= f(zs^{-1},(st^{-1})t)\cdot[f(y,s^{-1})]h^{s}\cdot b_1g \\ &= f(zt^{-1},t)\cdot[f(zs^{-1},st^{-1})]h^{t}\cdot[f(y,s^{-1})]h^{s}\cdot b_1g \\ &\in f(zt^{-1},t)\cdot[f(ys^{-1},st^{-1})]h^{t}\cdot W_1\cdot[f(y,s^{-1})]h^{s}\cdot b_1g \\ &\subset f(zt^{-1},t)\cdot[f(ys^{-1},st^{-1})]h^{t}\cdot[f(y,s^{-1})]h^{s}\cdot W_0b_1g \\ &= f(zt^{-1},t)\cdot\{f(ys^{-1},st^{-1})\cdot[f(y,s^{-1})]h^{st^{-1}}\}h^{t}\cdot W_0b_1g. \end{aligned}$$

The expression in curly brackets coincides with $f(y,t^{-1})$. Indeed

$$f(y,t^{-1}) = f(y,s^{-1}(st^{-1})) = f(ys^{-1},st^{-1})\cdot[f(y,s^{-1})]h^{st^{-1}}.$$

Hence

$$b \in f(zt^{-1},t)\cdot[f(y,t^{-1})]h^{t}\cdot Wg,$$

and consequently $(z,b) \in A(V,W,t)$.

Thus the family $\mathcal{B}(y,g)$ does not depend on the choice of $t \in T$ with $yt^{-1} \in U$. It is not difficult to verify that the totality of all families $\mathcal{B}(y,g)$ for all $(y,g) \in Y \times G$ determines a topology on $X = Y \times G$ such that $\mathcal{B}(y,g)$ is a neighbourhood basis at (y,g).

Let X be the topological space obtained by providing the set $Y \times G$ with the above topology. Define an action of T on X as follows:

$$(y,g)\pi^t = (y\rho^t, f(y,t)(gh^t)), \qquad (t \in T).$$

An elementary but tedious computation shows that the map π: $X \times T \to X$ is continuous. We do not present here the details but only note that, as it follows from the very definition of the topology on X, it suffices to verify the continuity only on $U^* \times G$, where $U^* = \{(y,t) | y \in U, y\rho^t \in U\}$. The latter follows easily from the continuity of h^t and $f|_{U^*}$. Thus (X,T,π) is a topological transformation group.

Let us define an action of G on X. If $(y,g) \in X$ and $a \in G$, then we set $(y,g)\sigma^a = (y,ga)$. Clearly, (X,G,σ) is a topological transformation group and moreover G acts freely on X.

Define a map $\varphi: X \to Y$ by $x\varphi = y$ whenever $x = (y,g)$. It is clear that φ is continuous and open and that the set $x\varphi\varphi^{-1} \equiv \{(y,g) | g \in G\}$ is compact. It follows directly from the definition of the topology on X that the topology of the subspace $U \times G \subset X$ coincides with the topology of the Cartesian product of $U \subset Y$ and G. In fact, if $y \in U$ and $V(y)$ is a neighbourhood with $V(y) \subset U$ then, in the capacity of an element $t \in T$ with $yt^{-1} \in U$, we can take $t = e$, where e is the identity of T. Then

$$A(V,W,g) = \{(z,h) | z \in V(y), h \in W(e)g\}.$$

Now it follows easily that the space X is Hausdorff. Since Y is compact and each fibre $x\varphi\varphi^{-1}$ is compact, the space X is also compact by the continuity of $\varphi: X \to Y$. If $y \in Y$, $g \in G$, $a \in G$, $t \in T$, then

$$(y,g)\sigma^a\pi^t = (y,ga)\pi^t = (y\rho^t, f(y,t)\cdot[(ga)h^t])$$

$$= (y\rho^t, f(y,t)\cdot(gh^t)\cdot(ah^t)) = (y,g)\pi^t\sigma^{ah^t}.$$

Thus (X,T,π,G,σ) is a normal group extension and, moreover, the induced homomorphism coincides with the given one, $h: T \to \mathrm{Aut}G$.

If we take the section $u: Y \to X$ defined by $u(y) = (y,e)$, where e is the identity of G, then the cocycle $f: Y \times T \to G$ corresponding to this section coincides with the given function f. In fact,

$$u(y)\pi^t = (y,e)\pi^t = (y\rho^t, f(y,t)) = (y\rho^t, e)\sigma^{f(y,t)} = u(y\rho^t)\sigma^{f(y,t)}.$$

The extension (X,T,π,G,σ) is locally trivial by Lemma 3·16.10, since the aforementioned section $u: Y \to X$ is continuous on the set U.

Thus (X,T,π,G,σ) is the required extension.∎

3·16.14 In the remainder of this section all topological spaces are assumed to be connected locally pathwise connected and semi-locally simply connected (unless specified otherwise) (Hu [1]).

3·16.15 Let (X,T,π,K,σ) be a free normal locally trivial extension of (Y,T,ρ). Then the map $\varphi:X \to Y$ is a locally trivial fibre bundle and since Y is compact and Hausdorff, $\varphi:X \to Y$ satisfies the covering path axiom (Hu [*1*]).

Let $x_0 \in X$, $y_0 = x_0\varphi$. Let $A = E(X,x_0)$ and $B = E(Y,y_0)$ be the universal covering spaces of X and Y respectively. Further, let $D_1 = \pi_1(Y,y_0)$, $D_2 = \pi_1(K)$ and $D = \pi_1(X,x_0)$. Let us denote by G the universal covering group of the group K. It is known that D_2 is a discrete (consequently, a central) normal subgroup of G and $K = G/D_2$.

The group D_1 acts in a natural way on $B = E(Y,y_0)$ (from the right). Let $[b] \in B$, *i.e.*, $[b]$ is a class of equivalent paths in $[Y;y_0,Y]$, and let $d_1 \in D_1$, *i.e.*, $d_1 = [\alpha]$, where α is a loop at the point y_0. Define a map $\delta_1:B \times D_1 \to B$ by $([b],d_1)\delta_1 = [\alpha^{-1}b]$. Clearly, (B,D_1,δ_1) is a transformation group. We define the transformation group (A,D,δ) in a similar way.

The transformation group (X,K,σ) induces the universal covering transformation group $(E(X,x_0),G,\tilde{\sigma})$ as follows. Let $x \in X$, $I = [0,1]$ and let $a:I \to X$ be a path satisfying $a(0) = x_0$, $a(1) = x$. If $g \in G$, then $g = [\tau]$, where $\tau(0) = e$ (e is the identity of the group K). The path $\xi:I \to X$, defined by $\xi(i) = a(1)\sigma^{\tau(i)}$ ($i \in I$) has x as its initial point. Define a map $\tilde{\sigma}:A \times G \to A$ by $([a],g)\tilde{\sigma} = [a\xi]$. It is easy to verify that this map is well defined and that $(A,G,\tilde{\sigma})$ is a topological transformation group.

Let us show that the actions of G and D on A just defined commute. In fact, let $[a] \in A$, $g = [\tau] \in G$ and $d = [\alpha] \in D$. Then $[a]\tilde{\sigma}^g = [a\xi]$ where $\xi:I \to X$ is already defined. Further, $[a\xi]\delta^d = [\alpha^{-1}a\xi]$. On the other hand, $([a]\delta^d)\tilde{\sigma}^g = [\alpha^{-1}a]\tilde{\sigma}^g = [\alpha^{-1}a\xi]$. ∎

3·16.16 Let $a:I \to X$ and $\xi:I \to X$ be the previously defined paths. Define a path $\lambda:I \to X$ by $\lambda(i) = x_0\sigma^{\tau(i)}$ ($i \in I$). Let $\tau(1) = k$. Let us denote by a_k the path $a_k:I \to X$, defined by $a_k(i) = a(i)\sigma^k$ ($i \in I$).

3·16.17 LEMMA: $[a\xi] = [\lambda a_k]$.

Proof: We need to show that the paths $a\xi$ and λa_k are homotopic. Set

$$h_j(i) = \begin{cases} x_0\sigma^{\tau(2i)}, & 0 \leq 2i \leq j \leq 1, \\ a(2i - j)\sigma^{\tau(j)}, & 0 \leq j < 2i \leq 1 + j \leq 2, \\ a(1)\sigma^{\tau(2i-1)}, & 0 \leq j < 2i - 1 \leq 1. \end{cases}$$

Evidently h_j ($j \in I$) is a homotopy with $h_0 = a\xi$, $h_1 = \lambda a_k$. ∎

3·16.18 COROLLARY: *If* $[a] \in D$ *and* $[\tau] \in D_2$, *then* $[a\xi] = [\xi a]$, *where* $\xi(i) = x_0\sigma^{\tau(i)}$ ($i \in I$).

Proof: Indeed, in this case $\xi = \lambda$ and $k = e$; hence $a_k = a$. ∎

3·16.19 Let $F = (x_0\varphi)\varphi^{-1} \subset X$ be the fibre over the point $y_0 = x_0\varphi$.

As $\varphi : X \to Y$ is a fibre bundle, we have the exact homotopy sequence

$$\ldots \to \pi_2(Y,y_0) \to \pi_1(F,x_0) \to \pi_1(X,x_0) \to \pi_1(Y,y_0) \to$$
$$\to \pi_0(F,x_0) \to \pi_0(X,x_0) \to \pi_0(Y,y_0), \qquad (16.20)$$

(*see* Hu [*1*]). Since the fibre F is homeomorphic to the group K, $\pi_1(F,x_0) = \pi_1(K) = D_2$. Since the group K is pathwise connected, $\pi_0(F,x_0) = 0$.

Suppose that, moreover, $\pi_2(Y,y_0) = 0$. Then we get from (16.20) a short exact sequence

$$0 \to D_2 \to D \to D_1 \to 0,$$

which means that the group D is an extension of the group D_2 by means of D_1. Corollary 3·16.18 says that this extension is central, *i.e.*, D_2 is a central subgroup of D. Such an extension is determined uniquely up to equivalence by an element $\{\varphi\}$ of the second cohomology group $H^2(D_1,D_2)$ (MacLane [*1*]). Let $\varphi : D_1 \times D_1 \to D_2$ be some cocycle from $\{\varphi\}$. Then the group D can be represented as a crossed product $D_1 \cdot D_2$. The elements of $D_1 \cdot D_2$ are of the form (d_1,d_2), where $d_1 \in D_1$, $d_2 \in D_2$, and the multiplication is given by

$$(c_1,c_2)\cdot(d_1,d_2) = (c_1d_1, \varphi(c_1,d_1)c_2d_2). \qquad (16.21)$$

3·16.20 From now on we shall assume that the universal covering space $B = E(Y,y_0)$ of Y is contractible. Let $\tilde{\varphi} : E(X,x_0) \to E(Y,y_0)$ be the map induced by the principal K-fibre bundle $\varphi : X \to Y$. Clearly $\tilde{\varphi} : A \to B$ is a principal G-fibre bundle and since B is contractible, $\tilde{\varphi} : A \to B$ is isomorphic as a principal fibre bundle to the Cartesian product $B \times G$ with the action of G determined by

$$(b,g)\hat{\sigma}^h = (b,gh), \qquad (b \in B, g \in G, h \in G).$$

Thus there exists a homeomorphism $\xi : B \times G \to A$ such that

$$(b,gh)\xi = [(b,g)\xi]\tilde{\sigma}^h, \qquad (b \in B, g \in G, h \in G).$$

The homeomorphism ξ is uniquely determined by the choice of a continuous section $u : B \to A$.

The homeomorphism ξ induces an action $\hat{\delta}$ of D on $B \times G$. Let $b \in B$, $g \in G$, $d = (d_1,d_2) \in D_1 \cdot D_2$. Since the group D_2 can be embedded in $D_1 \cdot D_2$ as the subgroup (e_1,D_2), where e_1 is the identity of D_1, it is clear that $((b,g),(e_1,d_2))\hat{\delta} = (b,d_2^{-1}g)$. It is also clear that

$$((b,g),(d_1,d_2))\hat{\delta} = (b\delta_1^{d_1}, g'),$$

where g' is some element of G. This follows from the fact that $D_1 = D/D_2$.

Define a map $f:B \times D_1 \to G$ by

$$((b,e),(d_1,e))\hat{\delta} = (b\delta_1{}^{d_1}, f(b,d_1)). \tag{16.22}$$

Since (A,D,δ) is a group of homeomorphisms, f is continuous.

It was shown above that the actions of D and G on A commute; therefore

$$((b,g),(d_1,e))\hat{\delta} = ((b,e)\partial^g,(d_1,e))\hat{\delta} = (((b,e),(d_1,e))\hat{\delta})\partial^g$$

$$= (b\delta_1{}^{d_1}, f(b,d_1)g).$$

Further, as $(d_1,d_2) = (e,d_2)(d_1,e)$ we have $((b,g),(d_1,d_2))\hat{\delta} = (((b,g),(e,d_2))\hat{\delta}, (d_1,e))\hat{\delta} = ((b,d_2{}^{-1}g),(d_1,e))\hat{\delta} = (b\delta_1{}^{d_1}, f(b,d_1) d_2{}^{-1}g)$. Thus we obtain a general formula

$$(b,g)(d_1,d_2) = (bd_1, f(b,d_1)d_2{}^{-1}g), \tag{16.23}$$

where bd_1 has been written to replace $b\delta_1{}^{d_1}$ and $(b,g)(d_1,d_2)$ denotes $((b,g),(d_1,d_2))\hat{\delta}$.

From (16.21) it follows that

$$(b,g)((c_1,c_2)\cdot(d_1,d_2)) = (b,g)(c_1d_1,\varphi(c_1,d_1)c_2d_2)$$

$$= (bc_1d_1, f(b,c_1d_1)\cdot[\varphi(c_1,d_1)c_2d_2]^{-1}g). \tag{16.24}$$

On the other hand, since the group D_2 is contained in the centre of G, we have:

$$((b,g)(c_1,c_2))(d_1,d_2) = (bc_1,f(b,c_1)c_2{}^{-1}g)(d_1,d_2)$$

$$= (bc_1d_1, f(bc_1,d_1)d_2{}^{-1}f(b,c_1)c_2{}^{-1}g)$$

$$= (bc_1d_1, f(bc_1,d_1)f(b,c_1)c_2{}^{-1}d_2{}^{-1}g). \tag{16.25}$$

Since $(B \times G, D, \hat{\delta})$ is a transformation group we obtain by comparing (16.24) and (16.25) that

$$f(b,c_1d_1) = \varphi(c_1,d_1)\cdot f(bc_1,d_1)\cdot f(b,c_1). \tag{16.26}$$

Let $\tilde{f}:B \times D_1 \to G/D_2 = K$ denote the map defined by $\tilde{f}(b,d_1) = D_2f(b,d_1)$. Since $\varphi(c_1,d_1) \in D_2$, it follows from (16.26) that $\tilde{f}$ is a continuous two-dimensional cocycle of the transformation group (B,D_1,δ_1) over the group K with the trivially induced homomorphism.

Equality (16.26) means that the function

$$\varphi(c_1,d_1) = f(b,c_1d_1)\cdot[f(b,c_1)]^{-1}\cdot[f(bc_1,d_1)]^{-1}, \quad (16.27)$$

is a three-dimensional cochain of the transformation group (B,D_1,δ_1) over the group G, which does not depend on $b \in B$ and takes its values in D_2. The function $f:B\times D_1 \to G$ is not unique, for it depends on the choice of the continuous section $u:B \to A$. Let $\ell:B \to G$ be a continuous map, let $b \in B$, $bu = [a]$, $b\ell = [\tau]$ and let $\nu:I \to X$ be the path defined by

$$\nu(i) = \begin{cases} a(2i), & 0 \leqslant i \leqslant \frac{1}{2}, \\ a(1)\sigma^{\tau(2i-1)}, & \frac{1}{2} < i \leqslant 1. \end{cases}$$

The map $u_1:B \to A$, defined by $bu_1 = [\nu]$, is a continuous section. It is easy to verify that the replacement of u by u_1 results in the substitution of an equivalent cochain for the cochain $f:B\times D_1 \to G$. Therefore the fibre bundle $\varphi:X \to X/K$ determines a class $\{\tilde{f}\}$ of (cohomological) continues two-dimensional cochains of the transformation group (B,D_1,δ_1) over the group K.

Let $C_c^2[(B,D_1,\delta_1),K]$ denote the set of all classes of equivalent (cohomological) continuous two-dimensional cocycles of the transformation group (B,D_1,δ_1) over the group K.

Two *PRINCIPAL K-FIBRE BUNDLES* (X_1,K,σ_1) and (X_2,K,σ_2) are said to be *ISOMORPHIC*, if there exists a homeomorphism Φ of X_1 onto X_2 with

$$(x_1\sigma_1{}^k)\Phi = (x_1\Phi)\sigma_2{}^k, \qquad (x_1 \in X_1, k \in K).$$

It follows from the above considerations that there exists a map ϑ of $\mathrm{Ext}[Y,K]$ into the set $C_c^2[(B,D_1,\delta_1),K]$.

3·16.21 THEOREM: *If the universal covering space of Y is contractible, then ϑ is a bijective map of* $\mathrm{Ext}[Y,K]$ *onto* $C_c^2[(B,D_1,\delta_1),K]$.

Proof: Let $\{\tilde{f}\} \in C_c^2[(B,D_1,\delta_1),K]$. Choose some cocycle $\tilde{f} \in \{\tilde{f}\}$. Then $\tilde{f}:B\times D_1 \to K$ is a continuous map. Define a map $\tilde{f}_{d_1}:B \to K$ by $\tilde{f}_{d_1}(b) = \tilde{f}(b,d_1)$ $(b \in B)$. There exists a map $f_{d_1}:B \to G$ which covers the map $\tilde{f}_{d_1}$. Thus the map $f:B\times D_1 \to G$, defined by $f(b,d_1) = f_{d_1}(b)$, is continuous.

As $\tilde{f}$ is a two-dimensional cocycle,

$$f(b,c_1d_1) = k(b,c_1,d_1)\cdot f(bc_1,d_1)f(b,c_1), \quad (16.28)$$

where $k(b,c_1,d_1) \in D_2$ $(b \in B; c_1,d_1 \in D_1)$. Since f is continuous, B is pathwise connected and the group D_2 is discrete, it follows that $k(b,c_1,d_1)$ does not depend on $b \in B$, and hence $k(b,c_1,d_1) = \psi(c_1,d_1)$ where ψ maps $D_1 \times D_1$ into D_2. Thus ψ is a two-dimensional cochain of the group D_1 over D_2. The cochain ψ is not uniquely determined, but it is easily seen that the element $\{\tilde{f}\} \in C_c^2[(B,D_1,\delta_1),K]$ determines a unique class $\{\psi\}$ of cohomological cochains from $Z^2[D_1,D_2]$.

Let us show that ψ is in fact a two-dimensional cocycle, *i.e.*,

$$\psi(d_1,h_1)\psi(c_1,d_1h_1) = \psi(c_1,d_1)\psi(c_1d_1,h_1) \qquad (16.29)$$

for all $c_1,d_1,h_1 \in D_1$. Reducing first the left side and then the right side of the last equality, we get

$$\begin{aligned}\psi(d_1,h_1)\psi(c_1,d_1h_1) &= \psi(d_1,h_1)f(b,c_1d_1h_1)[f(b,c_1)]^{-1}[f(bc_1,d_1h_1)]^{-1}\\ &= \psi(d_1,h_1)f(b,c_1d_1h_1)[f(b,c_1)]^{-1}[f(bc_1,d_1]^{-1}\\ &\quad\cdot[f(bc_1d_1,h_1)]^{-1}[\psi(d_1,h_1)]^{-1}\\ &= f(b,c_1d_1h_1)\cdot[f(bc_1d_1,h_1)f(bc_1,d_1)f(b,c_1)]^{-1}.\end{aligned}$$

Further:

$$\begin{aligned}\psi(c_1,d_1)\psi(c_1d_1,h_1) &= \psi(c_1,d_1)f(b,c_1d_1h_1)[f(bc_1d_1,h_1)f(b,c_1d_1)]^{-1}\\ &= \psi(c_1,d_1)f(b,c_1d_1h_1)\\ &\quad\cdot[f(bc_1d_1,h_1)\psi(c_1,d_1)f(bc_1,d_1)f(b,c_1)]^{-1}\\ &= f(b,c_1d_1h_1)[f(bc_1d_1,h_1)f(bc_1,d_1)f(b,c_1)]^{-1}.\end{aligned}$$

Thus (16.29) is proved.

Let D be the crossed product of D_1 and D_2 by means of the cocycle ψ and let A be the Cartesian product of the topological spaces B and G. Define an action $\hat{\delta}$ of D on A by (16.23) and an action $\hat{\sigma}$ of G on A by $(b,g)\hat{\sigma}^h = (b,gh)$ $(b \in B; g,h \in G)$. As $f(b,c_1d_1) = \psi(c_1,d_1)f(bc_1,d_1)f(b,c_1)$, it is not hard to verify that $(A,D,\hat{\delta})$ and $(A,G,\hat{\sigma})$ are commuting topological transformation groups. Set $X \equiv A/D$ and let φ be the natural map of X onto $Y = B/D_1$. It follows from the continuity of φ and from the compactness of Y and of the group $K = G/D_2$ that the space X is compact. The canonical projection of A onto X is a covering map, hence X is a principal K-fibre bundle with the base Y and the action of K on X induced by $\hat{\sigma}$. ▮

3·16.22 REMARK: If the group K is commutative then the set $C_c^2 = C_c^2[(B,D_1,\delta_1),K]$ can be provided with a natural group structure which makes C_c^2 into a commutative group $H_c^2[(B,D_1,\delta_1),K]$. It follows from the proof of the preceding theorem that there exists a natural map ν of the group $H_c^2[(B,D_1,\delta_1),K]$ into the group $H^2(D_1,D_2)$ which assigns to an element $\{f\} \in H_c^2$ the element $\{\psi\} \in H^2(D_1,D_2)$ satisfying the condition $f(b,c_1d_1) = \psi(c_1,d_1)f(bc_1,d_1)$ $f(b,c_1)$.

3·16.23 Let (X,T,π,K,σ) be a free central locally trivial group extension of (Y,T,ρ). Let S be the universal covering group of T and let $D_3 = \pi_1(T)$.

The transformation groups (X,T,π) and (Y,T,ρ) induce universal covering transformation groups $(A,S,\tilde{\pi})$ and $(B,S,\tilde{\rho})$, where $A = E(X,x_0)$, $B = E(Y,y_0)$. We shall show that $(A,S,\tilde{\pi},G,\tilde{\sigma})$ is a central group extension.

Let $s \in S$. Then $s = [\alpha]$, where $\alpha{:}I \to T$ is a path with $\alpha(0) = e$ (e is the identity of T). Further, let $a{:}I \to X$ be a path with $a(0) = x_0$, $a(1) = x \in X$. The action $\tilde{\pi}$ is defined in the following way

$$([a],s)\tilde{\pi} \equiv [a]\tilde{\pi}^s = [a\lambda],$$

where the path $\lambda{:}I \to X$ is defined by $\lambda(i) = a(1)\pi^{\alpha(i)}$ $(i \in I)$. It is easy to show that the transformation groups $(A,S,\tilde{\pi})$ and (A,D,δ) commute (similarly, the transformation groups $(B,S,\tilde{\rho})$ and (B,D_1,δ_1) commute).

Let $g \in G$. Then $g = [\tau]$, where the path $\tau{:}I \to K$ satisfies the conditions $\tau(0) = e$ (e is the identity of K) and $\tau(1) = k \in K$. We recall that $[a]\tilde{\sigma}^g = [a\xi]$, where $\xi{:}I \to K$ is defined by $\xi(i) = a(1)\sigma^{\tau(i)}$ $(i \in I)$. Then

$$([a]\tilde{\sigma}^g)\tilde{\pi}^s = [a\xi\mu],$$

where $\mu(i) = a(1)\sigma^{\tau(1)}\pi^{\alpha(i)}$ $(i \in I)$. Let us show that

$$([a]\tilde{\sigma}^g)\tilde{\pi}^s = ([a]\tilde{\pi}^s)\tilde{\sigma}^g.$$

Since $([a]\tilde{\pi}^s)\tilde{\sigma}^g = [a\lambda\varkappa]$, where $\varkappa{:}I \to X$ is defined by $\varkappa(i) = a(1)$ $\pi^{\alpha(1)}\sigma^{\tau(i)}$ $(i \in I)$, we need only show that $[a\xi\mu] = [a\lambda\varkappa]$. It will suffice to show that the paths $\xi\mu$ and $\lambda\varkappa$ are homotopic. Define H_j by

$$H_j(i) = \begin{cases} x\pi^{\alpha(2i)}, & 0 \leqslant 2i \leqslant j \leqslant 1, \\ x\pi^{\alpha(j)}\sigma^{\tau(2i-j)}, & 0 \leqslant j < 2i \leqslant 1+j \leqslant 2, \\ x\pi^{\alpha(j)}\sigma^{\tau(1)}\pi^{\alpha(2i-1)}, & 0 \leqslant j < 2i-1 \leqslant 1. \end{cases}$$

Then H_j is a homotopy with $H_0 = \xi\mu$, $H_1 = \lambda\varkappa$. Thus $(A,S,\tilde{\pi},G,\tilde{\sigma})$ is a central free group extension.

The transformation group $(A,S,\tilde{\pi})$ induces a transformation group $(B \times G,S,\hat{\pi})$ by means of the homeomorphism $\xi{:}B \times G \to A$. Let

$F:B \times S \to G$ be the map determined by the condition

$$(b,e)\hat{\pi}^s = (b\tilde{\rho}^s, F(b,s)), \qquad (b \in B, s \in S). \qquad (16.30)$$

Clearly, F is continuous. Since $(B \times G, S, \hat{\pi}, G, \hat{\sigma})$ is a central group extension,

$$(b,g)\hat{\pi}^s = ((b,e)\hat{\sigma}^g)\hat{\pi}^s = ((b,e)\hat{\pi}^s)g = (b\tilde{\rho}^s, F(b,s)g). \qquad (16.31)$$

We shall show that $F:B \times S \to G$ is a two-dimensional cocycle in $Z^2[(B,S,\tilde{\rho}),G]$. Indeed,

$$(b,g)\hat{\pi}^{sr} = (b\tilde{\rho}^{sr}, F(b,sr)g). \qquad (16.32)$$

On the other hand,

$$(b,g)\hat{\pi}^{sr} = ((b,g)\hat{\pi}^s)\hat{\pi}^r = (b\tilde{\rho}^s, F(b,s)g)\hat{\pi}^r$$

$$= (b\tilde{\rho}^{sr}, F(b\tilde{\rho}^s,r)F(b,s)g). \qquad (16.33)$$

Comparing (16.32) with (16.33) we get

$$F(b,sr) = F(b\tilde{\rho}^s,r)F(b,s).$$

But this just means that $F \in Z^2[(B,S,\tilde{\rho}),G]$.

The cocycle F is not unique; it depends on the choice of the section $u:B \to A$. But it is easy to show that the extension $(B \times G,S,\hat{\pi},G,\hat{\sigma})$ determines a unique element $\{F\} \in C_c^2[(B,S,\tilde{\rho}),G]$, *i.e.*, a certain class of equivalent continuous cocycles.

By $\text{Ext}_\ell[(Y,T,\rho),K]$ we shall *denote the set of all classes of congruent locally trivial free central group extensions* (X,T,π,K,σ) *of the transformation group* (Y,T,ρ). As it was shown above, to each element of $\text{Ext}_\ell[(Y,T,\rho),K]$ there can be assigned a uniquely defined element $\{F\} \in C_c^2[(B,S,\tilde{\rho}),G]$.

Recall that the actions of the groups S and D on $B \times G$ commute (and similarly the actions of S and D_1 on B commute). Hence

$$((b,g)(d_1,d_2))\hat{\pi}^s = ((b,g)\hat{\pi}^s)(d_1,d_2).$$

Reduce the left and right sides of this equality as follows:

$$((b,g)(d_1,d_2))\hat{\pi}^s = (bd_1, f(b,d_1)d_2^{-1}g)\hat{\pi}^s$$

$$= (bd_1\tilde{\rho}^s, F(bd_1,s)f(b,d_1)d_2^{-1}g); \qquad (16.34)$$

$$((b,g)\hat{\pi}^s)(d_1,d_2) = (b\tilde{\rho}^s, F(b,s)g)(d_1,d_2)$$

$$= (b\tilde{\rho}^s d_1, f(b\tilde{\rho}^s,d_1)d_2^{-1}F(b,s)g). \qquad (16.35)$$

Comparing (16.34) and (16.35), we obtain

$$F(b,s) = [f(b\tilde{\rho}^s,d_1)]^{-1}F(bd_1,s)f(b,d_1). \qquad (16.36)$$

Since $F(b,s)$ is a two-dimensional cocycle, $F(bd_1,s) \in Z^2[(B,S,\tilde{\rho}),G]$.

Equality (16.36) shows that the cocycles $F(b,s)$ and $F(bd_1,s)$ are equivalent, *i.e.*, they determine the same element $\{F(b,s)\} = \{F(bd_1,s)\} \in C^2_c[(B,S,\tilde{\rho}),G]$.

Let $\Phi(b,s) \in Z^2[(B,S,\tilde{\rho}),G]$, $d_1 \in D_1$. The element d_1 acts on Z^2, sending $\Phi(b,s)$ to $\Phi(bd_1,s)$. Moreover, this action induces an action of D_1 on $C^2_c[(B,S,\tilde{\rho}),G]$. Let $K^2_c = K^2_c[(B,S,\tilde{\rho}),G]$ denote the set of all elements $\Phi \in C^2_c$ fixed under the action of D_1. Equality (16.36) shows that $\{F\} \in K^2_c$.

Thus we have defined a map of the set $\mathrm{Ext}_\ell[(Y,T,\rho),K]$ into the set $K^2_c[(B,S,\tilde{\rho}),G]$.

3·16.24 Let $\{F\} \in K^2_c[(B,S,\tilde{\rho}),G]$. Choose some element $F \in \{F\}$. Then F is a continuous cocycle in $Z^2[(B,S,\tilde{\rho}),G]$. Since $\{F\} \in K^2_c$, we have $\{F(b,s)\} = \{F(bd_1,s)\}$ for each element $d_1 \in D_1$. Hence there exists a continuous map $f:B \times D_1 \to G$ with

$$F(b,s) = [f(b\tilde{\rho}^s,c_1)]^{-1}F(bc_1,s)f(b,c_1), \qquad (c_1 \in D_1). \quad (16.37)$$

Define a map $\tilde{f}:B \times D_1 \to K \equiv G/D_2$ by $\tilde{f}(b,d_1) = D_2 f(b,d)$. Set

$$\tilde{\lambda}(b,c_1,d_1) \equiv \tilde{f}(b,c_1d_1)[\tilde{f}(bc_1,d_1)\tilde{f}(b,c_1)]^{-1}.$$

It follows directly from the definition that $\tilde{\lambda}$ is a three-dimensional cochain of the transformation group $(B \times G,D_1,\delta_1)$ over the group K. The cochain $\tilde{\lambda}$ will be called an *OBSTRUCTION to the extension*. The obstruction $\tilde{\lambda}$ is not uniquely determined. It depends on the choice of the cocycle F in $\{F\}$ and on the choice of the map satisfying (16.37). As was already shown, if there exists an extension (X,T,π,K,σ), then the cochain $F:B \times S \to G$ and the function $f:B \times D_1 \to G$ can be chosen so that $\tilde{f}$ will be a cocycle in $H^2_c[(B,D_1,\delta_1),K]$ and consequently the obstruction $\tilde{\lambda}$ will 'vanish', *i.e.*, $\tilde{\lambda}(b,c_1,d_1) \equiv e$, where e is the identity of the group K.

Suppose in addition that the group T is simply connected. Then $S = T$.

3·16.25 LEMMA: *Let $\tilde{f}$ be a cocycle. Then there exists an extension (X,T,π,K,σ) such that the corresponding element of $K^2_c[(B,T,\tilde{\rho}),G]$ coincides with $\{F\}$.*

Proof: It follows from $\tilde{f} \in C^2_c[(B,D_1,\delta_1),K]$ that $f(b,c_1d_1) = k(b,c_1d_1)f(bc_1,d_1)f(b,c_1)$, where $k(b,c_1,d_1) \in D_2$. As is seen from the proof of Theorem 3·16.21, $k(b,c_1,d_1) \equiv \Psi(c_1,d_1)$ and moreover $\{\Psi\} \in H^2[D_1,D_2]$.

Let D be the cross product of D_1 and D_2 by means of the cocycle ψ and let A be the Cartesian product of B and G. Define an action $\hat{\delta}$ of the group D on A by (16.23). Further define an action $\hat{\pi}$ of the group $T = S$ on A by (16.31). In view of (16.36), the transformation groups $(B \times G, D, \hat{\delta})$ and $(B \times G, T, \hat{\pi})$ commute. It follows from the proof of Theorem 3·16.21 that $(B \times G, D, \hat{\delta})$ and $(B \times G, G, \hat{\sigma})$ also commute.

Let $X \equiv (B \times G)/D$ be the (quotient) space of the orbits of the transformation group $(B \times G, D, \hat{\delta})$. As was shown above, the space X is compact and the canonical map of X onto $Y = B/D_1$ is a principal locally trivial K-fibre bundle. Since the actions of $T = S$ and D on $B \times G$ commute, the transformation group $(B \times G, T, \hat{\pi})$ induces a transformation group (X, T, π). Similarly, the transformation group $(B \times G, G, \hat{\sigma})$ induces a transformation group (X, K, σ). It is easy to verify that (X, T, π, K, σ) is a free central group extension. ∎

3·16.26 EXAMPLE: Let $\mathbb{R}$ denote, as usual the additive group of real numbers, $B = \mathbb{R} \times \mathbb{R}$, $D_1 = \{(m,n) | m,n \in \mathbb{Z}\}$, $D_2 = \{p | p \in \mathbb{Z}\}$. Define an action of the group D_1 on B by $(x,y)(m,n) = (x - m, y - n)$ and an action of $\mathbb{R}$ on B by

$$((x,y),t)\tilde{\rho} = (x + \alpha t, y + \beta t), \qquad (t \in \mathbb{R}),$$

where α and β are certain real numbers.

Let $Y = B/D_1$, let $(Y, \mathbb{R}, \rho)$ be the flow induced by $(B, \mathbb{R}, \tilde{\rho})$ and let $\mathbb{K} = \mathbb{R}/D_2$.

Our goal is to describe all central locally trivial group extensions $(X, \mathbb{R}, \pi, \mathbb{K}, \sigma)$ of the flow $(Y, \mathbb{R}, \rho)$ by means of the group $\mathbb{K}$.

It is easy to show that each cross product of D_1 and D_2 is isomorphic to one of the groups $D^{(k)}$ $(k \in \mathbb{Z})$, where $D^{(k)}$ is characterized as follows. The set $D^{(k)}$ consists of all triples (m,n,p) of integers and the multiplication is defined by

$$(m_1,n_1,p_1)(m_2,n_2,p_2) = (m_1 + m_2, n_1 + n_2, p_1 + p_2 + km_1n_2).$$

Thus the cocycle corresponding to the group $D^{(k)}$ is of the form $\varphi_k(m_1,n_1,m_2,n_2) = km_1n_2$. The action of $D^{(k)}$ on $A = B \times \mathbb{R} = \mathbb{R}^3$ is determined by some cocycle $f_k(x,y,m,n)$ satisfying the condition

$$f_k(x,y,m_1 + m_2,n_1 + n_2) = km_1n_2 + f_k(x + m_1,y + n_1,m_2,n_2) + f_k(x,y,m_1,n_1).$$

For f_k we can take

$$f_k(x,y,m,n) = - knx. \qquad (16.38)$$

Then, by (16.23), the action of the group $D^{(k)}$ on A can be written in the following way: $(x,y,z)(m,n,p) = (x - m, y - n, z - knx - p)$. Set $X^{(k)} = A/D^{(k)}$. Evidently each space $X^{(k)}$ is a three-dimensional nilmanifold (*see* Subsection 2·10.10). As the space B is contractible, it follows from Theorem 3·16.21 that each principal $\mathbb{K}$-fibre bundle with base Y is isomorphic to one of the spaces $X^{(k)}$. The formula

$$((x,y,z),s)\hat{\sigma} = (x,y,z + s), \qquad (s \in \mathbb{R}),$$

determines an action of the group $\mathbb{R}$ on A, which induces an action σ of $\mathbb{K}$ on $X^{(k)}$.

Now we shall describe all the central locally trivial extensions $(X^{(k)},\mathbb{R},\pi,\mathbb{K},\sigma)$ of the transformation group $(Y,\mathbb{R},\rho)$. The universal covering flow $(A,\mathbb{R},\hat{\pi})$ is determined by means of some cocycle $F_k(x,y,t)$ which satisfies the condition

$$F_k(x,y,t) = F_k(x + m,y + n,t) + f_k(x,y,m,n) - f_k(x + \alpha t,y + \beta t,m,n),$$

which in view of (16.38) can be rewritten as

$$F_k(x,y,t) = F_k(x + m,y + n,t) + kn\alpha t. \tag{16.39}$$

Besides that, by the definition of a cocycle,

$$F_k(x,y,t + s) = F_k(x + \alpha t,y + \beta t,s) + F_k(x,y,t). \tag{16.40}$$

It follows immediately that for the cocycle F_k we may take

$$F_k(x,y,t) = - k(\alpha yt + \tfrac{1}{2}\alpha\beta t^2).$$

Any other cocycle $F_k{}'$ satisfying (16.39) and (16.40) can be obtained from F_k by adding an arbitrary continuous function $\Phi(x,y,t)$ satisfying the identities

$$\Phi(x,y,t) = \Phi(x + m,y + n,t), \tag{16.41}$$

$$\Phi(x,y,t + s) = \Phi(x + \alpha t,y + \beta t,s) + \Phi(x,y,t) \tag{16.42}$$

In particular, we can take

$$\Phi(x,y,t) = \int_0^t \Psi(x + \alpha\tau,y + \beta\tau)d\tau,$$

where $\Psi(x,y)$ is an arbitrary continuous periodic function with period unity in each variable, *i.e.*, satisfying $\Psi(x,y) \equiv \Psi(x + m,y + n)$. The flow $(A,\mathbb{R},\hat{\pi}_k)$ corresponding to the cocycle F_k is determined by

$$((x,y,z),t)\hat{\pi}_k = (x + \alpha t,y + \beta t,z - k(\alpha yt + \tfrac{1}{2}\alpha\beta t^2)).$$

It is easy to verify that the flow $(X^{(k)},\mathbb{R},\pi_k)$ induced by the transformation group $(A,\mathbb{R},\hat{\pi}_k)$ on $X^{(k)} = A/D^{(k)}$ is a nil-flow.

In the general case, a flow $(A,\mathbb{R},\hat{\pi}_k)$ which corresponds to an arbitrary cocycle $F_k{}'$ is determined by

$$((x,y,z),t)\tilde{\pi}_k = (x + \alpha t, y + \beta t, z - k(\alpha y t + \tfrac{1}{2}\alpha\beta t^2) + \Phi(x,y,t)),$$

where $\Phi(x,y,t)$ is a continuous function satisfying (16.41) and (16.42). In this case the flow induced on $X^{(k)}$ by $(A,\mathbb{R},\tilde{\pi}_k)$ will be called a *GENERALIZED NIL-FLOW.*

REMARKS AND BIBLIOGRAPHICAL NOTES

Central group extensions were considered by Ellis [*7-13*] and Furstenberg [*2*]. The general concept of a group extension is introduced by Bronšteĭn [*8*]. The results in this section are taken from the papers by Bronšteĭn [*8,9*]. A different approach to group extensions based on 'crossed representations' is developed by Knapp [*2*] (*see also* Ellis [*13*], pp. 196-198).

§(3·17): RELATIONSHIPS BETWEEN GROUP EXTENSIONS AND EQUICONTINUOUS EXTENSIONS; A STRENGTHENED STRUCTURE THEOREM FOR DISTAL EXTENSIONS

The considerations in Subsection 3·15.7 show that each equicontinuous minimal extension is subordinated to some central group extension with a compact topological structure group. This fact combined with the well known theorem on the structure of compact topological groups permits one to study in detail the structure of equicontinuous and consequently distal minimal extensions. This will be done in the present section.

3·17.1 Let (X,T,π,G,σ) be a central group extension, F a closed subgroup of G and $G/F = \{hF \mid h \in G\}$ the quotient space. Define two maps

$$\mu: X \times (G/F) \times T \to X \times (G/F), \delta: X \times (G/F) \times G \to X \times (G/F),$$

by

$$((x,hF),t)\mu = (x\pi^t, hF), \tag{17.1}$$

$$((x,hF),g)\delta = (xg, g^{-1}hF). \tag{17.2}$$

Then $(X \times (G/F),T,\mu,G,\delta)$ is a central group extension. Let $W \equiv (X \times (G/F))/G$ be the quotient space of the G-orbits and let (W,T,γ) be the transformation group induced by $(X \times (G/F),T,\mu)$.

3·17.2 THEOREM: *Let (X,T,π,G,σ) be a central group extension with (X,T,π) minimal and let $p:X \to X/G$ denote the canonical projection. Further, let (Z,T,ν) be a transformation group and let $\varphi:(X,T,\pi) \to (Z,T,\nu)$ and $\psi:(Z,T,\nu) \to (X/G,T,\rho)$ be* *homomorphisms with $\varphi \circ \psi = p$. Then there exist a closed subgroup F of the group G and a continuous bijective map ϑ of $W = (X \times (G/F))/G$ onto Z which is a homomorphism of (W,T,γ) onto (Z,T,ν). If, moreover, the space X is assumed to be compact, then ϑ is a homeomorphism.*

Proof: Set

$$F_x = \{g \mid g \in G, x\sigma^g\varphi = x\varphi\}, \qquad A_x = \{u \mid u \in X, u\varphi = x\varphi\}, \qquad (x \in X).$$

Henceforth we shall write xg instead of $x\sigma^g$.

Let us show that the set F_x does not depend on the choice of $x \in X$ and that it is a subgroup of G. Let $g \in F_x$, *i.e.*, $xg\varphi = x\varphi$. Then $x\pi^t g\varphi = xg\pi^t\varphi = xg\varphi\nu^t = x\varphi\nu^t = x\pi^t\varphi$. Hence it follows that $g \in F_{x\pi^t}$, *i.e.*, $F_x \subset F_{x\pi^t}$, $(x \in X, t \in T)$. But then $F_x = F_{x\pi^t}$. Let $x_1, x_2 \in X$. Since X is a minimal set of (X,T,π), there exists a net $\{t_i \mid i \in I\}$ $(t_i \in T)$ with $\{x_1\pi^{t_i} \mid i \in I\} \to x_2$. If $g \in F_{x_1}$, then it follows from the above that $x_1\pi^{t_i}\sigma^g\varphi = x_1\pi^{t_i}\varphi$.

Since σ^g and φ are continuous maps, we get by taking the limit that $x_2 g\varphi = x_2\varphi$, *i.e.*, $g \in F_{x_2}$. Thus $F_{x_1} \subset F_{x_2}$. By symmetry we finally obtain $F_{x_1} = F_{x_2}$ $(x_1, x_2 \in X)$.

Set $F = F_x$ $(x \in X)$. We shall show that F is a subgroup of G. If $g_1, g_2 \in F$, then $xg_1\varphi = x\varphi$, $xg_2\varphi = x\varphi$ for every point $x \in X$. Hence $(xg_1)g_2\varphi = (xg_1)\varphi = x\varphi$, *i.e.*, $g_1g_2 \in F$. Further, $xg_1\varphi = x\varphi = (xg_1)g_1^{-1}\varphi$, and therefore $g_1^{-1} \in F$. Thus, F is a subgroup of G. As the map $\varphi: X \to Z$ is continuous, the set F is closed in G.

Let $X/F = \{xF \mid x \in X\}$ be the quotient space of X with respect to the relation which is defined to hold between x_1 and x_2 iff $x_1F = x_2F$.

Further let $G/F \equiv \{hF \mid h \in G\}$ be the homogeneous space and let $R = X \times (G/F)$. Define a central group extension $(X \times (G/F), T, \mu, G, \sigma)$ by (17.1,2). Let $W = R/G \equiv \{\{(xg, g^{-1}hF) \mid g \in G\} \mid x \in X, h \in G\}$ be the space of orbits $\{(xg, g^{-1}hF) \mid g \in G\}$ of the transformation group (R,G,δ). Since the transformations $\mu^t: R \to R$ $(t \in T)$ and $\delta^g: R \to R$ $(g \in G)$ commute, the transformation group (R,T,μ) induces a transformation group (W,T,γ), where

$$(\{(xg, g^{-1}hF) \mid g \in G\}, t)\gamma = \{(x\pi^t g, g^{-1}hF) \mid g \in G\}.$$

Define a map $\Phi: W \to X/F$ by

$$(\{(xg, g^{-1}hF) \mid g \in G\})\Phi = xhF, \qquad (x \in X, h \in G),$$

and a map: $\Phi^*: X/F \to W$ by

$$(xF)\Phi^* = \{(xg, g^{-1}F) \mid g \in G\}, \qquad (x \in X).$$

Let us show that Φ^* is well defined. If $x_1F = xF$, then $x_1 = xf$ for some $f \in F$. Then

$$\{(x_1 g, g^{-1}F) | g \in G\} = \{xfg, g^{-1}F) | g \in G\}$$

$$= \{(xfg, (fg)^{-1}F) | g \in G\} = \{(xg, g^{-1}F) | g \in G\},$$

because F is a group. Hence $(x_1F)\Phi^* = (xF)\Phi^*$. It is easy to verify that Φ^* is the inverse to the map Φ.

Let us prove that Φ is continuous. Let $\lambda: R \to R/G \equiv W$ be the canonical projection and let $\Psi = \lambda \circ \Phi$. It suffices to verify that the composition $\Psi: R \to X/F$ is continuous.

Let $(x, hF) \in R$ and let S be an arbitrary neighbourhood of the point $xhF \equiv (x, hF)\Psi$ in X/F. There exists a neighbourhood S_1 of xh such that $S = \{yF | y \in S_1\}$. For the neighbourhood S_1 of $xh \in X$ we can choose a neighbourhood U of $x \in X$ and a neighbourhood V_1 of $h \in G$ so that $UV_1 \subset S_1$. Set $V = \{gF | g \in V_1\}$. Since the canonical map of X onto X/F is open, V is a neighbourhood of hF in G/F, and hence $U \times V$ is a neighbourhood of $(x, hF) \in R$. Evidently $(U \times V)$ $\Psi \subset S$.

Let us prove that Φ is open. Since the map $\lambda: R \to W$ is continuous, it will suffice to show that the composed map $\Psi = \lambda \circ \Phi$ is open. Let $(x, hF) \in R$ and let S be any neighbourhood of the point (x, hF) in R. There exist open neighbourhoods U of x and V of $h \in G$ such that

$$S_1 \equiv \{(y, gF) | y \in U, g \in V\} \subset S.$$

Then $S\Psi \supset S_1\Psi = \{ygF | y \in U, g \in V\}$. Since the set UVF is open in X, the set $S_1\Psi$ is open in X/F, and hence the map Ψ is open. We conclude that the map $\Phi: W \to X/F$ is a homeomorphism.

Define a map $\xi: X/F \to Z$ by $(xF)\xi = x\varphi$. The map ξ is well defined since $xF = x_1F$ implies $x_1 = xg$ for some element $g \in F$, and $x_1\varphi = xg\varphi = x\varphi$ $(x \in X)$ by definition of $F \equiv F_x$ $(x \in X)$. The map ξ is bijective. In fact, if $(xF)\xi = (x_1F)\xi$, then $x\varphi = x_1\varphi$, *i.e.*, $x_1 \in A_x = xF_x = xF$, and hence there exists an element $g \in F$ with $x_1 = xg$. But then $x_1F = xgF = xF$. The map ξ is continuous because the composed map $\varphi: X \to Z$ is continuous. Thus $\xi: X/F \to Z$ is a bijective continuous map.

Set $\vartheta = \Phi \circ \xi$. It follows from the preceding considerations that $\vartheta: W \to Z$ is bijective and continuous.

Clearly ϑ is a homomorphism of the transformation group (W, T, γ) onto (Z, T, ν).

Suppose now that moreover the space X is compact. The space X/F is Hausdorff since the map $\xi: X/F \to Z$ is continuous and bijective and Z is Hausdorff. From this and the compactness of X we deduce that the space X/F is also compact. Hence ξ is a homeomorphism. As was previously shown, $\Phi: W \to X/F$ is also a homeomorphism. Thus, if X is compact, then $\vartheta = \Phi \circ \xi$ is a homeomorphism. ▌

3·17.3 Let X be a compact space, $\varphi:(X,T) \to (Y,T)$ a distal minimal extension, G the Galois group of φ and $(M,T) \equiv (M(X,Y)$ $T)$ (*see* 3·15.3). There exists a subgroup K of G such that (X,T) $(M/K,T)$. Provide the group G (and its subgroup K) with the discrete topology. As G is a group of automorphisms of the transformation group (M,T) which is transitive on the fibres of φ, the canonical map $\varkappa:(M,T) \to (M/G,T) \approx (Y,T)$ is a central group extension with the discrete structure group G.

Let $G/K = \{hK \mid h \in G\}$. Define a central group extension $(M \times (G/K),T,\mu,G,\delta)$ in a way similar to (17.1,2). Let $W \equiv (M \times (G/K))/G$ be the quotient space of orbits of the transformation group $(M \times (G/K),G,\delta)$ and let (W,T,γ) be the transformation group induced by $(M \times (G/K),T,\mu)$. It follows from Theorem 3·17.2 that the transformation groups (X,T) and (W,T) are isomorphic.

3·17.4 Suppose in particular that $\varphi:(X,T) \to (Y,T)$ is an equicontinuous minimal extension. Then G can be provided in a natural way with a compact Hausdorff topology. We shall say in this case that (M,T) is the *CENTRAL GROUP EXTENSION ADJOINT TO THE EQUICONTINUOUS EXTENSION* $\varphi:(X,T) \to (Y,T)$. According to the terminology adopted in the theory of fibre bundles (Husemoller [*1*]), $X \approx (M \times (G/K))/G$ is the *TOTAL SPACE of a fibre bundle with base Y and fibre G/K associated with the principal G-fibre bundle* $\xi:M \to Y$. The group G is called the *STRUCTURE GROUP* of $\varphi:X \to Y$.

Let N be an arbitrary closed normal subgroup of G, $M/N = \{wN \mid w \in M\}$ and let $\varkappa:M \to M/N$ be the canonical projection. As $N \subset G$, (M,T) induces a transformation group $(M/N,T)$. Clearly $(M/N,T)$ is a central group extension of $(M/G,T)$ with the compact structure group G/N. Since N is a normal subgroup and K is a subgroup of G, NK is a closed subgroup of G. Consider the following commutative diagram:

$$\begin{array}{ccccc}
 & & (M,T) & \xrightarrow{\ \varkappa\ } & (M/N,T) \\
 & & \downarrow{\scriptstyle\psi} & & \downarrow{\scriptstyle\tilde\psi} \\
(X,T) & \approx & (M/K,T) & \xrightarrow{\ \tilde\varkappa\ } & (M/NK,T) \approx ((M/N)/(NK/N),T) \\
\downarrow{\scriptstyle\varphi} & & \downarrow & \swarrow{\scriptstyle\tilde\varphi} & \\
(Y,T) & \approx & (M/G,T) & \approx ((M/N)/(G/N),T). &
\end{array} \tag{17.3}$$

By Corollary 3·13.4, $\tilde\varkappa$ and $\tilde\varphi$ are equicontinuous extensions. Each fibre of $\tilde\varphi$ is homeomorphic to the homogeneous space $(G/N)/(NK/N) \approx G/NK$.

3·17.5 THEOREM: *Let X be a compact space and let $\varphi:(X,T) \to (Y,T)$ be a minimal equicontinuous extension. Then there exists a transfinite sequence*

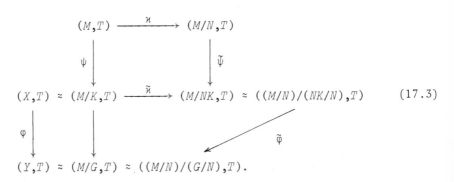

$$R_\varphi \equiv R_0 \supset R_1 \supset \cdots \supset R_\alpha \supset R_{\alpha+1} \supset \cdots \supset R_\vartheta = \Delta(X)$$

of closed invariant equivalence relations R_α $(0 < \alpha < \vartheta)$ *satisfying the conditions*

(1): *If* $\alpha < \vartheta$, *then the canonical projection* $\varphi_\alpha^{\alpha+1}: X/R_{\alpha+1} \to X/R_\alpha$ *is an equicontinuous locally trivial fibre bundle with a compact Lie group* G_α *as the structure group and with fibre homeomorphic to* G_α/H_α, *where* H_α *is a closed subgroup of* G_α*;*

(2): *If* β *is a limit ordinal,* $\beta < \vartheta$, *then* $R_\beta = \bigcap_{\alpha<\beta} R_\alpha$.

Proof: Let $\xi:(M,T) \to (Y,T)$ be the central group extension adjoint to the minimal equicontinuous extension $\varphi:(X,T) \to (Y,T)$ and let G be the structure group of ξ. Then $(M/G,T) \approx (Y,T)$ and there exists a closed subgroup K of G such that $(M/K,T) \approx (X,T)$. Note that if φ is not a homeomorphism then K is a proper subgroup of G.

Since G is a compact topological group, each neighbourhood of the identity of the group G contains a normal subgroup $N \subset G$ such that G/N is a Lie group (Pontrjagin [*1*], p. 330). Consider the commutative diagram (17.3) with N chosen to be the normal subgroup just defined. As G/N is a Lie group, the principal (G/N)-fibre bundle $\tilde\psi \circ \tilde\varphi$ is locally trivial by Gleason's theorem (*see* Montgomery and Zippin [*1*], p. 219). Hence $\tilde\psi$ is also a locally trivial fibre bundle with fibre $(G/N)/(NK/N)$ associated to the principal (G/N)-fibre bundle $\tilde\psi \circ \tilde\varphi$. If $K \neq G$, then we may suppose that $NK \neq G$, and therefore $\tilde\varphi$ is not a homeomorphism.

Since the extension $\tilde\varkappa$ is equicontinuous, we can repeat the above considerations with $\tilde\varkappa$ instead of φ. The proof is concluded by transfinite induction.∎

3·17.6 We may assume in Theorem 3·17.5 that the fibres of all $\varphi_\alpha^{\alpha+1}$ are either connected or finite. Indeed, consider the diagram (17.3), assuming that G is a compact Lie group and N is the component of the identity of G. Then the group G/N is finite, and hence the fibre $(G/N)/(NK/N)$ of $\tilde\varphi$ is also finite. The fibre of the extension $\tilde\varkappa$ is homeomorphic to the homogeneous space $NK/N \approx N/(K \cap N)$ of the connected group N. Thus each equicontinuous extension with a Lie structure group can be decomposed into a composition of two homomorphisms so that the first has connected fibres and the second has finite fibres.

3·17.7 Combining Theorem 3·17.5 with Theorem 3·14.23 we obtain the following proposition, which improves Theorem 3·14.23 on the structure of distal minimal extensions.

3·17.8 THEOREM: *Let* X *be a compact topological space,* T *a topological group and* $\varphi:(X,T) \to (Y,T)$ *a distal minimal ex-*

tension. Suppose that, moreover, at least one of the following conditions is satisfied:

(a): *The space X is metrizable;*

(b): *The group T is σ-compact.*

Then there exists a transfinite sequence:

$$R_\varphi \equiv R_0 \supset R_1 \supset \cdots \supset R_\alpha \supset R_{\alpha+1} \supset \cdots \supset R_\vartheta = \Delta(X),$$

of closed invariant equivalence relations R_α such that:

(1): *If $\alpha < \vartheta$, then the canonical projection $\varphi_\alpha^{\alpha+1}: X/R_{\alpha+1} \to X/R_\alpha$ is an equicontinuous locally trivial fibre bundle whose structure group G_α is a compact Lie group and the fibre is a homogeneous space of the form G_α/H_α, where H_α is a closed subgroup of G_α;*

(2): *If β is a limit ordinal, $\beta \leqslant \vartheta$, then $R_\beta = \bigcap_{\alpha<\beta} R_\alpha$.*

3·17.9 COROLLARY: *Suppose all the conditions of the preceding theorem are satisfied. If at least one fibre of the extension $\varphi: X \to Y$ is a finite-dimensional locally connected space, then the number ϑ in Theorem* 3·17.8 *is finite.*

3·17.10 COROLLARY: *Every compact metrizable locally connected finite-dimensional distal minimal set is a manifold (i.e., each point has a neighbourhood homeomorphic to an open ball of a Euclidean space).*

3·17.11 THEOREM: *Let X be a compact metrizable connected locally connected finite-dimensional space and (X,T) be a distal minimal transformation group. Then the fundamental group $\pi_1(X)$ contains a nilpotent subgroup of normal index (Kuroš [1]).*

Proof: We shall be using the notation of Theorem 3·17.8, where it is assumed that (Y,T) is a trivial transformation group (*i.e.*, $R_0 = X \times X$). By Corollary 3·17.9, the ordinal $\vartheta \equiv n$ is finite. We prove that for each i, $i = 0,1,\ldots,n$, the fundamental group of X/R_i contains a nilpotent subgroup of normal index. The required result corresponds to $i = n$.

The proof is by induction. If $i = 0$, then the space X/R_0 is a single point and the statement is trivial.

Suppose that the fundamental group of the space X/R_i, where $i < n$, contains a nilpotent subgroup of normal index. The space X/R_{i+1} is the total space of some locally trivial fibre bundle with base X/R_i, whose fibre is a homogeneous space of some compact Lie group G_i. Let $\xi:(W_{i+1},T) \to (X/R_i,T)$ be the group extension adjoint to the equicontinuous extension $\varphi_i^{i+1}:(X/R_{i+1},T) \to (X/R_i,T)$. The structure group of φ_i^{i+1} is equal to G_i.

We shall prove that $\pi_1(W_{i+1})$ contains a nilpotent subgroup of normal index. First consider the case when G_i is connected. Since the space X is connected, X/R_i is also connected. It is also clear that the spaces X/R_i and G_i are locally pathwise connected and locally simply connected. Therefore the induced homomorphism maps $\pi_1(W_{i+1})$ onto $\pi_1(X/R_i)$ and moreover the kernel of this homomorphism is contained in the centre of $\pi_1(W_{i+1})$ (Corollary 3·16.18). It follows from this and from the inductive hypothesis that $\pi_1(W_{i+1})$ contains a nilpotent subgroup of normal index. If the group G_i is finite, then ξ is a covering map, and hence the induced homomorphism ξ^* is a monomorphism of $\pi_1(W_{i+1})$ into $\pi_1(X/R_i)$. Thus $\xi^*(\pi_1(W_{i+1}))$ is a subgroup of normal index of the group $\pi_1(X/R)$. It is known (*see* Kuroš [*1*]) that the intersection of two subgroups of normal index is also a subgroup of normal index. Thus it follows from the induction assumption that $\pi_1(W_{i+1})$ contains a nilpotent subgroup of normal index. For the general case we note that the component of the identity of a compact Lie group is an invariant subgroup with a finite quotient-group. Therefore, combining the above arguments we get that in any case $\pi_1(W_{i+1})$ contains a nilpotent subgroup of normal index. In view of Theorem 2·11.26, it is clear that $\pi_1(X/R_{i+1})$ contains a subgroup of normal index which is a homomorphic image of $\pi_1(W_{i+1})$. Using this fact and the aforementioned property of $\pi_1(W_{i+1})$, we conclude that $\pi_1(X/R_{i+1})$ contains a nilpotent subgroup of normal index.∎

3·17.12 THEOREM: *Let T be a topological group such that for some compact subset $K \subset T$ every subgroup of T, both closed and open, containing K coincides with T. Let X be a compact space and let $\varphi:(X,T) \to (Y,T)$ be a minimal extension. Assume further that (Y,T) is equicontinuous, φ is an open map and at least one fibre X_y consists of a finite number of points. Then the transformation group (X,T) is also equicontinuous.*

Proof: By Theorem 3·13.17, the extension $\varphi:(X,T) \to (Y,T)$ is equicontinuous. Therefore (X,T) is distal and each fibre X_y consists of the same number of points. There exists a symmetric index $\gamma_0 \in \mathfrak{U}[X]$ such that $R_\varphi \cap \gamma_0{}^3 = \Delta(X)$. Since (Y,T) is equicontinuous there is a uniformity base $\mathfrak{U}_1$ of the uniformity $\mathfrak{U}[Y]$ consisting of indices that are invariant under the action of T.

Let $\alpha \in \mathfrak{U}[X]$. Choose $\delta \in \mathfrak{U}[X]$ so that $\delta K \subset \gamma_0$, $\delta^2 \subset \gamma_0$, $\delta \subset \alpha$. For this δ there exists an index $\varepsilon \in \mathfrak{U}[X]$ such that for every x_1, $x_2 \in X$ with $(x_1\varphi, x_2\varphi) \in \varepsilon$, either $(x_1,x_2) \in \delta$ or $(x_1,x_2) \in (X \times X) \setminus \gamma_0$. Set $A = \{(x_1,x_2) \mid (x_1\varphi, x_2\varphi) \in \varepsilon, (x_1,x_2) \in \gamma_0\}$. Then $A \subset \delta \subset \alpha$ and $AK \subset A$. Denote $\{t \mid t \in T, At \subset A\}$ by T_1. Clearly, the set T_1 is both open and closed, and, moreover, it follows from the above alternative that $T_1 = \{t \mid t \in T, At = A\}$. Hence T_1 is a subgroup of T with $K \subset T_1$, and therefore $T_1 = T$, because of the assumption made about T. If the index $\beta \in \mathfrak{U}[X]$ is chosen so that $\beta \subset \gamma_0$ and $\beta(\varphi \times \varphi) \subset \varepsilon$, then $\beta \subset A$ and therefore $\beta T \subset \alpha$.∎

3·17.13 THEOREM: *Let T be a σ-compact topological group satisfying the condition formulated in Theorem* 3·17.12, *let X be a compact space and let $\varphi:(X,T) \to (Y,T)$ be a distal minimal extension. Assume further that the transformation group (Y,T) is equicontinuous and φ has at least one zero-dimensional fibre. Then (X,T) is also equicontinuous and hence almost periodic.*

Proof: By Theorem 3·17.8 the extension φ can be decomposed into a transfinite sequence of open extensions with finite fibres. Applying transfinite induction in combination with Theorem 3·17.12 and Lemma 2·9.16 we see that the assertion is true.∎

3·17.14 LEMMA: *Let the group T satisfy the same conditions as in the preceding theorem and let X be a compact space. Suppose further that $\varphi:(X,T) \to (Y,T)$ is a distal minimal extension with at least one connected one-dimensional fibre. Then φ can be represented as a composition of an equicontinuous extension and a finite-to-one extension.*

Proof: We shall use the notation of Theorem 3·17.8. It follows from the restrictions put on φ that the fibres of the equicontinuous extension φ_0^1 are connected and of the form G_0/H_0, where G_0 is a compact Lie group and H_0 is a closed subgroup of G_0 with $\dim G_0/H_0 = 1$. Moreover, φ_1^ϑ is an extension with finite fibres.∎

REMARK: Arguments similar to that used in the proof of Theorem 3·17.12 together with a generalization of Lemma 1·6.26 to the case of extensions show that the extension φ is in fact equicontinuous whenever the hypotheses of Lemma 3·17.14 are fulfilled.

3·17.15 LEMMA: *If a minimal flow $(X,\mathbb{R},\pi)$ is an equicontinuous extension of some periodic flow $(Y,\mathbb{R})$, then $(X,\mathbb{R})$ is almost periodic.*

Proof: Let φ be a homomorphism of $(X,\mathbb{R})$ onto $(Y,\mathbb{R})$, $y_0 \in Y$ *and* $X_0 = \{x \mid x \in X, x\varphi = y_0\}$. Let τ denote the least positive period of the flow $(Y,\mathbb{R})$. Then $X_0\pi^\tau = X_0$ and $X_0\pi^t \cap X_0 = \emptyset$ whenever t is not a multiple of τ. Thus X_0 is a global section of the flow $(X,\mathbb{R})$. As $\varphi:(X,\mathbb{R}) \to (Y,\mathbb{R})$ is an equicontinuous extension, the homeomorphism $\pi^\tau : X_0 \to X_0$ determines an equicontinuous cascade. Hence it follows that the flow $(X,\mathbb{R},\pi)$ is equicontinuous and consequently almost periodic.∎

3·17.16 THEOREM: *Let X be a locally connected finite-dimensional compact metric space of dimension at least* 2, *and let $(X,\mathbb{R})$ be a distal minimal flow. Then there exists a homomorphism of $(X,\mathbb{R})$ onto some almost periodic flow $(X_0,\mathbb{R})$ with* $\dim X_0 \geqslant 2$.

Proof: Let

$$Q_1 = Q \equiv \bigcap_{\alpha \in \mathfrak{U}[X]} \overline{\alpha\mathbb{R}}.$$

Then $(X/Q_1,\mathbb{R})$ is a non-trivial almost periodic flow, and hence X/Q_1 is provided with a structure of a connected commutative compact topological Lie group (Theorem 3·17.8). Consequently X/Q_1 is a torus (Pontrjagin [*1*]) with $\dim X/Q_1 \geqslant 1$. Let us prove that $\dim X/Q_1 \geqslant 2$. Suppose the contrary is true. Then $(X/Q_1,\mathbb{R})$ is a periodic flow. Let

$$Q_2 = \cap\,[\overline{\alpha\mathbb{R}\cap Q_1}\,|\,\alpha \in \mathfrak{U}].$$

The flow $(X/Q_2,\mathbb{R})$ is a non-trivial equicontinuous extension of $(X/Q_1,\mathbb{R})$. It may be supposed that the natural map of X/Q_2 onto X/Q_1 is locally trivial, therefore $\dim X/Q_2 \geqslant 2$. It follows from Lemma 3·17.15 that the flow $(X/Q_2,\mathbb{R})$ is equicontinuous, but this contradicts the fact that $(X/Q_1,\mathbb{R})$ is the greatest equicontinuous factor of the flow $(X,\mathbb{R})$. ▌

3·17.17 THEOREM: *A distal minimal flow* $(X,\mathbb{R})$ *with a two-dimensional locally connected compact metric phase space is almost periodic. Hence* X *is homeomorphic to the two-dimensional torus.*

Proof: By the preceding theorem and Theorem 3·17.8, the flow $(X,\mathbb{R})$ covers an almost periodic flow; therefore it is itself almost periodic. ▌

3·17.18 LEMMA: *Let* G *be a compact connected group and* H *be a closed subgroup of* G *such that the space* G/H *is one-dimensional. Then* H *contains the commutator of* G*; hence* H *is a normal subgroup and the group* G/H *is commutative.*

Proof: This assertion is proved in the book by Hoffman and Mostert [*1*] (p. 303) under the additional hypothesis that H is connected. As to the general case, let H_0 denote the component of the identity of the group H; then the space G/H_0 is also one-dimensional, hence G/H_0 is a commutative group, and therefore G/H is also commutative. ▌

3·17.19 THEOREM: *Let* $(X,\mathbb{R})$ *be an arbitrary distal minimal flow with a three-dimensional locally connected metric phase space. Then there exists a generalized nil-flow* $(X',\mathbb{R})$ *covering* $(X,\mathbb{R})$.

Proof: By Theorem 3·17.8 and Corollary 3·17.9, there exists a finite sequence

$$X \times X \equiv R_0 \supset R_1 \supset \cdots \supset R_n = \Delta(X)$$

of closed invariant equivalence relations R_i such that φ_i^{i+1}: $(X/R_{i+1},\mathbb{R}) \to (X/R_i,\mathbb{R})$ is a locally trivial equicontinuous extension with the structure group a Lie group G_i and with a fibre of the form G_i/N_i, where N_i is a closed subgroup of G_i $(i = 0,1,\ldots,$ $n - 1)$. Denote X/R_i by X_i. It follows from Theorem 3·17.16 that we may suppose $(X_1,\mathbb{R})$ to be an almost periodic flow on the two-dimensional torus. Let $\xi:(M,\mathbb{R}) \to (X_1,\mathbb{R})$ be the central group extension adjoint to $\varphi_1^2:(X_2,\mathbb{R}) \to (X_1,\mathbb{R})$. The structure group of ξ is equal to G_1. Let H be the component of the identity of the group G_1. Consider the following commutative diagram:

$$\begin{array}{ccc}
 & (M,\mathbb{R}) & \\
 \swarrow^{\lambda} & & \searrow^{\mu} \\
(X_2,\mathbb{R}) \approx (M/N_1,\mathbb{R}) & \xleftarrow{\ \psi\ } & (M/(N_2\cap H),\mathbb{R}) \\
\downarrow{\scriptstyle \varphi_1^2} & & \downarrow{\scriptstyle \nu} \\
(X_1,\mathbb{R}) \approx (M/G_1,\mathbb{R}) & \xleftarrow{\ \eta\ } & (M/H,\mathbb{R}).
\end{array}$$

As G_1/H is a finite group, η is a covering map. Hence it follows from Theorem 3·17.12 that the transformation group $(M/H,\mathbb{R})$ is almost periodic and M/H is homeomorphic to the two-dimensional torus. Similarly, ψ is also a covering map and the fibres of ν are homeomorphic to those of φ_1^2.

Without loss of generality we may suppose that $\dim(G_1/N_1) \geqslant 1$.

Since $\dim X_2 \leqslant 3$ and $\dim X_1 = 2$, we conclude that $\dim(G_1/N_1) = 1$ and $\dim X_2 = 3$. Hence it follows from the previous considerations that $\dim(H/(N_1\cap H)) = 1$. By Lemma 3·17.18, $N_1\cap H$ is a normal subgroup of G and $H/(N_1\cap H)$ is a one-dimensional connected compact commutative Lie group; hence $H/(N_1\cap H)$ is isomorphic to the group $\mathbb{K} = \{\xi \mid \xi \in \mathbb{C}, |\xi| = 1\}$. Thus ν is a central group extension of an almost periodic flow on a two-dimensional torus M/H with $H/(N_1\cap H) \approx \mathbb{K}$ as structure group. By the considerations in Subsection 3·16.26, $(M/(N_1\cap H),\mathbb{R})$ is a generalized nil-flow.

As $\dim X_2 = \dim X_3 = \cdots = \dim X = 3$, the flow $(X,\mathbb{R})$ covers $(X_2,\mathbb{R})$. Since $(M/(N_1\cap H),\mathbb{R})$ also covers $(X_2,\mathbb{R})$, there exists a minimal flow $(X',\mathbb{R})$ with a compact phase space which covers both $(X,\mathbb{R})$ and $(M/(N_1\cap H),\mathbb{R})$. Thus $(X',\mathbb{R})$ is the required flow.∎

3·17.20 We give an example of a distal minimal cascade for which not every endomorphism is an automorphism. This example may be also regarded as an illustration of the theory of group extensions.

Let $\mathbb{K} = \{\xi \mid \xi \in \mathbb{C}, |\xi| = 1\}$. Consider a homeomorphism $F:\mathbb{K}^\infty \to \mathbb{K}^\infty$ of the form $F(z,w_1,\dots,w_n,\dots) = (\alpha z, \varphi(z)w_1,\dots,\varphi(\beta^{n-1}z)w_n,\dots)$, where the numbers $\alpha,\beta \in \mathbb{K}$ and the continuous map $\varphi:\mathbb{K} \to \mathbb{K}$ will be defined below in such a way that the cascade $(\mathbb{K}^\infty, F)$ is minimal. The map $S:\mathbb{K}^\infty \to \mathbb{K}^\infty$, where

$$S(z,w_1,\dots,w_n,\dots) = (\beta z, w_2,\dots,w_{n+1},\dots),$$

commutes with F, and hence S is an endomorphism of the cascade $(\mathbb{K}^\infty, F)$. But the map S is not injective.

Let $\gamma \in \mathbb{K}$ be a certain transcendental number, *i.e.*, $p(\gamma) \neq 0$ for each polynomial function p with integer coefficients. Choose a number $\beta \in \mathbb{K}$ such that $\beta^n \neq 1$ for all $n \in \mathbb{Z}$. Then the set $\{\beta^n | n = 1,2,\ldots\}$ is dense in $\mathbb{K}$. Hence we can choose an increasing sequence $N = \{n(k) | k = 1,2,\ldots\}$ of positive integers so that

$$|\beta^{n(k)} - \gamma| < 1/k, \qquad (k = 1,2,\ldots).$$

We shall now prove the existence of a number $\alpha \in \mathbb{K}$ such that $\alpha^n \neq 1$ for all $n \in \mathbb{Z}$, and

$$|\alpha^{m(\ell)} - 1| < C[m(\ell)]^{-2}$$

for some subsequence $\{m(\ell)\} = \{n(k_\ell) | \ell = 1,2,\ldots\}$ of the sequence N and some constant number c. The number $\alpha \in \mathbb{K}$ can be represented in the form $\alpha = \exp(2\pi i a)$, where $a \in [0,1]$, i is the imaginary number $\sqrt{-1}$. Therefore it suffices to find a number $a \in [0,1]$ with

$$\left|a - \frac{p(\ell)}{m(\ell)}\right| < \frac{1}{[m(\ell)]^3}, \qquad (\ell = 1,2,\ldots),$$

where $p(\ell)$ is a certain integer.

Set $m(1) = n(1)$,

$$I_1(j) = \left[\frac{j}{m(1)} - \frac{1}{[m(1)]^3}, \frac{j}{m(1)} + \frac{1}{[m(1)]^3}\right], \qquad E_1 = \bigcup_{j=1}^{m(1)-1} I_1(j).$$

Suppose that the numbers $m(k)$ and the sets $E_k \subset [0,1]$ $(k = 1,\ldots,\ell)$ have been already defined. Choose a number $m(\ell + 1) \equiv n(k_{\ell+1})$ so that

$$\frac{2}{m(\ell + 1)} < \frac{1}{[m(\ell)]^3},$$

and define

$$I_{\ell+1}(j) = \left[\frac{j}{m(\ell + 1)} - \frac{1}{[m(\ell + 1)]^3}, \frac{j}{m(\ell + 1)} + \frac{1}{[m(\ell + 1)]^3}\right],$$

$$E_{\ell+1} = \bigcup_{j=1}^{m(\ell+1)-1} I_{\ell+1}(j).$$

Clearly, each line segment of the form $I_\ell(j)$ contains at least two segments of the form $I_{\ell+1}(k)$, and therefore the set $E = \bigcap_{\ell=1}^{\infty} E_\ell$ is non-countable. Each number a in E satisfies the required inequality.

Thus we have chosen numbers α and β with the necessary properties. Now we define the function $\varphi:\mathbb{K}\to\mathbb{K}$. For $n\in\mathbb{Z}$ we put:

$$a_n = \begin{cases} \alpha^n - 1, & \text{whenever } n\in N \text{ or } (-n)\in N; \\ 0 & \text{otherwise.} \end{cases}$$

Since

$$\sum_{n\in\mathbb{Z}} |a_n| \leqslant 2c \sum_{n=1}^{\infty} \frac{1}{n^2} < \infty,$$

and $a_{-n} = \bar{a}_n$ $(n = 1,2,\ldots)$, the function $H(z) \equiv \sum_{n\in\mathbb{Z}} a_n z^n$ is continuous on $\mathbb{K}$ and real valued. Define:

$$\varphi(z) = \exp[2\pi i H(z)], \qquad (z\in\mathbb{K}).$$

It remains to show that the cascade $(\mathbb{K}^\infty,F)$ is minimal for the above $\alpha,\beta\in\mathbb{K}$ and $\varphi:\mathbb{K}\to\mathbb{K}$. Observe that $(\mathbb{K}^\infty,F)$ can be regarded as the projective limit of the projective system $\{(\mathbb{K}^{n+1},F_n)\,|\,n = 0,1,2,\ldots\}$, where

$$F_0(z) = \alpha z,$$

$$F_n(z,w_1,\ldots,w_n) = (\alpha z,\varphi(z)w_1,\ldots,\varphi(\beta^{n-1}z)w_n), \qquad (n = 1,2,\ldots).$$

with canonical homomorphisms. Therefore it will suffice to show the minimality of $(\mathbb{K}^{n+1},F_n)$ for all $n = 0,1,2,\ldots$.

Since $\alpha^n \neq 1$ $(n\in\mathbb{Z})$, $(\mathbb{K},F_0)$ is a minimal cascade. Suppose that $(\mathbb{K}^{r+1},F_r)$ is minimal. Let us prove the minimality of $(\mathbb{K}^{r+2}, F_{r+1})$. Suppose the contrary. Notice that $(\mathbb{K}^{r+2},F_{r+1})$ is a group extension of $(\mathbb{K}^{r+1},F_r)$ with the structure group $\mathbb{K}$. Since every closed subgroup of $\mathbb{K}$ is either finite or coincides with $\mathbb{K}$, the following alternative holds: either the cascade $(\mathbb{K}^{r+2},F_{r+1})$ is minimal or there exists a positive integer m such that the cascade $(\mathbb{K}^{r+2},F^*_{r+1})$, where

$$F^*_{r+1}(z,w_1,\ldots,w_{r+1}) = (\alpha z,\varphi(z)w_1,\ldots,\varphi(\beta^{r-1}z)w_r,[\varphi(\beta^r z)]^m w_{r+1})$$

is congruent as a group extension to the cascade $(\mathbb{K}^{r+1}\times\mathbb{K},\Phi_{r+1})$ defined by $(x,\xi)\Phi_{r+1} = (xF_r,\xi)$ $(x\in\mathbb{K}^{r+1},\xi\in\mathbb{K})$. The second case can happen iff there exists a continuous map $\Psi:\mathbb{K}\to\mathbb{K}$ satisfying the equality

$$\Psi(\alpha z)\cdot[\Psi(z)]^{-1} = [\varphi(\beta^r z)]^m, \qquad (z\in\mathbb{K}). \tag{17.4}$$

This follows from Theorem 3·16.13.

Every continuous map $\Psi:\mathbb{K} \to \mathbb{K}$ can be represented in the form

$$\Psi(z) = z^{\ell}\exp[2\pi i\psi(z)],$$

where $\ell \in \mathbb{Z}$ is the degree of Ψ and $\psi:\mathbb{K} \to \mathbb{R}$ is a continuous map. Then (17.4) implies that

$$\exp(2\pi i\ell a)\cdot\exp[2\pi i\psi(\alpha z)]\cdot\exp[-2\pi i\psi(z)] = \exp[2\pi imH(\beta^2 z)].$$

Hence there exists a number $q \in \mathbb{Z}$ with

$$\ell a + \psi(\alpha z) - \psi(z) = mH(\beta^r z) + q, \qquad (z \in \mathbb{K}).$$

The function $\psi:\mathbb{K} \to \mathbb{R}$ can be expanded in a Fourier series which converges to ψ in L_2:

$$\psi(z) \sim \sum_{n\in\mathbb{Z}} b_n z^n, \qquad (z \in \mathbb{K}).$$

Then

$$\ell a + \sum b_n\alpha^n z^n - \sum b_n z^n = m \sum a_n z^n \beta^{rn} + q.$$

Comparing the coefficients at z^n we obtain

$$(\alpha^n - 1)b_n = ma_n\beta^{rn}, \qquad (n \in \mathbb{Z}).$$

If $n \in N = \{n(k)|k = 1,2,\ldots\}$, then $b_n = m\beta^{rn} = p(\beta^n)$, where $p(z) = mz^n$ $(z \in \mathbb{C})$. Therefore

$$\lim_{k\to\infty} b_{n(k)} \equiv \lim_{k\to\infty} p(\beta^{n(k)}) = p(\gamma) \neq 0, \qquad (17.5)$$

because γ is a transcendental number. But it follows from (17.5) that

$$\sum_{n\in\mathbb{Z}} |b_n|^2 = \infty,$$

which leads to a contradiction. Thus it is proved that all cascades $(\mathbb{K}^{n+1},F_n)$ are minimal, and hence $(\mathbb{K}^\infty,F)$ is also minimal.

REMARKS AND BIBLIOGRAPHICAL NOTES

Theorem 3·17.2 is a slight improvement of a proposition due to Ellis [*10,11*]. The construction in 3·17.1 is well known from the theory of fibre bundles (*cf.* Husemoller [*1*]). Statements 3·17.5-11 are taken from the paper by Bronšteĭn [*14*]. Theorem 3·17.12 (in the case $T = \mathbb{R}$) was proved by Žikov [*2*]. His proof is similar to some constructions by Amerio [*1,2*] (*see also* Demidovič [*2*]). The case $T = \mathbb{Z}$ can easily be reduced to the case $T = \mathbb{R}$ by embedding the cascade in a flow. Theorem 3·17.12 was proved by Ellis ([*13*], p. 56) under some additional assumptions (for example that the group T is connected or finitely-generated). The proof of Theorem 3·17.12 given here is new (it was independently found by Sacker and Sell *[*1*], *[*2*]). Theorem 3·17.13 is an extension of a result obtained by Žikov [*2,5*] for the case $T = \mathbb{R}$. The assertion contained in the remark after Lemma 3·17.14 is proved by Žikov for $T = \mathbb{R}$. Theorems 3·17.16,17,19 are due to Bronšteĭn [*14*]. Relations between distal, equicontinuous and group extensions were also considered by Ellis [*10,11*]. Example 3·17.20 is taken from a paper by Parry and Walters [*1*].

§(3·18): A METHOD FOR CONSTRUCTING MINIMAL SETS

3·18.1 Let X be a compact topological space and let T and G be topological groups. Moreover, assume G to be compact. Let (X,T,π,G,σ) be a central group extension. Then the transformation group (X,T,π) induces a transformation group $(X/G,T,\rho)$ on the quotient space $Y \equiv X/G = \{xG \mid x \in X\}$ of G-orbits. Let φ denote the canonical projection of X onto X/G.

Let $f: Y \times T \to G$ be a continuous map. The image of $(y,t) \in Y \times T$ under f will be denoted by $f(y,t)$. Suppose that f satisfies the identity

$$f(y,ts) = f(y,t)f(y\rho^t,s), \qquad (y \in Y; t,s \in T). \tag{18.1}$$

It is easy to verify that the map $\mu: X \times T \to X$, where

$$(x,t)\mu = x\pi^t\sigma^{f(x\varphi,t)}, \tag{18.2}$$

determines a topological transformation group (X,T,μ) and that (X,T,μ,G,σ) is a tame group extension. The map $\varphi: X \to X/G$ is a homomorphism of (X,T,μ) onto $(X/G,T,\rho)$. In what follows we shall write xg instead of $x\sigma^g$.

Suppose that for the given central group extension (X,T,π,G,σ) the transformation group $(X/G,T,\rho)$ is minimal. It will be shown in this section that under some restrictions on $Y \equiv X/G$, G and T one can choose a map f satisfying (18.1) in such a way that φ:

$(X,T,\mu) \to (Y,T,\rho)$ is a distal minimal extension. This fact is of considerable interest in connection with the improved version, Theorem 3·17.8, of the theorem on the structure of distal minimal extensions. It can also be used to construct particular examples. For instance, the above mentioned construction enables us to produce a distal minimal cascade on the Klein bottle and on the Cartesian product of the n-sphere and the circle, also to construct a distal minimal flow on the Cartesian product of the Klein bottle and the circle. In the same way, a minimal cascade of the class V is constructed so that some of its factors do not belong to V nor even to PD. Extensions of weakly mixing minimal cascades are also considered in this section. It is proved that most extensions of a given weakly mixing minimal cascade obtained by means of the aforementioned construction are also minimal and weakly mixing.

3·18.2 LEMMA: *If (X,T,π,G,σ) is a central group extension, the spaces X and G are compact and $f:Y \times T \to G$ is a continuous map satisfying the identity* (18.1), *then the transformation group (X,T,μ), where μ is defined by* (18.2), *is a distal extension of (Y,T,ρ) by the homomorphism $\varphi:X \to Y$.*

Proof: Let $x_1,x_2 \in X$, $x_1 \neq x_2$, but $x_1\varphi = x_2\varphi$. We shall prove that the points x_1 and x_2 are distal under (X,T,μ). Suppose the contrary holds. As X is a compact space, there exist a net $\{t_i \mid i \in I\}$ $(t_i \in T)$ and a point $x \in X$ such that

$$\{x_1\mu^{t_i} \mid i \in I\} \to x, \qquad \{x_2\mu^{t_i} \mid i \in I\} \to x.$$

It follows from the definition of μ that there exist elements $g_i \in G$ $(i \in I)$, for which $x_k\mu^{t_i} = x_k\pi^{t_i}g_i$ $(k = 1,2)$. Without loss of generality we may suppose that the net $\{g_i \mid i \in I\}$ converges to some element $g \in G$. Since the space X is compact and $\{x_1\mu^{t_i}\} = \{x_1\pi^{t_i}g_i\} \to x$, we have $\{x_1\pi^{t_i}\} \to xg^{-1}$. Similarly, $\{x_2\pi^{t_i}\} \to xg^{-1}$. Hence the points x_1 and x_2 are proximal under (X,T,π). But (X,T,π) is a central group extension of (Y,T,ρ), therefore it follows from Lemma 3·15.2 that $x_1 = x_2$ in spite of the assumption $x_1 \neq x_2$. ∎

3·18.3 Henceforth we shall assume that X is a compact metric space, G is a compact connected Lie group and G acts on X. The image of $x \in X$ under $g \in G$ will be denoted by xg. Let ψ be a homeomorphism of X onto X such that $xg\psi = x\psi g$ for all $g \in G$, $x \in X$, and let $Y \equiv X/G = \{xG \mid x \in X\}$. Clearly, the space Y is compact Hausdorff and second countable; hence Y is metrizable. In what follows we shall assume that the set Y is infinite. The canonical map $\pi:X \to X/G$ is open. Since $g\circ\psi = \psi\circ g$ for all $g \in G$, the homeomorphism $\psi:X \to X$ induces a homeomorphism φ of the space Y onto itself.

3·18.4 Let (X,ψ) and (Y,φ) be cascades determined by the homeomorphisms ψ and φ respectively, let $x_0 \in X$, $x_0\pi = y_0$ and let $O(x,\psi) = \{x\psi^n | n \in \mathbb{Z}\}$ be the orbit of $x \in X$ under (X,ψ).

Further, let $C(Y,G)$ be the space of all continuous maps of Y into G. If $u \in C(Y,G)$, then the image of $y \in Y$ under the map u will be denoted by $u(y)$. Provide the space $C(Y,G)$ with the metric

$$\rho(u,v) = \sup\{\rho[u(y),v(y)] \,|\, y \in Y\},$$

compatible with the topology of uniform convergence. Then $C(Y,G)$ is a complete metric space.

Let $u \in C(Y,G)$. Define a map $\bar{u}:X \to X$ by:

$$x\bar{u} = x\psi u(x\pi\varphi), \qquad (x \in X).$$

Note that $x\bar{u}\pi = x\psi\pi = x\pi\varphi$.

3·18.5 LEMMA: *Let* $u \in C(Y,G)$ and let $x \in X$. Then:

(1): $x\bar{u}^n = x\psi^n u(b\varphi)\cdots u(b\varphi^n)$, $(n \geqslant 1)$,
$x\bar{u}^n = x\psi^n[u(b\varphi)]^{-1}\cdots[u(b\varphi^{n+1})]^{-1}$, $(n \leqslant -1)$,
where $b = x\pi$;

(2): $\bar{u}$ *is a homeomorphism of* X *onto* X.

Proof: Statement (1) is immediate; the proof is by induction. Statement (2) follows from the fact that $\bar{u}$ us continuous and has a continuous inverse map determined by formula (1) for $n = -1$. ▌

3·18.6 Define a map $f_u:Y \times \mathbb{Z} \to G$ as follows:

$$f_u(y,n) = \begin{cases} u(y\varphi)\cdots u(y\varphi^n) & (n \geqslant 1) \\ e \text{ (where } e \text{ is the identity of } G) & (n = 0), \\ [u(y)]^{-1}\cdots[u(y\varphi^{n+1})]^{-1} & (n \leqslant -1). \end{cases}$$

This map satisfies the identity

$$f_u(y,n+m) = f_u(y,n)f_u(y\varphi^n,m),$$

which is similar to (18.1). Therefore it follows from Lemma 3·18.2 that the cascade $(X,\bar{u})$ is a distal extension of (Y,φ) by the map $\pi:X \to Y$.

3·18.7 LEMMA: *Let* $V_1,\ldots,V_n$ *be open non-empty subsets of* G. *There exists a positive integer* p *such that if* $W_i \in \{V_1,\ldots V_n\}$ $(i = 1,\ldots,p)$, *then* $W_1W_2\ldots W_p = G$.

Proof: There exist a neighbourhood V of the identity $e \in G$ and elements $g_1,\ldots,g_n \in G$ for which $Vg_i \subset V_i$ $(i = 1,\ldots,n)$. Since the group G is compact there exists a neighbourhood U of $e \in G$

such that $gUg^{-1} \subset V$ for every element $g \in G$. Let $N = \cap [gVg^{-1} | g \in G]$. Then $U \subset N \subset V$ and $gN = Ng$ $(g \in G)$.

We shall show that $U^p = G$ for some superscript p. Indeed, $G = \cup [U^i | i = 1,2,\ldots]$, as the group G is connected. Since G is compact, there is a superscript p with $G = \cup [U^i | i = 1,\ldots,p]$. But $U^k \subset U^{k+1}$ $(k = 1,2,\ldots)$, and hence $G = U^p$.

If $W_i \in \{V_1,\ldots,V_n\}$ $(i = 1,\ldots,p)$, then $W_1W_2\ldots W_p \supset Vh_1Vh_2\ldots \supset Vh_p$, where $h_i \in \{g_1,\ldots,g_n\}$ $(i = 1,\ldots,p)$. Then $G \supset W_1\ldots W_p$ $Nh_1\ldots Nh_p = N^p h_1\ldots h_p \supset U^p h_1\ldots h_p = G$. ▮

3·18.8 LEMMA: *Let $f \in C^p(Y,G)$ and $\varepsilon > 0$. There exists a number $\delta > 0$ such that if F is a finite subset of Y and u is a map of F into G satisfying the condition $\rho(f(y),u(y)) < \delta$ $(y \in F)$, then there exists a map $v \in C(Y,G)$ for which $v(y) = u(y)$ $(y \in F)$ and $\rho(f(y),v(y)) < \varepsilon$ $(y \in Y)$.*

Proof: Let $f \in C(Y,G)$ and $\varepsilon > 0$. As G is a Lie group, each point $g \in G$ has a coordinate neighbourhood, *i.e.*, a neighbourhood homeomorphic to the interior of a ball of some Euclidean space. Since the group G is compact, there is a $\delta > 0$ such that the spherical δ-neighbourhood of each point $g \in G$ is contained in a coordinate neighbourhood $S(g)$ of diameter less than $\varepsilon/2$.

Let F be a finite subset of Y and let $u:F \to G$ satisfy the condition $\rho(f(y),u(y)) < \delta$ $(y \in F)$. For each $y \in F$ choose a neighbourhood U_y of y so that all the U_y are mutually disjoint and each point $y \in F$ is contained in exactly one neighbourhood. Also choose the neighbourhoods U_y small enough so that $f(U_y) \subset S(f(y),\delta)$. By Tietze's Extension Theorem (*see* Hu [*1*]) there exist continuous maps $u_y:\bar{U}_y \to S(f(y))$ such that $u_y(y) = u(y)$, if $y \in F$, and $u_y(z) = f(z)$ if $z \in \bar{U}_y \setminus U_y$. Set $v(z) = u_y(z)$ if $z \in U_y$, and $v(z) = f(z)$ if z does not belong to any U_y $(y \in F)$. Then v is the required map. ▮

3·18.9 THEOREM: *Suppose that all the conditions formulated in Subsection* 3·18.3 *are satisfied. Let $x_0 \in X$ and $x_0\pi = y_0$. If the orbit $O(y_0,\varphi)$ is dense in Y, then there exists a homeomorphism h of X onto X such that the orbit $O(x_0,h)$ is dense in X and $\pi:X \to Y$ is a homomorphism of the cascade (X,h) onto (Y,φ). If the group G is commutative, then (X,h) is a central group extension of (Y,φ).*

Proof: Let U be an open non-empty subset of X and let

$$E(U) = \{u | u \in C(Y,G), O(x_0,\bar{u}) \cap U \neq \emptyset\}.$$

We shall first show that $E(U)$ is an open dense subset of $C(Y,G)$.

Let $u \in E(U)$. Then there exists a superscript n such that $x_0 \bar{u}^n \in U$. The map $\bar{u}^n$ is a composition of continuous maps, and hence we can choose a number $\varepsilon > 0$ so that $\rho(u,v) < \varepsilon$ implies $x_0\bar{v}^n \in U$. Hence it follows that $E(U)$ is open.

Choose $\delta > 0$ which corresponds to u and ε according to Lemma 3·18.8. Since the space G is compact, there exist open sets V_i

$(i = 1,\ldots,n)$ such that $G = \bigcup_{i=1}^{n} V_i$ and such that the diameter of each V_i does not exceed δ. Let p be the positive integer corresponding to the system $\{V_1,\ldots,V_n\}$ according to Lemma 3·18.7.

Since the map $\pi: X \to Y$ is open and the set U is open in X, the set $U\pi$ is open in Y. By hypothesis, the orbit $O(y_0,\varphi)$ is dense in Y; hence there exists an $r \in \mathbb{Z}$ with $|r| \geqslant p$ and $y_0\varphi^r \in U\pi$. Suppose, for example, that $r > 0$. There exists an element $g_0 \in G$ with $x_0\psi^r g_0 \in U$. Further, since $G = \bigcup_{i=1}^{n} V_i$, there exist elements W_i of $\{V_1,\ldots,V_n\}$ such that

$$u(y_0\varphi^i) \in W_i, \qquad (i = 1,\ldots,r).$$

Then $r \geqslant p$ because $W_1\ldots W_r = G$, and hence the element g_0 can be represented in the form $g_0 = g_1\ldots g_r$, where g_i is some element of W_i $(i = 1,\ldots,r)$. We may suppose without loss of generality that the elements g_i $(i = 1,\ldots,r)$ are all distinct from each other.

Denote $\{y_0\varphi^i \mid i = 1,\ldots,r\}$ by F. Define a map $w: F \to G$ by $w(y_0\varphi^i) = g_i$; w is well defined because the points $y_0\varphi^i$ $(i = 1,\ldots,r)$ are distinct. For, were it not so, the set $O(y_0,\varphi)$ and consequently the set $Y = \overline{O(y_0,\varphi)}$ would be finite, contrary to our hypothesis. Observe that $\rho(u(y),w(y)) < \delta$ $(y \in F)$. Let v be the continuous extension of w as guaranteed by Lemma 3·18.8. Then $\rho(v(y),u(y)) < \varepsilon$ $(y \in Y)$ and $v \in E(U)$, by the very definition of the homeomorphism $\bar{v}^r$.

Thus we have proved that if U is non-empty and open in X, then $E(U)$ is open and dense in $C(Y,G)$.

Let $\mathfrak{B}$ be a countable base of open sets of the compact metric space X. According to Baire's Theorem, the set $\Phi \equiv \bigcap [E(U) \mid U \in \mathfrak{B}]$ is dense in the complete metric space $C(Y,G)$. If $u \in \Phi$, then $h = \bar{u}$ satisfies all the required conditions.

Suppose now that the group G is commutative. Since $x\bar{u} = x\psi u(b\varphi)$, where $b = x\pi$, we have $xg\bar{u} = xg\psi u(b\varphi) = x\psi g u(b\varphi) = x\psi u(b\varphi)g = x\bar{u}g$. Thus (X,h) is in this case a central group extension of (Y,φ). ▮

3·18.10 THEOREM: *Suppose that all the conditions stated in Subsection* 3·18.3 *are satisfied. Let* (Y,φ) *be a minimal cascade. Then there exists a homeomorphism* h *of* X *onto* X *such that* (X,h) *is a minimal cascade and* $\pi: X \to Y$ *is a distal homomorphism of* (X,h) *onto* (Y,φ).

Proof: Let $h \equiv \bar{u}$ be the homeomorphism resulting from the preceding theorem. Then the orbit $O(x_0,h)$ is dense in X. As was already noticed, Lemma 3·18.2 guarantees that (X,h) is a distal extension of (Y,φ) by the homomorphism π. According to Lemma 3·14.4, (X,h) is a minimal cascade. ▮

3·18.11 COROLLARY: *If the hypotheses of Theorem* 3·18.10 *are fulfilled and* (Y,φ) *is distal, then there exists a homeomorphism* h *of* X *onto* X *such that* (X,h) *is a distal minimal cascade.*

Proof: This follows from Theorem 3·18.10 and Lemma 3·18.2. ▮

3·18.12 THEOREM: *Let all conditions formulated in Subsection* 3·18.3 *be satisfied and assume that* (Y,φ) *is a minimal weakly mixing cascade. Then there exists a Baire subset* $W \subset C(Y,G)$ *such that for each map* $u \in W$ *the cascade* $(X,\bar{u})$ *is minimal weakly mixing and* $\pi:(X,\bar{u}) \to (Y,\varphi)$ *is a distal extension.*

Proof: Since (Y,φ) is a weakly mixing cascade, there is a point $(y_0,y_1) \in Y \times Y$ whose orbit is dense in $Y \times Y$ (see Subsection 3·13.14). Choose points $x_0,x_1 \in X$ so that $x_0\pi = y_0$, $x_1\pi = y_1$. Let $\{U_i \mid i = 1,2,\ldots\}$ be a countable base of the topology of the Cartesian product $X \times X$. Let E_1 denote the subset of $C(Y,G)$ consisting of all maps u such that $(x_0,x_1)\bar{u}^m \in U_i$ for some integer $m \in \mathbb{Z}$. Clearly each set E_i is open in $C(Y,G)$.

Let us prove that each set E_i is dense in $C(Y,G)$. Let $u \in C(Y,G)$ and $\varepsilon > 0$. Choose a number $\delta > 0$ according to Lemma 3·18 .8. Set $H = u(Y)$ and select an open covering $V_1,\ldots,V_n$ of the set $H \cup H^{-1}$ so that $\operatorname{diam} V_i < \delta$ $(i = 1,\ldots,n)$. Then choose a positive integer p satisfying the condition of Lemma 3·18.7. Since the canonical projection $\pi:X \to Y \equiv X/G$ is open, there is an integer $m \in \mathbb{Z}$ with $(y_0,y_1)\varphi^m \in U_i\pi$. Without loss of generality we may suppose that $|m| \geqslant p$. For definiteness let $m > 0$. Choose some sets $W_j^0,W_j^1 \in \{V_1,\ldots,V_n\}$ so that $u(y_0\varphi^j) \in W_j^0$, $u(y_1\varphi^j) \in W_j^1$ $(j = 1,\ldots,p)$. Since $(y_0,y_1)\varphi^m \in U_i\pi$, there exist elements $g_0,g_1 \in G$ satisfying the condition $(x_0g_0,x_1g_1)\psi^m \in U_i$. By the choice of p, $W_1^0\ldots W_m^0 = W_1^1\ldots W_m^1 = G$. Therefore the elements g_0 and g_1 can be represented in the form $g_0 = g_1^0\ldots g_m^0$ and $g_1 = g_1^1\ldots g_m^1$, where $g_j^\ell \in W_j^\ell$ $(j = 1,\ldots,m;\ \ell = 0,1)$. Let $F = \{y_0\varphi,\ldots,y_0\varphi^m\} \cup \{y_1\varphi,\ldots,y_1\varphi^m\}$. Define a map $v:F \to G$ by $v(y_\ell\varphi^i) = g_j^\ell$, $(j = 1,\ldots,m; \ell = 0,1)$. This map is well defined because we may suppose that F consists of $2m$ distinct points. By the choice of δ, there exists an extension $w:Y \to G$ of the function $v:F \to G$ which satisfies the condition $\rho(u,w) < \varepsilon$. It follows directly from the definition of $\bar{w}^m$ that $w \in U_i$.

Thus each set E_i is open and dense in $C(Y,G)$. Let $u \in W = \bigcap_{i=1}^{\infty} E_i$. Then $(X,\bar{u})$ is a weakly mixing cascade, and hence there are dense orbits in X. Since (Y,φ) is a minimal cascade and $\pi:(X,\bar{u}) \to (Y,\varphi)$ is a distal extension (Lemma 3·18.2), the cascade $(X,\bar{u})$ is also minimal. ▮

3·18.13 Let X be a compact metric space, G a compact connected Lie group and $(X,\mathbb{R},\pi,G,\sigma)$ a central group extension. Let $(Y,\mathbb{R},\rho) \equiv (X/G,\mathbb{R},\rho)$ denote the flow induced by $(X,\mathbb{R},\pi)$. Suppose that the flow $(Y,\mathbb{R},\rho)$ has a global section $(W,\mathbb{Z},\rho)$ and that the space W is infinite. Set $M = W\varphi^{-1}$, where $\varphi:X \to X/G$ is the canonical projection. Clearly, $(M,\mathbb{Z},\pi)$ is a global section of $(X,\mathbb{R},\pi)$ and $(M,\mathbb{Z},\pi,G,\sigma)$ is a central group extension. Hence we can apply Theorems 3·18.9,10 to this extension. In particular, assume that $(Y,\mathbb{R},\rho)$ is a minimal flow. Then $(W,\mathbb{Z},\rho)$ is a minimal cascade. By Theorem 3·18.10 we can determine a minimal cascade $(M,\mathbb{Z},\nu)$ so that the canonical map of M onto W is a homomorphism of $(M,\mathbb{Z},\nu)$ onto $(W,\mathbb{Z},\rho)$. Since G is a connected Lie group, it follows directly from the definition of $(M,\mathbb{Z},\nu)$ (see Subsection 3·18.4) that the map ν is isotopic to π^1. The cascade $(M,\mathbb{Z},\nu)$ can be embedded in a flow as a global section. Recall the corresponding construction. Given the Cartesian product $M \times \mathbb{R}$, we define an action of $\mathbb{Z}$ by

$$(b,r)n = (b\nu^n, r + n), \qquad (b \in M, r \in \mathbb{R}, n \in \mathbb{Z}),$$

and an action of the group $\mathbb{R}$ by

$$(b,r)s = (b, r + s), \qquad (b \in M; r,s \in \mathbb{R}).$$

Since these actions of $\mathbb{Z}$ and $\mathbb{R}$ commute, the group $\mathbb{R}$ acts in a natural way on the quotient space $X(M,\mathbb{Z},\nu) \equiv (M \times \mathbb{R})/\mathbb{Z}$. The latter space is compact because it can be obtained from the product $M \times [0,1]$ by the identifications $(y,1) = (y\nu,0)$ $(y \in M)$.

3·18.14 LEMMA: *If the cascades* $(M,\mathbb{Z},\nu)$ and $(M,\mathbb{Z},\lambda)$ *are such that the maps* ν *and* λ *are isotopic, then the phase spaces of the flows obtained from* $(M,\mathbb{Z},\nu)$ *and* $(M,\mathbb{Z},\lambda)$ *by means of the above construction are homeomorphic.*

Proof: Let ν and λ be isotopic. This means that there exists a continuous map $\Phi:M \times [0,1] \to M$ satisfying the following conditions: (1) $(b,0)\Phi = b\nu$, $(b,1)\Phi = b\lambda$ $(b \in M)$; (2) for each $\xi \in [0,1]$ the map $\varphi_\xi:M \to M$ defined by $b\varphi_\xi = (b,\xi)\Phi$ $(b \in M)$ is a homeomorphism of M onto itself.

Define a map $F:M \times [0,1] \to M \times [0,1]$ as follows: $(b,r)F = (b\varphi_r, r)$ $(b \in M, r \in [0,1])$. We conclude that F is a homeomorphism. Further,

$$(b,1)F = (b\lambda,1), \qquad (b\lambda,0)F = (b\lambda\varphi_0,0) = (b\lambda\nu,0).$$

Hence it follows that F induces a homeomorphism F_1 of $X(M,\mathbb{Z},\lambda)$ onto $X(M,\mathbb{Z},\nu)$. ∎

3·18.15 It follows from Lemma 3·18.14 that the phase space of the flow, which includes the cascade $(M,\mathbb{Z},\nu)$ as a global section, is homeomorphic to X, because the maps ν and π^1 are isotopic. Denote this flow by $(X,\mathbb{R},\delta)$. Clearly, $\varphi:(X,\mathbb{R},\delta) \to (Y,\mathbb{R},\rho)$ is a distal minimal extension.

3·18.16 We shall now give an example of a distal minimal cascade on the Klein bottle and an example of a distal minimal flow on the Cartesian product of the Klein bottle and a circle.

Let $\mathbb{K} = \{\xi \mid \xi \in \mathbb{C}, |\xi| = 1\}$. $\mathbb{Z}_2 = \{1,-1\}$ is a subgroup of the group $\mathbb{K}$. Let $G = \mathbb{Z}_2 \cdot \mathbb{K}$ denote a semidirect product of $\mathbb{Z}_2$ and $\mathbb{K}$. The group G consists of all pairs (ε,γ) $(\varepsilon \in \mathbb{Z}_2, \gamma \in \mathbb{K})$ and the multiplication is defined by

$$(\varepsilon_1,\gamma_1)\cdot(\varepsilon_2,\gamma_2) = (\varepsilon_1\varepsilon_2, \gamma_1^{\varepsilon_2}\gamma_2).$$

Define an action of the group G on the torus $\mathbb{K} \times \mathbb{K}$ by

$$(\alpha,\beta)(\varepsilon,\gamma) = (\alpha\varepsilon, \beta^{\varepsilon}\gamma), \qquad ((\alpha,\beta) \in \mathbb{K} \times \mathbb{K}, (\varepsilon,\gamma) \in G).$$

Define a homeomorphism $\tau : \mathbb{K} \times \mathbb{K} \to \mathbb{K} \times \mathbb{K}$ by $(\alpha,\beta)\tau = (\alpha e^{i},\beta)$ $(\alpha,\beta \in \mathbb{K})$ (i is the imaginary identity). It is easy to verify that the action of G commutes with τ. The set $H = \mathbb{Z}_2 \cdot 1 = \{(1,1),(-1,1)\}$ is a subgroup of G. Then τ induces a homeomorphism τ_1 of the quotient space $(\mathbb{K} \times \mathbb{K})/H$ and a homeomorphism τ_2 of $(\mathbb{K} \times \mathbb{K})/G$.

Let us prove that $(\mathbb{K} \times \mathbb{K})/H$ is the Klein bottle. To do this, we note that the fundamental domain of the covering $(\mathbb{K} \times \mathbb{K}) \to (\mathbb{K} \times \mathbb{K})/H$ is of the form

$$\{(\alpha,\beta) \mid \alpha = e^{it}, 0 < t \leqslant \pi; \beta \in \mathbb{K}\},$$

and that

$$(1,\beta) \sim (1,\beta)(-1,1) \equiv (-1,\beta^{-1}) \equiv (e^{i\pi},\beta^{-1})$$

is the corresponding identification. The space $(\mathbb{K} \times \mathbb{K})/G$ is a circle obtained from the circle $\mathbb{K} \times 1$ by identifying the points $(\alpha,1)$ and $(-\alpha,1) \equiv (\alpha e^{i\pi},1)$.

The homeomorphism τ_2 of $(\mathbb{K} \times \mathbb{K})/G$ satisfies

$$\{(\alpha,1),(-\alpha,1)\}\tau_2 = \{(\alpha e^{i},1),(-\alpha e^{i},1)\}, \qquad (\alpha \in \mathbb{K}).$$

Thus $((\mathbb{K} \times \mathbb{K})/G, \tau_2)$ is a minimal almost periodic cascade on the circle.

Observe that $N = 1\cdot\mathbb{K}$ is a subgroup of $\mathbb{Z}_2 \cdot \mathbb{K}$. Clearly N is isomorphic to the group $\mathbb{K}$. Since the action of G commutes with τ, we can define a homeomorphism τ_3 of the quotient space $(\mathbb{K} \times \mathbb{K})/N \approx \mathbb{K} \times 1$ onto itself:

$$(\alpha,1)\tau_3 = (\alpha e^{i},1), \qquad (\alpha \in \mathbb{K}).$$

We have the following commutative diagram

$$\begin{array}{ccc} \mathbb{K}\times\mathbb{K} & \xrightarrow{\lambda} & (\mathbb{K}\times\mathbb{K})/H \\ \mu\downarrow & & \downarrow\varphi \\ (\mathbb{K}\times\mathbb{K})/N & \xrightarrow{\psi} & (\mathbb{K}\times\mathbb{K})/G, \end{array}$$

where λ,φ,μ,ψ are canonical maps. Observe that ψ is a twofold covering of the circle.

Let $C_0(\mathbb{R},\mathbb{R})$ denote the family of all continuous 2π-periodic functions $\Phi:\mathbb{R}\to\mathbb{R}$ satisfying the conditions:

$$\Phi(0) = 0, \qquad \Phi(t+\pi) = -\Phi(t), \qquad (t\in\mathbb{R}). \tag{18.3}$$

Identify the group $\mathbb{K}$ with the quotient group $\mathbb{R}/M$, where $M = \{2k\pi \mid k\in\mathbb{Z}\}$. Then each function $\Phi\in C_0(\mathbb{R},\mathbb{R})$ determines a function $u:\mathbb{K}\to\mathbb{K}$ as follows:

$$u(e^{it}) = e^{i\Phi(t)}, \qquad (t\in\mathbb{R}). \tag{18.4}$$

It follows from (18.3) that

$$u(-1) = u(1) = 1, \qquad u(-\alpha) \equiv u(\alpha e^{i\pi}) = [u(\alpha)]^{-1}, \qquad (\alpha\in\mathbb{K}). \tag{18.5}$$

The totality of functions $u:\mathbb{K}\to\mathbb{K}$ of the form (18.4) with $\Phi\in C_0(\mathbb{R},\mathbb{R})$ will be denoted by $C_0(\mathbb{K},\mathbb{K})$.

We shall now apply the construction of Theorem 3·18.10 to the group extension $\mu:(\mathbb{K}\times\mathbb{K},\tau)\to((\mathbb{K}\times\mathbb{K})/N,\tau_3)$ with the structure group $N = 1\cdot\mathbb{K}\approx\mathbb{K}$, but unlike in Theorem 3·18.10 we shall look for a function u belonging to the class $C_0(\mathbb{K}\times 1, 1\cdot\mathbb{K})\approx C_0(\mathbb{K},\mathbb{K})$. This requires some modifications in the proof. In Theorem 3·18.9 it will be necessary to take the set

$$\{y_0\varphi^\ell \mid \ell = 1,\dots,r\}\cup\{-(y_0\varphi^\ell)\mid \ell = 1,\dots,r\},$$

at the place of F (here $y_0 = 1\in\mathbb{K}$, $y_0\varphi^\ell = 1\cdot e^{i\ell} = e^{i\ell}$) and to observe that the points $y_0\varphi^\ell$ $(\ell = 1,\dots,r)$ are not only all distinct and different from ± 1, but also that among them there are no two points with the same absolute value. If the function $u: F\to\mathbb{K}$ satisfies the condition of 'antisymmetry' $u(-\alpha) = [u(\alpha)]^{-1}$ $(\alpha\in F)$, then an extension of it $v:\mathbb{K}\to\mathbb{K}$ can be constructed which satisfies not only the conclusion of Lemma 3·18.8, but also the condition $v\in C_0(\mathbb{K},\mathbb{K})$ (where, of course, it is assumed that $f\in C_0(\mathbb{K},\mathbb{K})$).

Thus there is a function $u\in C_0(\mathbb{K},\mathbb{K})$ such that the homeomorphism $\bar{u}:\mathbb{K}\times\mathbb{K}\to\mathbb{K}\times\mathbb{K}$ defined by

$$x\bar{u} = (x\tau)u((x\mu)\tau_3), \qquad (x\in\mathbb{K}\times\mathbb{K}), \tag{18.6}$$

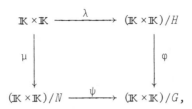

determines a distal minimal cascade $(\mathbb{K}\times\mathbb{K},\bar{u})$ and moreover $\mu:(\mathbb{K}\times\mathbb{K},\bar{u}) \to ((\mathbb{K}\times\mathbb{K})/N,\tau_3)$ is a group extension. Rewrite the formula (18.6) in more detail. Let $x = (\alpha,\beta)$, $x\mu = (\alpha,1) \approx \alpha$, then

$$(\alpha,\beta)\bar{u} = (\alpha e^{i},\beta)(1,u(\alpha e^{i})) = (\alpha e^{i},\beta u(\alpha e^{i})), \qquad (\alpha,\beta \in \mathbb{K}). \quad (18.7)$$

We show that the action of the group H commutes with the homeomorphism $\bar{u}$. In fact:

$$[(\alpha,\beta)\bar{u}](-1,1) = (\alpha e^{i},\beta u(\alpha e^{i}))(-1,1) = (-\alpha e^{i},\beta^{-1}\cdot[u(\alpha e^{i})]^{-1});$$

$$[(\alpha,\beta)(-1,1)]\bar{u} = (-\alpha,\beta^{-1})\bar{u} = (-\alpha e^{i},\beta^{-1}\cdot u(-\alpha e^{i})).$$

The required result follows now from (18.5). Thus the action of H commutes with $\bar{u}$, and hence $\bar{u}$ induces a homeomorphism δ of the Klein bottle onto itself. Clearly $((\mathbb{K}\times\mathbb{K})/H,\delta)$ is a distal minimal cascade.

Since $u \in C_0(\mathbb{K},\mathbb{K})$, there is a function $\Phi \in C_0(\mathbb{R},\mathbb{R})$ which satisfies the condition (18.4). Then (18.7) takes the form

$$(M + s,M + t)\bar{u} = (M + s + 1,M + t + \Phi(s + 1)) \qquad (s,t \in \mathbb{R}).$$

The action of the group H on $\mathbb{K}\times\mathbb{K} \approx (\mathbb{R}/M)\times(\mathbb{R}/M)$ can be written as $(M + s,M + t)(-1,1) = (M + s + \pi,M - t)$ $(s,t \in \mathbb{R})$. The map

$$(M + s,M + t)\varphi_{\xi} = (M + s + \xi,M + t + \xi\Phi(s + 1)),$$

$$(s,t \in \mathbb{R};0 \leqslant \xi \leqslant 1),$$

determines an isotopy of the identity map φ_0 to $\varphi_1 = \bar{u}$. Further, in view of (18.3), we obtain that

$$\begin{aligned}[(M + s,M + t)(-1,1)]\varphi_{\xi} &= (M + s + \pi,M - t)\varphi_{\xi} \\ &= (M + s + \pi + \xi,M - t + \xi\Phi(s + \pi + 1)) \\ &= (M + s + \xi + \pi,M - t - \xi\Phi(s + 1) \\ &= (M + s + \xi,M + t + \xi\Phi(s + 1))(-1,1) \\ &= [(M + s,M + t)\varphi_{\xi}](-1,1), \\ &\quad (s,t \in \mathbb{R},0 \leqslant \xi \leqslant 1).\end{aligned}$$

Thus the map $\delta:(\mathbb{K}\times\mathbb{K})/H \to (\mathbb{K}\times\mathbb{K})/H$ is isotopic to the identity map.

It follows from this by Lemma 3·18.14 that the phase space of the flow, which includes the cascade $((\mathbb{K}\times\mathbb{K})/H,\delta)$ as a global section, is homeomorphic to the Cartesian product of the Klein bottle $(\mathbb{K}\times\mathbb{K})/H$ and a circle. Thus there is a distal minimal flow $(S^1\times(\mathbb{K}\times\mathbb{K})/H,\mathbb{R})$.

3·18.17 We give an example of a minimal cascade of the class V which admits a homomorphic map onto some cascade which is not of the class PD. This example demonstrates at the same time that there exist almost distal minimal extensions which do not belong to the class of V-extensions. This fact is remarkable, in view of Theorem 3·14.36. First we shall prove several lemmata.

3·18.18 An extension $\varphi:(X,T)\to(Y,T)$ is said to be prime if each transformation group (Z,T) satisfying the commutative diagram

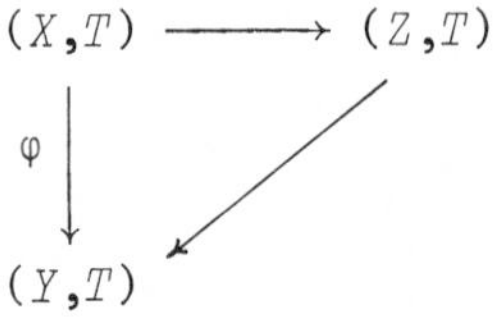

coincides with either (X,T) or (Y,T).

3·18.19 LEMMA: *Let X be a compact space, let (X,T) be a minimal transformation group and let S_1,S_2 and R be closed invariant equivalence relations satisfying the conditions S $S_1\subset R$, $S_2\subset R$. Let $\varphi:(X/S_1,T)\to(X/R,T)$ and $\psi:(X/(S_1\cap S_2),T)\to(X/S_2,T)$ denote the natural homomorphisms. Suppose that φ is a distal (proximal, almost automorphic, equicontinuous) extension. Then ψ is an extension from the same class.*

Proof: (a): If φ is a distal extension, then $S_1\supset P\cap R\supset P\cap S_2$; hence $S_1\cap S_2\supset P\cap S_2$, and therefore ψ is distal (Corollary 3·14.2) (b): If φ is a proximal extension, then $S_1\supset R\mathfrak{I}\supset S_2\mathfrak{I}$; hence $S_1\cap S_2\supset S_2\mathfrak{I}$, and therefore ψ is a proximal extension (Lemma 3·14.16). (c): Let φ be an almost automorphic extension, *i.e.*, such that there is a point $a\in X$ with $aS_1=aR\supset aS_2$. Then $aS_2\supset a(S_1\cap S_2)=aS_1\cap aS_2\supset aS_2$, and hence $a(S_1\cap S_2)=aS_2$. Thus ψ is almost automorphic. (d): If φ is an equicontinuous extension, then $S_1\supset Q(R)\supset Q(S_2)$; hence $S_1\cap S_2\supset Q(S_2)$ and ψ is equicontinuous by Corollary 3·13.3. ▮

3·18.20 LEMMA: *Let $\{(X_\lambda,T)\mid 0\leq\lambda\leq\vartheta\}$ be a PD-system with homomorphisms φ_μ^λ $(0\leq\mu\leq\lambda\leq\vartheta)$ and let (X_ϑ,T) be a minimal transformation group with a compact phase space. For each closed invariant equivalence relation $S\subset R(\varphi_0^\vartheta)$ there exists a PD-system $\{(Z_\lambda,T)\mid 0\leq\lambda\leq\vartheta\},\{\psi_\mu^\lambda\mid 0\leq\mu\leq\lambda\leq\vartheta\}$*

and a homomorphism $\xi_\lambda:(Z_\lambda,T) \to (X_\lambda,T)$ $(0 \leqslant \lambda \leqslant \vartheta)$ *such that*

(1): $(Z_0,T) = (X_\vartheta/S,T)$;

(2): ξ_0 *is the canonical map;*

(3): *the diagram*

$$\begin{array}{ccccccccc} Z_0 & \leftarrow \cdots \leftarrow & Z_\lambda & \xleftarrow{\psi_\lambda^{\lambda+1}} & Z_{\lambda+1} & \leftarrow \cdots \leftarrow & Z_\vartheta \\ \downarrow{\scriptstyle \xi_0} & & \downarrow{\scriptstyle \xi_\lambda} & & \downarrow{\scriptstyle \xi_{\lambda+1}} & & \downarrow{\scriptstyle \xi_\vartheta} \\ X_0 & \leftarrow \cdots \leftarrow & X_\lambda & \xleftarrow[\varphi_\lambda^{\lambda+1}]{} & X_{\lambda+1} & \leftarrow \cdots \leftarrow & X_\vartheta \end{array}$$

is commutative;

(4): $\xi_\vartheta:(Z_\vartheta,T) \to (X_\vartheta,T)$ *is an isomorphism.*

Similar statements hold for PE-systems and V-systems.

Proof: Construct inductively a projective system $\{(Z_\lambda,T)\ 0 \leqslant \lambda \leqslant \vartheta\}$, $\{\psi_\mu^\lambda | 0 \leqslant \mu \leqslant \lambda \leqslant \vartheta\}$.

Set $(Z_0,T) = (X_\vartheta/S,T)$ and denote the canonical projection of $Z_0 \equiv X_\vartheta/S$ onto $X_0 = X_\vartheta/R(\varphi_0^\vartheta)$ by ξ_0. We get the commutative diagram

$$\begin{array}{ccc} X_1 & \xleftarrow{\varphi_1^\vartheta} & X_\vartheta \\ {\scriptstyle \varphi_0^1}\downarrow & & \downarrow{\scriptstyle \sigma_0} \\ X_0 & \xleftarrow[\xi_0]{} & Z_0 . \end{array}$$

Suppose that the transformation groups (Z_λ,T) and homomorphisms $\sigma_\lambda:(X_\vartheta,T) \to (Z_\lambda,T)$, $\xi_\lambda:(Z_\lambda,T) \to (X_\lambda,T)$ have been already defined for all ordinals λ smaller than some ordinal $\mu \leqslant \vartheta$ and the diagrams

$$\begin{array}{ccc} X_{\lambda+1} & \xleftarrow{\varphi_{\lambda+1}^\vartheta} & X_\vartheta \\ {\scriptstyle \varphi_\lambda^{\lambda+1}}\downarrow & \swarrow{\scriptstyle \varphi_\lambda^\vartheta} & \downarrow{\scriptstyle \sigma_\lambda} \\ X_\lambda & \xleftarrow[\xi_\lambda]{} & Z_\lambda \end{array}$$

are commutative.

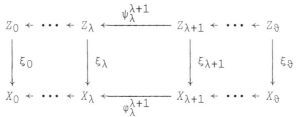

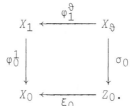

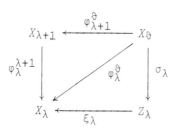

First consider the case when $\mu = \lambda + 1$. Define:

$$(Z_{\lambda+1},T) = (X_\vartheta/(R(\varphi_{\lambda+1}^\vartheta) \cap R(\sigma_\lambda)),T).$$

There exist homomorphisms $\sigma_{\lambda+1}$, $\xi_{\lambda+1}$ and $\psi_\lambda^{\lambda+1}$ making the diagram

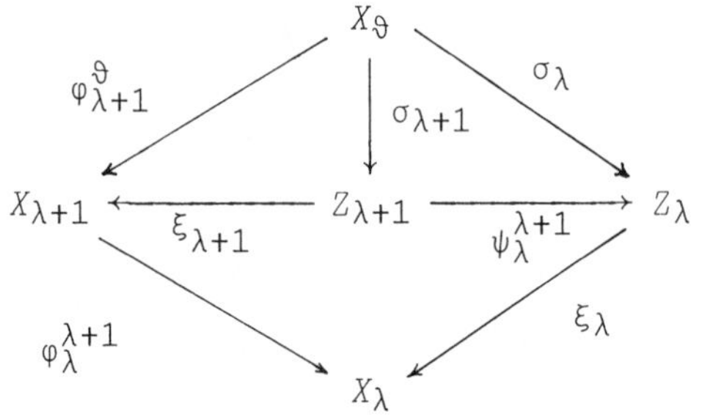

commute. It follows from Lemma 3·18.19 that if $\varphi_\lambda^{\lambda+1}$ is distal (proximal, equicontinuous, almost automorphic), then $\psi_\lambda^{\lambda+1}$ has the same property.

If μ is a limit ordinal, then define (Z_μ,T) to be the projective limit of the system $\{(Z_\lambda,T)\,|\,0 \leqslant \lambda < \mu\}$. Clearly, there exist homomorphisms $\xi_\mu:(Z_\mu,T) \to (X_\mu,T)$ and $\sigma_\mu:(X_\vartheta,T) \to (Z_\mu,T)$ with $\varphi_\mu^\vartheta = \sigma_\mu \circ \xi_\mu$.

Thus we have constructed a projective system $\{(Z_\lambda,T)\,|\,0 \leqslant \lambda \leqslant \vartheta\}$ with homomorphisms $\psi_\mu^\lambda\,|\,0 \leqslant \mu \leqslant \lambda \leqslant \vartheta\}$ and defined homomorphisms $\xi_\lambda:(Z_\lambda,T) \to (X_\lambda,T)$, $\sigma_\lambda:X_\vartheta \to Z_\lambda$ which satisfy the condition $\varphi_\lambda^\vartheta = \sigma_\lambda \circ \xi_\lambda$ $(0 < \lambda \leqslant \vartheta)$. It is easily seen that $\{(Z_\lambda,T)\,|\,0 \leqslant \lambda \leqslant \vartheta\}$ is the required system. ∎

3·18.21 LEMMA: *Let*

$$
\begin{array}{ccc}
(X,T) & \xleftarrow{\ \lambda\ } & (Z,T) \\
\varphi \downarrow & & \downarrow \mu \\
(Y,T) & \xleftarrow{\ \nu\ } & (W,T)
\end{array}
$$

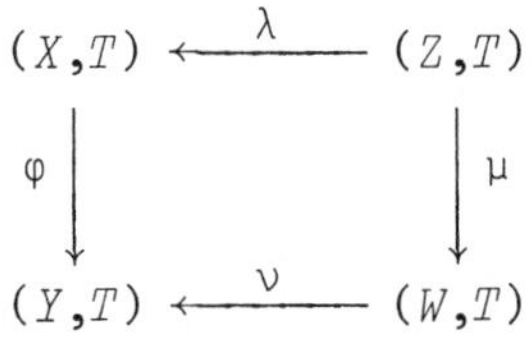

be a commutative diagram, (Z,T) a minimal transformation group with a compact phase space, φ a prime proximal extension, λ distal, μ proximal and $(Y,T) = (Z/P^*(Z,T),T)$. If φ is not subordinated to the extension ν, then (W,T) does not belong to the class PD.

Proof: By Lemma 3·14.18, φ is the largest proximal extension subordinated to $\lambda\circ\varphi \equiv \mu\circ\nu$. Since $Y = Z/P^*(Z,T)$, we have $(W/P^*(W,T),T) = (Y,T)$.

Suppose that (W,T) belongs to the class *PD*. By Lemma 3·18.20, there exists a *PD*-system $\{(W_\lambda,T)|0 \leqslant \lambda \leqslant \vartheta\}$ with $(W_0,T) = (Y,T) \equiv (W/P^*(W,T),T),(W_\vartheta,T) = (W,T)$. We may assume that all elements of the projective system are distinct. Since $W_0 = W/P^*(W,T)$, the extension $\psi_0^1:(W_1,T) \to (W_0,T)$ is proximal. Since it is subordinated to the map $\mu\circ\nu = \lambda\circ\varphi$, there exists a homomorphism $\xi:(X,T) \to (W_1,T)$ subordinated to φ. As φ is a prime extension, either $(W_1,T) = (Y,T)$ or $(W_1,T) = (X,T)$. The first case cannot happen since all transformation groups (W_λ,T) are assumed to be distinct, whereas $(W_0,T) = (Y,T)$. The second case is also impossible, because it contradicts the fact that φ is not subordinated to ν. ∎

3·18.22 Now we pass directly to the construction of the promised example. We take $\mathbb{Z}$ as the phase group. Let $p:(X,\sigma) \to (\mathbb{K},\tau)$ be the extension produced in Subsection 3·14.37. Recall that X is a minimal set, $(\mathbb{K},\tau)$ is an almost periodic minimal cascade and there exists a point $k_0 \in \mathbb{K}$ such that kp^{-1} consists of a single point for all $k \notin O(k_0) \equiv \{k_0\tau^n|n \in \mathbb{Z}\}$, but

$$(k_0\tau^n)p^{-1} = \{x_0\sigma^n,\bar{x}_0\sigma^n\}, \qquad (n \in \mathbb{Z}),$$

where x_0 and $\bar{x}_0$ are two fixed points from X. Clearly, p is a prime proximal extension.

Let S^2 be the two-dimensional sphere, let $SO(3)$ be the group of all rotations of S^2 and let $f:X \to SO(3)$ be some continuous map. Define a homeomorphism τ_1 of the space $Y = X \times S^2$ onto itself by

$$(x,s)\tau_1 = (x\sigma,sf(x)), \qquad (x \in X, s \in S^2),$$

where $sf(x)$ is the image of the point $s \in S^2$ under the rotation $f(x) \in SO(3)$ (note that we use the left-sided notation for f).

Let us show that there exists a function $f \in C(X,SO(3))$ which satisfies the following four conditions:

(1): $s_0 \in S^2:\{f(x_0),f(\bar{x}_0)\} \subset H(s_0) \equiv \{A|A \in SO(3),s_0A = s_0\}$;

(2): $f(x_0) = Bf(\bar{x}_0)$, where B is an element of $SO(3)$ such that its powers are dense in the group $H(s_0)$ of rotations of the sphere S^2 around an axis passing through s_0;

(3): $f(x_0\sigma^n) = f(\bar{x}_0\sigma^n)$, if $n \neq 0$, $n \in \mathbb{Z}$;

(4): (Y,τ_1) is a minimal cascade.

In order to prove this assertion denote by d some metric on S^2 and define $C_\varepsilon = \{f|f \in C(X,SO(3))$ for every $\varepsilon > 0$; f satisfies (1) and (3); $d(f(x_0),f(\bar{x}_0)) \geqslant \varepsilon\}$. For all sufficiently small $\varepsilon > 0$ the set C_ε is a non-empty closed subset of the complete space $C(X,SO(3))$. Modifying slightly the proof of Theorem 3·18.9, we can show that there exists a residual subset C_ε^* of functions $f \in C_\varepsilon$ such that the orbit of the point (x_0,s_0) is dense in $X \times S^2 \equiv Y$. By Theorem 3·18.10, (Y,τ_1) is a minimal cascade. Thus, if $f \in C_\varepsilon^*$, then conditions (1), (3) and (4) are satisfied. Since the condition (2) also determines a residual subset of C_ε, we can choose a function $f \in C_\varepsilon$ so that all four conditions are satisfied.

Let $\lambda:(X \times S^2,\tau_1) \to (X,\sigma)$ be the canonical projection. It follows from Lemma 3·18.2 that $\lambda:(Y,\tau_1) \to (X,\sigma)$ is distal. Clearly (Y,τ_1) belongs to the class PD (and even to the class V).

Define an invariant equivalence relation R on Y as follows: $(y_1,y_2) \in R$ and $(y_2,y_1) \in R$ iff either $y_1 = y_2$ or $y_1 = (x_0,s_0)\tau_1^n$, $y_2 = (\bar{x}_0,s_0)\tau_1^n$ for some $n \in \mathbb{Z}$. Let us prove that R is closed. In fact, suppose that

$$\{(x_0,s_0)\tau_1^{n_k}\} \to (x_1,s_1), \qquad \{(\bar{x}_0,s_0)\tau_1^{n_k}\} \to (x_2,s_2), \qquad (18.8)$$

where $\{n_k\}$ is a sequence of distinct integers. Since the points x_0 and $\bar{x}_0$ are asymptotic in both directions, we have $x_1 = x_2$. We shall prove that $s_1 = s_2$. For every positive integer n, define a map $f_n:X \to SO(3)$ by

$$f_0(x) = e \text{ (where } e \text{ is the identity of } SO(3));$$

$$f_n(x) = f(x)f(x\sigma)\cdot \ldots \cdot f(x\sigma^{n-1}), \quad (n = 1,2,\ldots);$$

$$f_n(x) = [f(x\sigma^{-1})]^{-1}\cdot \ldots \cdot[f(x\sigma^{n})]^{-1}, \quad (n = -1,-2,\ldots).$$

Then

$$(x,s)\tau_1^{n} = (x\sigma^{n},sf_n(x)), \qquad (n \in \mathbb{Z}).$$

It follows from (1) and (3) that $s_0 f_n(x_0) = s_0 f_n(\bar{x}_0)$ $(n \neq 1)$. Taking (18.8) into account, we conclude that $s_1 = s_2$.

Thus R is a closed invariant equivalence relation. Let $\varphi:Y \to Y/R \equiv W$ be the canonical map. It can now be seen directly that there exists a natural homomorphism $\nu:W \to \mathbb{K}$.

Let us show that the minimal set W does not belong to the class PD. It suffices to verify that the commutative diagram

$$\begin{array}{ccc} X & \xleftarrow{\ \lambda\ } & Y \\ \rho\big\downarrow & & \big\downarrow\varphi \\ \mathbb{K} & \xleftarrow{\ \nu\ } & W \end{array}$$

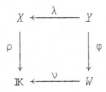

satisfies all the conditions of Lemma 3·18.21. It has already been noted that Y is a compact minimal set, ρ is a prime proximal extension, λ is a distal extension and φ is a proximal extension. Clearly, there is no homomorphism $\psi : W \to X$ with $\psi \circ \rho = \nu$, *i.e.*, the extension ρ is not subordinated to ν. It remains to prove that

$$Y/P^*(Y) = \mathbb{K}.$$

Since $(\mathbb{K},\tau) = (X/P^*(X),\sigma)$, it will suffice to show that

$$((x,s_1),(x,s_2)) \in P^*(Y), \qquad (x \in X; s_1,s_2 \in S^2).$$

We see easily that

$$((x_0,s),(\bar{x}_0,sB)) \in P(Y), \qquad ((\bar{x}_0,sB),(x_0,sB)) \in P(Y), \qquad (s \in S^2).$$

Hence

$$((x_0,s),(x_0,sB)) \in P(Y) \circ P(Y), \qquad (s \in S^2).$$

We note by the way, that the points (x_0,s) and (x_0,sB) are distal, and therefore $P \circ P \neq P$ for (Y,τ_1).

Thus $((x_0,sB),(x_0,s)) \in P^*(Y)$ $(s \in S^2)$. Hence it follows that

$$((x_0,s),(x_0,sB^n)) \in P^*(Y), \qquad (s \in S^2),$$

for all integers n. By condition (2),

$$((x_0,s),(x_0,sA)) \in P^*(Y), \qquad (A \in H(s_0), s \in S^2). \quad (18.10)$$

Set

$$H(s) = \{A \mid A \in SO(3), sA = s\}, \qquad (s \in S^2).$$

Clearly

$$H(s_0)C = CH(s_0C), \qquad (C \in SO(3)). \quad (18.11)$$

Let A_n denote the element $f_n(x_0) \in SO(3)$ and let $\varkappa$ denote the natural map $\varkappa : Y \to Y/P^*(Y)$. Using (18.11), we obtain:

$$(x_0\sigma^n, sA_n)\varkappa = (x_0,s)\tau_1^n\varkappa = (x_0,sA)\tau_1^n\varkappa = (x_0\sigma^n, sAA_n)\varkappa$$

$$= (x_0\sigma^n, sA_nA^*)\varkappa, \qquad (A \in H(s_0), s \in S^2),$$

where

$$A^* = A_n^{-1}AA_n.$$

Taking into account the fact that $(S^2)A_n = S^2$ and letting $C = A_n$, we get by (18.10) that

$$((x_0\sigma^n,s),(x_0\sigma^n,sA)) \in P^*(Y), \qquad (A \in H(s_0A_n), s \in S^2). \quad (18.12)$$

Let a_1 and a_2 be any two points of the sphere S^2. Choose a point $a_3 \in S^2$ so that $a_2 \in a_1H(a_3)$. Since the orbit of $(x_0,s_0) \in Y$ is dense in $Y = X \times S^2$, for each point $x \in X$ there exists a sequence $\{n_k\}$ such that

$$\lim_{k\to\infty} \{x_0\sigma^{n_k}\} = x, \qquad \lim_{k\to\infty} \{s_0A_{n_k}\} = a_3.$$

Thus it follows from (18.12) that

$$((x,a_1),(x,a_1A)) \in P^*(Y), \qquad (A \in H(a_3)),$$

and since $a_2 \in a_1H(a_3)$,

$$((x,a_1),(x,a_2)) \in P^*(Y), \qquad (x \in X; a_1,a_2 \in S^2).$$

Thus the diagram (18.9) satisfies all the conditions of Lemma 3·18.21. Therefore W does not belong to the class PD. Recall that there is a homomorphism $\varphi: Y \to W$, where Y belongs to the class V and consequently to PD. Thus, the class PD of minimal transformation groups is not closed under homomorphisms. Clearly W contains a residual subset of distal points.

3·18.23 Let us show that the Cartesian product of the n-sphere S^n and the circle can support a distal minimal cascade. First consider a more general construction. Let Y be an infinite compact metric space and let G be a connected compact Lie group which acts transitively on a compact metric space A.

Let X denote the Cartesian product of Y and A and let (Y,φ) be a minimal cascade. Define a homeomorphism $\psi: X \to X$ by

$$(y,a)\psi = (y\varphi,a), \qquad (y \in Y, a \in A),$$

and an action of G on X by $(y,a)g = (y,ag)$ $(y \in Y, a \in A, g \in G)$, where ag denotes the image of the point a under the action of $g \in G$ on A. Then (X,ψ) is a central group extension of (Y,φ). By Theorem 3·18.10, there exists a homeomorphism h of X onto X such that (X,h) is a minimal cascade.

In particular, let $Y = S^1$; $\varphi: S^1 \to S^1$ be the rotation through the angle of one radian; $A = S^n$; $G = SO(n + 1)$. Corollary 3·18 .11 shows that there exists a distal minimal cascade $(S^1 \times S^n, h)$.

REMARKS AND BIBLIOGRAPHICAL NOTES

The method of constructing minimal cascades presented in Subsections 3·18.3-11 is due to Ellis [*9*]. Corollary 3·18.11 is a particular case of a more general result obtained by Ellis [*9*], but Theorem 3·18.10 is not completely covered by the results of Ellis because the group G in Theorem 3·18.10 is not assumed to be commutative. The last fact seems to be important in connection with the theorem on the structure of distal minimal extensions, since it shows that even in the case $T = \mathbb{Z}$ there exist minimal equicontinuous extensions whose structure group is an arbitrary compact connected Lie group. Theorem 3·18.12 is due to Peleg [*1*]. The examples presented in Subsection 3·18.16 were constructed by Ellis [*8*,*9*], but the proofs are new and, as seems to us, simpler. Lemmas 3·18.19,20 are, possibly, new. Lemma 3·18.21 and example 3·18.22 are due to Shapiro [*2*]. A similar example was found by Martin [*1*]. Example 3·18.23 is taken from the paper by Ellis [*9*].

§(3·19): DISJOINTNESS

In this section we shall study the disjointness of minimal sets as well as that of extensions. Throughout the section it will be assumed that the phase spaces of transformation groups are compact.

3·19.1 Let $\varphi_1:(X_1,T) \to (Y,T)$ and let $\varphi_2:(X_2,T) \to (Y,T)$ be two extensions. Recall that $X_1 \cdot X_2$ denotes the closed invariant subset $\{(x_1,x_2) | x_1 \in X_1, x_2 \in X_2, x_1\varphi_1 = x_2\varphi_2\}$ of the Cartesian product of the transformation groups (X_1,T) and (X_2,T). The map $\varphi_1 \cdot \varphi_2:(X_1 \cdot X_2,T) \to (Y,T)$ defined by

$$(x_1,x_2)(\varphi_1 \cdot \varphi_2) = x_1\varphi_1 \equiv x_2\varphi_2, \qquad ((x_1,x_2) \in X_1 \cdot X_2),$$

is a homomorphism of $X_1 \cdot X_2$ onto Y. The extension $\varphi_1 \cdot \varphi_2$ is called the *WHITNEY SUM* of the extensions φ_1 and φ_2. The transformation group $(X_1 \cdot X_2,T)$ will sometimes be called the *WHITNEY SUM OF THE TRANSFORMATION GROUPS* (X_1,T) *AND* (X_2,T) *RELATIVE TO THE EXTENSIONS* φ_1 *AND* φ_2.

Two minimal extensions φ_1 and φ_2 are said to be *DISJOINT* if their Whitney sum $\varphi_1 \cdot \varphi_2$ is also minimal. Disjointness of φ_1 and φ_2 is denoted by $\varphi_1 \perp \varphi_2$. Minimal transformation groups (X_1,T) and (X_2,T) are called disjoint, $((X_1,T) \perp (X_2,T))$, if their Cartesian product $(X_1,T) \times (X_2,T)$ is also minimal. The second definition is really a particular case of the first if we identify a transformation group (X,T) with the extension $\varphi:(X,T) \to (Y,T)$ where (Y,T) is the trivial transformation group (*i.e.*, Y is a single point).

3·19.2 LEMMA: *If* φ *and* ψ *are disjoint minimal extensions and* λ *is subordinated to* φ*, then* $\lambda \perp \psi$.

Proof: This is immediate from the definitions. ▮

3·19.3 Let

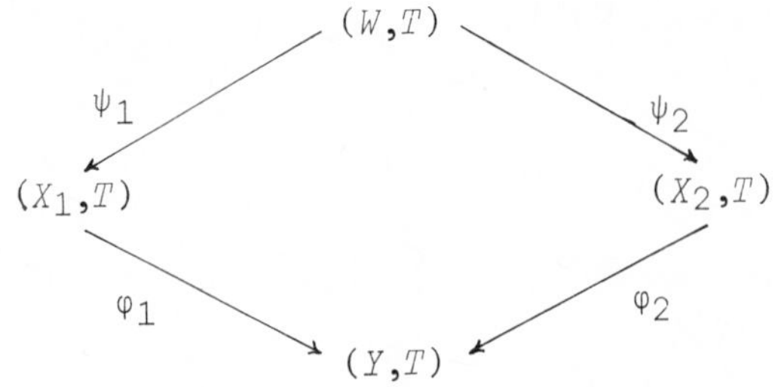

(19.1)

be a commutative diagram of minimal transformation groups. Let p_1 and p_2 denote the canonical projections of $X_1 \cdot X_2$ onto X_1 and X_2 respectively. There exists a unique homomorphism $\tau:(W,T) \to (X_1 \cdot X_2, T)$ for which the following diagram is commutative:

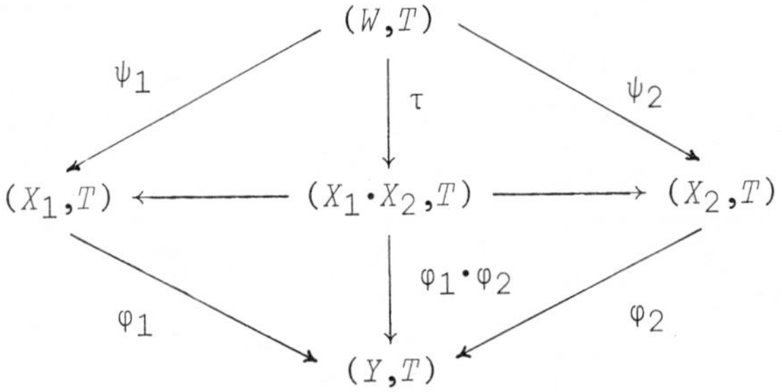

(19.2)

That homomorphism is given by

$$w\tau = (w\psi_1, w\psi_2), \qquad (w \in W).$$

3·19.4 LEMMA: *Minimal extensions* φ_1 *and* φ_2 *are disjoint iff* $W\tau = X_1 \cdot X_2$ *for any transformation group* (W,T) *which satisfies the commutative diagram* (19.1).

Proof: If $\varphi_1 \perp \varphi_2$, then $X_1 \cdot X_2$ is a minimal set. Since $W\tau$ is a non-empty closed invariant set, $W\tau = X_1 \cdot X_2$. Conversely, suppose that $X_1 \cdot X_2$ contains a proper closed invariant subset, say W. Let τ be the inclusion map of W into $X_1 \cdot X_2$ and let $\psi_i = \tau o p_i$ $(i = 1,2)$. By hypothesis, $W\tau = X_1 \cdot X_2$, *i.e.*, $W = X_1 \cdot X_2$, a contradiction. ▮

3·19.5 LEMMA: *Let* φ_1 *and* φ_2 *be minimal extensions. If there exist a transformation group* (Z,T) *and homomorphisms* ξ_1, ξ_2, λ, *satisfying the commutative diagram*

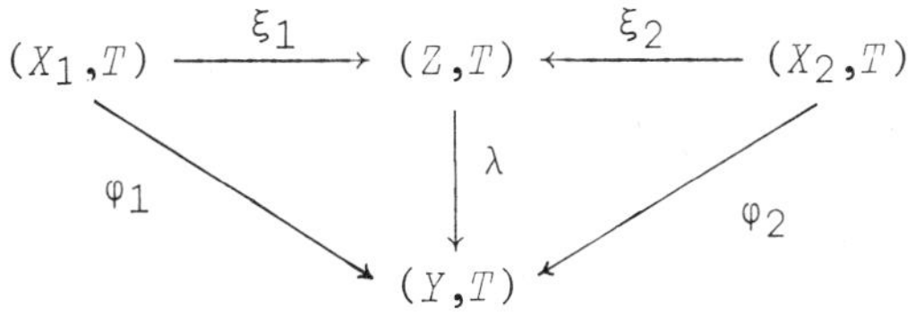

$$\begin{array}{ccccc} (X_1,T) & \xrightarrow{\xi_1} & (Z,T) & \xleftarrow{\xi_2} & (X_2,T) \\ & \searrow^{\varphi_1} & \downarrow \lambda & \swarrow_{\varphi_2} & \\ & & (Y,T) & & \end{array} \tag{19.3}$$

and λ *is not an isomorphism, then the extensions* φ_1 *and* φ_2 *are not disjoint.*

Proof: Denote $\{(z_1,z_2)|z_1,z_2 \in Z, z_1\lambda = z_2\lambda\}$ by R. Define a homomorphism $\mu:(X_1 \cdot X_2,T) \to (R,T)$ by

$$(x_1,x_2)\mu = (x_1\xi_1,x_2\xi_2), \qquad ((x_1,x_2) \in X_1 \cdot X_2).$$

Let us prove that $(X_1 \cdot X_2)\mu = R$. Let $(z_1,z_2) \in R$. Since $z_1,z_2 \in Z$ and Z is a minimal set, there exist points $x_1 \in X_1$ and $x_2 \in X_2$ such that $x_1\xi_1 = z_1$, $x_2\xi_2 = z_2$. Since $z_1\lambda = z_2\lambda$, it follows that $x_1\varphi_1 = x_1\xi_1\lambda = x_2\xi_2\lambda = x_2\varphi_2$. Hence $(x_1,x_2) \in X_1 \cdot X_2$ and $(x_1,x_2)\mu = (z_1,z_2)$.

Since λ is not an isomorphism, $\Delta(Z) \equiv \{(z,z)|z \in Z\}$ is a proper subset of R. Therefore $(X_1 \cdot X_2)\mu = R$ implies that the transformation group $(X_1 \cdot X_2,T)$ is not minimal. ∎

3·19.6 COROLLARY: *If there exist a non-trivial transformation group* (Z,T) *and homomorphisms* $\xi_i:(X_i,T) \to (Z,T)$ *(*$i = 1,2$*), then* (X_1,T) *and* (X_2,T) *are not disjoint.*

3·19.7 Lemma 3·19.5 can be rephrased as follows: if the extensions $\varphi_1:(X_1,T) \to (Y,T)$ and $\varphi_2:(X_2,T) \to (Y,T)$ are disjoint, then (Y,T) is a 'maximal common factor' of the transformation groups (X_1,T) and (X_2,T). The converse does not hold in general. The next example shows that there exist two equicontinuous minimal transformation groups with no non-trivial common factors, but such that their Cartesian product is not minimal.

3·19.8 Let G be a compact topological group, H a closed subgroup of G, $G/H \equiv \{Ha|a \in G\}$ the coset space and $(G/H,G)$ the equicontinuous minimal transformation group defined by

$$(Ha,g) \mapsto Hag, \qquad (Ha \in G/H, g \in G).$$

Let H_1 and H_2 be closed subgroups of G. The following assertions (a) and (b) hold:

(a): $(G/H_1,G) \perp (G/H_2,G)$ iff $H_1H_2 = G$.

If $(G/H_1,G) \perp (G/H_2,G)$, then the transformation group $(G/H_1 \times G/H_2,G)$, where $(H_1a,H_2b)g = (H_1ag,H_2bg)$ $(a,b,g \in G)$, is minimal. Since the group G is compact, this means that the orbit of the point (H_1,H_2) coincides with $(G/H_1) \times (G/H_2)$. Hence for every element $a \in G$ there is an element $g \in G$ such that $H_1g = H_1a, H_2g =$

H_2. It follows that $g \in H_2$ and $a \in H_1H_2$. Since a is an arbitrary element of G, $G = H_1H_2$. The proof of the second half of the statement is similar.

(b): The transformation groups $(G/H_1,G)$ and $(G/H_2,G)$ have no non-trivial common factors iff for each element $g \in G$ the smallest closed subgroup of G, containing H_1 and gH_2g^{-1} coincides with G.

Indeed, if (Z,G) is a homomorphic image of both $(G/H_1,G)$ and $(G/H_2,G)$, then there exist a closed subgroup H_3 of G and an element $g \in G$ such that $(Z,G) \approx (G/H_3,G)$, $H_3 \supset H_1$, $H_3 \supset gH_2g^{-1}$.

Now we are ready to construct the promised example. Let G be the symmetric group of degree four, *i.e.*, S_4. Let H_1 denote the subgroup generated by a three-cycle and H_2 be the subgroup generated by a four-cycle. Then for each element $g \in G$ the subgroup gH_2g^{-1} is also generated by a four-cycle. It is easy to show that each subgroup of G containing H_1 and gH_2g^{-1} coincides with G. On the other hand, the set H_1H_2 contains not more than twelve elements, therefore $H_1H_2 \neq G$. Thus $(G/H_1,G)$ and $(G/H_2,G)$ have no non-trivial common factors, although they are not disjoint.

The next lemma shows that for equicontinuous minimal transformation groups with a commutative phase group, disjointness is equivalent to the absence of non-trivial common factors.

3·19.9 LEMMA: *Let T be an Abelian group. If the equicontinuous minimal transformation groups (X_1,T) and (X_2,T) have no non-trivial common factors, then* $(X_1,T) \perp (X_2,T)$.

Proof: According to Theorem 2·8.5, there exist a compact commutative topological group G and a continuous group homomorphism $\varphi : T \to G$ such that $\overline{T\varphi} = G$ and $(X_1,T) \approx (G,T,\sigma)$, where $(g,t)\sigma = g(t\varphi)$ $(g \in G, t \in T)$.

Suppose that the Cartesian product $(X_1 \times X_2,T)$ is not minimal. Then $X_1 \times X_2$ contains a proper minimal set M. The group G acts on $X_1 \times X_2 \approx G \times X_2$ in a natural way. Set $K = \{g \mid g \in G, Mg = M\}$. Clearly, K is a closed subgroup of G which acts on M and, moreover, $(X_2,T) \approx (M/K,T)$. There exists a natural homomorphism of (M,T) onto (X_1,T), which induces a homomorphism of the transformation group $(M/K,T) \approx (X_2,T)$ onto $(X_1/K,T)$. It is given by $(x_1K,x_2) \mapsto x_1K$ $(x_1 \in X_1 \approx G, x_2 \in X_2)$. Thus, the transformation group $(X_1/K,T)$ is a common factor of (X_1,T) and (X_2,T). By hypothesis, the space $X_1/K \approx G/K$ consists of a single point, *i.e.*, $K = G$. Hence $Mg = M$ for all $g \in G$. It follows from this that $M = X_1 \times X_2$, a contradiction.▌

3·19.10 Throughout the remainder of this section a group extension will mean a central group extension.

3·19.11 Let G be a compact topological group and let (X,T,G) be a group extension. Moreover let M be a compact topological space and let (M,G) be a transformation group. Define actions of the groups T and G on $X \times M$ as follows:

$$(x,m)t = (xt,m), \qquad (x,m)g = (xg,mg), \qquad (x \in X, m \in M, t \in T, g \in G).$$

We obtain a group extension $(X \times M,T,G)$ which is said to be *M-ASSOCIATED WITH* (X,T,G).

3·19.12 LEMMA: *Let $(X \times M,T,G)$ be an M-associated group extension. Consider the commutative diagram*

$$\begin{array}{ccc} (Y,T) \equiv ((X \times M)/G,T) & \xrightarrow{\;\mu\;} & (X/G,T), \\ \downarrow \psi & \swarrow \lambda & \\ (W,T) & & \end{array}$$

where $((x,m)G)\mu = xG$ $(x \in X, m \in M)$. The transformation group $(Y,T) \equiv ((X \times M)/G,T)$ is minimal iff $(X/G,T)$ and $(Y/Q^*(\psi),T)$ are minimal.

Proof: The relation $Q^*(\psi) \equiv Q^*(R_\psi)$ was defined in Subsection 3·13.5. Since the necessity is obvious, we prove only the sufficiency. Suppose $(X/G,T)$ and $(Y/Q^*(\psi),T)$ are minimal. Let Z denote some minimal subset of X. Clearly $K \equiv \{g \mid g \in G, Zg = Z\}$ is a closed subgroup of the compact topological group G, which acts on Z in such a way that (Z,T,K) is a group extension. Let $(Z \times M,T,K)$ be the M-associated group extension. Define a map $\varphi:(Z \times M)/K \to (X \times M)/G$ by

$$((z,m)p)\varphi = (z,m)p_1, \qquad (z \in Z, m \in M),$$

where $p:Z \times M \to (Z \times M)/K$ and $p_1:X \times M \to (X \times M)/G$ are the canonical projections. Let $x \in X$, $m \in M$. There exists an element $g \in G$ with $xg \in Z$. Hence $((xg,mg)p)\varphi = (xg,mg)p_1 = (x,m)p_1$. Thus φ maps $(Z \times M)/K$ onto $(X \times M)/G$. The map φ is injective. Indeed, if $(z,m)p_1 = (z',m')p_1$, then there is an element $g \in G$ such that $z' = zg$, $m' = mg$, and hence $Zg = Z$, as $z,z' \in Z$. Consequently $g \in K$, and therefore $(z,m)p = (z',m')p$. It is easy to verify that the map φ is continuous, and since X,M,G and K are compact, φ is a homeomorphism. Thus φ is an isomorphism of $((Z \times M)/K,T)$ onto (Y,T).

Consider the transformation group $(W \times (M/K),T)$, where

$$(w,mK)t = (wt,mK), \qquad (w \in W, m \in M, t \in T).$$

Define a homomorphism $\rho:((Z \times M)/K,T) \to (W \times (M/K),T)$ by

$$((z,m)p)\rho = ((z,m)p\varphi\psi,mK), \qquad (z \in Z, m \in M).$$

We shall show that ρ is surjective. In fact, let $w \in W, m \in M$. There exists a point $z \in Z \subset X$ such that $(zG)\lambda = w$. Then

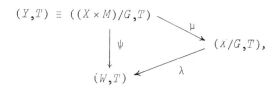

$$((z,m)p)\rho = ((z,m)p\varphi\psi,mK) = ((z,m)p_1\psi,mK)$$

$$= ((z,m)p_1\mu\lambda,mK) = ((zG)\lambda,mK) = (w,mK).$$

Let $p_2: W\times(M/K) \to W$ denote the canonical projection map. The extension p_2 is, of course, equicontinuous. Since there exists a homomorphism $\psi = \varphi^{-1}\circ\rho\circ p_2$ of (Y,T) onto (W,T), it follows from the definition of $Q^*(\psi)$ and from the equicontinuity of the extension p_2 that there is a homomorphism of the minimal transformation group $(Y/Q^*(\psi),T)$ onto $(W\times(M/K),T)$ (see Subsection 3•13.5).

Thus, the transformation group $(W\times(M/K),T)$ is minimal, but this can happen only when M/K consists of a single point, *i.e.*, K acts transitively on M. It follows from this that the transformation group $(Y,T) \equiv ((Z\times M)/K,T)$ is minimal. In fact, let $(z,m),(z',m') \in Z\times M$. Since K is transitive on M, $m = m'k$ for some element $k \in K$. Then

$$(z',m')K = (z',mk^{-1})K = ((z'k)k^{-1},mk^{-1})K = (z'k,m)K.$$

As (Z,T) is a minimal transformation group, there exists a net $\{t_n\}$, $t_n \in T$, such that $\{zt_n\} \to z'k \in Z$. Therefore

$$\{(z,m)Kt_n\} = \{(zt_n,m)K\} \to (z'k,m)K = (z',m')K,$$

but this just means that $((Z\times M)/K,T)$ is minimal.∎

3•19.13 COROLLARY: *Let $(X\times M,T,G)$ be the M-associated group extension and let $(Y,T) = ((X\times M)/G,T)$. The transformation group (Y,T) is minimal iff $(X/G,T)$ and $(Y/Q^*(Y,T),T)$ are minimal.*

Proof: Corollary 3•19.13 is the particular case of Lemma 3•19.12 when W consists of a single point.∎

3•19.14 COROLLARY: *Let (X,T,G) be a group extension. Then the transformation group (X,T) is minimal iff $(X/G,T)$ and $(X/Q^*(X,T),T)$ are minimal.*

Proof: Take $M = G$ and define a transformation group (G,G) by $(a,g) \mapsto g^{-1}a$ $(a,g \in G)$. Now apply Corollary 3•19.13. It is easy to prove that the map

$$\lambda:((X\times G)/G,T) \to (X,T),$$

defined by

$$[(x,a)G]\lambda \equiv \{(xg,g^{-1}a)\,|\,g \in G\}\lambda = xa,$$

is an isomorphism; hence (X,T) is minimal.∎

3•19.15 Let $\varphi:(X,T) \to (Y,T)$ be an extension and let $R = \{(x_1,x_2)\,|\, x_1,x_2 \in X, x_1\varphi = x_2\varphi\}$. Recall that $Q^*(R) \equiv Q^*(\varphi)$ denotes the smallest closed invariant equivalence relation containing

$$Q(R) \equiv Q(\varphi) = \bigcap_{\alpha\in\mathfrak{U}[X]} \overline{\alpha T \cap R}.$$

The relation $Q^*(R)$ can be obtained by means of the following inductive process.

Set $Q_0 = Q(R)$ and suppose that we have already defined closed symmetric reflexive invariant relations $Q_\alpha(\varphi)$ for all ordinals α smaller than some ordinal β. If β is a limit ordinal, then set

$$Q_\beta(\varphi) = \overline{\bigcup_{\alpha<\beta} Q_\alpha(\varphi)}.$$

If $\beta = \gamma + 1$, then define

$$Q_\beta \equiv Q_{\gamma+1} = Q_\gamma \circ Q_\gamma.$$

Thus the $Q_\alpha(\varphi)$ are closed symmetric reflexive invariant relations defined for all ordinals α and

$$Q(R) \equiv Q_0 \subset Q_1 \subset \cdots \subset Q_\alpha \subset Q_{\alpha+1} \subset \cdots \subset Q^*(R).$$

By Zorn's Lemma, there is an ordinal ϑ such that $Q_{\vartheta+1} = Q_\vartheta$, *i.e.*, $Q_\vartheta = Q_\vartheta \circ Q_\vartheta$. Evidently, $Q_\vartheta = Q^*(R)$.

3·19.16 LEMMA: *Let $\varphi:(X,T) \to (W,T)$ and $\psi:(Y,T) \to (W,T)$ be homomorphisms. Suppose that ψ is an open map. If $(a_1,a_2) \in Q^*(\varphi)$ and the point $y \in Y$ satisfies the condition $y\psi = a_1\varphi$, then*

$$((a_1,y),(a_2,y)) \in Q^*(\varphi\cdot\psi).$$

Proof: Let $(a_1,a_2) \in Q(\varphi)$. Then $a_1\varphi = a_2\varphi$. For each index $\alpha \in \mathfrak{U}[X]$ there are points $a_{1\alpha}, a_{2\alpha} \in X$ and an element $t_\alpha \in T$ such that

$$(a_1,a_{1\alpha}) \in \alpha, \qquad (a_2,a_{2\alpha}) \in \alpha, \qquad a_{1\alpha}\varphi = a_{2\alpha}\varphi,$$

$$(a_{1\alpha}t_\alpha, a_{2\alpha}t_\alpha) \in \alpha. \tag{19.4}$$

Then $\{a_{1\alpha}\varphi \mid \alpha \in \mathfrak{U}[X]\} \to a_1\varphi = y\psi$. As ψ is open, we can choose points $y_\alpha \in Y$ $(\alpha \in \mathfrak{U}[X])$ such that $\{y_\alpha\} \to y$, $y_\alpha\psi = a_{1\alpha}\varphi$. Hence

$$\{(a_{1\alpha},y_\alpha)\} \to (a_1,y), \qquad \{(a_{2\alpha},y_\alpha)\} \to (a_2,y),$$
$$a_{1\alpha}\varphi = a_{2\alpha}\varphi = y_\alpha\psi. \tag{19.5}$$

It follows from (19.4) and (19.5) that

$$((a_1,y),(a_2,y)) \in Q(\varphi\cdot\psi).$$

Define the relations $Q_\alpha(\varphi)$ and $Q_\alpha(\varphi\cdot\psi)$ as above. Suppose that the following assertion holds: if $(a_1,a_2) \in Q_\alpha(\varphi)$ and $y\psi = a_1\varphi$, then $((a_1,y),(a_2,y)) \in Q_\alpha(\varphi\cdot\psi)$ for all $\alpha < \beta$.

If $\beta = \alpha + 1$ and $(a_1,a_2) \in Q_{\alpha+1}(\varphi)$, $y\psi = a_1\varphi$, then it follows from $Q_{\alpha+1}(\varphi) = Q_\alpha(\varphi) \circ Q_\alpha(\varphi)$ that there is a point $a_3 \in X$ with (a_1,a_3), $(a_3,a_2) \in Q_\alpha(\varphi)$. Clearly, $a_1\varphi = a_2\varphi = a_3\varphi = y\psi$. Therefore $((a_1,y),(a_3,y)) \in Q_\alpha(\varphi\cdot\psi)$, $((a_3,y),(a_2,y)) \in Q_\alpha(\varphi\cdot\psi)$, and consequently $((a_1,y),(a_2,y)) \in Q_{\alpha+1}(\varphi\cdot\psi)$. The case when β is a limit ordinal can be dealt with similarly. Thus the above assertion holds for all ordinals α. ▮

3·19.17 LEMMA: *Let $\varphi:(X,T) \to (W,T)$ and $\psi:(Y,T) \to (W,T)$ be minimal extensions, where ψ is equicontinuous. Then*

$$Q^*(\varphi\circ\psi) = \{((x_1,y),(x_2,y)) \mid (x_1,x_2) \in Q^*(\varphi), y \in Y, x_1\varphi = y\psi\}.$$

Proof: If $((x_1,y_1),(x_2,y_2)) \in Q(\varphi\cdot\psi)$, then $x_1\varphi = y_1\psi = x_2\varphi = y_2\psi$ and $(x_1,x_2) \in Q(\varphi)$, $(y_1,y_2) \in Q(\psi)$. Since ψ is equicontinuous, $y_1 = y_2 = y$. Hence $Q(\varphi\cdot\psi) \subset \{((x_1,y),(x_2,y)) \mid (x_1,x_2) \in Q(\varphi), y \in Y, x_1\varphi = x_2\varphi = y\psi\}$. Therefore the inclusion $Q^*(\varphi\cdot\psi) \subset \{((x_1,y),(x_2,y) \mid (x_1,x_2) \in Q^*(\varphi), y \in Y, x_1\varphi = x_2\varphi = y\psi\}$ holds. The reverse inclusion follows from Lemma 3·19.16 (if we recall that an equicontinuous minimal extension is open). ▮

3·19.18 LEMMA: *Let $\varphi:(X,T) \to (W,T)$, let $\psi:(Y,T) \to (W,T)$ be a minimal extension and assume further that ψ is equicontinuous. Then the extensions φ and ψ are disjoint iff $\varphi_0 \perp \psi$, where $\varphi_0:(X/Q^*(\varphi),T) \to (W,T)$ is the canonical homomorphism.*

Proof: The necessity follows from Lemma 3·19.2. Let $\varphi_0 \perp \psi$, *i.e.*, $(X/Q^*(\varphi)\cdot Y,T)$ is a minimal transformation group. Let $\mu: X\cdot Y \to X$ be the canonical map and let $\xi:(Z,T) \to (W,T)$ be the principal group extension adjoint to the equicontinuous extension $\psi: (Y,T) \to (W,T)$. Let G denote the structure group of the extension ξ; then $(W,T) = (Z/G,T)$. There exists a closed subgroup K of G such that $(Y,T) = (Z/K,T)$. Let $\nu:X\cdot Z \to X$ be the canonical map, *i.e.*, $(x,z)\nu = x$ $(x \in X, z \in Z, x\varphi = z\ )$. The fibre $x\nu^{-1}$ over the point $x \in X$ has the form

$$x\nu^{-1} = \{(x,z) \mid z \in Z, z\xi = x\varphi\}.$$

Since the group G acts on Z, there is a natural action of G on $X\cdot Z$, namely $(x,z)g = (x,zg)$ $((x,z) \in X\cdot Z, g \in G)$. This action is well defined, because $zg\xi = z\xi$ $(z \in Z, g \in G)$. The fibre $(z\xi)\xi^{-1}$ coincides with zG. Hence it follows that the fibres of the extension ν coincide with the orbits of $(X\cdot Z,G)$. In other words, $\nu:X\cdot Z \to X$ is a group extension and $(X,T) \approx ((X\cdot Z)/G,T)$.

Set $M = G/K = \{hK \mid h \in G\}$. Consider the M-associated extension $((X\cdot Z) \times M,T,G)$:

$$((x,z),hK)g = ((x,zg),g^{-1}hK), \qquad ((x,z) \in X\cdot Z; h,g \in G),$$

$$((x,z),hK)t = ((xt,zt),hK), \qquad ((x,z) \in X\cdot Z; h \in G, t \in T).$$

The transformation group $(X\cdot Y,T)$ can be identified with $(((X\cdot Z) \times M)/G,T)$. Apply Lemma 3·19.12 to the commutative diagram:

$$\begin{array}{ccccc} (((X\cdot Z)\times M)/G,T) \approx (X\cdot Y,T) & & \xrightarrow{\ \mu\ } & & (X,T) \approx ((X\cdot Z)/G,T) \\ & \varphi\cdot\psi \searrow & & \swarrow \varphi & \\ & & (W,T) & & \end{array} .$$

In order to prove that $(X\cdot Y,T)$ is minimal it will suffice to show that the transformation group $((X\cdot Y)/Q^*(\varphi\cdot\psi),T)$ is minimal. By Lemma 3·19.17,

$$Q^*(\varphi\cdot\psi) = \{((x_1,y),(x_2,y)) \mid (x_1,x_2) \in Q^*(\varphi), y \in Y, x_1\varphi = x_2\varphi = y\psi\}.$$

Therefore

$$((X\cdot Y)/Q^*(\varphi\cdot\psi),T) \approx (X/Q^*(\varphi)\cdot Y,T),$$

and the latter transformation group is minimal because $\varphi_0 \perp \psi$. ∎

3·19.19 COROLLARY: *If* (X,T) *and* (Y,T) *are minimal transformation groups and* (Y,T) *is equicontinuous, then*

$$(X,T) \perp (Y,T) \text{ iff } (X/Q^*(X),T) \perp (Y,T).$$

3·19.20 LEMMA: *Suppose that the following diagram*

$$\begin{array}{ccccc} X & \xrightarrow{\ \mu\ } & X_2 & & Z \\ & \varphi_1 \searrow & \downarrow \varphi_2 & \swarrow \varphi_3 & \\ & & W & & \end{array}$$

of minimal transformation groups and homomorphisms is commutative and moreover that

(1): φ_3 *is an open map;*

(2): μ *is an equicontinuous extension;*

(3): $Q^*(\varphi_1) \supset R_\mu$.

Then $\varphi_1 \perp \varphi_3$ *holds iff* $\varphi_2 \perp \varphi_3$.

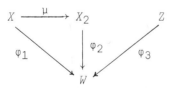

Proof: Necessity follows from Lemma 3·19.2. Let $\varphi_2 \perp \varphi_3$. We shall prove that $\varphi_1 \perp \varphi_3$. Let $\lambda: X \to X_2$ denote the principal group extension adjoint to the equicontinuous extension $\mu: X_1 \to X_2$ and let G be its structure group. Then $(X_2,T) = (X/G,T)$ and there is a closed subgroup K of G such that $(X_1,T) = (X/K,T)$. By Theorem 3·17.2, $(X_1,T) \approx ((X\times M)/G,T)$, where $M = G/K \equiv \{hK|h \in G\}$ and $(X\times M,T,G)$ is an M-associated group extension.

By hypothesis, $(X_2\cdot Z,T) = ((X/G)\cdot Z,T)$ is a minimal transformation group. Let G act trivially on Z, *i.e.*, $zg = z$, $(z \in Z, g \in G)$. Then there exist natural isomorphisms between $((X/G)\cdot Z,T)$ and $((X\cdot Z)/G,T)$ as well as between $(((X\times M)/G)\cdot Z,T)$ and $(((X\cdot Z)\times M)/G,T)$. Consider the commutative diagram

$$((X\cdot Z)\times M)/G \approx ((X\times M)/G)\cdot Z \xrightarrow{\ \mu_1\ } (X/G)\cdot Z \approx (X\cdot Z)/G$$

$$\varphi_1\cdot\varphi_2 \searrow \qquad \swarrow \varphi_2\cdot\varphi_3$$

$$W_1 \ ,$$

where $((x,m)G,z)\mu_1 = (xG,z)$. Apply Lemma 3·19.12 to the M-associated group extension $((X\cdot Z)\times M,T,G)$. By hypothesis, the transformation group $((X\cdot Z)/G,T) \approx ((X/G)\cdot Z,T) = (X_2\cdot Z,T)$ is minimal. In order to prove that $\varphi_1 \perp \varphi_3$ it will suffice to show that the transformation group

$$((X_1\cdot Z)/Q^*(\varphi_1\cdot\varphi_3),T) \approx (((X\times M)/G\cdot Z)/Q^*(\varphi_1\cdot\varphi_3),T),$$

is minimal. We shall first prove the inclusion

$$R_{\mu_1} \equiv R(\mu_1) \subset Q^*(\varphi_1\cdot\varphi_3). \tag{19.6}$$

Let $((a_1,z_1),(a_2,z_2)) \in R(\mu_2)$, where $a_1,a_2 \in (X\times M)/G$ and $z_1,z_2 \in Z$. This means that $(a_1,z_1)\mu_1 = (a_2,z_2)\mu_2$, *i.e.*, $(a_1\mu,z_1) = (a_2\mu,z_2)$, hence $(a_1,a_2) \in R(\mu)$, $z_2 = z_1 \equiv z$. By condition (3), $(a_1,a_2) \in Q^*(\varphi_1)$. Applying Lemma 3·19.16 to the extensions φ_1 and φ_3, we get $((a_1,z_1),(a_2,z_2)) \equiv ((a_1,z),(a_2,z)) \in Q^*(\varphi_1\cdot\varphi_3)$. Thus, the inclusion (19.6) is proved. Hence there exists a homomorphism of the minimal transformation group $(X_2\cdot Z,T) \approx ((X/G)\cdot Z,T)$ onto $((X_1\cdot Z)/Q^*(\varphi_1\cdot\varphi_3),T)$. ∎

3·19.21 COROLLARY: *Let (X_1,T), (X_2,T) and (Z,T) be minimal transformation groups and let $\mu:(X_1,T) \to (X_2,T)$ be an equicontinuous extension with $R_\mu \subset Q^*(X_1,T)$. Then:*

$$(X_1,T)\perp(Z,T) \textit{ iff } (X_2,T)\perp(Z,T).$$

3·19.22 LEMMA: *Let $\varphi:(X,T) \to (Z,T)$, $\mu:(X,T) \to (Y,T)$ and $\psi:(Y,T) \to (Z,T)$ be minimal extensions with $\varphi = \mu\circ\psi$ and $Q^*(\varphi) \supset R(\mu)$. Then $Q^*(\psi) = Q^*(\varphi)(\mu\times\mu)$.*

Proof: The condition $Q^*(\varphi) \supset R(\mu)$ implies that $Q^*(\varphi)(\mu\times\mu)$ is a closed invariant equivalence relation. By Theorem 3·13.2,

$Q(\psi) = Q(\varphi)(\mu \times \mu)$. Since $Q(\varphi)(\mu \times \mu) \subset Q^*(\varphi)(\mu \times \mu)$, we conclude that

$$Q^*(\psi) \equiv [Q(\varphi)(\mu \times \mu)]^* \subset Q^*(\varphi)(\mu \times \mu).$$

Further, it follows easily from

$$Q(\varphi)(\mu \times \mu) = Q(\psi) \subset Q^*(\psi),$$

by applying the inductive process of constructing the relation $Q^*(\varphi)$ that

$$Q^*(\varphi)(\mu \times \mu) \subset Q^*(\psi). \blacksquare$$

The next theorem is a generalization of Lemma 3·19.20.

3·19.23 THEOREM: *Consider the commutative diagram*

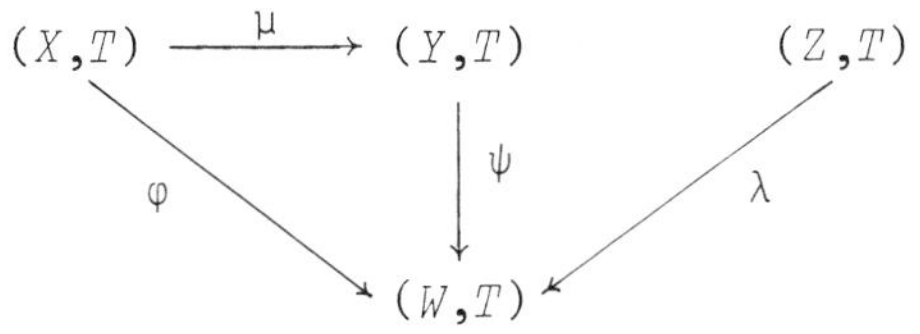

of minimal transformation groups and homomorphisms, where

(1): λ *is an open map;*

(2): μ *is an F-extension;*

(3): $Q^*(\varphi) \supset R_\mu$.

Under these conditions, the extensions φ *and* λ *are disjoint iff* ψ *and* λ *are disjoint.*

Proof: We assume that $\psi \perp \lambda$ and prove that $\varphi \perp \lambda$. By hypothesis (2), μ can be decomposed into a transfinite sequence of equicontinuous extensions $\varphi_\alpha^{\alpha+1}$ $(0 \leqslant \alpha < \vartheta)$:

$$(Y,T) \equiv (Y_0,T) \leftarrow (Y_1,T) \leftarrow \dots \leftarrow (Y_\alpha,T) \xleftarrow{\varphi_\alpha^{\alpha+1}} (Y_{\alpha+1},T) \leftarrow$$

$$\leftarrow \dots \leftarrow (Y_\vartheta,T) \equiv (X,T).$$

Since $\psi \perp \lambda$, we have that $(Y_0 \cdot Z,T)$ is a minimal transformation group. Suppose that $(Y_\gamma \cdot Z,T)$ are minimal for all ordinals smaller than some ordinal β. If β is a limit ordinal, then $(Y_\beta \cdot Z,T)$ is minimal by Lemma 2·9.16. Let $\beta = \alpha + 1$. Consider the commutative diagram

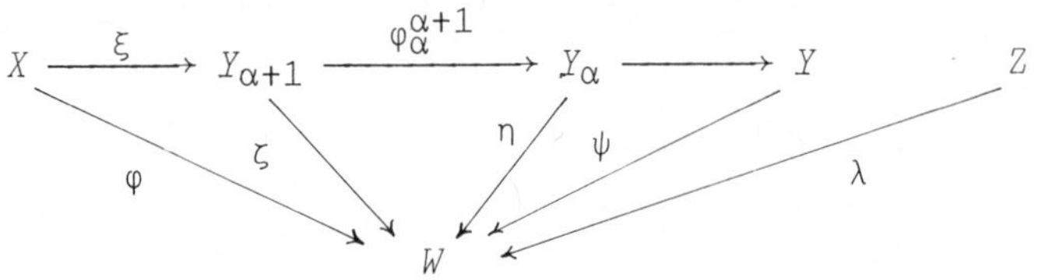

where ξ, η and ζ are canonical. Let us prove the inclusion

$$R(\varphi_\alpha^{\alpha+1}) \subset Q^*(\zeta).$$

By Lemma 3·19.22, $Q^*(\varphi)(\xi \times \xi) = Q^*(\zeta)$, therefore

$$Q^*(\zeta) = Q^*(\varphi)(\xi \times \xi) \supset R_\mu(\xi \times \xi) = R(Y_{\alpha+1} \to Y) \supset R(\varphi_\alpha^{\alpha+1}).$$

By hypothesis, $\varphi_\alpha^{\alpha+1}$ is an equicontinuous extension. Hence it follows from Lemma 3·19.20 and the condition $\eta \perp \lambda$ that $\zeta \perp \lambda$, *i.e.*, the transformation group $(Y_{\alpha+1} \cdot Z, T)$ is minimal.

Thus it is shown by induction that $(Y_\beta \cdot Z, T)$ is minimal for all β, $\beta \leqslant \vartheta$. Letting $\beta = \vartheta$ we get the required result. ▮

3·19.24 REMARK: If the space X is metrizable or if the group T is σ-compact, then condition (2) in Theorem 3·19.23 can be replaced by requiring the distality of μ (Theorem 3·14.24).

3·19.25 THEOREM: *Let $\varphi:(X,T) \to (W,T)$ and $\psi:(Y,T) \to (W,T)$ be minimal extensions. Suppose that φ is open and ψ is an F-extension. Then φ and ψ are disjoint iff $\varphi_0 \perp \psi_0$, where φ_0: $(X/Q^*(\varphi),T) \to (W,T)$ and $\psi_0:(Y/Q^*(\psi),T) \to (W,T)$ are the canonical homomorphisms.*

Proof: Let $(X_0,T) = (X/Q^*(\varphi),T)$, $(Y_0,T) = (Y/Q^*(\psi),T)$ and $\varphi_0 \perp \psi_0$. By Corollary 3·19.21, $\varphi \perp \psi_0$. Apply Theorem 3·19.23 to the following commutative diagram:

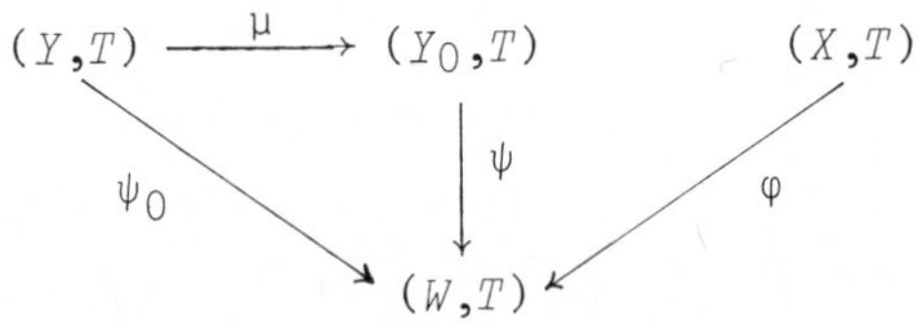

It is clear that μ is an F-extension. Since $R(\mu) = Q^*(\psi)$, condition (3) of Theorem 3·19.23 is satisfied. By hypothesis, φ is an open map. Since $\varphi \perp \psi_0$, it follows from Theorem 3·19.23 that $\varphi \perp \psi$. ▮

3·19.26 COROLLARY: *Let (X,T) and (Y,T) be minimal transformation groups and let (Y,T) belong to the class F. Then*

$$(X,T)\perp(Y,T)\ \textit{iff}\ (X/Q^*(X,T),T)\perp(Y/Q^*(Y,T),T).$$

Proof: Apply Theorem 3·19.25 with W assumed to be a single point. ▮

3·19.27 LEMMA: *If $\varphi:(X,T)\to(Y,T)$ is a proximal extension and (Y,T) is minimal, then X contains exactly one minimal set.*

Proof: Let M_1 and M_2 be minimal subsets of X. Since $M_1\varphi = M_2\varphi = Y$, there exist points $m_1 \in M_1$ and $m_2 \in M_2$ such that $m_1\varphi = m_2\varphi$. By hypothesis, the points m_1 and m_2 are proximal. Therefore $m_1\xi = m_2\xi$ for some element ξ from the enveloping semigroup $E(X,T)$. Hence it follows that $M_1 \cap M_2 \neq \emptyset$, and consequently $M_1 = M_2$. ▮

3·19.28 LEMMA: *Consider the commutative diagram:*

$$\begin{array}{ccccc} (X,T) & \xrightarrow{\ \mu\ } & (Y,T) & & (Z,T) \\ & \searrow^{\varphi} & \downarrow{\psi} & \swarrow_{\lambda} & \\ & & (W,T) & & \end{array}$$

where all transformation groups are minimal and μ is a proximal extension. If $\psi\perp\lambda$ and the set of almost periodic points is dense in $(X\cdot Z,\varphi,\lambda) \equiv X\hat{\cdot}Z$, then $\varphi\perp\lambda$.

Proof: Define a homomorphism $\zeta:X\cdot Z \to Y\cdot Z$ by $(x,z)\zeta = (x\mu,z)$ $((x,z)\in X\cdot Z)$. Clearly, the extension ζ is proximal. If $Y\cdot Z$ is a minimal set, then $X\cdot Z$ contains a unique minimal set M (Lemma 3·19.27); hence $M = X\cdot Z$, because the set of almost periodic points is dense in $X\cdot Z$. Thus, if $\psi\perp\lambda$, then $\varphi\perp\lambda$. ▮

3·19.29 COROLLARY: *If (X,T), (Y,T) and (Z,T) are minimal transformation groups, $\mu:(X,T)\to(Y,T)$ is a proximal extension and the set of almost periodic points is dense in $X\times Z$, then $(Y,T)\perp(Z,T)$ implies $(X,T)\perp(Z,T)$.*

3·19.30 We shall say that a minimal transformation group (X,T) belongs to the class $\mathfrak{K}$ if there exist a point $x_0 \in X$ and an idempotent $u\in E(X,T)$ such that uE is a minimal right ideal of the semigroup $E(X,T)$ and

$$x_0utu = x_0ut, \qquad (t\in T).$$

If the group T is commutative, then each minimal transformation group (X,T) belongs to the class $\mathfrak{K}$. If the point $x_0\in X$ is distal from all other points $x\in X$, then $(X,T)\in\mathfrak{K}$. Indeed, in this case $x_0u = x_0$ and $(x_0t)u = x_0t$ for all $t\in T$ and all idempotents $u\in E(X,T)$. Hence

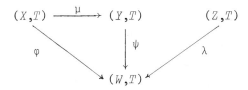

$$x_0utu = (x_0t)u = x_0t = x_0ut, \qquad (t \in T, u \in E, u^2 = u).$$

3·19.31 LEMMA: *If $(X,T) \in \mathcal{K}$ and (Y,T) is a minimal transformation group, then the set of almost periodic points of $(X,T) \times (Y,T)$ is dense in $X \times Y$.*

Proof: Let $p:(X \times Y) \to X$ be the canonical projection and let $\vartheta:E(X \times Y,T) \to E(X,T)$ be the homomorphism of the enveloping semigroups induced by p. By Lemma 1·4.20, there exist a minimal right ideal K of $E(X \times Y,T)$ and an idempotent $v \in K$ such that $v\vartheta = u$. Let $q:X \times Y \to Y$ denote the projection map onto the second coordinate, let $\xi:E(X \times Y,T) \to E(Y,T)$ be the semigroup homomorphism induced by $q:(X \times Y,T) \to (Y,T)$ and let $w = v\xi$. We specify some point $y_0 \in Y$ and proceed to prove that $(x_0ut, y_0w\tau)$ is almost periodic for all $t,\tau \in T$. In fact, $\tau^{-1}v\tau$ is an idempotent belonging to the minimal right ideal $\tau^{-1}K$ of $E(X \times Y,T)$ and

$$(x_0ut, y_0w\tau)\tau^{-1}v\tau = (x_0(ut\tau^{-1}u)\tau, y_0w\tau)$$

$$= (x_0(ut\tau^{-1})\tau, y_0w\tau) = (x_0ut, y_0w\tau),$$

but this just means that $(x_0ut, y_0w\tau)$ is an almost periodic point. Clearly, the set of these points for all $t,\tau \in T$ is dense in $X \times Y$. ∎

3·19.32 COROLLARY: *Let (X,T), (Y,T) and (Z,T) be minimal transformation groups and let $\mu:(X,T) \to (Y,T)$ be a proximal extension. Assume moreover that either $(X,T) \in \mathcal{K}$ or $(Z,T) \in \mathcal{K}$. Then $(Y,T) \perp (Z,T)$ implies $(X,T) \perp (Z,T)$.*

Proof: This follows from Lemma 3·19.31 and Corollary 3·19.29. ∎

3·19.33 THEOREM: *Let $\varphi:(X,T) \to (W,T)$ and $\psi:(Y,T) \to (W,T)$ be minimal extensions such that φ is open and ψ is a PE-extension. If the set of almost periodic points is dense in $X \cdot Y$, then*

$$\varphi \perp \psi \text{ iff } \varphi_0 \perp \psi_0,$$

where φ_0 and ψ_0 are the same as in Theorem 3·19.25.

Proof: As ψ is a *PE*-extension, there exists a projective system $\{(Y_\alpha,T) \mid 0 \leq \alpha \leq \vartheta\}$ such that $(Y_0,T) = (Y/Q^*(\psi),T)$, $(Y_\vartheta,T) = (Y,T)$, $\psi = \varphi_0^\vartheta$ and each extension $\varphi_\alpha^{\alpha+1}:(Y_{\alpha+1},T) \to (Y_\alpha,T)$ is either equicontinuous or proximal (Lemma 3·18.20). Let $\varphi_0 \perp \psi_0$. Then $\varphi \perp \psi_0$ by Corollary 3·19.21.

Suppose that $\varphi_0 \perp \psi_\gamma$, where $\psi_\gamma:(Y_\gamma,T) \to (W,T)$ is the natural map, for all ordinals γ smaller than some ordinal β. If β is a limit ordinal, then $\psi_\beta \perp \varphi$, since $(Y_\beta \cdot X,T)$ is equal to the projective limit of the system $\{(Y_\gamma \cdot X,T) \mid \gamma < \beta\}$ of minimal transformation groups. Let $\beta = \alpha + 1$ and let ξ denote the natural map of

Y onto $Y_{\alpha+1}$. Then by Lemma 3·19.22,

$$Q^*(\psi_{\alpha+1}) = Q^*(\psi)(\xi \times \xi) = [R(Y \to Y_0)](\xi \times \xi)$$

$$= R(Y_{\alpha+1} \to Y_0) \supset R(\varphi_\alpha^{\alpha+1}). \qquad (19.7)$$

Suppose first that $\varphi_\alpha^{\alpha+1}$ is an equicontinuous extension. Apply Theorem 3·19.23 to the diagram

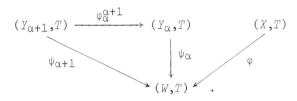

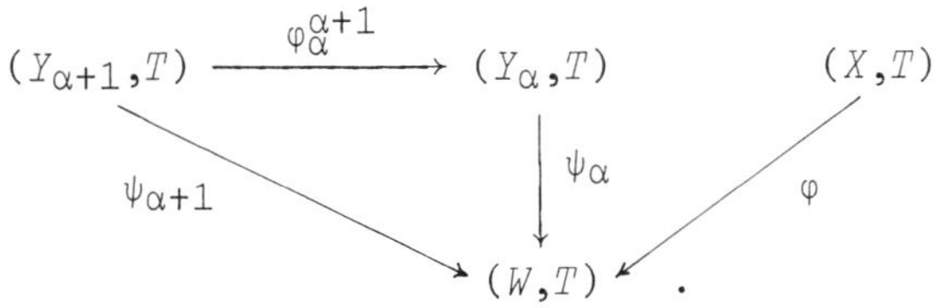

Condition (3) holds in view of (19.7). The other conditions are also satisfied. Hence $\psi_{\alpha+1} \perp \varphi$.

Now consider the case when $\varphi_\alpha^{\alpha+1}$ is a proximal extension. By hypothesis, the set of almost periodic points is dense in $X \cdot Y$; hence the same is valid for $X \cdot Y_{\alpha+1}$. Thus $\psi_{\alpha+1} \perp \varphi$ by Lemma 3·19.28.

We have proved by induction that $\psi_\alpha \perp \varphi$ for all ordinals α, $\alpha \leqslant \vartheta$. Letting $\alpha = \vartheta$, we get $\psi \perp \varphi$. ∎

3·19.34 COROLLARY: *Let (X,T) and (Y,T) be minimal transformation groups. Assume that (Y,T) belongs to the class PE and the set of almost periodic points is dense in $(X \times Y,T)$ (in particular, $(X,T) \in \mathfrak{K}$ or $(Y,T) \in \mathfrak{K}$). Then*

$$(X,T) \perp (Y,T) \text{ iff } (X/Q^*(X,T),T) \perp (Y/Q^*(Y,T),T).$$

3·19.35 COROLLARY: *Suppose that T is a commutative group, (X,T) and (Y,T) are minimal transformation groups and $(Y,T) \in$ PE. Then the following statements are mutually equivalent:*

(1): $(X,T) \perp (Y,T)$;

(2): $(X/Q^*(X,T),T) \perp (Y/Q^*(Y,T),T)$;

(3): *(X,T) and (Y,T) have no non-trivial common (equicontinuous) factors.*

Proof: Statements (1) and (2) are equivalent by Lemma 3·19.31 and Corollary 3·19.34 because T is commutative. If (3) is satisfied, then $(X/Q^*(X,T),T)$ and $(Y/Q^*(Y,T),T)$ have no non-trivial common factors, and hence (2) holds by Lemma 3·19.9. Thus (3) implies (2). But (1) implies (3) according to Lemma 3·19.5. Thus, all three statements are equivalent to each other. ∎

3·19.36 LEMMA: *Consider the commutative diagram*

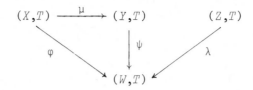

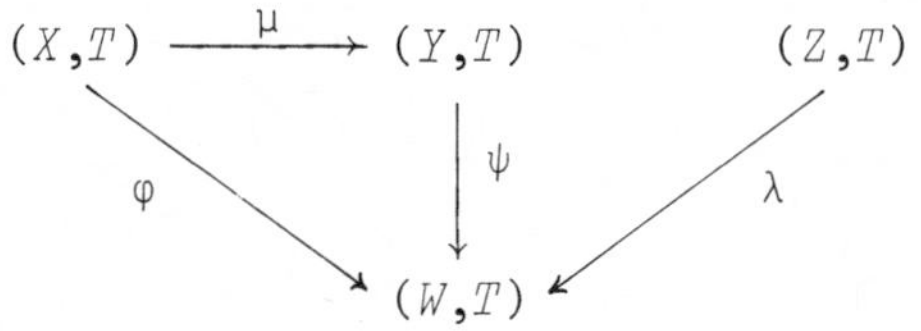

of minimal transformation groups, where μ *is an almost automorphic extension and* λ *is an open map. Then* $\psi \perp \lambda$ *implies* $\varphi \perp \lambda$.

Proof: Define a map $\mu_1 : X \cdot Z \to Y \cdot Z$ by $(x,z)\mu_1 = (x\mu, z)$ $((x,z) \in X \cdot Z)$. Suppose $\psi \perp \lambda$, *i.e.*, $(Y \cdot Z, T)$ is a minimal transformation group. Denote $\{x \mid x \in X, (x\mu)\mu^{-1} = \{x\}\}$ by X_0. By hypothesis, $X_0 \neq \emptyset$; since X is a minimal set and X_0 is invariant, the set X_0 is dense in X.

Clearly μ_1 is injective at all points of the set $A \equiv \{(x,z) \mid (x,z) \in X \cdot Z, x \in X_0\}$. Let us prove that A is dense in $X \cdot Z$. Let $(a,b) \in X \cdot Z$, let U be a neighbourhood of $a \in X$, and let V be a neighbourhood of b in Z. By hypothesis, $a\varphi = b\lambda$. Since the map λ is open, we can choose a small enough neighbourhood V_1 of $b\lambda$ so that $z\lambda \in V_1$ implies $(z\lambda)\lambda^{-1} \cap V \neq \emptyset$. As X_0 is dense in X, φ is continuous and $a\varphi = b\lambda$, there is a point $x_0 \in X_0$ such that $x_0 \in U$ and $x_0\varphi \in V_1$. Choose a point $z_0 \in Z$ so that $z_0\lambda = x_0\varphi$. Then $(z_0\lambda)\lambda^{-1} \cap V \neq \emptyset$. Let $z_1 \in (z_0\lambda)\lambda^{-1} \cap V$. Then $z_1\lambda = z_0\lambda = x_0\varphi$. Thus $(x_0, z_1) \in (X \cdot Z) \cap (U \times V)$ and hence A is dense in $X \cdot Z$.

Since $Y \cdot Z$ is a compact minimal set and $\mu_1 : X \cdot Z \to Y \cdot Z$ is injective at all points of A, the set of almost periodic points is dense in $X \cdot Z$. Thus $\varphi \perp \lambda$ by Lemma 3·19.28. ∎

3·19.37 COROLLARY: *If* (X,T), (Y,T) *and* (Z,T) *are minimal transformation groups and* $\varphi : (X,T) \to (Y,T)$ *is an almost automorphic extension, then*

$$(X,T) \perp (Z,T) \text{ iff } (Y,T) \perp (Z,T).$$

3·19.38 THEOREM: *Suppose that* $\varphi : (X,T) \to (W,T)$ *and* $\psi : (Y,T) \to (W,T)$ *are minimal extensions and moreover that* φ *is open and* ψ *is a V-extension. Then*

$$\varphi \perp \psi \text{ iff } \varphi_0 \perp \psi_0,$$

where φ_0 *and* ψ_0 *are as in Theorem* 3·19.25.

Proof: The proof is similar to that of Theorem 3·19.33 with the only difference that Lemma 3·19.36 should be used for the almost automorphic extension $\varphi_\alpha^{\alpha+1}$. ∎

3·19.39 COROLLARY: *Let* (X,T) *be an arbitrary minimal transformation group and let* (Y,T) *be a minimal transformation group belonging to the class V. Then*

$$(X,T) \perp (Y,T) \text{ iff } (X/Q^*(X,T),T) \perp (Y/Q^*(Y,T),T).$$

3·19.40 Let A be any class of minimal transformation groups with compact phase spaces and a fixed phase group T. We shall denote by $A^{\perp}$ the family of all minimal transformation groups (X,T) satisfying the condition $(X,T) \perp (Y,T)$ for all $(Y,T) \in A$.

Let D denote the class of all distal minimal transformation groups and let W denote the class of all minimal transformation groups (X,T) with $Q^*(X,T) = X \times X$.

3·19.41 THEOREM: *Suppose that T is a σ-compact topological group. Then $D^{\perp} = W$.*

Proof: This is an immediate consequence of the Theorems 3·19.25 and 3·14.24.∎

REMARKS AND BIBLIOGRAPHICAL NOTES

The notion of disjointness of minimal sets was introduced and first studied by Furstenberg [*2*]. Many papers have been devoted to this concept: Bronšteĭn [*12*], Ellis [*13*], Keynes [*6*], Knapp [*3*], Peleg [*2*], Petersen [*1*], Shapiro [*2*,*4*], Wu [*5*,*6*,*7*], and others. In particular, Shapiro [*2*] has introduced the notion of disjointness for extensions. The material presented in Subsections 3·19.3-6 is due to Shapiro [*2*]. Example 3·19.8 was invented by Knapp [*3*]. Lemma 3·19.12 was obtained by the author as a generalization of Corollary 3·19.13 proved by Keynes [*6*], which is in turn a strengthening of the Corollary 3·19.14, proved by Bronšteĭn [*12*]. In this section, an attempt was made to give an account of results concerning disjointness of extensions. Most of the results are new. Even some corollaries on disjointness of minimal sets are a strengthening of previously known results. (*See* for example, Corollaries 3·19.26,34). Corollary 3·19.35 can be found in the paper by Keynes [*6*]. The class $\mathcal{K}$ of transformation groups was introduced in the book by Ellis [*13*].

CHAPTER IV

EXTENSIONS AND EQUATIONS

§(4·20): SHIFT DYNAMICAL SYSTEMS

We shall define shift dynamical systems in various function spaces. Several general (although simple) lemmas will be stated which will permit us to simplify the verification of the continuity axiom and will allow us to significantly shorten the proofs. Moreover, we shall consider the question of completing a transformation group (*i.e.*, extending it by continuity onto the completion of the phase space).

4·20.1 LEMMA: *Let X and Y be uniform spaces, let T be a locally compact topological space, and let $\sigma: X \times T \to Y$ be a map satisfying the following conditions:*

CONDITION (1): *For every point $x \in X$, every index $\alpha \in \mathfrak{U}[Y]$ and every compact subset $K \subset T$ there exists an index $\beta \in \mathfrak{U}[X]$ such that*

$$((x,t)\sigma,(x',t)\sigma) \in \alpha \qquad (x' \in x\beta, t \in K);$$

CONDITION (2): *The map $\sigma_x: T \to Y$, defined by $t\sigma_x = (x,t)\sigma$ $(t \in T)$, is continuous for each point x in some dense subset X_0 of X. Then $\sigma: X \times T \to Y$ is continuous.*

Proof: Let $x \in X$, $t \in T$, $\alpha \in \mathfrak{U}[Y]$. Choose an index $\beta \in \mathfrak{U}[Y]$ so that $\beta^4 \subset \alpha, \beta = \beta^{-1}$. There exists a neighbourhood V of the point $t \in T$ with $\bar{V}$ compact. According to Condition (1), we can choose an index $\gamma \in \mathfrak{U}[X]$ such that

$$((x,\tau)\sigma,(x',\tau)\sigma) \in \beta, \qquad (x' \in x\gamma, \tau \in \bar{V}).$$

Let $x^* \in x\gamma \cap X_0$. By Condition (2), there exists a neighbourhood W of $t \in T$ with

$$((x^*,t)\sigma,(x^*,\tau)\sigma) \in \beta, \qquad (\tau \in W).$$

If $\tau \in W \cap V$, then $((x,t)\sigma,(x,\tau)\sigma) = ((x,t)\sigma,(x^*,t)\sigma)\circ((x^*,t)\sigma,(x^*,\tau)\sigma)\circ((x^*,\tau)\sigma,(x,\tau)\sigma) \in \beta\circ\beta\circ\beta = \beta^3$.

Let $x' \in x\gamma$, $t' \in W \cap V$; then

$$((x,t)\sigma,(x',t')\sigma) = ((x,t)\sigma,(x,t')\sigma)\circ((x,t')\sigma,(x',t')\sigma) \in \beta^3\circ\beta \subset \alpha. \blacksquare$$

4·20.2 LEMMA: *Let X and Y be uniform spaces, let T be a locally compact topological space and let $\sigma:X \times T \to Y$ be a map such that*

CONDITION (1): *if F is a Cauchy filter of the space $(X,\mathfrak{U}[X])$, $\alpha \in \mathfrak{U}[Y]$ and K is a compact subset of T, then there exists a subset $A \in F$ such that*

$$((x_1,t)\sigma,(x_2,t)\sigma) \in \alpha, \qquad (t \in K; x_1,x_2 \in A);$$

CONDITION (2): *the map $\sigma_x:T \to Y$ is continuous for all $x \in X$. Then σ is continuous. If X_1 is the completion of $(X,\mathfrak{U}[X])$ and Y_1 is the completion of $(Y,\mathfrak{U}[Y])$, then there exists a continuous map $\sigma_1:X_1 \times T \to Y_1$, which extends σ.*

Proof: The continuity of $\sigma:X \times T \to Y$ follows from Lemma 4·20.1, if F is taken to be the neighbourhood filter of any point $x \in X$. Condition (1) means that for each fixed $t \in T$ the map $\sigma^t:X \to Y$ defined by $x\sigma^t = (x,t)\sigma$, $(x \in X)$, sends Cauchy filters to Cauchy filters; therefore we can define a map $\sigma_1:X_1 \times T \to Y_1$. Moreover, it is easily seen from Condition (1) and from the definition of completion that for every point $x_1 \in X_1$, every index $\alpha_1 \in \mathfrak{U}[Y_1]$ and an arbitrary compact subset $K \subset T$ there exists an index $\beta_1 \in \mathfrak{U}[X_1]$ with

$$((x_1,t)\sigma_1,(x_1',t)\sigma_1) \in \alpha_1, \qquad (x_1' \in x_1\beta_1, t \in K).$$

The set X is dense in X_1. Since each map $\sigma_x:T \to Y$, $(x \in X)$ is continuous and $\mathfrak{U}[Y_1]$ induces on Y the initial uniformity $\mathfrak{U}[Y]$, the required result follows from Lemma 4·20.1. $\blacksquare$

We obtain, as a corollary, the following statement about the completion of a topological transformation group.

4·20.3 LEMMA: *Let T be a locally compact topological group, $(X,\mathfrak{U})$ a uniform space and $((X,\mathfrak{U}),T,\pi)$ a topological transformation group. Further, let $\mathfrak{V}$ be a uniformity on X, weaker than $\mathfrak{U}$, and satisfying the following condition: for every Cauchy filter F of the space $(X,\mathfrak{V})$, arbitrary index $\alpha \in \mathfrak{V}$ and compact subset $K \subset T$, there exists a subset $A \in F$ such that*

$$((x_1,t)\pi,(x_2,t)\pi) \in \alpha, \qquad ((x_1,x_2) \in A \times A, t \in K).$$

Then the transformation group (X,T,π) can be extended by continuity to the completion $(X_1,\mathfrak{V}_1)$ of the space $(X,\mathfrak{V})$.

Proof: Since $\mathcal{V}$ is weaker than $\mathcal{U}$, the map $\pi_x : T \to (X,\mathcal{V})$ is continuous for all $x \in X$. By Lemma 4·20.2, the map $\pi : (X,\mathcal{V}) \times T \to (X,\mathcal{V})$ is continuous and it can be extended by continuity to $(X_1,\mathcal{V}_1) \times T$. Thus we obtain a map $\pi_1 : X_1 \times T \to X_1$. Since X is dense in X_1, the identity axiom and the homomorphism axiom for (X_1,T,π_1) follow from the fact that (X,T,π) is a transformation group. ∎

4·20.4 LEMMA: *Let* $\mathbb{R}^+$ *be the semigroup of non-negative real numbers,* (X,ρ) *a locally compact metric space and* Ω *a topological space. Suppose that* $f : \Omega \times \mathbb{R}^+ \to X$ *is a map satisfying the following conditions:*

CONDITION (1): *the map* $f_w : \mathbb{R}^+ \to X$ *defined by* $f_w(t) = f(\omega,t)$ *is continuous for all* $\omega \in \Omega$;

CONDITION (2): f *is continuous at every point of the form* $(\omega,0)$ $(\omega \in \Omega)$;

CONDITION (3): *if* $\omega_0 \in \Omega$, $T > 0$ *and* G *is a neighbourhood of the set* $f(\omega_0[0,T])$ *with a compact closure* $\bar{G}$, *then there exists a neighbourhood* $U = U(\omega_0)$ *such that for every* $\varepsilon > 0$ *we can choose a number* $\delta = \delta(\varepsilon) > 0$ *so that* $\rho(f(\omega,t_1), f(\omega,t_2)) < \varepsilon$ *whenever* $\omega \in U$, $f(\omega,t_1) \in G$, $f(\omega,t_2) \in G$, $|t_1 - t_2| < \delta$;

CONDITION (4): *if* $\omega_0 \in \Omega$, $T > 0$, $\{\omega_\alpha\} \to \omega_0$ *and* $\{f\omega_\alpha : [0,T] \to X\}$ *converges to* $\varphi : [0,T] \to X$ *uniformly on* $[0,T]$, *then* $\varphi(t) = f\omega_0(t)$ $(t \in [0,T])$.

Then f *is continuous.*

Proof: Let $\omega_0 \in \Omega$, $T > 0$. By Condition (1), the set $f(\omega_0,[0,T])$ is compact. Since X is locally compact, there exists a number $\varepsilon_0 > 0$ such that the closure of the spherical ε_0-neighbourhood of the set $f(\omega_0,[0,T])$ is compact.

Let us prove that for every $\varepsilon \in (0,\varepsilon_0)$ there exists a neighbourhood $U = U(\omega_0)$ such that $\rho(f(\omega_0,t),f(\omega,t)) < \varepsilon (t \in [0,T], \omega \in U)$. By Lemma 4·20.1, this will imply the continuity of f.

Suppose that, to the contrary, for some $\varepsilon_1 \in (0,\varepsilon_0)$ there exist nets $\{t_\alpha\}$, $t_\alpha \in [0,T]$, and $\{\omega_\alpha\} \to \omega_0$ with

$$\left.\begin{aligned} &\rho(f(\omega_0,t),f(\omega_\alpha,t)) < \varepsilon_1 < \varepsilon_0 \qquad (0 \leqslant t < t_\alpha), \\ &\rho(f(\omega_0,t_\alpha),f(\omega_\alpha,t_\alpha)) = \varepsilon_1. \end{aligned}\right\} \tag{20.1}$$

By Condition (2), there exist a number $h > 0$ and an index α_0 such that $t_\alpha > h$ for all $\alpha > \alpha_0$. Without loss of generality we may suppose that $\{t_\alpha\} \to t^*$. Then $t^* \geqslant h > 0$. For the point $\omega_0 \in \Omega$, the number $T > 0$ and the ε_0-neighbourhood of the set $f(\omega_0,[0,T])$ choose a neighbourhood U of ω_0 and a number $\delta = \delta(\varepsilon_1/3)$ in accordance with Condition (3). If $0 < \tau < t^*$ and $t^* - \tau < \delta$, then by Lemma 4·20.1 we have that

$$\rho(f(\omega_0,\tau),f(\omega_0,t_\alpha)) < \frac{\varepsilon_1}{3}, \tag{20.2}$$

$$\rho(f(\omega_\alpha,\tau),f(\omega_\alpha,t_\alpha)) < \frac{\varepsilon_1}{3}, \qquad (20.2)$$

for all α greater than some index $\alpha_1 > \alpha_0$.

By Condition (3), the net $\{f\omega_\alpha:[0,\tau] \to X\}$ of maps can be assumed to be uniformly convergent on $[0,\tau]$. By Condition (4), the limit is equal to $f\omega_0:[0,\tau] \to X$. Corresponding to $\varepsilon_1/3$ choose an index $\alpha_2 > \alpha_1$ so that $\rho(f(\omega_0,\tau),f(\omega_\alpha,\tau)) < \varepsilon_1/3$ for all $\alpha > \alpha_2$. Then it follows from Lemma 4·20.2 that

$$\rho(f(\omega_0,t_\alpha),f(\omega_\alpha,t_\alpha)) < \varepsilon_1.$$

This contradicts the second part of Lemma 4·20.1.▮

4·20.5 Let T be a locally compact topological group, X a topological space and let $C(T,X)$ be the set of all continuous maps $\varphi:T \to X$ provided with the compact-open topology, which is defined as follows. Let $\mathfrak{K} = \mathfrak{K}(T)$ be the class of all compact subsets of the group T, and let $G(X)$ be the totality of open subsets of X. The family of all finite intersections of sets of the form $G(K,A) = \{\varphi \mid \varphi \in C(T,X), \varphi(K) \subset A\}$, where $K \in \mathfrak{K}$, $A \in G$, is a base for some topology in $C(T,X)$ which is called the compact-open topology.

Define a map $\sigma:C(T,X) \times T \to C(T,X)$ as follows: if $\varphi \in C(T,X)$, $t \in T$, then $(\varphi,t)\sigma = \varphi_t$, where $\varphi_t(\tau) = \varphi(t\tau)$ $(\tau \in T)$. Let us show that σ is continuous. Let $\varphi \in C(T,X)$, $t \in T, K \in \mathfrak{K}, A \in G(X)$ and $\varphi_t(K) \equiv \varphi(tK) \subset A$. There exists a neighbourhood V of $t \in T$ such that the closure $\bar{V}$ is compact and $\varphi(\bar{V}K) \subset A$. The set $U \equiv \{\psi \mid \psi \in C(T,X), \psi(\bar{V}K) \subset A\}$ is a neighbourhood of φ with $(U,V)\sigma \subset G(K,A)$. Thus σ is a continuous map. It is easy to verify that σ satisfies the identity axiom and the homomorphism axiom; hence $(C(T,X),T,\sigma)$ is a topological transformation group.

If, in particular, $(X,\mathfrak{U})$ is a uniform space, then the set $C(T,X)$ can be provided with a uniformity $\mathfrak{V}$, compatible with the compact-open topology. The family $\{V(\alpha,K) \mid \alpha \in \mathfrak{U}, K \in \mathfrak{K}(T)\}$, where $V(\alpha,K) = \{(\varphi,\psi) \mid \varphi,\psi \in C(T,X), (\varphi(t),\psi(t)) \in \alpha, (t \in K)\}$, is a base for this uniformity.

4·20.6 Let T be a locally compact topological group, X a topological space, Y a uniform space and $C(X \times T,Y)$ the family of all continuous maps $\varphi:X \times T \to Y$. If $\mathfrak{K}(X)$ is the set of all compact subsets of the space X, $\alpha \in \mathfrak{U}[Y]$, $K \in \mathfrak{K}(T)$ and $L \in \mathfrak{K}(X)$, then the family of all sets of the form

$$V(\alpha,K,L) = \{(\varphi,\psi) \mid \varphi,\psi \in C(X \times T,Y),$$
$$(\varphi(x,t),\psi(x,t)) \in \alpha, (t \in K, x \in L)\}$$

is a base for some uniformity on $C(X \times T,Y)$.

Define a map $\sigma:C(X \times T,Y) \times T \to C(X \times T,Y)$ by $(\varphi,t)\sigma = \varphi_t$, where

$\varphi_t(x,\tau) = \varphi(x,t\tau)(x \in X, \tau \in T)$. In other words, σ is a 'shift' of the functions φ in the second variable.

Let us show that $(C(X \times T,Y),T,\sigma)$ is a topological transformation group. It will suffice to verify the continuity of σ. We use Lemma 4·20.1 to show this. We first prove that for every function $\varphi \in C(X \times T,Y)$ the map $\sigma_\varphi : T \to C(X \times T,Y)$ defined by $\sigma_\varphi(t) = \varphi_t$ $(t \in T)$ is continuous. Let $t \in T$, $K \in \mathfrak{K}(T)$, $L \in \mathfrak{K}(X)$ and $\alpha \in \mathfrak{U}[Y]$. Choose a neighbourhood V of the element t so that $\bar{V}$ is compact. By the uniform continuity of the function φ on the compact subset $L \times \bar{V}K$, there exists for $\alpha \in \mathfrak{U}[Y]$ a neighbourhood W of the identity of T such that $Wt \subset V$ and $(\varphi(x,s),\varphi(x,\tau s)) \in \alpha$ $(x \in L,\ s \in tK,\ \tau \in W)$. In other words, $(\varphi_t,\varphi_{\tau t}) \in V(\alpha,K,L)(\tau \in W)$. Now let $L \in \mathfrak{K}(X)$, $K_1 \in \mathfrak{K}(T)$, $K_2 \in \mathfrak{K}(T)$. Then $K_1K_2 \in \mathfrak{K}(T)$. If $(\varphi,\psi) \in V(\alpha,K_1K_2,L)$, then $(\varphi(x,s),\psi(x,s)) \in \alpha(s \in K_1K_2, x \in L)$; hence $(\varphi_\tau(x,s),\psi_\tau(x,s)) \in \alpha(x \in L, s \in K_2, \tau \in K_1)$. By Lemma 4·20.1, σ is continuous.

4·20.7 A function $\varphi \in C(X \times T,Y)$ is said to be a *COMPACT FUNCTION* if for every compact subset $L \subset X$ the closure of the set $\{\varphi(x,t) | x \in L, t \in T\}$ is compact. A function $\varphi \in C(X \times T,Y)$ is called a *LAGRANGE STABLE FUNCTION* if the closure of the family of shifts $\{\varphi_\tau | \tau \in T\}$ is compact in the space $C(X \times T,Y)$.

By using Ascoli's theorem (see Kelley[*1*]) it is not hard to prove that a function $\varphi \in C(X \times T,Y)$ is Lagrange stable iff it is compact and for every compact subset $L \subset X$ the restriction of φ to $L \times T$ is uniformly continuous.

A function $\varphi \in C(X \times T,Y)$ is called a *FUNCTION RECURRENT IN THE SENSE OF BIRKHOFF* if the closure of the orbit of φ in $(C(X \times T,Y),T,\sigma)$ is a compact minimal set. As follows from §(5), a function φ is recurrent iff it is Lagrange stable, and for every $\alpha \in \mathfrak{U}[Y]$ and arbitrary compact subsets $L \subset X$ and $K \subset T$ there exists a syndetic subset $A \subset T$ such that

$$(\varphi(x,t),\varphi(x,\tau t)) \in \alpha \qquad (x \in L, t \in K, \tau \in A).$$

In this case we also say that the function $\varphi = \varphi(x,t)$ is recurrent in the variable t uniformly with respect to x on every compact set $L \subset X$.

A function $\varphi \in C(X \times T,Y)$ is called *ALMOST PERIODIC IN THE SENSE OF BOHR* if the orbit closure of the point φ in $(C(X \times T,Y),T,\sigma)$ is an almost periodic minimal set (or, equivalently, if φ is a compact function and for every $\alpha \in \mathfrak{U}[Y]$ and every $L \in \mathfrak{K}(X)$ there exists a syndetic subset $A \subset T$ with

$$(\varphi(x,t),\varphi(x,\tau t)) \in \alpha \qquad (x \in L, t \in T, \tau \in A).$$

Let S be a subsemigroup of the group T. A function $\varphi \in C(X \times T,Y)$ is said to be *POISSON S-STABLE* if for every $\alpha \in \mathfrak{U}[Y]$, $L \in \mathfrak{K}(X)$, $K \in \mathfrak{K}(T)$ and $s \in S$ there exists an element $q \in S$ such that $(\varphi(x,t),\varphi(x,sqt)) \in \alpha(x \in L, t \in K)$.

4·20.8 It is easy to transfer the definitions in Subsection 4·20.7 to the case of a dynamical system $(C(T,Y),T,\sigma)$, which is a particular case of Example 4·20.6 when X consists of one point.

On the other hand, let us provide each of the spaces $C(X\times T,Y)$, $C(X,Y)$, and $C(T,C(X,Y))$ with the uniformity corresponding to the uniform convergence on compact subsets (see Subsection 4·20.6). Then it is easy to verify that the map which assigns to each function $\varphi \in C(X\times T,Y)$ the function $\phi \in C(T,C(X,Y))$, where $\phi(t)(x) = \varphi(x,t)\,(x\in X, t\in T)$, is bijective and *equimorphic* (*i.e.*, it is uniformly continuous and the inverse map is also uniformly continuous). Therefore the transformation group $(C(X\times T,Y),T,\sigma)$ from Subsection 4·20.6 can be reduced to $(C(T,Y'),T,\sigma)$, where $Y' = C(X,Y)$.

4·20.9 Let us give an example illustrating Lemma 4·20.3 on the completion of transformation groups.

Let $\mathbb{R}^n$ be the n-dimensional Euclidean space, $C(\mathbb{R},\mathbb{R}^n)$ the family of all continuous functions $\varphi:\mathbb{R}\to\mathbb{R}^n$ provided with the uniformity described in Subsection 4·20.5, and let $(C(\mathbb{R},\mathbb{R}^n),\mathbb{R},\sigma)$ be the shift transformation group. Specify a number p, $1 < p < \infty$. Let I be an arbitrary segment of the real line $\mathbb{R}$ and let $\varepsilon > 0$. Let $V_p(I,\varepsilon)$ denote the family of all pairs of functions $\varphi,\psi \in C(\mathbb{R},\mathbb{R}^n)$ such that

$$\int_I |\varphi(s) - \psi(s)|^p ds < \varepsilon^p,$$

where $|\cdot|$ denotes the Euclidean norm in $\mathbb{R}^n$. It is easy to verify that the totality of sets of the form $V_p(I,\varepsilon)$ forms a base for some uniformity in $C(\mathbb{R},\mathbb{R}^n)$, which is weaker than the uniformity discussed in Subsection 4·20.5. Denote by $L_p(\mathbb{R},\mathbb{R}^n)$ the completion of the uniform space obtained in this manner. It is not difficult to show, by using Lemma 4·20.3, that the transformation group $(C(\mathbb{R},\mathbb{R}^n),\mathbb{R},\sigma)$ can be extended by continuity to $L_p(\mathbb{R},\mathbb{R}^n)$. In fact, let I and K be two segments of $\mathbb{R}$ and $\varepsilon > 0$. If $(\varphi,\psi) \in V_p(I + K,\varepsilon)$, then $(\varphi_\tau,\psi_\tau) \in V_p(I,\varepsilon)$ $(\tau \in K)$.

REMARKS AND BIBLIOGRAPHICAL NOTES

The shift dynamical system defined in the space of continuous functions $\varphi:\mathbb{R}\to\mathbb{R}$ was first studied by Bebutov [*1*]. More general shift transformation groups were considered by Gottschalk and Hedlund [*2*], Sell [*4*] and Scerbakov [*6*]. Lemma 4·20.4 is similar to the well known Kamke Lemma (see, for example, Hartman [*1*]).

§(4·21): EXTENSIONS ASSOCIATED WITH CERTAIN CLASSES OF EQUATIONS

In this section we shall show that extensions of dynamical systems can be associated in a natural way with several classes of differential, integral, difference, and algebraic equations. In order to give the first and the simplest examples demonstrating the usefulness of considering relations between equations and the corresponding extensions, we prove the existence of recurrent or Poisson stable solutions of differential equations with recurrent or Poisson stable, respectively, right sides.

4·21.1 Let W be an open subset of the space $\mathbb{R}^n$. We shall consider differential equations of the form

$$\frac{dx}{dt} = f(x,t), \qquad (21.1)$$

where the right side f is assumed to be defined and continuous for all $x \in W$ and $t \in \mathbb{R}$. Consider the set $C(W \times \mathbb{R}, \mathbb{R}^n)$ of all such functions f and determine the flow $(C(W \times \mathbb{R}, \mathbb{R}^n), \mathbb{R}, \sigma)$, described in Subsection 4·20.6. For brevity denote $(C(W \times \mathbb{R}, \mathbb{R}^n), \mathbb{R}, \sigma)$ by $(Y, \mathbb{R}, \sigma)$. The family of all pairs (f, φ), where $f \in C(W \times \mathbb{R}, \mathbb{R}^n)$, and $\varphi \in C(\mathbb{R}, W)$ is such that $\varphi: \mathbb{R} \to W$ is a solution of Equation (21.1), is denoted by X. It is easy to see that X is an invariant subset of the Cartesian product of the transformation groups $(C(W \times \mathbb{R}, \mathbb{R}^n), \mathbb{R}, \sigma)$ and $(C(\mathbb{R}, W), \mathbb{R}, \sigma)$. Therefore we have defined a dynamical system $(X, \mathbb{R}, \sigma)$, for which the canonical projection $p: X \to Y$, $(\varphi, f)p = f$, is a homomorphism. The fibre of the map $p: X \to Y$ over the point $f \in Y$ coincides, in principle, with the set of all solutions $\varphi: \mathbb{R} \to W$ of Equation (21.1).

4·21.2 Let Y_0 be the subset of Y consisting of all functions $f \in C(W \times \mathbb{R}, \mathbb{R}^n)$ such that for each point $\alpha \in W$ there exists a unique solution $\varphi: \mathbb{R} \to W$ of Equation (21.1) with the initial condition $\varphi(0) = a$. Clearly, Y_0 is invariant under shifts.

Let $X_0 = Y_0 \times W$. Define a map $\pi: X_0 \times \mathbb{R} \to X_0$ by $(f, a, t)\pi = (f_t, \varphi(f, a, t))$ $(f \in Y_0, a \in W, t \in \mathbb{R})$, where $\varphi(f, a, \cdot)$ is the solution of Equation (21.1) with $\varphi(f, a, 0) = a$. It follows from the basic properties of solutions of differential equations and from Lemma 4·20.4 that $\varphi: Y_0 \times W \times \mathbb{R} \to W$ is a continuous map. Therefore π is also continuous. Clearly $(f, a, 0)\pi = (f, a)$ $(f \in Y_0, a \in W)$. Let s, $t \in \mathbb{R}$. Since the solutions of Equation (21.1) with $f \in Y_0$ are uniquely determined by the initial conditions, we have the equality

$$\varphi(f, a, t + s) = \varphi(f_t, \varphi(f, a, t), s),$$

which implies that the homomorphism axiom is satisfied. Thus $(X_0, \mathbb{R}, \pi)$ is a flow, and the canonical projection $p: X_0 \to Y_0$ determines an extension $p: (X_0, \mathbb{R}, \pi) \to (Y_0, \mathbb{R}, \sigma)$. We note (although it is evident) that the map p is open in this case.

Let $X_1 = \{(f,\varphi) | (f,\varphi) \in X, f \in Y_0\}$. It is easy to see that the dynamical systems $(X_0,\mathbb{R},\pi)$ and $(X_1,\mathbb{R},\sigma)$ are isomorphic. The required isomorphism of X_1 onto X_0 is of the form

$$(f,\varphi) \mapsto (f,\varphi(0)).$$

4·21.3 The considerations in Subsection 3·21.1 are valid for equations of the form (4·21.1) defined in a Banach space, *i.e.*, in the case when $\mathbb{R}^n$ is replaced by an arbitrary Banach space. As to Subsection 4·21.2, it is necessary to assume in addition the continuity of the map $\varphi: Y_0 \times W \times \mathbb{R} \to W$. The latter condition holds if the function f satisfies, for example, the Lipschitz condition: for every $B > 0$ there exists a number $L(B) > 0$ such that $\|f(x,t) - f(y,t)\| \leqslant L(B)\|x - y\|$ whenever $\|x\| \leqslant B$, $\|y\| \leqslant B$, $|t| \leqslant B$.

4·21.4 Now we return to Equation (21.1) in Euclidean n-space $\mathbb{R}^n$. Let us specify some solution $\varphi: \mathbb{R} \to \mathbb{R}^n$ of this equation and construct an extension of transformation groups as follows. Let $H(f)$ denote the closure of the family of shifts $\{f_\tau | \tau \in \mathbb{R}\}$ in the topology of uniform convergence on compact subsets of $\mathbb{R}^n \times \mathbb{R}$. In other words, $H(f)$ is the closure of the orbit of the point f in the shift dynamical system $(C(\mathbb{R}^n \times \mathbb{R},\mathbb{R}^n),\mathbb{R},\sigma)$. Hence we get a flow $(H(f),\mathbb{R},\sigma)$. Similarly, let $H(f,\varphi)$ denote the orbit closure of the point (f,φ) in the Cartesian product of flows $(C(\mathbb{R}^n \times \mathbb{R},\mathbb{R}^n), \mathbb{R},\sigma)$ and $(C(\mathbb{R},\mathbb{R}^n),\mathbb{R},\sigma)$.

Let $p: H(f,\varphi) \to H(f)$ be the natural projection. We shall say that the extension $p:(H(f,\varphi),\mathbb{R},\sigma) \to (H(f),\mathbb{R},\sigma)$ is associated with the solution $\varphi: \mathbb{R} \to \mathbb{R}^n$ of Equation (21.1).

It is not difficult to prove that if $(g,\psi) \in H(f,\varphi)$, then $\psi: \mathbb{R} \to \mathbb{R}^n$ is a solution of the equation

$$\frac{\mathrm{d}x}{\mathrm{d}t} = g(x,t).$$

In fact, if $(g,\psi) \in H(f,\varphi)$, then there exists a sequence $\{\tau_n\}$ such that $\{f_{\tau_n}\} \to g$ in $C(\mathbb{R}^n \times \mathbb{R},\mathbb{R}^n)$ and $\{\varphi_{\tau_n}\} \to \psi$ in $C(\mathbb{R},\mathbb{R}^n)$. The set K, consisting of all points $\varphi_{\tau_n}(t) (t \in [-T,T], n = 1,2,\ldots)$ and $\psi(t) (t \in [-T,T])$, is a compact subset of $\mathbb{R}^n$ for every $T > 0$. Hence it follows from the inequality

$$\|f_{\tau_n}(\varphi_{\tau_n}(s),s) - g(\psi(s),s)\| < \|f_{\tau_n}(\varphi_{\tau_n}(s),s) - g(\varphi_{\tau_n}(s),s)\|$$
$$+ \|g(\varphi_{\tau_n}(s),s) - g(\psi(s),s)\|$$

that $\{f_{\tau_n}(\varphi_{\tau_n}(\cdot),\cdot)\} \to g(\psi(\cdot),\cdot)$ in the space $C(\mathbb{R},\mathbb{R}^n)$.

In the case of an equation of the form (21.1) defined in a Banach space, a similar statement holds under the additional hypothesis that the set $\{\varphi(s) | s \in \mathbb{R}\}$ is compact in E (in this case the solution $\varphi:\mathbb{R} \to E$ is called compact).

4·21.5 Fix a number $h > 0$. Denote by Φ the space of all continuous maps $\varphi:[-h,0] \to \mathbb{R}^n$ with the norm:

$$\|\varphi\| = \sup_{-h\leqslant\vartheta\leqslant 0} |\varphi(\vartheta)|,$$

where $|\cdot|$ denotes the Euclidean norm in $\mathbb{R}^n$. If $\psi:[-h,A) \to \mathbb{R}^n$ is a continuous map, then $\psi^t (0 \leqslant t < A)$ denotes the element of the space Φ defined by:

$$\psi^t(\vartheta) = \psi(t + \vartheta) \qquad (-h \leqslant \vartheta \leqslant 0).$$

Let $F:\mathbb{R}^+ \times \Phi \to \mathbb{R}^n$ be a continuous map. An equation of the form

$$x'(t) = F(t,x^t), \tag{21.2}$$

where $x'(t)$ is the right derivative of $x(t)$, is called a *FUNCTIONAL-DIFFERENTIAL EQUATION*. A continuous function $x:[-h,A) \to \mathbb{R}^n$ is said to be a solution of this equation with the initial condition $\varphi \in \Phi$, if the function $x = x(t)$ satisfies Equation (21.2) for all $t \in [0,A)$ and $x(\vartheta) = \varphi(\vartheta)$, if $-h \leqslant \vartheta \leqslant 0$.

Let $Y = C(\mathbb{R}^+ \times \phi,\mathbb{R}^n)$ be the space of all continuous maps $F:\mathbb{R}^+ \times \Phi \to \mathbb{R}^n$ provided with the uniformity corresponding to the uniform convergence on compact subsets of $\mathbb{R}^+ \times \Phi$. Then we have a semigroup shift dynamical system $(C(\mathbb{R}^+ \times \Phi,\mathbb{R}^n),\mathbb{R}^+,\sigma)$.

Let X be the family of all pairs of the form (F,φ), where $F \in C(\mathbb{R}^+ \times \Phi,\mathbb{R}^n)$ and $\varphi:[-h,\infty) \to \mathbb{R}^n$ is a solution of Equation (21.2). We can define a shift transformation semigroup $(X,\mathbb{R}^+,\sigma)$ so that the canonical projection $p:X \to Y$ is a homomorphism.

4·21.6 Consider the *NON-LINEAR INTEGRAL VOLTERRA EQUATION*

$$x(t) = f(t) + \int_0^t a(t,s)g(x(s),s)ds \qquad (t \geqslant 0), \tag{21.3}$$

where $f:\mathbb{R}^+ \to \mathbb{R}^n$, $g:\mathbb{R}^n \times\mathbb{R}^+ \to \mathbb{R}^n$ and $a:\mathbb{R}^+ \times\mathbb{R}^+ \to M^n$ are given continuous functions (M^n denotes the space of all real $(n \times n)$-matrices) and $x = x(t)$ is the unknown function.

Let $C \equiv C(\mathbb{R}^+,\mathbb{R}^n)$, $A \equiv C(\mathbb{R}^+ \times\mathbb{R}^+,M^n)$ and let $G \equiv C(\mathbb{R}^n \times\mathbb{R}^+,\mathbb{R}^n)$ be the spaces of continuous maps provided with the uniformity of uniform convergence on compact subsets. Let G_0 denote the subspace of G consisting of the functions $g \in G$ satisfying the following Lipschitz condition: for every compact subset $K \subset \mathbb{R}^n$ and every segment $I = [0,T], 0 < T < \infty$ there exists a number $L = L(g,K,I)$ such that

$$|g(x,s) - g(y,s)| < L|x - y|, \qquad (x,y \in K; s \in I).$$

If $f \in C$, $a \in A$ and $g \in G_0$, then the contraction map theorem ensures the existence and uniqueness of a continuous solution $x(t) = x(f,a,g,t)$ of Equation (21.3) defined on some segment $[0,h]$, $h > 0$. It is also easy to verify that the solution depends continuously on $f \in C$, $a \in A$, $g \in G_0$ and $t \in [0,h]$ (moreover, the continuity with respect to f,a,g is uniform on $[0,h]$).

Let X_0 denote the subspace of the uniform space $C \times A \times G_0$ consisting of all triples (f,a,g) such that the solution of (21.3) is defined on the half-line $\mathbb{R}^+$. Let $(f,a,g) \in X_0$. Then we can associate with an equation of the form (21.3) an extension

$$p:(X_0,\mathbb{R}^+,\pi) \to (Y_0,\mathbb{R}^+,\sigma) \tag{21.4}$$

as follows. Define

$$(f,a,g,\tau)\pi = (S^\tau f, a_\tau, g_\tau), \qquad ((f,a,g) \in X_0, \tau \geqslant 0), \tag{21.5}$$

where

$$(S^\tau f)(t) = f(t + \tau) + \int_0^\tau a(t + \tau,s)g(x(s),s)\mathrm{d}s, \tag{21.6}$$

$$a_\tau(t,s) = a(t + \tau, s + \tau), \tag{21.7}$$

$$g_\tau(x,s) = g(x,s + \tau), \qquad (t,s,\tau \in \mathbb{R}^+; x \in \mathbb{R}^n), \tag{21.8}$$

and $x(s) = x(f,a,g,s)$ is the solution of Equation (21.3).

Let Y_0 be the image of X_0 under the canonical projection $p:C \times A \times G_0 \to A \times G_0$. It is clear that the equalities (21.7) and (21.8) determine a transformation semigroup $(Y_0,\mathbb{R}^+,\sigma)$.

Let us show that (21.5) determines a transformation semigroup $(X_0,\mathbb{R}^+,\pi)$. The identity axiom is evidently satisfied. The homomorphism axiom follows from the equalities

$$(S^{\tau_2}(S^{\tau_1}f))(t) = f(t + \tau_1 + \tau_2) + \int_0^{\tau_1} a(t + \tau_1 + \tau_2, s)g(x(s),s)\mathrm{d}s$$

$$+ \int_0^{\tau_2} a(t + \tau_1 + \tau_2, s + \tau_1)g(x(s + \tau_1), s + \tau_1)\mathrm{d}s$$

$$= f(t + \tau_1 + \tau_2) + \int_0^{\tau_1} a(t + \tau_1 + \tau_2, s)g(x(s),s)\mathrm{d}s$$

$$+ \int_{\tau_1}^{\tau_1+\tau_2} a(t + \tau_1 + \tau_2, \xi)g(x(\xi),\xi)\mathrm{d}\xi = (S^{\tau_1+\tau_2}f)(t).$$

To prove the continuity axiom, we shall make use of Lemma 4·20.4. The continuity of the map (21.6) and consequently of the map π at $\tau \in \mathbb{R}^+$ is evident. Let $x(s) = x(f,a,g,s)$, $T > 0$ and let K be the closure of the spherical neighbourhood with radius one of the set $\{x(s) | 0 < s < 2T\}$. If the point $(f^*,a^*,g^*) \in X_0$ is sufficiently near (f,a,g), then $x^*(s) \in K$ for $s \in [0,2T]$ (here $x^*(s) \equiv x(f^*,a^*,g^*,s)$). It is now a standard exercise to show that the function $S^\tau f$ is continuous at $(f,a,g) \in X_0$ uniformly with respect to τ in every segment $[0,T]$. By Lemma 4·20.4, the map is continuous. Thus formulae (21.5-8) determine a transformation semigroup $(X_0,\mathbb{R}^+,\pi)$.

Observe that

$$(S^\tau t)(0) = f(\tau) + \int_0^\tau a(\tau,s)g(x(s),s)ds \equiv x(f,a,g,\tau) \equiv x(\tau).$$

This equality allows us to compare motions of the dynamical system $(X_0,\mathbb{R}^+,\pi)$ with the solutions of the Equation (21.3).

4·21.7 Let T be a locally compact connected locally pathwise connected simply connected topological group and let $a_k:T \to \mathbb{C}$ be continuous functions ($k = 1,\ldots,n$). Consider *ALGEBRAIC EQUATIONS* of the form:

$$\lambda^n + a_1(t)\lambda^{n-1} + \ldots + a_{n-1}(t)\lambda + a_n(t) = 0. \qquad (21.9)$$

A continous function $\lambda:T \to \mathbb{C}$ which satisfies Equation (21.9) identically is called a *SOLUTION OF THE ALGEBRAIC EQUATION* (21.9). If the discriminant $D(t) \equiv D(a\ (t),\ldots,a_n(t))$ of Equation (21.9) is not equal to zero for all $t \in T$, then there are exactly n solutions of this equation (*cf.* Gorin and Lin [1]).

Let Y be the space of all continuous functions $a:T \to \mathbb{C}^n$, $a(t) = (a_1(t),\ldots,a_n(t))$ $(t \in T)$ with

$$D(a_1(t),\ldots,a_n(t)) \neq 0 \qquad (t \in T).$$

Denote by X the set of all functions $b:T \to \mathbb{C}^{n+1}$, $b(t) = (d_1(t),\ldots,d_n(t),\lambda(t))$ such that $D(d_1(t),\ldots,d_n(t)) = 0$ and $\lambda = \lambda(t)$ is a solution of the equation

$$\lambda^n + d_1(t)\lambda^{n-1} + \ldots + d_{n-1}(t)\lambda + d_n(t) = 0.$$

We define shift dynamical systems on X and Y. Then the canical projection $p:X \to Y$, $(d_1(t),\ldots,d_n(t),\lambda(t)) \to (d_1(t),\ldots,d_n(t))$ determines an extension $p:(X,T,\sigma) \to (Y,T,\sigma)$. Each fibre of this extension consists of n points.

4·21.8 Let E be a Banach space and let $F:E \times \mathbb{Z} \to E$ be a continuous map. The *DIFFERENCE EQUATION*

$$x(n+1) = F(x(n),n), \tag{21.10}$$

where $x:\mathbb{Z} \to E$ is the unknown function, is similar to the differential Equation (21.1). Define a cascade $(C(X \times \mathbb{Z}, E), \mathbb{Z}, \sigma)$ following the method of Subsection 4•20.6. Let X be the space of pairs (Φ, ξ), where $\Phi \in C(E \times \mathbb{Z}, E)$ and $\xi:\mathbb{Z} \to E$ is a solution of Equation (21.10) with right side Φ. Define a cascade on X by the shift of the variable $n \in \mathbb{Z}$ by unity. Then we obtain an extension $p:(X \mathbb{Z}, \sigma) \to (C(E \ \mathbb{Z}, E), \mathbb{Z}, \sigma)$.

4•21.9 Let E be a Banach space, W an open subset of E, and $f:W \times \mathbb{R} \to E$ a continuous map. Consider the differential equation

$$\frac{dx}{dt} = f(x,t). \tag{21.11}$$

Let $\varphi:\mathbb{R} \to W$ be a compact solution of this equation (*i.e.* the set $K \equiv \overline{\varphi(\mathbb{R})} \subset W$ is compact.)

Let $H_K(f)$ denote the closure of the set of shifts $\{f_\tau:K \times \mathbb{R} \to E \mid \tau \in \mathbb{R}\}$ in the space $C(K \times \mathbb{R}, E)$. According to Subsection 4•20.6 we have a flow $(H_K(f), \mathbb{R}, \sigma)$. Similarly, let $H_K(f,\varphi)$ denote the closure of the family of shifts $\{(f_\tau, \varphi_\tau) \mid \tau \in \mathbb{R}\}$ of the pair (f,φ) in the Cartesian product of the spaces $C(K \times \mathbb{R}, E)$ and $C(\mathbb{R}, E)$ and let $(H_K(f,\varphi), \mathbb{R}, \sigma)$ denote the corresponding shift dynamical system. The canonical projection $p:H_K(f,\varphi) \to H_K(f)$ is a homomorphism. Sometimes we shall also use the flow $(H(\varphi), \mathbb{R}, \sigma)$ defined on the orbit closure of the function $\varphi:\mathbb{R} \to W$.

4•21.10 LEMMA: *Let φ be a compact solution of the Equation (21.1) and let $K \equiv \overline{\varphi(\mathbb{R})}$. If the set $H_K(f)$ is compact, then the set $H_K(f,\varphi)$ is also compact.*

Proof: Let $\{(g_n, \psi_n)\}$ be a sequence of points in $H_K(f,\varphi)$. Without loss of generality we may assume that $\{g_n\} \to g$ in $H_K(f)$. The family $\{\psi_n \mid n = 1,2,\ldots\}$ is equicontinuous on each segment of the real line $\mathbb{R}$. Indeed, this follows from the equality

$$\psi_n(t) = \psi_n(0) + \int_0^t g_n(\psi_n(\tau),\tau)d\tau,$$

if we take into account that $\overline{\psi_n(\mathbb{R})} \subset K$ and $\{g_n\} \to g$ in $C(K \times \mathbb{R}, E)$.

4•21.11 THEOREM: *If the function $f \in C(W \times \mathbb{R}, E)$ is recurrent (in the sense of Birhoff) and the differential equation (21.1) has a compact solution $x = \varphi(t)$, then it has at least one recurrent solution.*

Proof: Let $K = \{\varphi(t) \mid t \in \mathbb{R}\}$. Since the function f is recurrent, $H_K(f)$ is a compact minimal set. By Lemma 4•21.10, the set $H_K(f,\varphi)$ is also compact. Thus Lema 3•14.14 asserts that there is a pair $(f,\psi) \in H_K(f,\varphi)$ such that $H_K(f,\psi)$ is a compact minimal set.

By the remark at the end of Subsection 4·21.4, ψ is a solution of Equation (21.1). Clearly $H(\psi)$ is a compact minimal set, and hence ψ is a recurrent function.

4·21.12 Let (X,T) be a transformation group and let P be a subsemigroup of the group T. A point $x \in X$ is said to be *POISSON P-STABLE* if for every neighbourhood U of x and every element $p \in P$

$$xpP \cap U \neq \emptyset$$

(in other words, $x \in \overline{xpP}$). In the case when $T = \mathbb{R}$ and $P = \mathbb{R}^+$, this definition agrees with the well-known definition of Poisson stability in the positive direction.

4·21.13 LEMMA: *Let $\varphi:(X,T) \to (Y,T)$ be an extension and the point $y_0 \in Y$ be Poisson P-stable for some subsemigroup P of T. Suppose that there exists a point $x_0 \in X$ such that $x_0\varphi = y_0$ and each net $\{x_\alpha\}$, $x_\alpha \in x_0P$, which satisfies the condition $\{x_\alpha\varphi\} \to y_0$, contains a convergent subnet. Then there is a Poisson p-stable point $a_0 \in \overline{x_0P} \cap y_0\varphi^{-1}$.*

Proof: Let $\mathcal{A}$ be the class of all closed P-invariant subsets A of the set $\overline{x_0P}$ for which $A \cap y_0\varphi^{-1} \neq \emptyset$. Clearly, $\overline{x_0P} \in \mathcal{A}$. By hypothesis, the set $y_0\varphi^{-1} \cap \overline{x_0P}$ is compact. Therefore the system $\mathcal{A}$ is inductive. By Zorn's lemma, there exists a minimal element $A_0 \in \mathcal{A}$. Let $a_0 \in A_0 \cap y_0\varphi^{-1}$. Let us prove that the point a_0 is Poisson P-stable. Suppose that to the contary, there exist an open neighbourhood U_0 of the point a_0 and an element $p_0 \in P$ such that $a_0p_0P \cap U_0 = \emptyset$. Hence

$$\overline{a_0p_0P} \cap U_0 \neq \emptyset. \tag{21.11}$$

Since the set A_0 is closed and P-invariant, we have $\overline{a_0p_0P} \subset A_0$ and moreover $\overline{a_0p_0P} \neq A_0$ by (21.11). We shall prove that $\overline{a_0p_0P} \cap y_0\varphi^{-1} \neq \emptyset$. Indeed, $y_0 \in \overline{y_0p_0P}$ because y_0 is Poisson P-stable. Hence there exists a net $\{p_\alpha\}$, $p_\alpha \in P$ such that $\{y_0p_0p_\alpha\} \to y_0$. Since $a_0\varphi = y_0$, we may suppose without loss of generality that the net $\{a_0p_0p_\alpha\}$ converges to some point $B \in Y$. Clearly, $B\varphi = y_0$, $B \in \overline{a_0p_0P}$. Thus $\overline{a_0p_0P} \in \mathcal{A}$, $\overline{a_0p_0P} \subset A_0$ and $\overline{a_0p_0P} \neq A_0$. But this contradicts the minimality of the element $A_0 \in \mathcal{A}$.

4·21.14 THEOREM: *If the function $f \in C(W \times \mathbb{R}, E)$ is Poisson stable in the positive direction and Equation (21.1) has a compact solution $x = \varphi(t)$, then $H(\varphi)$ contains a solution of (21.1), which is Poisson stable in the positive direction.*

Proof: Since $\varphi(t)$ is a compact solution, the extension $p:H_K(f,\varphi) \to H_K(f)$, where $K = \overline{\{\varphi(t) \mid t > 0\}}$ and P is taken to be the semigroup of positive real numbers satisfies all the hypotheses of Lemma 4·21.13 (the requirement that φ be compact can be relaxed; the details are left to the reader). Therefore there is a point $(f,\psi) \in H_K(f,\varphi)$ which is Poisson stable in the positive direction. Clearly, ψ is the desired solution.

REMARKS AND BIBLIOGRAPHICAL NOTES

Dynamical systems associated with non-autonomous differential equations have first appeared in the papers by Miller [*1*], Millionščikov [*1*,*2*], Seifert [*1-6*], Sell [*1-5*] and Ščerbakov [*1-6*]. Žikov [*1-9*] has used extensions of dynamical systems in the investigation of almost periodic solutions of differential equations. Dynamical systems associated with functional differential equations were considered by Dubolar' [*1*]. Similar questions for difference equations were studied by Štefanitse [*1*]. Relations between nonlinear integral Volterra equations and dynamical systems were considered by Miller and Sell [*1*]. The construction presented in Subsection 4•21.7 was proposed by Žikov [*2*]. Theorem 4•21.11 is proved in the following papers: Bender [*1*], Miller [*1*], Millionščikov [*1*,*2*] and Ščerbakov [*1*,*2*]. Theorem 4•21.14 is due to Ščerbakov [*4*] (see also Lam [*1*]).

The theory of differential equations in a Banach space is presented in the books by Daletskiĭ and Kreĭn [*1*] and Dieudonné [*1*]. There are many papers devoted to the existence, uniqueness and extendibility of solutions and their continuous dependence on the right side.

Among these, the paper by Lasota and Yorke [*1*] is especially worth mentioning because it shows that the existence of solutions and their continuous dependence on the right side is a generic property.

§(4.22): ALMOST AUTOMORPHIC EXTENSIONS AND SYNCHRONOUS SOLUTIONS OF DIFFERENTIAL EQUATIONS

We shall investigate presently the relationship between almost automorphic extensions and solutions of differential equations synchronous with the right side. We shall give some sufficient conditions for an extension of a minimal transformation group to be almost automorphic. Relations between 'two-sided' and 'one-sided' properties of extenstions will also be considered. Synchronous and uniformly synchronous solutions will be defined and some simple conditions ensuring the existence of such solutions will be presented. We shall investigate in some detail the case of linear differential equations. Conditions ensuring the existence of a single uniformly synchronous solution of a quasi-linear equation will also be given. At the end of the section we shall consider scalar differential equations.

4•22.1 Let T be a topological group and let S be a subsemigroup of T satisfying the conditions

$$SS^{-1} = T, \qquad sS = Ss \qquad (s \in S). \tag{22.1}$$

Assume further that S contains the identity e of the group T. Let X be uniform space, Y a compact space, (X,T,π) and (Y,T,σ) transformation groups: and $p\ (X,T,\pi) \to (Y,T,\sigma)$ a homomorphism. Assume that the set xS is compact for every point $x \in X$. Let $y_0 \in Y$, X_0 $\{x|x \in X, xp = y_0\}$. Let $E(X,S)$ denote, as usually, (see §(1.4)) the closure of the subset $\{\pi^s|s \in S\}$ of the space X^X provided with the Tikhonov topology. It follows from our assumptions that $E(X,S)$ is a compact subsemigroup of X^X. Put

$$E_0 = \{\xi|\xi \in E(X,S), X_0\xi \subset X_0\}.$$

Clearly, E_0 is a compact subsemigroup of the semigroup $E(X,S)$ and $\overline{xS} \cap X_0 = xE_0$ for all $x \in X$. Identify two elements ξ and η of E, if $x\xi = x\eta$ for all $x \in X_0$, and denote the quotient semigroup by P_0. The topology in P_0 coincides with the topology of a subspace of the space $X_0^{X_0}$. Since $\overline{xS} \cap X_0 = xP_0$ $(X \in X_0)$ the map $p{:}\overline{xS} \to Y$ is injective at a point $x \in X_0$ if x is a fixed point for all transformations $p{:}X_0 \to X_0$ belonging to the semigroup P_0.

4·22.2 LEMMA: *Let T be a topological group, S a subsemigroup of T satisfying conditions* (22.1) *and let (X,T,π) be a transformation group. Then a compact set $M \subset X$ is minimal if it is S-minimal.*

Proof: Suppose that the set M is S-minimal. Since $(Ms)S \subset MSs \subset Ms$ for all $s \in S$, the set Ms is S-invariant. Moreover, this set is closed and is contained in the S-minimal set M. Hence $M = Ms$ for all $s \in S$. Since $T = SS^{-1}$, the set M is invariant. Let $x \in M$. Then $M \supset \overline{xT} \supset \overline{xS} = M$, and hence $M = xT (x \in M)$. Thus, the set M is minimal.

Conversely, let M be a compact minimal set. Clearly, M is S-invariant. Hence there exists an S-minimal set $A \subset M$. According to the first part of the proof, the set A is minimal, whence $A = M$. Thus, M is an S-minimal set.

4·22.3 THEOREM: *Let T be a topological group, S a subsemigroup of T satisfying conditions* (22.1), *X a uniform space, Y a compact space, $p{:}(X,T,\pi) \to (Y,T,\sigma)$ an extension, $y_0 \in Y$ and $X_0 = \{x|x \in X, xp = y_0\}$. Suppose further that the following conditions are satisfied:*

CONDITION (1): *(Y,T) is a minimal transformation group;*

CONDITION (2): *For each point $x \in X$ the set xS is compact.*

If a point $x_0 \in X_0$ is fixed under all transformations of the semigroup P_0, then the map $p\ x_0T, \to Y$ is injective at the point x_0, x_0T is a compact minimal set and the extension $p{:}(x_0T,T) \to (Y,T)$ is proximal.

Proof: By Lemma 4·22.2, $\overline{y_0S} = \overline{y_0T}$. If $p{:}\overline{x_0S} \to \overline{y_0S}$ is injective at the point $x_0 \in X$, then $\overline{x_0S}$ is evidently S-minimal and hence T-minimal. The fact that the extension $p{:}(x_0T,T) \to (Y,T)$ is proximal follows from Lemma 3·12.7.

4·22.4 LEMMA: *Let T be a topological group; S a subsemigroup of T satisfying the conditions* (22.1)*; X a compact space and let φ $(X,T,\pi) \to (Y,T,\rho)$ be an extension where (Y,T,ρ) is a minimal transformation group. Then the extension $\varphi:(X,T) \to (Y,T)$ is distal if the extension $\varphi:(X,S) \to (Y,S)$ is distal.*

Proof: By Theorem 3·14.7, the extension $\varphi:(X,T) \to (Y,T)$ is distal iff the orbit closure of each pair $(x_1,x_2) \in R_\varphi$ in $(X \times X,T)$ is a compact minimal set. Lemma 4·22.2 shows that (Y,S) is a minimal transformation semigroup. Theorem 3·14.7 can be extended to the case of transformation semigroups (*cf.* Gerko [2]). Therefore the required result follows from Lemma 4·22.2.

4·22.5 LEMMA: *Let T be a topological group; S a subsemigroup of T satisfying the conditions* (22.1)*; X a compact space, and let $p:(X,T) \to (Y,T)$ be a minimal extension. Then $p:(X,S) \to (Y,S)$ is equicontinuous iff the extension $p:(X,T)$ (Y,T) is equicontinuous.*

Proof: Let $p:(X,S) \to (Y,S)$ be an equicontinuous extension. Then $p:(X,S^{-1}) \to (Y,S^{-1})$ is distal. By Lemma 4·22.4, the extension $p:(X,T) \to (Y,T)$ is also distal. It will suffice to show that $p:(X,S^{-1}) \to (Y,S^{-1})$ is an equicontinuous extension. Suppose that to the contrary, for some index $\beta_0 \in \mathfrak{U}[X]$ and every index $\alpha \in \mathfrak{U}[X]$ we can choose points $a_\alpha, B_\alpha \in X$ and an element $s_\alpha \in S$ so that

$$(a_\alpha,b_\alpha) = \alpha, \qquad a_\alpha p = b_\alpha p, \qquad (a_\alpha s_\alpha{}^{-1}, b_\alpha s_\alpha{}^{-1}) = \beta_0. \qquad (22.2)$$

Hence $\overline{(a_\alpha,b_\alpha,S^{-1}} \cap (X \times X \setminus \beta_0) \neq \emptyset$. Since the extension $p:(X,T) \to (Y,T)$ is distal, $\overline{(a_\alpha,b_\alpha)T}$ is a compact minimal set for every $\alpha \in U[X]$, and hence $\overline{(a_\alpha,b_\alpha)S^{-1}} = \overline{(a_\alpha,b_\alpha)T} = \overline{(a_\alpha,b_\alpha)S}$ by Lemma 4·22.2. Thus

$$\overline{(a_\alpha,b_\alpha)S} \cap (X \times X \setminus \beta_0) \neq \emptyset, \qquad (\alpha \in \mathfrak{U}(X))$$

but in view of (22.2) this contradicts the equicontinuity of p $(X,S) \to (Y,S)$.

4·22.6 Let X and Y be uniform spaces, T a topological group, (X,T,π) and (Y,T,ρ) transformation groups, and $x \in X$, $y \in Y$. We shall say that the point y is *SYNCHRONOUS WITH THE POINT* x if for every neighbourhood U of y there exists a neighbourhood V of x such that $x\pi^t \in V$ implies $y\rho^t \in U$. This definition is equivalent to the following: there exists a continuous map $p:xT \to yt$ satisfying the condition $(x\pi^t)p = y\rho^t (t \in T)$. If such a map exists, it is uniquely determined. If the map $p:xT \to yT$ is moreover uniformly continuous, then y is said to be *UNIFORMLY SYNCHRONOUS WITH* x. The points x and y are called *ISOCHRONOUS* (*UNIFORMLY ISOCHRONOUS*) if they are mutually synchronous (uniformly synchronous).

4·22.7 LEMMA: *Let* $\mathcal{A}$ *be a family of subsets of the group* T. *If a point* $x \in X$ *is* $\mathcal{A}$*-recursive and the point* $y \in Y$ *is synchronous with* x, *then* y *is also* $\mathcal{A}$*-recursive.*

4·22.8 LEMMA: *If* $\overline{xT}$ *is a compact minimal set (almost periodic minimal set) and the point* $y \in Y$ *is uniformly synchronous with* x, *then* $\overline{yT}$ *has the corresponding property.*

4·22.9 In accordance with the general definition, a compact solution $\varphi \in C(\mathbb{R},E)$ of the equation

$$\frac{dx}{dt} = f(x,t) \qquad (x \in W \subset E, t \in \mathbb{R}) \tag{22.3}$$

is called *A SOLUTION SYNCHRONOUS WITH THE RIGHT SIDE* f (or, in short, *SYNCHRONOUS*), if for the compact set $K = \varphi(\mathbb{R})$ and every neighbourhood U of the point $\varphi \in C(\mathbb{R},K)$ there exists a neighbourhood V of the point $f \in C(K\times\mathbb{R},E)$ such that $f_\tau \in V$ implies $\varphi_\tau \in U$. In other words a compact solution φ is called synchronous if for every $\varepsilon < 0$ and every (bounded) segment $I_1 \subset \mathbb{R}$ there are $\delta > 0$ and a segment $I_2 \subset \mathbb{R}$ such that

$$\|f(x,t) - f(x,t+\tau)\| < \delta \qquad (x \in K,\ t \in I_2)$$

implies

$$\|\varphi(t) - \varphi(t+\tau)\| < \varepsilon \qquad (t \in I_1).$$

4·22.10 LEMMA: *A compact solution* φ *is synchronous iff the map* $p:H_K(f,\varphi) \to H_K(f)$ *is injective at the point* (f,φ).

Proof: The necessity is obvious. Suppose that the projection P is injective at the point (f,φ), but that the solution φ is not synchronous. Then there exists a sequence $\{\tau_n | n = 1,2,\ldots\}$ such that $\{f_{\tau_n}\} \to f$, but $\{\varphi_{\tau_n}\}$ does not converge to φ. Recall that $K \equiv \varphi(\mathbb{R})$ is a compact set and $\varphi_{\tau_n}(t) = \varphi(t+\tau_n) \in K\ (t \in \mathbb{R})$. Since the sequence $\{f_{\tau_n}(x,t)\}$ converges to $f(x,t)$ uniformly on subsets of the form $K\times I$, where I is an arbitrary bounded segment of the real line $\mathbb{R}$, and

$$\varphi_{\tau_n}(t) = \varphi_{\tau_n}(0) + \int_0^t \tau_n(\varphi_{\tau_n}(\xi),\xi)d\xi,$$

the family $\{\varphi_{\tau_n}:\mathbb{R} \to E | n = 1,2,\ldots\}$ is equicontinuous on every segment I. By Ascoli's theorem, we may assume that $\{\varphi_{\tau_n}\} \to \psi$ in the space $C(\mathbb{R},E)$. Thus $\psi \neq \varphi$ and $(f,\psi) \in H_K(f,\varphi)$. But this contradicts the hypothesis that p is injective at the point (f,φ). ∎

4·22.11 In accordance with the general definition, a compact solution φ is called a *UNIFORMLY SYNCHRONOUS SOLUTION*. if for every index α of the uniform space $C(\mathbb{R},E)$ there exists an index β of the uniform space $C(K \times \mathbb{R},E)$ where $K = \varphi(\mathbb{R})$, such that $(f_\tau, f_\xi) \in \beta$ implies $(\varphi_\tau, \varphi_\xi) \in \alpha$ for any $\tau\xi \in \mathbb{R}$. In other words, a compact solution φ is uniformly synchronous, if for every $\varepsilon > 0$ and every segment $I_1 \subset \mathbb{R}$ there exist a number $\delta > 0$ and a segment I_2 $\mathbb{R}$ such that

$$\|f(x,t + \tau) - f(x,t + \xi)\| < \delta \qquad (x \in K,\ t \in I_1)$$

implies

$$\|\varphi(t + \tau) - \varphi(t + \xi)\| < \varepsilon \qquad (t \in I\ ).$$

4·22.12 LEMMA: *Let φ be a compact solution of equation* (22.3), $K = \varphi(\mathbb{R})$ *and let* $f:K\ \mathbb{R} \to E$ *be Lagrange stable. Then the following statements are mutually equivalent:*

(1) *The solution φ is uniformly synchronous with f;*

(2) $p:H_k(f,\varphi) \to H_k(f)$ *is bijective;*

(3) $p:H_k(f,\varphi) \to H_k(f)$ *is an equimorphism*

Proof: Clearly (1) implies (2). Since the orbit closure of f in $(C(K \times \mathbb{R}, E), \mathbb{R}, \sigma)$ is compact, the proof of the implication (2) => (1) is essentially the same as the proof of Lemma 4·22.10. Evidently (1) and (3) are equivalent. ∎

Now we shall present some simple conditions that ensure the existence of synchronous and uniformly synchronous solutions of the equation (22.3).

4·22.13 LEMMA: *If φ is a compact solution of* (22.2) *and* $H(\varphi)$ *does not contain other compact solutions of this equation (in particular, if there is no solution ψ distinct from φ and satisfying the condition* $\overline{\psi(\mathbb{R})} \subset \overline{\varphi(\mathbb{R})}$*), then φ is a synchronous solution.*

Proof: The fibre of the map $p:H_k(f,\varphi) \to H_k(f)$ over the point f consists of pairs (f,ψ) such that $f = \lim f_{\tau_n}$ and $\psi = \lim \varphi_{\tau_n}$ for some sequence $\{\tau_n | n = 1,2,\ldots\}$. Therefore ψ is a compact solution of the equation (22.3). The hypothesis of the lemma implies that the map p is one-to-one at the point (f,φ). Hence φ is synchronous by Lemma 4·22.10. ∎

4·22.14 COROLLARY: *If φ is the unique compact solution of* (22.3), *then it is synchronous.*

4·22.15 LEMMA: *Let φ be a compact solution of* (22.3), $K = \overline{\varphi(\mathbb{R})}$ *and let* $f:K \times \mathbb{R} \to E$ *be Lagrange stable. If for every function* $g \in H_k(f)$ *the equation*

$$\frac{dx}{dt} = g(x,t) \qquad\qquad (22.4)$$

has no more than one solution ψ *with* $\psi \in H(\varphi)$ *(in particular if this equation has no more than one compact solution), then* φ *is a uniformly synchronous solution.*

Proof: It follows from the hypotheses of the lemma that $p: H_K(f,\varphi) \to H_K(f)$ is a bijective map. To conclude the proof we refer to Lemma 4·22.12. ▮

4·22.16 We shall say that the compact solutions of the equation (22.3) are *POSITIVELY SEPARATED SOLUTIONS* if for each compact solution φ and every solution ψ satisfying the conditions $\psi \in H(\varphi)$, $\psi \not\equiv \varphi$, there exists a number $\ell > 0$, such that $\|\varphi(t) - \psi(t)\| \geqslant \ell$ for all $t \geqslant 0$.

In the case of non-homogeneous linear differential equations the above property is equivalent to the following condition on positive separation from zero of the non-trivial compact solutions of the corresponding homogeneous equation: for every such solution $\varphi: \mathbb{R} \to E$ there exists a number $\ell > 0$ such that $\|\varphi(t)\| \geqslant \ell$ for all $t \geqslant 0$.

4·22.17 THEOREM: *Let* $\varphi: \mathbb{R} \to E$ *be a compact solution of Equation* (22.3), $K = \varphi(\mathbb{R})$ *and let the function* $f \in C(K \times \mathbb{R}, E)$ *be recurrent in the sense of Birkhoff. If* φ *is the only compact solution of the equation* (22.3) *and every equation of the form* (22.4) *with* $g \in Y \equiv H_K(f)$ *satisfies the condition on separation of compact solutions in the positive direction, then* φ *is a uniformly synchronous recurrent (in the sense of Birkhoff) solution.*

Proof: By hypothesis, Y is a compact minimal set. According to Lemma 4 21.10, the space $X \equiv H_K(f,\varphi)$ is also compact. Therefore X contains some minimal set M. Let us prove that $X = M$. Since $Mp = Y$, there is a pair $(f,\psi) \in M$. Clearly ψ is a compact solution of the equation (22.3), and hence $\psi = \varphi$ by hypothesis. Thus $X = M$ and the map $p: X \to Y$ is injective at the point (f,φ) Hence it follows from Lemma 3·12.7 that p is a proximal extension. On the other hand, the positive separation of compact solutions implies that the extension $p:(X,\mathbb{R}^+) \to (Y,\mathbb{R}^+)$ is distal. By Lemma 4·22.4 the extension $p:(X,\mathbb{R}) \to (Y,\mathbb{R})$ is also distal. This is possible only when $p: X \to Y$ is a bijective map. By Lemma 4·22.12, φ is a uniformly synchronous solution. ▮

4·22.18 COROLLARY: *Let* $\varphi: \mathbb{R} \to E$ *be a compact solution of the equation* (22.3), $K = \varphi(\mathbb{R})$ *and let the function* $f: K \times \mathbb{R} \to E$ *be almost periodic in* t *(uniformly with respect to* $x \in K$*). Suppose further that* φ *is the only compact solution of* (22.3) *and every equation of the form* (22.4) *with* $g \in H_K(f)$ *satisfies the condition on positive separation of compact solutions. Then* φ *is a uniformly synchronous almost periodic solution.*

4·22.19 Consider the Linear non-homogeneous differential equation

$$x' = A(t)x + f(t), \tag{22.5}$$

where $A \in C(\mathbb{R}, L(E,E))$, ($L(E,E)$ is the normed ring of continuous linear maps of E into E, and $f \in C(\mathbb{R}, E)$.

It is useful to consider at the same time the corresponding homogeneous equation

$$x' = A(t)x. \tag{22.6}$$

For each point $x_0 \in E$ there exists a unique solution $\varphi: \mathbb{R} \to E$ of the equation (22.5) with the initial condition $\varphi(0) = x_0$, and this solution depends continuously on (A, f, x_0). Therefore we can construct an extension of dynamical systems associated with Equation (22.5) in the way it has been done in Subsections 4·21.2. Let Y denote the closure of the family $\{(A_\tau, f_\tau)\ \tau \in \mathbb{R}\}$ in the cartesian product $C(\mathbb{R}, L(E,E)) \times C(\mathbb{R}, E)$. Set $X = Y \times E$. The mapping

$$(B,g,x)\pi^\tau \quad (B_\tau, g_\tau, \varphi(\tau; B,g,x) \qquad (\tau \in \mathbb{R}\ ,\ (B,g) \in Y,\ x \in E),$$

where $\varphi(t; B,g,x)$ is the solution of the equation

$$x' = B(t)x + g(t),$$

with the initial condition $\varphi(0; B,g,x) = x$ determines a dynamical system $(X, \mathbb{R}, \pi)$. The projection $p: X \to Y$, $(B,g,x) \to (B,g)$, is a homomorphism.

4·22.20 THEOREM: *If* (22.5) *has a compact solution* $\varphi: \mathbb{R} \to E$ *and* (22.6) *does not have non-trivial compact solutions, then* φ *is a synchronous solution*

Proof: This follows from Corollary 4·22.14, since φ is the only compact solution of the equation (22.5). ▮

4·22.21 THEOREM: *If the functions* $A \in C(\mathbb{R}, L(E,E))$ *and* $f \in C(\mathbb{R}, E)$ *are Lagrange stable and Equation* (22.5) *has a compact solution* $\varphi: \mathbb{R} \to E$, *whilst each equation of the form*

$$x' = B(t)x, \tag{22.8}$$

with $B \in H(A)$ *has no non-trivial compact solutions, then* φ *is a uniformly synchronous solution.*

Proof: This follows directly from Lemma 4·22.15. ▮

4·22.22 Let $U(t)$ be the Cauchy operator of the Equation (22.6) and let $U(t,\tau) = U(t)U^{-1}(\tau)$ be the evolution operator.

Following Daletskiĭ and Kreĭn [*1*], we shall say the equation (22.6) satisfies the condition of [*REGULAR*] *EXPONENTIAL DICHOTOMY* on the real line $\mathbb{R}$ (in brief, Equation (22.6) is *e-DICHOTOMIC* on $\mathbb{R}$) if for some (and hence for every) number $t_0 \in \mathbb{R}$ the space E can be decomposed into a direct sum of closed linear subspaces

$$E = E_1(t_0) \oplus E_2(t_0) \tag{22.9}$$

so that

(a) if $x_1(t) = U(t,t_0)x_1^0$ is a solution of (22.6) with $x_1^0 \equiv x_1(t_0) \in E_1(t_0)$, then

$$\|x_1(t)\| < N\, e^{-\nu_1(t-s)}\|x\ (s)\| \qquad (t > s;\ t,s \in \mathbb{R})$$

for some numbers $N_1 > 0$, $V_1 > 0$;

(b) if $x_2(t) = U(t,\ _0)x_2^0, tx_2(t_0) = x_2^0 \in E_2(t_0)$, then

$$\|x_2(t)\| < N_2 e^{-\nu_1(s-t)}\|x_2(s)\| \qquad (t < s;\ t,s \in \mathbb{R})$$

for some $N_2 > 0$, $V_2 > 0$;

(c) if $E_k(t) = U(t,t_0)E_k(t_0)$ $(k = 1,2)$, then the mutual inclination $\mathrm{Sn}(E_1(t),E_2(t))$ of the subspaces $E_1(t)$ and $E_2(t)$ satisfies the inequality:

$$\mathrm{Sn}(E_1(t),E_2(t)) \geqslant \gamma > 0 \qquad (t \in \mathbb{R}).$$

We remind the reader that the mutual inclination of two subspaces A_1 and A_2 of a Banach space E is defined by

$$\mathrm{Sn}(A_1,A_2) = \inf\{\|x_1 + x_2\|\ \ |x_k \in A_k, \|x_k\|=1, k = 1,2\}$$

Clearly, in the case of a constant operator $A(t) \equiv A$, the Equation (22.6) is e-dichotomic iff the spectrum of A does not meet the imaginary axis.

If Equation (22.6) is e-dichotomic on $\mathbb{R}$, then for each continuous bounded function $f\colon \mathbb{R} \to E$ there is exactly one bounded (on the whole axis) solution of Equation (22.5). This solution can be represented in the form:

$$x(t) = \int_{\mathbb{R}} G(t,\tau)f(\tau)\ \tau,$$

where $G(t,\tau)$ is the (principal) Green's function of (22.6):

$$G(t,\tau) = \begin{cases} U(t)P_1U^{-1}(\tau) & (t > \tau), \\ U(t)P_2U^{-1}(\tau) & (t > \tau), \end{cases}$$

and P_1, P_2 are the projectors corresponding to the direct sum (22.9). The function $G(t,\tau)$ is continuous on $\mathbb{R}\times\mathbb{R}$ except on the line $t = \tau$, where it has a jump. Green's function satisfies the inequality

$$\|G(t,\tau)\| \leqslant Ne^{-\nu|t-\tau|} \qquad (t,\tau \in \mathbb{R}), \qquad (22.10)$$

where N and ν are some positive constants.

4·22.23 THEOREM: *Let the functions* $A \in C(\mathbb{R},L(E,E))$ *and* $f \in C(\mathbb{R},E)$ *be Lagrange stable and let the homogeneous Equation* (22.6) *be a e-dichotomic on* $\mathbb{R}$. *Then Equation* (22.5) *has a unique Lagrange stable uniformly synchronous solution.*

Proof: Since the Equation (22.6) is e-dichotomic, there exists a unique bounded (on $\mathbb{R}$) solution $\varphi: \mathbb{R} \to E$ of Equation (22.5). Since A and f are Lagrange stable functions, it will suffice to show that φ is a uniformly synchronous solution.

Let us prove that for every $\varepsilon > 0$ and $\ell > 0$ there exist numbers $\delta > 0$ and $L > 0$ such that if for some $\lambda,\xi \in \mathbb{R}$

$$\sup_{|t|<L} \|A(t + \lambda) - A(t + \xi)\| < \delta$$
$$\sup_{|t|<L} \|f(t + \lambda) - f(t + \xi)\| < \delta, \qquad (22.11)$$

then

$$\sup_{|t|<} \|\varphi(t + \lambda) - \varphi(t + \xi)\| < \delta.$$

Let $G(t,\tau)$ be the Green's function of Equation (22.6). Then

$$\varphi(t + \lambda) = \int_{\mathbb{R}} G(t + \lambda,\tau)\, f(\tau)d\tau = \int_{\mathbb{R}} G(t + \lambda,\tau + \lambda)f(\tau + \lambda)d\tau,$$
$$\varphi(t + \xi) = \int_{\mathbb{R}} G(t + \xi,\tau + \xi)f(\tau + \xi)d\tau. \qquad (22.12)$$

By virture of (22.10), we have for all $s \in \mathbb{R}$ that

$$\|G(t + s,\tau + s)\| < Ne^{-\nu[t-\tau]}, \qquad (t,\tau \in \mathbb{R}). \qquad (22.13)$$

It follows from this and from the boundedness of f on $\mathbb{R}$ that for $\varepsilon > 0$ and $\ell > 0$ we can choose a number $L > \ell > 0$ so that

$$\sup_{|t|<\ell} \left\|\int_{\mathbb{R}} F(t,\tau,\xi,\lambda)d\tau - \int_{-L}^{L} F(t,\xi,\lambda)d\tau\right\| < \varepsilon_1 \qquad (22.14)$$

where

$$F(t,\tau,\xi,\lambda) = G(t + \lambda,\tau + \lambda)f(\tau + \lambda) - G(t + \xi,\tau + \xi)f(\tau + \xi), \qquad (22.15)$$

$$\varepsilon_1 = \varepsilon(1 + 2LM + 2LN)^{-1}, \qquad M = \sup_{t\in\mathbb{R}} \|f(t)\|. \tag{Contd) (22.15}$$

Under our assumptions the Cauchy operator, and consequently the Green's function, depend continuously on $A \in C(\mathbb{R},L(E,E))$ (*cf.* Daletskiĭ and Kreĭn [*1*], Lemma III 2.3). Hence for $\varepsilon > 0$ and $L > 0$ we can choose a number $\gamma > 0$ so that

$$\sup_{|t|\leqslant L} \|A(t + \lambda) - A(t + \xi)\| < \gamma$$

implies

$$\|G(t + \lambda,\tau + \lambda) - G(t + \xi,\tau + \xi)\| < \varepsilon_1, \qquad (|t| \leqslant L,\ |\tau| \leqslant L). \tag{22.16}$$

Let $\delta = \min\{\delta,\varepsilon_1\}$ and let (22.11) be satisfied. Then it follows from (22.13-16) that

$$\sup_{|t|\leqslant \ell} \|\varphi(t + \lambda) - \varphi(t + \xi)\| < \varepsilon_1 + \left\|\int_{-L}^{L} F(t,\tau,\xi,\lambda)d\tau\right\|$$

$$\leqslant \varepsilon_1 + \left\|\int_{-L}^{L} [G(t + \lambda,\tau + \lambda) - G(t + \xi,\tau + \xi)]f(\tau + \lambda)d\tau\right\|$$

$$+ \left\|\int_{-L}^{L} G(t + \xi,\tau + \xi)[f(\tau + \lambda) - f(\tau + \xi)]d\tau\right\|$$

$$< \varepsilon_1 + 2\varepsilon_1 LM + 2\varepsilon_1 LN = \varepsilon. \blacksquare$$

4·22.24 Let us now turn our attention to quasi-linear equations in a Banach space E. Let $A \in C(\mathbb{R},L(E,E))$ and $F \in C(E\times\mathbb{R},E)$. Consider the differential equation

$$x' = A(t)x + F(x,t). \tag{22.17}$$

4·22.25 THEOREM: *Let the function A be Lagrange stable and let Equation* (22.6) *be e-dichotomic on* $\mathbb{R}$. *Suppose that the function F satisfies the following conditions:*

(1): *For each $x \in E$ the function F is uniformly continuous in $t \in \mathbb{R}$ and the set $F(x,\mathbb{R}) \equiv \{F(x,t)|t \in \mathbb{R}\}$ is compact in E;*

(2): *The function F satisfies the Lipschitz condition:*

$$\|F(x_1,t) - F(x_2,t)\| < L\|x_1 - x_2\|, \qquad (x_1,x_2 \in E, t \in \mathbb{R}),$$

where $0 < L < L_0$ *and* L_0 *is a sufficiently small number, depending only on the function* A.

Then equation (22.17) *has a unique Lagrange stable uniformly synchronous solution.*

Proof: Let Q be any compact subset of the space E. It is not hard to deduce from the conditions (1,2) that the function F is uniformly continuous on $Q \times \mathbb{R}$ and the set $F(Q,\mathbb{R}) = \{F(x,t) \mid x \in Q, t \in \mathbb{R}\}$ is compact in E.

Consider the space Φ consisting of all continuous Lagrange stable functions $\varphi: \mathbb{R} \to E$ which are synchronous with $(A,F) \in C(\mathbb{R}, L(E,E)) \times C(E \times \mathbb{R}, E)$ and provide Φ with the metric of uniform convergence on $\mathbb{R}$. It is easy to verify that the metric space Φ is complete.

Let $\varphi \in \Phi$. Define a function $f \in C(\mathbb{R},E)$ by

$$f(t) = F(\varphi(t),t), \qquad (t \in \mathbb{R}). \tag{22.18}$$

Let us show that f is Lagrange stable. In fact, $f(\mathbb{R}) \subset F(Q,\mathbb{R})$, where $Q = \overline{\varphi(\mathbb{R})}$. As has been pointed out above, the set $F(Q,\mathbb{R})$ is compact in E. Hence the function f is compact. It follows from the uniform continuity of F on $Q \times \mathbb{R}$ and the uniform continuity of φ on $\mathbb{R}$ that the function $f: \mathbb{R} \to E$ is also uniformly continuous.

Since the equation (22.6) is e-dichotomic on $\mathbb{R}$, it follows from Theorem 4·22.23 that the equation

$$x' = A(t)x + F(\varphi(t),t) \tag{22.19}$$

has a unique Lagrange stable solution $\psi: \mathbb{R} \to E$. Let us prove that $\psi \in \Phi$. If $\lambda, \xi \in \mathbb{R}$, the function A_λ is sufficiently close to A_ξ, F_λ is sufficiently close to F_ξ and $\varphi \in \Phi$, then the function f_λ, where $f_\lambda(t) = F(\varphi(t+\lambda), t+\lambda)$ $(t \in \mathbb{R})$, is arbitrarily close to the function f_ξ. Hence ψ_λ can also be made arbitrarily close to ψ_ξ, as can be seen from the proof of Theorem 4·22.23. Thus, the function $\psi: \mathbb{R} \to E$ is uniformly synchronous with $(A,F) \in C(\mathbb{R}, L(E,E)) \times C(E \times \mathbb{R}, E)$.

The above considerations enable us to define a map $B: \Phi \to \Phi$ as follows: to each function $\varphi \in \Phi$ we assign the unique Lagrange stable solution ψ of Equation (22.19). Clearly $\varphi \in \Phi$ is a solution of (22.17) iff φ is a fixed point of the map $B: \Phi \to \Phi$.

Let us show that if the Lipschitz constant of the function $F \in C(E \times \mathbb{R}, E)$ is sufficiently small, then B is a contraction map. If $\varphi_1, \varphi_2 \in \Phi$, then

$$B(\varphi_k)(t) = \int_{\mathbb{R}} G(t,\tau) F(\varphi_k(\tau),\tau)\,d\tau, \qquad (k = 1,2; t \in \mathbb{R}),$$

where $G(t,\tau)$ is the Green's function of the Equation (22.6). Hence it follows from (22.10) that

$$\sup_{t \in \mathbb{R}} \|B(\varphi_1)(t) - B(\varphi_2)(t)\| < 2NL\nu^{-1} \sup_{t \in \mathbb{R}} \|\varphi_1(t) - \varphi_2(t)\|.$$

Thus, if $L < L_0 \equiv \nu/2N$, then B is a contraction, and hence there exists a unique fixed point φ of the mapping B. In other words, Equation (22.17) has a unique Lagrange stable solution $\varphi \in \Phi$. By the definition of the space Φ, the solution $\varphi \in C(\mathbb{R},E)$ is uniformly synchronous with the right side of Equation (22.17).▮

4·22.26 Now we shall investigate equations of the form (22.5), where $A \in C(\mathbb{R},L(E,E))$ and $f \in C(\mathbb{R},E)$ are Lagrange stable functions. Suppose that the corresponding homogeneous equation (22.6) is e-dichotomic on $\mathbb{R}$. Let φ_0 be the unique Lagrange stable solution of Equation (22.5) and let $Q = \overline{\varphi_0(\mathbb{R})}$. The the following theorem is valid.

4·22.27 THEOREM: *Let W be a bounded neighbourhood of the set Q. Suppose that the function $F \in C(\bar{W} \times \mathbb{R},E)$ satisfies the following conditions:*

(1): *For each point $x \in \bar{W}$, the function F is uniformly continuous in t and the set $F(X,\mathbb{R})$ is compact in E;*

(2): *F satisfies the Lipschitz condition in $x \in \bar{W}$.*

Under these conditions there exists a number $\lambda_0 > 0$ such that for all λ, $|\lambda| < \lambda_0$, the equation

$$x' = A(t)x + f(t) + F(x,t) \qquad (22.20)$$

has in $\bar{W}$ a unique Lagrange stable uniformly synchronous solution which tends to φ_0 uniformly on $\mathbb{R}$ when $\lambda \to 0$.

Proof: For any compact set $Q \subset \bar{W}$, the function F is uniformly continuous on $Q \times \mathbb{R}$ and the set $F(Q,\mathbb{R})$ is compact in E. Hence F is Lagrange stable. It is also easy to show that F is bounded on $\bar{W} \times \mathbb{R}$. Let Φ denote the set of all Lagrange stable functions $\varphi: \mathbb{R} \to W$ which are uniformly synchronous with (A,f,F) provided with the metric of uniform convergence on $\mathbb{R}$. Following the proof of Theorem 4·22.25, define a map $B_\lambda: \Phi \to \Phi$ which to each function $\varphi \in \Phi$ assigns the unique Lagrange stable solution of the equation

$$\frac{dx}{dt} = A(t)x + f(t) + \lambda F(\varphi(t),t).$$

If λ is sufficiently small, then B_λ is a contraction map. We may now conclude the proof by an argument similar to that used in the proof of Theorem 4·22.25. The details are left to the reader.▮

4·22.28 THEOREM: *Consider the scalar differential equation*

$$\frac{dx}{dt} = f(x,t), \qquad (x \in \mathbb{R}, t \in \mathbb{R}). \qquad (22.21)$$

Suppose that it has a bounded solution $\varphi: \mathbb{R} \to \mathbb{R}$ and that all equations of the form

$$\frac{\mathrm{d}x}{\mathrm{d}t} = g(x,t), \tag{22.22}$$

with $g \in H(f)$ are such that their solutions are uniquely determined by the initial conditions. Set $K = \overline{\varphi(\mathbb{R})}$. Then there exists a dense subset $L \subset H_K(f)$ such that, for each function $g \in L$, Equation (22.22) has a synchronous solution $\mu:\mathbb{R} \to \mathbb{R}$ belonging to $H(\varphi)$.

Proof: Define a map $s:H_K(f) \to \mathbb{R}$ by

$$s(g) = \sup\{\lambda(0) \mid (g,\lambda) \in H_K(f,\varphi)\}.$$

Clearly, s is an upper semi-continuous function, and hence the set of its continuity points is dense in $H_K(f)$ (Kuratowski [*1*]). There exists a solution μ of (22.22) with the initial condition $\mu(0) = s(g)$ which is defined on the real line $\mathbb{R}$ and satisfies the condition $(g,\mu) \in H_K(f,\varphi)$. Since all equations of the form (22.22) satisfy the uniqueness condition, the transition along the solutions is a monotonic operator. Hence it follows by integral continuity that the set L is invariant and, moreover, $s(g_t) = \mu(t)$ $(t \in \mathbb{R},\ g \in L)$. Consequently the mapping $p:H_K(f\ \varphi) \to H_K(f)$ is injective at all points of the form (g,μ), where $g \in L$ and $\mu(0) = s(g)$. By Lemma 4·22.10, $\mu:\mathbb{R} \to \mathbb{R}$ is a synchronous solution of Equation (22.22).∎

4·22.29 LEMMA: *Let T be a topological group, S a subsemigroup of T satisfying the conditions* (22.1), *X and Y compact spaces and $p:(X,T) \to (Y,T)$ an extension of a minimal transformation group (Y,T). Further, let $x \in X$ and let M be a minimal set contained in $\overline{xS}$. Then there exists a point $m \in M$ such that $xp = mp$ and $(x,m) \in P(X,S)$ (i.e., for every $\alpha \in \mathfrak{U}[X]$ there is an element $s \in S$ with $(xs,ms) \in \alpha$).*

Proof: Given a minimal set $M \subset \overline{xS}$, choose a minimal right ideal I of the semigroup $E(X,S)$ so that $xI = M$. Let $y_0 = xp$, and let X_0 and E_0 be the same as in Subsection 4·22.1. Since Y is a minimal set, $M \cap X_0 \neq \emptyset$. Hence it follows that $I \cap E_0 \neq \emptyset$. By Lemma 4·4.11, there is an idempotent $u \in I \cap E_0 \subset E(X,S)$. Then m $xu \in M$ and $(x,xu) \in P(X,S)$. Since $u \in E_0$, we have $xu \in X_0$ and hence $(xu)p = xp = y_0$.∎

4·22.30 THEOREM: *Suppose all the hypotheses of Theorem* 4·22.28 *are satisfied and furthermore that the function f: $K \times \mathbb{R} \to \mathbb{R}$ is recurrent in the sense of Birkhoff, i.e., $H_K(f)$ is a compact minimal set. Assume, further, that the bounded solutions of all equations of the form* (22.22) *with $g \in H(f)$ are positively separated. Then the solution $\varphi:\mathbb{R} \to \mathbb{R}$ of Equation* (22.21) *is uniformly synchronous.*

Proof: The closure of the positive semiorbit of the point $(f,\varphi) \in H_K(f,\varphi)$ contains a compact minimal set M. According to Theorem 4·22.28, the extension $p:M \to H_K(f)$ is almost automorphic and consequently proximal. On the other hand, it follows from the positive separation condition that this extension is distal in the positive direction. By Lemma 4·22.4, the extension $p:(M,\mathbb{R}) \to (H_K(f),\mathbb{R})$ is distal. Hence $p:M \to H_K(f)$ is a bijective map. Let us prove that $(f,\varphi) \in M$. In fact, by Lemma 4·22.28, there is a point $(f,\psi) \in M$ such that (f,φ) and (f,ψ) are proximal in the positive direction. Since the bounded solutions are positively separated, we get $\varphi = \psi$. Hence $(f,\varphi) \in M$. Since $p:M \to H_K(f)$ is a homeomorphism, φ is a uniformly synchronous solution.∎

4·22.31 THEOREM: *Let all the hypotheses of Theorem 4·22.28 be satisfied and let $f:K \times \mathbb{R} \to \mathbb{R}$ be recurrent in the sense of Birkhoff. If the function f is monotonic in x (in the same sense for all $t \geq 0$, for example, $f(x_1,t) \geq f(x_2,t)$ for all $t \geq 0$, whenever $x_1 \geq x_2$), then each bounded solution $\varphi:\mathbb{R} \to \mathbb{R}$ of (22.21) is uniformly sunchronous.*

Proof: It is easy to verify that the monotonicity of the function f, and consequently of all functions $g \in H_K(f)$, ensures the distality of the extension $p:H_K(f,\varphi) \to H_K(f)$ in at least one direction. To conclude the proof, we refer to Theorem 4·22.30.∎

4·22.32 If the function f is strictly monotonic, then Equation (22.21) has at most one bounded solution. If the hypotheses of Theorem 4·22.31 are fulfilled, then the difference of any two bounded solutions (which are uniformly synchronous, by Theorem 4·22.31) is constant. The proof of these statements is left to the reader.

4·22.33 We shall now present an example which shows that, even in the situation of Theorem 4·22.28, uniformly synchronous solutions need not necessarily exist. Consider the differential equation (14.7) in Example 3·14.38. Since the winding number of this equation equals $\alpha = 1/2\pi$, for every number ϑ_0, the function of φ

$$w(\varphi,\vartheta_0) \equiv f(\varphi,\vartheta_0) - \alpha\varphi,$$

is bounded on $\mathbb{R}$. Let us introduce a new dependent variable $w = \vartheta - \alpha\varphi$. Then w satisfies the equation

$$\frac{dw}{d\varphi} = B(\varphi,w), \tag{22.23}$$

where

$$B(\varphi,w) = A(\varphi,w + \alpha\varphi) - \alpha,$$

and all the solutions of (22.23) are bounded. It is easy to see that the function $B = B(\varphi,w)$ is almost periodic in φ uniformly with respect to $w \in \mathbb{R}$.

Let us prove that (22.23) has no almost periodic solutions. Suppose that, to the contrary, there exists a number $\vartheta_0 \in \mathbb{R}$ such that $w = w(\varphi,\vartheta_0)$ is an almost periodic solution. Let ϑ_1 and ϑ_2 be the end-points of some interval complementary to the set $X \subset S^1$. There exist sequences of integers $\{p_n\}$ and $\{q_n\}$ such that

$$\{f(\varphi + 2\pi p_n,\vartheta_0) + 2\pi q_n\} \to f(\varphi,\vartheta_1)$$

uniformly on each segment of the real line $-\infty < \varphi < +\infty$. Since the function $w = w(\varphi,\vartheta_0) = f(\varphi,\vartheta_0) - \alpha\varphi$ is almost periodic, we may suppose without loss of generality that the sequence

$$\{(\varphi + 2\pi p_n,\vartheta_0)\}$$

is uniformly convergent to a certain almost periodic function $\lambda(\varphi,\vartheta_1)$. It is easy to see that the difference $\lambda(\varphi,\vartheta_1) - w(\varphi,\vartheta_1)$ is constant. Hence $w(\varphi,\vartheta_1)$ is an almost periodic function. Of course, the same is true for $w(\varphi,\vartheta_2)$. But

$$w(\varphi,\vartheta_2) - w(\varphi,\vartheta_1) = f(\varphi,\vartheta_2) - f(\varphi,\vartheta_1),$$

and the difference $f(\varphi,\vartheta_2) - f(\varphi,\vartheta_1)$ tends to zero as $\varphi \to \infty$, by (14.8). Since the functions $w(\varphi,\vartheta_2)$ and $w(\varphi,\vartheta_1)$ are almost periodic, this is possible only when $w(\varphi,\vartheta_1) \equiv w(\varphi,\vartheta_2)$, which contradicts the hypothesis $\vartheta_1 \not\equiv \vartheta_2$. This contradiction shows that Equation (22.23) does not have almost periodic solutions. Note that the orbit closure of each point $(\varphi,\vartheta) \in S^1 \times S^1$ contains a Levitan almost periodic point. Hence it follows that for each solution $w = w(\varphi,\vartheta_0)$ of (22.23) the set $H(w)$ contains at least one Levitan almost periodic function $\lambda = \lambda(\varphi)$, but perhaps there do not exist any Levitan almost periodic solutions of the Equation (22.23).

REMARKS AND BIBLIOGRAPHICAL NOTES

Lemma 4·22.2 is taken from the paper by Gerko [*1*]. Lemma 4·22.4 is essentially due to Žikov [*2*]. The notions of synchronous and uniformly synchronous solutions of differential equations are introduced by Ščerbakov [*6*]; the statements 4·22.13-21 can also be found there. The relationship between synchronous solutions and almost automorphic extensions has been studied here for the first time (although it is implicitly used also in the book by Ščerbakov [*6*]). The usage of extensions enables us to essentially shorten and simplify the proofs. Theorems 4·22.23,25,27 (in the case of a constant function $A(t) \equiv A$) were obtained by Ščerbakov [*6*] (see also: Birjuk [*1*], Daletskiĭ and Kreĭn [*1*]). Theorem 4·22.28 was proved by Žikov [*2,8*]. Scalar (one-dimensional) differential equations with an almost periodic right side were investigate by many authors (see: Barbălat [*1*], Borukhov [*1*], Demidovič [*1*], Gheorgiu [*1*], Halanay [*1*], Opial [*1,2*] and others). Many results obtained in these papers are particular cases of Theorem 4·22.31. Example 4·22.33 is due to Opial [*2*].

As was noted in the proof of Theorem 4·22.28, the transition operator along the solutions of a one-dimensional differential equation is monotonic. Differential equations with some montonicity properties were studied by Žikov [*2*,*8*,*9*] and Čeresiz [*1*,*2*].

§(4·23): EXTENSIONS WITH ZERO-DIMENSIONAL FIBRES AND ALMOST PERIODIC SOLUTIONS

We shall give here several general conditions for the existence of minimal extensions with zero-dimensional fibres. These conditions are related to the existence of a fixed point of a certain transformation group. By using some of the results from the theory of compact transformation groups, we obtain sufficient conditions for the existence of almost periodic solutions of systems of two or three scalar differential equations with almost periodic right sides. We shall investigate the motions of the points $x_0 \in X$ which are uniformly asymptotically stable relative to the extension $p:(X,T) \to (Y,T)$. It will be proved that a uniformly asymptotically stable solution of a differential equation with an almost periodic right side is asymptotically almost periodic. Algebraic equations with recurrent or almost periodic coefficients will also be considered, and sufficient conditions for the existence of recurrent (respectively, almost periodic) solutions of such equations will be presented.

4·23.1 Let T be a topological group, let S be a subsemigroup of T satisfying conditions (22.1), X a uniform space, Y a compact space, $p:(X,T,\pi) \to (Y,T,\rho)$ an extension, $y_0 \in Y$ and $X_0 = \{x \mid x \in X, xp = y_0\}$. Suppose, moreover, that the following three conditions are satisfied:

(1): (Y,T,ρ) is a minimal transformation group;

(2): For each point $x \in X$, the set $\overline{xS}$ is compact;

(3): The extension $p:(X,S) \to (Y,S)$ is equicontinuous, *i.e.*, for every $\alpha \in \mathfrak{U}[X]$ there exists a $\beta \in \mathfrak{U}[X]$ such that $(x_1 s, x_2 s) \in \alpha$ for all $s \in S$ whenever $(x_1,x_2) \in \beta$ and $x_1 p = x_2 p$.

Let $E(X,S)$, E_0 and P_0 be the same as in Subsection 4·22.1. It follows from Condition (3) that P_0 is a compact topological semigroup of continuous maps of X_0 into X_0. Let I be a minimal right ideal of the semigroup $E(X,S)$. Then for each point $x \in X_0$ the set xI is S-minimal and $(xI)p = \overline{y_0 S} = \overline{y_0 T} = Y$. Hence $xI \cap X_0 \neq \emptyset$. Since $xI \subset \overline{xS}$ and $\overline{xS} \cap X_0 = xE_0$, we conclude that $I \cap E_0 \neq \emptyset$.

Denote $I \cap E_0$ by M. It is easy to verify that M is a right ideal of the semigroup E_0. Let $R = \{(\xi,\eta) \mid \xi,\eta \in M; x\xi = x\eta\ (x \in X_0)\}$ and $q:M \to M/R$ be the canonical projection. Let us prove that the semigroup M/R is in fact a group. It will suffice to show that M/R does not contain more than one idempotent (note that the semigroup M/R is compact and contains at least one idempotent, by Lemma 1·4.11). If u is an idempotent of the semigroup M/R, then there exists an idempotent u_1 of the semigroup M such that $u_1 q = u$. Hence it will suffice to verify that if u and v are idempotents of M, then $uq = vq$. *i.e.*, $xu = xv$ ($x \in X_0$). Let $x \in X_0$ and $a = xu$. Then $(a,av) \in P(X,S)$, where $P(X,S)$ is the family of all pairs of points that are proximal under (X,S). Since $u \in M \subset I$, the set $\overline{aS} \equiv \overline{xuS}$ is S-minimal. By Lemma 4·22.2, $\overline{aS} = \overline{aT}$. By hypothesis, $p:(\overline{aT},S) \to (Y,S)$ is an equicontinuous extension. Thus $p:(\overline{aT},T) \to (Y,T)$ is also equicontinuous (Lemma 4·22.5). Recall that $x \in X_0$, $u \in M \subset E_0$. Hence $a \equiv xu \in X_0$, and similarly $av \in X_0$. As was pointed out above, the points a and ab are proximal under (X,S); but since $p:(\overline{aT},T) \to (Y,T)$ is a distal extension, this can happen only when $a = av$, *i.e.*, $xu = xuv$ for all $x \in X_0$. We remind the reader that $u,v \in I$. By Lemma 1·4.2(2), $uv = v$, and hence $xu = xv$ ($x \in X_0$). Thus we have proved that $G \equiv M/R$ is a group. Since M is a minimal right ideal of the semigroup E_0, G is a (minimal) right ideal of the semigroup $P_0 \equiv E_0/R$ (see 4·22.1).

Let ω be the identity of the compact topological group G. The continuous mapping $\omega: X_0 \to X_0\omega$ is a retraction, and therefore the set $X_0\omega = \{x \mid x \in X_0, x\omega = x\}$ is closed. Since G is a minimal right ideal of the semigroup P_0, we have $G = \omega P_0$. If $x \in X_0$ and $z = x\omega$, then $zP_0 = x\omega P_0 = (x\omega)\omega P_0 = zG$ and $zG = x\omega G = xG\omega \subset X_0\omega$. Thus $(X_0\omega, G)$ is a topological transformation group and $zP_0 = zG$ for every point $z \in X_0\omega$.

4·23.2 THEOREM: *Suppose that the conditions stated at the beginning of Subsection* 4·23.1 *are satisfied. Then for every point* $y_0 \in Y$ *the semigroup* P_0 *contains a minimal right ideal* G, *which is a compact topological group. Let* ω *be the identity of* G. *Then* $\omega: X_0 \to X_0\omega$ *is a retraction and* $(X_0\omega, G)$ *is a topological transformation group.*

If $z \in X_0\omega$ *is a fixed point of the transformation group* $(X_0\omega, G)$, *then* $p:\overline{zT} \to Y$ *is a homeomorphism.*

If G_0 *denotes the identity component of* G *and* $z \in X_0\omega$ *is a fixed point of the transformation group* $(X_0\omega, G)$, *then* $p:(\overline{zT},T) \to (Y,T)$ *is an equicontinuous extension with zero-dimensional fibres. In particular, if* (Y,T) *is an almost periodic transformation group and the group* T *satisfies the hypotheses of Theorem* 3·17.13, *then* $\overline{zT}$ *is also an almost periodic minimal set.*

Proof: Let $z \in X_0\omega$ be a fixed point of the transformation group $(X_0\omega, G)$. Then $zP_0 = zG = \{z\}$. By Theorem 4·22.3, $p:(\overline{zT},T) \to (Y,T)$ is a proximal extension. On the other hand, it has already been shown that $p:(\overline{zT},T) \to (Y,T)$ is an equicontinuous and, moreover, distal extension. Thus p is an isomorphism.

Let G_0 be the identity component of G. Then G is a normal subgroup and the quotient group G/G_0 is totally disconnected and compact, and therefore zero-dimensional. If a point $z \in X_0\omega$ is fixed under all transformations in the group G_0, then $zP_0 = zG \approx G/G_0$. Therefore the set zP_0 is zero-dimensional. Since the extension $p:(\overline{zT},T) \to (Y,T)$ is equicontinuous, all the fibres of this extension are zero-dimensional.

If Y is an almost periodic minimal set and the group T satisfies the hypotheses of Theorem 3·17.13, then the transformation group $(\overline{zT},T)$ is also almost periodic by Theorem 3·17.13.∎

4·23.3 We state (without proof) the following facts from the theory of compact topological transformation groups:

(A): THEOREM: (Montgomery and Zippin [*1*], p. 260): *Let G be a connected compact group and let $(\mathbb{R}^3,G)$ be an effective group of transformations of three-dimensional Euclidean space. Then G is either a group of rotations about an axis or a group of orthogonal transformations. Hence in either of these cases there are fixed points under the action of G.*

(B): THEOREM: (Hofmann and Mostert [*2*], p. 225): *Let X be a locally compact connected subspace of an n-dimensional manifold and let G be a compact topological group which acts effectively, but not transitively on X in such a way that the orbit of some point is $(n - 1)$-dimensional. Then G is a Lie group.*

(C): THEOREM: (Hofmann and Mostert [*1*], p. 310): *If G is a compact topological group, H is a closed subgroup of G and the space G/H is contractible, then $G = H$.*

Theorem (B) will be used in what follows only when X is contractible. Under this condition the group G acts on X transitively iff X consists of a single point. Indeed, if $X = xG$ $(x \in X)$ and $H = \{g|xg = x\}$, then the space G/H is homeomorphic to X and therefore contractible. Then $G = H$ by Theorem (C), and hence X consists of a single point.

4·23.4 THEOREM: *Assume that all the conditions stated at the beginning of Subsection* 4·23.1 *are satisfied. Then the following assertions hold:*

(1): *If the fibre X_0 is homeomorphic to a subset of the real line, then there is a point $x_0 \in X_0$ such that the map $p:\overline{x_0T} \to Y$ is a homeomorphism. For all other points $x \in X_0$ the map $p:xT \to Y$ is two-to-one (i.e., each fibre consists of two points);*

(2): *If the fibre X_0 is homemorphic to the plane $\mathbb{R}^2$, then there exists a point $x_0 \in X_0$ such that the fibres of the map $p:\overline{x_0T} \to Y$ are zero-dimensional;*

(3): *If the retract $X_0\omega$ contains a subset X_1, which is invariant under the action of the group G and homeomorphic to three-dimensional Euclidean space, then there is a point $x_0 \in X_1$ such that $p:(\overline{x_0T},T) \to (Y,T)$ is an equicontinous extension with zero-dimensional fibres.*

Proof: (1): We may suppose that $X_0\omega$ is a connected closed subset of the real line, *i.e.*, a segment. Each transformation g: $X_0\omega \to X_0\omega$ $(g \in G)$ is a homeomorphism, and therefore g is either a monotonically increasing or a monotonically decreasing function.

Let H denote the totality of monotonically increasing maps $g \in G$. Clearly, H is a subgroup of the group G of normal index not greater than two, and therefore H is a normal subgroup of G. Since the G-orbit of each point is compact, H acts trivially on $X_0\omega$. Hence we can define an action of the group G/H on $X_0\omega$. If $H \neq G$, then there exists a monotonically decreasing map $g \in G$. This map has exactly one fixed point, therefore the group G/H, and consequently the group G, has a unique fixed point. The orbit of any other point consists of exactly two points.

(2): The orbit xG of each point $x \in X_0\omega$ is either zero-dimensional or one-dimensional, then this orbit is open in X_0 (and at the same time compact); which is impossible. Let us prove that there is always a point whose G-orbit is zero-dimensional. Suppose the contrary holds. By Theorem (B), G is a Lie group, and therefore all orbits are locally connected. Let G_0 be the component of the identity of the group G. Since $\dim(xG) = 1$, we have $\dim(xG_0) = 1$ for all $x \in X_0\omega$, and the orbit xG_0 is homeomorphic to the space G_0/H_0, where $H = \{g | g \in G, xg = x\}$. By Lemma 3·17.18, G_0/H_0 is a compact one-dimensional commutative Lie group, *i.e.*, $G_0/H_0 \approx \mathbb{K}$. By our assumption, the orbit xG_0 of each point $x \in X_0\omega$ is a non-degenerate Jordan curve. Let D be the part of the plane inside of the curve xG_0. Since $X_0\omega$ is a retract of the plane X_0, $X_0\omega$ can be contracted in itself to a point. Therefore $\bar{D} \subset X_0\omega$. By Jordan's Theorem the set $\bar{D}$ is invariant under the action of the connected group G_0.

Let $d(x)$ denote the diameter of the set xG_0 $(x \in X_0\omega)$. Since the group G_0 is compact and acts on $X_0\omega$ continuously, the function $d = d(x)$ is continuous. Let $d_0 = \inf\{d(x) | x \in \bar{D}\}$ and let the point $x_0 \in X_0\omega$ be such that $d_0 = d(x_0)$. By our assumption, x_0G_0 is a Jordan curve distinct from a point, hence $d(x_0) > 0$. If x_1 is a point inside of the curve x_0G_0, then the orbit x_1G_0 lies inside x_0G_0; therefore $d(x_1) < d(x_0)$, which is a contradiction.

(3): Let G_0 denote, as before, the identity component of the group G. Without loss of generality we may suppose that G_0 acts effectively on X_1. By Theorem (A) the transformation group (X_1, G_0) has at least one fixed point. Therefore the required result follows from Theorem 4·23.2. ∎

4·23.5 Consider the differential equation

$$\frac{dx}{dt} = f(x,t), \qquad (x \in \mathbb{R}^n, t \in \mathbb{R}). \tag{23.1}$$

Suppose that for all $g \in H(f) \subset C(\mathbb{R}^{n+1}, \mathbb{R}^n)$ the solutions of the equation

$$\frac{dx}{dt} = g(x,t) \tag{23.2}$$

are defined for all $t \in \mathbb{R}$ and uniquely determined by the initial condition $\varphi(0) = x_0 \in \mathbb{R}^n$. We shall say that Equation (23.1) is *UNIFORMLY POSITIVELY STABLE* if for every closed bounded set $K \subset \mathbb{R}^n$ and every $\varepsilon > 0$ there is a number $\delta > 0$ such that if $\|x_1 - x_2\| < \delta$, $x_1, x_2 \in K$ and $g \in H(f)$, then the solutions $\varphi_1(t)$, $\varphi_2(t)$ of the Equation (23.2) with the initial conditions $\varphi_i(0) = x_i$ $(i = 1,2)$ satisfy the inequality $\|\varphi_1(t) - \varphi_2(t)\| < \varepsilon$ for all $t \geqslant 0$. Let $X = H(f) \times \mathbb{R}^n$. Define a dynamical system $(X, \mathbb{R})$ as was described in Subsection 4·21.2. Then the natural homomorphism $p:(X, \mathbb{R}^+) \to (H(f), \mathbb{R}^+)$ is an equicontinuous extension.

4·23.6 THEOREM: *Consider Equation* (23.1) *in the case when* $n = 2$. *Suppose that this equation is uniformly positively stable and its right side* $f \in C(\mathbb{R}^n \times \mathbb{R}, \mathbb{R})$ *is almost periodic. If this equation has at least one solution which is bounded on* $\mathbb{R}^+$, *then it has an almost periodic solution* $\varphi(t)$ *(which is not necessarily synchronous).*

Proof: If the hypotheses of the theorem are satisfied, then all solutions are bounded for $t \geqslant 0$ and the fibres of the extension are homemorphic to the plane $\mathbb{R}^2$. The required result follows now from Theorem 4·23.4(2) and Subsection 4·23.5.▮

4·23.7 THEOREM: *Consider the Equation* (23.1) *with* $n = 3$. *Suppose that this equation is uniformly positively stable and that the right side* $f(x,t)$ *is almost periodic in* t *(uniformly on* x *varying in every bounded set). If all the solutions of* (23.1) *are bounded on the entire axis, then there is at least one almost periodic solutoin of this equation.*

Proof: This follows directly from Theorem 4·23.4(3).▮

4·23.8 THEOREM: *Let* E *be a Banach space,* W *an open subset of* E, $f: W \times \mathbb{R} \to E$ *a continuous mapping,* $\varphi: \mathbb{R} \to W$ *a compact solution of the equation*

$$\frac{dx}{dt} = f(x,t), \qquad (x \in W, t \in \mathbb{R}), \tag{23.3}$$

and $K = \overline{\varphi(\mathbb{R})}$. *Assume further that the function* $f(x,t)$ *is almost periodic in* $t \in \mathbb{R}$ *(uniformly relative to* $x \in K$*),* $H(\varphi)$ *contains finitely many solutions of* (23.3) *and each equation of the form*

$$\frac{dx}{dt} = g(x,t), \qquad (x \in W, t \in \mathbb{R}), \tag{23.4}$$

where $g \in H_K(f)$, *satisfies the condition on postive separation of compact solutions* $\psi \in H(\varphi)$. *Then* φ *is an almost periodic solution.*

Proof: Apply Theorem 3·17.12 to the extension $p: H_K(f,\varphi) \to H_K(f)$. By hypothesis, $H_K(f)$ is an almost periodic minimal set, the fibre over the point f consists of a finite number of points and the extension p is distal in the positive direction. By

Lemma 4·22.4, p is distal in both directions. It follows that $H_K(f,\varphi)$ is a compact minimal set. By Theorem 3·17.12, $H_K(f,\varphi)$ is almost periodic, and hence the function φ is also almost periodic. ▮

4·23.9 COROLLARY: *Let* $\varphi:\mathbb{R}\to E$ *be a compact solution of the equation*

$$x' = A(t)x + f(t), \tag{23.5}$$

where $A \in C(\mathbb{R},L(E,E))$ *and* $f \in C(\mathbb{R},E)$ *are almost periodic functions. Assume further that each equation of the form*

$$x' = B(t)x,$$

where $B \in H(A)$, *satisfies the condition on positive separation from zero of non-trivial compact solutions and* $H(\varphi)$ *contains only finitely many solutions of Equation* (23.5). *Then* φ *is an almost periodic solution.*

4·23.10 Let X and Y be metric spaces, $(X,\mathbb{R})$ and $(Y,\mathbb{R})$ be flows, and let $p:(X,\mathbb{R})\to(Y,\mathbb{R})$ be an extension. We shall say that a set $A \subset X$ is *UNIFORMLY STABLE IN THE POSITIVE DIRECTION RELATIVE TO THE EXTENSION* p if for each number $\varepsilon > 0$ there exists a $\delta = \delta(\varepsilon) > 0$ such that $\rho(at,xt) < \varepsilon$ for all $t \geqslant 0$ whenever $a \in A$, $x \in X$, $\rho(a,x) < \delta$ and $ap = xp$ (where ρ denotes the metric in X).

4·23.11 LEMMA: *If the set A is uniformly stable in the posititive direction relative to the extension p and $p:X\to Y$ is open at all points of the set $(\bar{A}p)p^{-1}$, then $\bar{A}$ has the above stability property.*

Proof: Given $\varepsilon > 0$, choose a number $\delta = \delta(\varepsilon)$ according to the definition of uniform stability of the set A in the positive direction relative to p. Let us prove that if $a \in \bar{A}$, $x \in X$, $ap = xp$ and $\rho(a,x) < (1/3)\delta(\varepsilon/3)$, then $\rho(at,xt) < \varepsilon$ for all $t \geqslant 0$. Suppose that to the contrary there exist $a_0 \in \bar{A}$, $x_0 \in X$ and $t_0 > 0$ such that $a_0p = x_0p$, $\rho(a_0,x_0) < (1/3)\delta(\varepsilon/3)$ and $\rho(a_0t_0,x_0t_0) \geqslant \varepsilon$. Choose a number δ_1 so that $\rho(x_0t_0,x_1t_0) < \varepsilon/3$ whenever $\rho(x_0,x_1) < \delta_1$. Since the map p is open at the point $x_0 \in (\bar{A}p)p^{-1}$, there exists a number $\delta_2 > 0$ such that for every point a' in the δ_2-neighbourhood of a_0 there is a point $x' \in X$ with $x'p = a'p$, $\rho(x_0,x') < \delta_1$. Suppose that moreover δ_2 is small enough so that $\rho(a_0t_0, a't_0) < \varepsilon/3$, whenever $\rho(a_0,a') \leq \delta_2$. Suppose further that $\delta_2 < \delta_1 < (1/3)\delta(\varepsilon/3)$. Since $a_0 \in \bar{A}$, there is a point $a_1 \in A$ belonging to the δ_2-neighbourhood of a_0. Then $\rho(a_0t_0,a_1t_0) < \varepsilon/3$ and there exists a point $x_1 \in X$ for which $a_1p = a_1p$ and $\rho(x_1,x_0) < \delta_1$. Thus $\rho(x_1t_0,x_0t_0) < \varepsilon/3$. Since $\rho(x_1,a_1) < \rho(x_1,x_0) + \rho(x_0,a_0) + \rho(a_0,a_1) < \delta(\varepsilon/3)$, it follows that $\rho(x_1t_0,a_1t_0) < \varepsilon/3$, by the choice of $\delta(\varepsilon/3)$. Hence

$$\rho(x_0t_0,a_0t_0) \leqslant \rho(x_0t_0,x_1t_0) + \rho(x_1t_0,a_1t_0) + \rho(a_1t_0,a_0t_0) < \varepsilon.$$

But this contradicts our assumption. ▮

4·23.12 A set $A \subset X$ is said to be a *UNIFORM POSITIVE ATTRACTOR RELATIVE TO THE EXTENSION* p if there exists a number $\delta_0 > 0$ such that for every $\varepsilon > 0$ there is a number $L(\varepsilon) > 0$ with the property that $a \in A$, $z \in X$, $ap = zp$ and $\rho(a,z) < \delta_0$ imply $\rho(at,zt) < \varepsilon$ for all $t \geqslant L(\varepsilon)$.

4·23.13 LEMMA: *Let $A \subset X$. If the extension p is open at all points $x \in (\bar{A}p)p^{-1}$ and the set A is uniformly positively attracting relative to the extension p, then so is $\bar{A}$.*

Proof: The proof is similar to that of Lemma 4·23.11 and therefore it will be omitted. ▮

4·23.14 An extension $p:(X,\mathbb{R}) \to (Y,\mathbb{R})$ is said to be *POSITIVELY (NEGATIVELY) LOCALLY PROXIMAL IN THE FIBRE* $X_0 = \{x \mid x \in X, xp = y_0\}$ *OVER A POINT* $y_0 \in Y$ if for every point $x \in X_0$ there exists a number $\delta_x > 0$ such that $\rho(x,x') < \delta_x$ and $x'p = xp$ imply that the points x and x' are positively (negatively) proximal, *i.e.*, $\inf_{t>0} \rho(xt,x't) = 0$ $(\inf_{t<0} \rho(xt,x't) = 0)$.

4·23.15 THEOREM: *Let $M \subset X$ be a compact minimal set. If the extension $p:(M,\mathbb{R}) \to (Y,\mathbb{R})$ is distal in at least one direction and there exists a point $y_0 \in Y$ such that p is locally proximal in the fibre $M_0 = \{x \mid x \in M, xp = y_0\}$ in a certain direction, then $p:M \to Y$ is a finite-to-one locally homeomorphic map.*

Proof: By Lemma 4·22.4, the extension $p:(M,\mathbb{R}) \to (Y,\mathbb{R})$ is distal (in both directions). For each point $x \in M_0$ choose $\delta_x > 0$ according to the local proximality condition. Select a finite subcovering of the set M_0 from the covering $\{S(x,\delta_x) \mid x \in M_0\}$. Clearly each subset of the form $S(x,\delta_x) \cap M_0$ consists of the single point x. Hence M_0 is a finite fibre and since $p:(M,\mathbb{R}) \to (Y,\mathbb{R})$ is a distal minimal extension, all the fibres consist of the same number of points. Moreover, the map $p:M \to Y$ is open (Corollary 3·12.25). Consequently p is a local homeomorphism. ▮

4·23.16 THEOREM: *Let $p:(X,\mathbb{R}) \to (Y,\mathbb{R})$ be an extension, $(Y,\mathbb{R})$ an almost periodic minimal transformation group and $x_0 \in X$ a point such that $x_0\mathbb{R}^+$ is compact in X (i.e., $\overline{x_0\mathbb{R}^+}$ is compact) and uniformly stable in the positive direction relative to the extension p. If $p:(\overline{x_0\mathbb{R}^+},\mathbb{R}) \to (Y,\mathbb{R})$ is locally proximal in a certain direction at least in one fibre and $p:X \to Y$ is open at all points $x \in (\overline{x_0 p\mathbb{R}^+})p^{-1}$, then there exists a point $z_0 \in \overline{x_0\mathbb{R}^+}$ such that $\overline{z_0\mathbb{R}}$ is an almost periodic minimal set.*

Proof: The compactness of $\overline{x_0\mathbb{R}^+}$ ensures the existence of a point $z_0 \in \overline{x_0\mathbb{R}^+}$ such that $\overline{z_0\mathbb{R}}$ is a compact minimal set. By Lemma 4·23.11, the set $\overline{x_0\mathbb{R}^+}$ is uniformly stable in the positive direction relative to the extension p. This implies that the extension $p:(\overline{z_0\mathbb{R}},\mathbb{R}) \to (Y,\mathbb{R})$ is distal in the negative direction and thus, by

Lemma 4•22.4, it is distal in both directions. By Theorem 4•23.15, the fibres of this extension are finite. By Theorem 4•17.12, the transformation group $(\overline{z_0\mathbb{R}},\mathbb{R})$ is almost periodic.∎

4•23.17 A *POINT* $x \in X$ is called *UNIFORMLY ASYMPTOTICALLY STABLE RELATIVE TO THE EXTENSION* $p:(X,\mathbb{R}) \to (Y,\mathbb{R})$ if the set $x\mathbb{R}^+$ is uniformly stable in the positive direction relative to p and uniformly attracting relative to p. A point $x \in X$ is called a *POSITIVELY ASYMPTOTICALLY ALMOST PERIODIC POINT*, if there exist an almost periodic minimal set $M \subset X$ and a point $a \in M$ such that

$$\lim_{t \to +\infty} \rho(xt,at) = 0.$$

4•23.18 THEOREM: *Let* $p:(X,\mathbb{R}) \to (Y,\mathbb{R})$ *be an extension, where* $(Y,\mathbb{R})$ *is an almost periodic minimal transformation group. Let* $x_0 \in X$ *be a uniformly asymptotically stable point relative to* p, *such that the set* $\overline{x_0\mathbb{R}^+}$ *is compact and the map* p: $X \to Y$ *is open at all points of the set* $(\overline{x_0p\mathbb{R}^+})p^{-1}$. *Then the point* x_0 *is positively asymptotically almost periodic.*

Proof: By Lemmas 4•23.11 and 4•23.13, the set $\overline{x_0\mathbb{R}^+}$ is uniformly stable in the positive direction and uniformly attracting relative to p. Therefore the extension $p:(\overline{x_0\mathbb{R}^+},\mathbb{R}) \to (Y,\mathbb{R})$ is locally positively proximal in every fibre. By Theorem 4•23.16, there exists an almost periodic minimal set $M \subset \overline{x_0\mathbb{R}^+}$. By Lemma 4•22.29, there is a point $a \in M$ such that $ap = x_0p$ and the points a and x_0 are proximal. Let ε be an arbitrary positive number. Choose δ_0 and $L(\varepsilon)$ as required by the condition that $\overline{x_0\mathbb{R}^+}$ is uniformly attracting relative to p. Since the points a and x_0 are proximal, there is a number $t_0 \in \mathbb{R}$ with $\rho(x_0t_0,at_0) < \delta_0$. Then $\rho(x_0t,at) < \varepsilon$ for all $t \geqslant t_0 + L(\varepsilon)$. This means that x_0 is a positively asymptotically almost periodic point.∎

4•23.19 Consider the equation

$$\frac{dx}{dt} = f(x,t) \tag{23.6}$$

in a Banach space E, where f is assumed to be continuous on $W \times \mathbb{R}$ (W is an open subset of E). Suppose that for each equation of the form

$$\frac{dx}{dt} = g(x,t), \tag{23.7}$$

where $g \in H(f) \in C(W \times \mathbb{R},E)$, and for each point $z \in W$ there exists a unique solution $\varphi:\mathbb{R} \to W$ with the initial condition $\varphi(0) = z$ and that this solution depends continuously on the initial point and on the right side $g \in H(f)$. Define the dynamical system $(H(f) \times W,\mathbb{R})$ as was done in Subsection 4•21.3. Let $p:H(f) \times W \to H(f)$ be the canonical projection. Clearly p is open.

Recall the following definitions. A *solution* $\varphi:\mathbb{R}^+ \to W$ is said to be *uniformly stable in the positive direction*, if for every $\varepsilon > 0$ there exists a $\delta > 0$ such that if $t_0 \geqslant 0$ and a solution $\psi:\mathbb{R}^+ \to W$ of (23.6) satisfies the condition $\|\psi(t_0) - \varphi(t_0)\| < \delta$, then $\|\psi(t) - \varphi(t)\| < \varepsilon$ for all $t \geqslant t_0$.

A *SOLUTION* $\varphi:\mathbb{R}^+ \to W$ of Equation (23.6) is called *UNIFORMLY POSITIVELY ATTRACTING* if there exists a number $\delta_0 > 0$ such that for every $\varepsilon > 0$ we can choose an $L > 0$ so that if $\|\psi(t_0) - \varphi(t_0)\| < \delta_0$ for some $t_0 \in \mathbb{R}^+$, then $\|\psi(t) - \varphi(t)\| < \varepsilon$ for all $t \geqslant L + t_0$ (here ψ denotes an arbitrary solution of (23.6)).

A *SOLUTION* $\varphi:\mathbb{R}^+ \to W$ is called *UNIFORMLY ASYMPTOTICALLY STABLE IN THE POSITIVE DIRECTION* if it is uniformly stable and uniformly attracting (in the positive direction).

A *SOLUTION* $\varphi:\mathbb{R}^+ \to W$ is called *TOTALLY STABLE* if for every $t_0 \geqslant 0$ and every $\varepsilon > 0$ there exists a number $\delta > 0$ such that if a function $F:W \times \mathbb{R}^+ \to E$ is continuous and satisfies the condition $\|F(x,t)\| < \delta (x \in W,\ t \geqslant 0)$ and $y_0 \in W$ is any point with $\|\varphi(t_0) - y_0\| < \delta$, then for each solution $y(t)$ of the equation

$$y' = f(y,t) + F(y,t),$$

with the initial condition $y(t_0) = y_0$ the inequality $\|\varphi(t) - y(t)\| < \varepsilon$ holds for all $t \geqslant t_0$.

Let $\varphi:\mathbb{R}^+ \to W$ be a compact solution of Equation (23.6) and let $K = \overline{\varphi(\mathbb{R}^+)}$. The solution φ is called stable in the positive direction under the perturbations from $H_K(f)$ if for every $\varepsilon > 0$ there exist numbers $\delta > 0$ and $\ell > 0$ such that if $g \in H_K(f)$ and $\|g(x,t) - f(x,t + \tau)\| < \delta$ for some $\tau \geqslant 0$ and all $x \in K$, $|t| \leqslant \ell$, where $\psi(t)$ is a solution of (23.7) with $\|\psi(0) - \varphi(\tau)\| < \delta$, then $\|\psi(t) - \varphi(t + \tau)\| < \varepsilon$ for all $t \geqslant 0$.

It is easy to show that for equations with almost periodic right hand sides total stability implies stability under the perturbations from $H_K(f)$. Let the right hand side of (23.6) be almost periodic in $t \in \mathbb{R}$, uniformly on $x \in K$ and let $\varphi:\mathbb{R}^+ \to K$ be a solution which is stable in the positive direction under the perturbations from $H_K(f)$. It follows directly from the definitions that in this case the point (f,φ) is uniformly positively Ljapunov stable in the corresponding shift dynamical system, and hence it is positively asymptotically almost periodic. Thus φ is an asymptotically almost periodic solution.

4·23.20 THEOREM: *Suppose that the following conditions are satisfied:*

(1): *The function $f \in C(W \times \mathbb{R},E)$ is almost periodic (i.e., $f(x,t)$ is almost periodic in t uniformly on x when varying in an arbitrary compact set $K \subset W$);*

(2): *There exists a compact solution $\varphi:\mathbb{R}^+ \to W$ of Equation* (23.6), *which is uniformly stable in the positive direction;*

(3): *If ψ is a solution of* (23.7) *and* $\psi(\mathbb{R}) \subset \overline{\varphi(\mathbb{R}^+)} \equiv K_0$, *then there exists a number $\delta_0 > 0$ such that for every solution $\lambda:\mathbb{R} \to K_0$ of Equation (23.7) with $\|\lambda(0) - \psi(0)\| < \delta_0$ there is a sequence $\{t_k\}$, $\{t_k\} \to \infty$ for which $\|\psi(t_k) - \lambda(t_k)\| \to 0$.*

Then there exists an almost periodic solution $\xi:\mathbb{R} \to K_0$ of Equation (23.6).

Proof: By hypothesis (1), $(H(f),\mathbb{R})$ is an almost periodic minimal transformation group. It follows from Condition (2) that the closure of the positive semiorbit of the point $(f,\varphi(0)) \in H(f) \times W$ is compact and uniformly stable in the positive direction relative to the extension p. Condition (3) implies that the extension is positively locally proximal in the fibre over the point f. Thus by Theorem 4·23.16, $H(\varphi)$ contains an almost periodic solution $\xi:\mathbb{R} \to K_0$ of Equation (23.6). ∎

4·23.21 THEOREM: *Let $f(x,t)$ be almost periodic in t, uniformly in $x \in K$ for every compact set $K \subset W$. Suppose that $\varphi:\mathbb{R}^+ \to W$ is a compact uniformly asymptotically stable solution of Equation (23.6). Then the function φ is positively asymptotically almost periodic.*

Proof: This is an immediate consequence of Theorem 4·23.18. ∎

4·23.22 The following examples illustrate Theorems 4·23.20,21. Let $\{\omega_n\}$ be a sequence of positive numbers satisfying the condition $\sum_{n=1}^{\infty} n\omega_n < \infty$. The function

$$\varphi(t) = \sum_{n=1}^{\infty} n\omega_n \sin\omega_n t, \qquad (t \in \mathbb{R}),$$

is almost periodic, but its primitive

$$\psi(t) = \int_0^t \varphi(u)du = \sum_{n=1}^{\infty} n(1 - \cos\omega_n t) \geqslant 0,$$

is unbounded. Each non-trivial solution of the equation

$$\frac{dx}{dt} = - x\varphi(t), \qquad (x \in \mathbb{R}, t \in \mathbb{R}),$$

satisfies the hypotheses of Theorem 4·23.20, but it is not asymptotically almost periodic.

The differential equation

$$\frac{dz}{dt} = iz\varphi(t), \qquad (z \in \mathbb{C}, t \in \mathbb{R}), \tag{23.8}$$

where i is the imaginary identity, has no non-trivial almost periodic solutions, although all its solutions are bounded and uniformly Ljapunov stable. The extension of the dynamical systems associated with this equation is equicontinuous; therefore each non-trivial solution is distal and moreover recurrent in the sense of Birkhoff, but neither of them is Levitan almost periodic. The fibres of each minimal extension associated with a non-trivial solution of Equation (23.8) are homeomorphic to a circle. Thus there is an analogy between Equation (23.8) and Example 2•9.19 of a distal but not almost periodic minimal flow on the three-dimensional torus.

4•23.23 Let T be a locally compact connected locally pathwise connected simply connected topological group. Consider an algebraic equation of the form

$$\lambda^n + a_1(t)\lambda^{n-1} + \dots + a_n(t) = 0, \tag{23.9}$$

where the $a_k : T \to \mathbb{C}$ are continuous functions. According to the definition given in Section 4•21.7, a solution of (23.9) is a continuous function $\lambda : T \to \mathbb{C}$ which satisfies Equation (23.9) for all $t \in T$. In what follows, it will be assumed that the discriminant

$$D(t) \equiv D(a_1(t), \dots, a_n(t)),$$

of Equation (23.9) is non-zero for all $t \in T$. Under this condition there are exactly n solutions of (23.9).

4•23.24 THEOREM: *Let* $a : T \to \mathbb{C}^n$, $a(t) = (a_1(t), \dots, a_n(t))$ $(t \in T)$, *be recurrent in the sense of Birkhoff. Then each solution* $\lambda : T \to \mathbb{C}$ *of* (23.9) *is also recurrent.*

Proof: Since $a : T \to \mathbb{C}^n$ is recurrent, each solution of (23.9) is bounded and uniformly continuous. The orbit closure $H(a)$ of the point $a \in C(T, \mathbb{C}^n)$ in the shift dynamical system is a compact minimal set. Let $H(\lambda, a) = H(\lambda(t), a_1(t), \dots, a_n(t)) \subset C(T, \mathbb{C}^{n+1})$ and $p : H(\lambda, a) \to H(a)$ be the canonical projection. Let us prove that the extension p is distal in the fibre over the point a. Suppose the contrary holds. Then there exist points (μ, a) and (ν, a) from $H(\lambda, a)$ and a net $\{t_\alpha\}$ of elements of the (additive) group T such that $\{\mu(t + t_\alpha)\} \to \mu(t)$, $\{\nu(t + t_\alpha)\} \to \mu(t)$, $\{a_k(t + t_\alpha)\} \to a_k(t)$ $(k = 1, \dots, n)$ in the topology of uniform convergence on compact subsets of T. Setting $t = 0$, where 0 is the zero element of the group T, we obtain:

$$\{\mu(t_\alpha)\} \to \mu(0), \qquad \{\nu(t_\alpha)\} \to \mu(0),$$

$$\{a_k(t_\alpha)\} \to a_k(0), \qquad (k = 1, \dots, n).$$

Since $D(0) \neq 0$, this can happen only when $\mu(t_\alpha) = \nu(t_\alpha)$ for $\alpha > \alpha_0$. It follows from the conditions imposed on the group T that $\mu(t) \equiv \nu(t)$. Hence the extension p is distal in the fibre over the point a. By Lemma 3·14.4, $H(\lambda,a)$ is a compact minimal set. Hence the function $\lambda:T \to \mathbb{C}$ is recurrent in the sense of Birkhoff. ∎

4·23.25 THEOREM: *Suppose that the functions* $a_1(t),\ldots,a_n(t)$ *are almost periodic and the discriminant* $D(t)$ *is separated from zero (i.e.,* $\inf_{t\in T}|D(t)| > 0$*). Then each solution* $\lambda:T \to \mathbb{C}$ *of* (23.9) *is almost periodic.*

Proof: Since the discriminant is separated from zero, the extension $p:H(\lambda,a) \to H(a)$ is distal. Clearly, all the fibres of this extension are finite. To conclude the proof we refer to Theorem 3·17.12. ∎

Now we shall give an example which shows that in Theorem 4·23.25 the hypothesis $\inf|D(t)| > 0$ cannot be replaced by the condition $D(t) \neq 0 (t \in T)$. Moreover, we shall produce a Bohr almost periodic function $f:\mathbb{R} \to \mathbb{C}$ such that $|f(t)| > 0$ $(t \in R)$, $\inf |f(t)| = 0$, but the solutions of the equation $\lambda^2 - f(t) = 0$ are not Levitan almost periodic. We precede this with a proof of the following lemma.

4·23.26 LEMMA: *Let* $f:\mathbb{R} \to \mathbb{C}$ *be a Bohr almost periodic function with* $|f(t)| > 0$ $(t \in \mathbb{R})$, $\inf|f(t)| = 0$. *Suppose that each function* $g \in H(f)$ *vanishes at no more than one point. Then Levitan almost periodicity of the function* $u(t) = \sqrt{f(t)}$ $(t \in \mathbb{R})$ *is Bohr almost periodicity.*

Proof: Here $\sqrt{f(t)}$ means one of the two continuous solutions of the equation $u^2 - f(t) = 0$. Suppose that $u(t)$ is a Levitan almost periodic solution. This means that there exist a compact commutative group G, a continuous homomorphism $\alpha:\mathbb{R} \to G$ onto a dense subgroup $G_1 \equiv \{\alpha(t) | t \in \mathbb{R}\}$ of G and a continuous function $v:G_1 \to \mathbb{C}$ such that $v(\alpha(t)) = u(t)$. Moreover we may suppose that there exists a continuous function $F:G \to \mathbb{C}$ such that $F(\alpha(t)) = f(t)$ $(t \in \mathbb{R})$.

Let $\delta = |f(0)|/2 > 0$. Choose a sufficiently small neighbourhood U of the identity $e = \alpha(0)$ of the group G such that

(1) $|F(g)| > \delta$ $(g \in U)$;

(2) If $\Phi:U \to \mathbb{C}$ is a continuous branch of the function $\sqrt{F(g)}$, then $|\Phi(g_1) + \Phi(g_2)| \geq \delta$ for all $g_1,g_2 \in U$.

Since $u(t) = v(\alpha(t))$ $(t \in \mathbb{R})$ and the function $v:G_1 \to \mathbb{C}$ is continuous, there exists a neighbourhood $U_1 \subset U$ of the point $e \in G$, such that $|u(0) - u(t)| = |v(e) - v(\alpha(t))| < \delta/2$ whenever $\alpha(t) \in U_1$. It follows from Condition (2) that the function $v(\alpha(t))$, when restricted to U_1, agrees with one of the branches of $\sqrt{F(g)}$. To be precise, let:

$$v(\alpha(t)) = \Phi(\alpha(t)), \qquad (\alpha(t) \in U_1).$$

This means that the function $u(t) = v(\alpha(t))$ can be extended by continuity onto U_1.

Let us show that such an extension is possible on the whole space G. Let $g \in G$. We shall prove that if $\{\alpha(t_n)\} \to g$, then $\{v(\alpha(t_n))\}$ is also convergent. Suppose that, to the contrary, there exist two subnets $\{t_n'\}$ and $\{t_n''\}$ such that

$$\lim_{n\to\infty} v(\alpha(t_n')) \neq \lim_{n\to\infty} v(\alpha(t_n'')).$$

The function $F(g\alpha(t))$ belongs to $H(f)$, and therefore it vanishes at no more than one point. Hence we may suppose that $F(g\alpha(t)) \neq 0$ for $t > 0$. There exists a number $\tau > 0$ such that $g\alpha(\tau) \in U_1$. Since $u(t) = v(\alpha(t)) = \Phi(\alpha(t))$ for $\alpha(t) \in U_1$ and $\lim\{\alpha(\tau + t_n)\} = \lim\{\alpha(t_n)\cdot\alpha(\tau)\} = g\alpha(\tau)$, we have

$$\lim_{n\to\infty}\{v(\alpha(\tau + t_n))\} = \lim_{n\to\infty}\{v(\alpha(\tau + t_n'))\}$$

$$= \lim_{n\to\infty}\{v(\alpha(\tau + t_n''))\} = \Phi(g\alpha(\tau)).$$

Since $F(g\alpha(t)) \neq 0$ for $t > 0$, $v(\alpha(\tau + t_n')) \neq v(\alpha(\tau + t_n''))$ and there exists a number $\gamma > 0$ such that:

$$|v(\alpha(\tau + t_n')) - v(\alpha(\tau + t_n''))| \geq \gamma$$

for all sufficiently large subscripts n. But this leads to a contradiction. Thus the function $u(t) = v(\alpha(t))$ can be extended by continuity onto G. Hence $u(t)$ is a Bohr almost periodic function. ∎

4·23.27 We shall construct now a Bohr almost periodic function $f:\mathbb{R} \to \mathbb{C}$ which satisfies the hypotheses of Lemma 4·23.26 and which has the property that the function $u(t) = \sqrt{f(t)}$ is not Levitan almost periodic.

Let G be the two-dimensional torus, $\alpha:\mathbb{R} \to G$ a dense one-parameter subgroup and $(G,\mathbb{R},\sigma)$ the flow defined by

$$(g,t)\sigma = g\alpha(t), \qquad (t \in \mathbb{R}, g \in G).$$

Let g_1, g_2 be two points of the torus with distinct orbits. Each continuous function $F:G \to \mathbb{C}$ can be considered in a natural way as a vector field on G. Define a function $F:G \to \mathbb{C}$ so that: (1) $F(g_1) = F(g_2) = 0$; $F(g) \neq 0$, if $g \neq g_1$, $g \neq g_2$; (2) the degree of the singular point g_1 of the vector field F is equal to 1 and the degree of g_2 is equal to (-1). It is easy to see that such functions exist.

Let g_0 be a point of G outside the orbits of the points g_1, g_2 in the dynamical system $(G,\mathbb{R},\sigma)$. The function $u(t) = \sqrt{F(g_0\alpha(t))}$ is bounded and uniformly continuous. Let $X = H(F(g_0\alpha(t)),u(t))$, *i.e.*, X is the orbit closure of the pair of functions $F(g_0\alpha(t))$ and $u(t)$ in the corresponding shift transformation group. There exists a homomorphism p of the flow $(X,\mathbb{R})$ onto the flow $(G,\mathbb{R},\sigma)$ which sends the point $(F(g_0\alpha(t)),u(t))$ to the point g_0. The fibre X_0 over the point g_0 consists of no more than two points (which correspond to different branches of the function $\sqrt{F(g_0\alpha(t))}$. It follows from the choice of the point g_0 and from property (1) of the function F that if X_0 contains two points, then they are distal. Therefore X is a minimal set.

We shall prove that the fibre X_0 consists in fact of two points. Suppose the contrary holds. Then $p:X \to G$ is an almost automorphic extension; hence $(F(g_0\alpha(t)),u(t))$ is Levitan almost periodic and we deduce from Lemma 4·23.26 that $u(t)$ is Bohr almost periodic. Hence it follows that X is an almost periodic minimal set, each fibre consists of a single point, and the function $\sqrt{F(g_0\alpha(t))}$ can be continuously extended to the whole torus G. But this is impossible because the increase of the argument of the function F after a turn around the point g_1 along a small curve is equal to 2π. This contradiction shows that the fibre X_0 consists of two distal points, and therefore $(F(g_0\alpha(t)),-u(t)) \in X$. Since X is a minimal set, we conclude that $(k(t),\ell(t)) \in X$ implies $(k(t),-\ell(t)) \in X$. Identifying such pairs of points we obtain a minimal set Y^*. The canonical map $s:X \to Y^*$ is a twofold covering. The projection on the first coordinate determines a homomorphism $r:(Y^*,\mathbb{R}) \to (G,\mathbb{R},\sigma)$. Clearly, r is an almost automorphic extension. Since the function $u(t)$ is not Bohr almost automorphic, the map r is not injective. Thus the extension $p:(X,\mathbb{R}) \to (G,\mathbb{R})$ is decomposed into a twofold covering s and an almost automorphic extension r. The last statement is in accordance with the theorem on the structure of extensions with a finite fibre (Theorem 3·13.18 and Remark 3·13.21).

REMARKS AND BIBLIOGRAPHICAL NOTES

Subsections 4·23.1-9 contain an exposition of results due to Žikov [*2*] (with some generalizations and improvements). In connection with Theorems 4·23.6,7 it is natural to suggest the following conjecture: if an equation

$$\frac{dx}{dt} = f(x,t), \qquad (x \in \mathbb{R}^n, t \in \mathbb{R})$$

with right side almost periodic in t satisfies the conditions on uniform positive stability and positive dissipativity, then it has at least one almost periodic solution. As was shown by Fink and Frederickson [*1*], dissipation alone does not imply the existence of almost periodic solutions. At the same time it is well known that in the periodic case, dissipativity ensures the existence of a periodic solution. It follows from the dissipation condition that $X_{0\omega}$ is a retract of some ball of $\mathbb{R}^n$. By Theorem 4·23.2, in order to prove the above conjecture it will suffice to prove the following statement: each compact connected topological group G of homeomorphisms of a compact contractible subset A of the space $\mathbb{R}^n$ has a fixed point. It is easy to verify that this statement is true for commutative groups, but in the general case the answer is unknown to the author. Conner and Montgomery [*1*] have shown that the group $SO(3)$ can act on $\mathbb{R}^{12}$ without fixed points. This shows that if the assumption on compactness of A is dropped, the statement becomes false.

Theorem 4·23.8 was proved by Amerio [*1,2*]. The material presented in 4·23.10-18 is taken from the paper by Bronšteĭn and Černii [*1*]. The relationship between various types of stability and the existence of almost periodic solutions of differential equations was studied by many authors (Deysach and Sell [*1*], Fink [*1,2*], Fink and Seifert [*1*], Miller [*1*], Seifert [*1-6*], Yoshizawa [*1-3*] and others). Their proofs are based either on considerations similar to the one at the end of Subsection 4·23.19 or on the fact that uniform asymptotic stability implies (under some additional conditions) total stability (Barbašin [*3*], Kato and Yoshizawa [*1*]). The proofs of Theorems 4·23.20,21 were found independently by Žikov. Relations between almost periodic differential equations and almost periodic flows were investigated by Cartwright [*1-3*]. Examples presented in Subsection 4·23.22 are due to Bohr [*2*]. Asymptotically almost periodic functions are introduced by Fréchet [*1*]. Theorem 4·23.24 was proved by Bronšteĭn [*19*]. Theorem 4·23.25 is the main result of the papers by Walters [*1*] and Bohr and Flanders [*1*]. The proof of this theorem based on the theory of distal extensions is due to Žikov [*2*]. So are Lemma 4·23.26 and Example 4·23.27.

§(4·24): LINEAR EXTENSIONS OF DYNAMICAL SYSTEMS AND LINEAR DIFFERENTIAL EQUATIONS

We shall introduce here and study linear extensions of semigroup dynamical systems. We shall extend to such extensions the well known minimax method due to Favard. Several general theorems on the existence of proximal and almost periodic extensions will be proved. These lead to some corollaries on synchronous solutions of linear differential equations. We investigate conditions under which the primitive of a recurrent (in particular, almost periodic) function is isochronous with the given function. Results due to Bohl, Bohr and Kadets on this subject will be presented.

4·24.1 Let X be a topological space, E a linear topological space over the field of real or complex numbers and $\sigma: X \times E \to X$ a continuous action of the additive group E on X. Henceforth it is assumed that the action σ is free, *i.e.*, $x\sigma^a \neq x$ for $a \neq 0$ and all $x \in X$.

Denote $R = \{(x, x\sigma^a) \mid x \in X, a \in E\}$. If $x_1, x_2 \in R$, then there is a uniquely defined element $\eta(x_1, x_2) \in E$ satisfying the condition

$$x_1\sigma^{\eta(x_1,x_2)} = x_2.$$

The function $\eta:R \to E$ is called the shift map. Let $Y \equiv X/R = \{xE|x \in X\}$ be the space of E-orbits, *i.e.*, the quotient space of X with respect to the equivalence relation R, and let $p:X \to X/R$ denote the canonical projection. Then (X,p,Y) is called an *E-EXTENSION*. If the shift map $\eta:R \to E$ is continuous, then the E-extension is called a *PRINCIPAL E-EXTENSION* . Each fibre $X_y = \{x|x \in X, xp = y\}$ $(y \in Y)$ is in this case homeomorphic to the space E. Indeed, if x_0 is any fixed point of the fibre X_y, then the map $x \mapsto \eta(x_0,x)(x \in X_y)$ is a homeomorphism.

Let (X,p,Y) be a principal E-extension, S a topological semigroup with identity e and (X,S,π) a transformation semigroup satisfying the condition

$$xE\pi^s = x\pi^s E, \qquad (x \in X, s \in S). \tag{24.1}$$

Since $p:X \to Y$ is an open map, (X,S,π) induces a transformation semigroup (Y,S,ρ). For $x \in X$, $s \in S$ and $a \in E$ there exists by (24.1) an element $\xi(x,s,a) \in E$ such that

$$x\sigma^a\pi^s = x\pi^s\sigma^{\xi(x,s,a)}. \tag{24.2}$$

Indeed, $\xi(x,s,a) = \eta(x\pi^s, x\sigma^a\pi^s)$. It is easily seen that $\xi:X \times S \times E \to E$ is a continuous map.

For each element $b \in E$ we have by (24.2) that

$$x\sigma^{a+b}\pi^s = x\pi^s\sigma^{\xi(x,s,a+b)}. \tag{24.3}$$

On the other hand,

$$x\sigma^{a+b}\pi^s = x\sigma^a\pi^s\sigma^{\xi(x\sigma^a,s,b)} = x\pi^s\sigma^{\xi(x,s,a)}\sigma^{\xi(x\sigma^a,s,b)}. \tag{24.4}$$

Since the action σ is free, equating (24.3) and (24.4) yields the identity

$$\xi(x,s,a + b) = \xi(x,s,a) + \xi(x\sigma^a,s,b), \quad (x \in X, s \in S; a,b \in E). \tag{24.5}$$

Consider the special case when

$$\xi(x,s,a) = \xi(x\sigma^b,s,a), \qquad (x \in X, s \in S; a,b \in E).$$

Then we can define a function $\xi_1:Y \times S \times E \to E$:

$$\xi_1(y,s,a) = \xi(x,s,a), \qquad (y \in Y, s \in S, a \in E, x \in X_y).$$

It follows from the continuity of ξ and the openness of the map $p:X \to Y$ that ξ_1 is also continuous. For $y \in Y$ and $s \in S$ define a map $A(y,s):E \to E$ by

$$A(y,s)a = \xi_1(y,s,a), \qquad (a \in E).$$

(24.5) implies the following identity:

$$A(y,s)(a + b) = A(y,s)a + A(y,s)b, \qquad (a,b \in E; y \in Y, s \in S). \tag{24.6}$$

If E is a linear space over $\mathbb{R}$, then it follows from the continuity of $A(y,s):E \to E$ and from (24.6) that $A(y,s):E \to E$ is a linear map $(y \in Y, s \in S)$. If E is considered over the field of complex numbers, we also have to require that

$$A(y,s)(ia) = iA(y,s)a, \qquad (y \in Y, s \in S, a \in E),$$

where i is the imaginery identity. Then $A(y,s)$ is a continuous linear map and

$$x\sigma a\pi^s = x\pi^s\sigma A(y,s)a, \qquad (y \in Y, x \in X_y, s \in S, a \in E), \tag{24.7}$$

in view of (24.2). Hence it follows that

$$A(y,s_1s_2) = A(y\rho^{s_1},s_2)\circ A(y,s_1), \qquad (y \in Y; s_1,s_2 \in S). \tag{24.8}$$

If S is a group, then also the following identities hold:

$$A(y,e) = I, \qquad [A(y,s)]^{-1} = A(y\rho^s,s^{-1}), \qquad (y \in Y, s \in S), \tag{24.9}$$

where e is the identity of the group S and I is the identity map of E.

Let $L(E,E)$ denote the set of all continuous linear maps of E into E. If there is a map $A:Y \times S \to L(E,E)$ satisfying the identity (24.8), then $p:(X,S,\pi) \to (Y,S,\rho)$ is called an *E-LINEAR EXTENSION*.

Since the map $(y,s,a) \mapsto A(y,s)a$ is continuous and satisfies (24.8), the map

$$(y,a)\pi_1^s = (y\rho^s, A(y,s)a), \qquad (y \in Y, s \in S, a \in E) \tag{24.10}$$

determines a transformation semigroup $(Y \times E,S,\pi_1)$. The extension $q:(Y \times E,S,\pi_1) \to (Y,S,\rho)$, where $q:Y \times E \to Y$ denotes the canonical projection, is E-linear and it is called the *HOMOGENEOUS LINEAR EXTENSION CORRESPONDING TO THE E-LINEAR EXTENSION* $p:(X,S,\pi) \to (Y,S,\rho)$. This agrees with the terminology adopted in the theory of linear differential equations. Yet, it would be more precise to use the term *affine* for linear extensions and the term *linear* for homogeneous linear extensions.

4·24.2 We shall now consider an important particular case of E-extensions. Let (Y,S,ρ) be a given dynamical system, $X = Y\times E$ and let $p:X\to Y$ be the canonical projection. Suppose that there is defined an E-linear extension $p:(X,S,\pi)\to(Y,S,\rho)$ with the natural action of the additive group E on $X = Y\times E$:

$$(y,a)\sigma^b = (y,a+b), \qquad (y\in Y; a,b\in E).$$

The equality

$$(y,0)\pi^s = (y\rho^s, f(y,s)), \qquad (y\in Y, s\in S) \tag{24.11}$$

determines a continuous map $f:Y\times S\to E$ such that the following identity holds:

$$f(y,s_1s_2) = f(y\rho^{s_1},s_2) + A(y\rho^{s_1},s_2)f(y,s_1). \tag{24.12}$$

It follows from (24.11) and (24.12) that

$$(y,a)\pi^s = (y\rho^s, f(y,s) + A(y,s)a), \qquad (y\in Y; a\in E, s\in S). \tag{24.13}$$

Conversely, let $f:Y\times S\to E$ and $A:Y\times S\to L(E,E)$ be functions satisfying the equalities (24.12) and (24.8). If the maps $(y,s)\mapsto f(y,s)$ and $(y,s,a)\mapsto A(y,s)a$ $(y\in Y, s\in S, a\in E)$ are continuous, then the formula (24.13) determines a transformation semigroup $(Y\times E,S,\pi)$ such that $q:(Y\times E,S,\pi)\to(Y,S,\rho)$ is an E-linear extension. The latter will be called a *SKEW PRODUCT EXTENSION*.

4·24.3 LEMMA: *Let E be a quasi-complete linear topological space, $p:(X,S,\pi)\to(Y,S,\rho)$ an E-linear extension and let $x_0\in X$. Suppose that there exists a non-empty compact set $K\subset Y$ such that $Kp^{-1}\cap\overline{x_0S}$ is compact. Let $F(y)$ denote the closed convex hull of the set $\overline{x_0S}\cap X_y$ in the affine space X_y. Then the set $F = \bigcup_{y\in K} F(y)$ is compact and $\overline{F(y)S}\cap X_y\subset F(y)$ $(y\in K\cap\overline{x_0Sp})$.*

Proof: Let $y\in K$ and let U be an arbitrary neighbourhood of the set $F(y)$. Let us prove that there exists a neighbourhood V of y such that $F(z)\subset U$ $(z\in V\cap K)$.

Suppose that, to the contrary, then there exists a net $\{y_\alpha\}$ such that $\{y_\alpha\}\subset K$, $\{y_\alpha\}\to y$, but

$$F(y_\alpha)\cap(X\setminus V_0) \neq \emptyset \tag{24.14}$$

for some fixed neighbourhood U_0 of $F(y)$. Choose arbitrary points

$x_\alpha \in \overline{x_0S} \cap X_{y_\alpha} \subset \overline{x_0S} \cap Kp^{-1}$. Without loss of generality we may assume that the net $\{x_\alpha\}$ converges to some point $x^* \in \overline{x_0S} \cap X_y$.

Let $N(x) = \{\eta(x,z) \mid z \in \overline{x_0S}, xp = zp\}$ $(x \in \overline{x_0S} \cap Kp^{-1})$. The set $N(x)$ is compact because it is homeomorphic to the compact set $\overline{x_0S} \cap X_{xp}$. Since the space E is quasi-complete the closed convex hull $\overline{\text{co}}N(x)$ of $N(x)$ in E is compact (Bourbaki [*1*]). Further, it follows from the definition of the set $F(y)$ that

$$(x, \overline{\text{co}}N(x))\sigma = F(xp), \qquad (x \in \overline{x_0S} \cap Kp^{-1}).$$

For the neighbourhood U_0 of $F(y)$ we can find a convex neighbourhood W of the zero element of the space E and a neighbourhood V_0 of the point x^* such that

$$(z, \overline{\text{co}}N(x^*) + W + W)\sigma \subset U_0, \qquad (z \in V_0 \cap \overline{x_0S} \cap Kp^{-1}).$$

Hence for sufficiently large α:

$$(x_\alpha, \overline{\text{co}}N(x^*) + W + W)\sigma \subset U_0.$$

Since $\{x_\alpha\} \to x^*$ and the set $\overline{x_0S} \cap Kp^{-1}$ is compact, we have $N(x_\alpha) \subset N(x^*) + W$ for all sufficiently large α. Hence it follows that $\text{co}N(x_\alpha) \subset \text{co}N(x^*) + W$. Therefore

$$\overline{\text{co}}N(x_\alpha) \subset \overline{\text{co}}N(x^*) + W + W.$$

Thus there exists an index α_0 such that

$$F(y_\alpha) = F(x_\alpha p) = (x_\alpha, \overline{\text{co}}N(x_\alpha))\sigma \subset U_0$$

for all $\alpha > \alpha_0$. But this contradicts (24.14).

Thus for every point $y \in K$ and every neighbourhood U of $F(y)$ there exists a neighbourhood V of y such that $F(z) \subset U$ for all $z \in V \cap K$. Since the set K is compact and each fibre $F(y)$ of the continuous map $p: F \to K$ is compact, the set F is also compact. Since the set $\overline{x_0S}$ is invariant and the semigroup S acts affinely on the fibres X_y, we have $F(y)\pi^s \subset F(y\rho^s)$ $(y \in \overline{x_0Sp} \cap K)$. It follows from the above considerations that $\overline{F(y)S} \cap X_y \subset F(y)$ for all $y \in K \cap \overline{x_0Sp}$. ∎

4·24.4 COROLLARY: *Let E be a quasi-complete linear topological space, and let $p:(X,S,\pi) \to (Y,S,\rho)$ be an E-linear extension. Let $x_0 \in X$ be such that the set x_0S is compact. Then the set $F = \cup[F(y) \mid y \in \overline{x_0pS}]$ is compact and invariant.*

Proof: This follows directly from Lemma 4·24.3, if we take K to be equal to $\overline{x_0pS}$. ∎

4·24.5 THEOREM: *Let E be a quasi-complete linear topological space, $p:(X,S,\pi) \to (Y,S,\rho)$ an E-linear extension, $x_0 \in X$ and $y_0 = x_0p$. Suppose further that for some neighbourhood V_0 of y_0, the set $\bar{V}_0p^{-1} \cap \overline{x_0S}$ is compact. Then there exists a* *point* $z \in F(y_0) = \overline{co}(\overline{x_0 S} \cap X_{y_0})$ *such that the extension* p: $(\overline{zS}, S, \pi) \to (\overline{y_0 S}, S, \rho)$ *is proximal in the fibre over the point* y_0.

Proof: Apply Lemma 4·24.3 with K equal to the set $(\bar{V}_0 p^{-1} \cap \overline{x_0 S})p$. Then the set $F = \cup [F(y) | y \in (\bar{V}_0 p^{-1} \cap \overline{x_0 S})p]$ is compact. Let $\mathcal{A}$ denote the family of all non-empty convex compact subsets $D \subset F(y_0)$ which satisfy the condition

$$\overline{DS} \cap X_{y_0} \subset D.$$

By Lemma 4·24.3, $F(y_0) \in \mathcal{A}$. It is easy to verify that the set $\mathcal{A}$ is inductive. Hence, by Zorn's Lemma, there exists a minimal element $D_0 \in \mathcal{A}$.

We shall prove that the extension $p:(\overline{D_0 S}, S, \pi) \to (\overline{y_0 S}, S, \rho)$ is proximal in the fibre over y_0. Let $x_1, x_2 \in \overline{D_0 S} \cap X_{y_0}$. Then $x_1 x_2 \in D_0$. Let us prove that the points x_1 and x_2 are S-proximal. Suppose the contrary holds. Put $z \equiv \frac{1}{2}(x_1 + x_2)$. Let w be an arbitrary point in $\overline{zS} \cap X_{y_0} \subset D_0$. Then there exists a net $\{s_\alpha\}$ such that $w = \lim\{z\sigma^{s_\alpha}\}$. Hence $\{z\sigma^{s_\alpha} p\} \to y_0$. Since the set F is compact, $x_1, x_2 \in D_0 \subset F(y_0)$ and $\{x_1 s_\alpha p\} = \{x_2 s_\alpha p\} = \{z s_\alpha p\} \to y_0$, we have $x_1 s_\alpha \in F$, $x_2 s_\alpha \in F$ for all sufficiently large α. Therefore we may assume without loss of generality that $\{x_1 s_\alpha\} \to z_1$, $\{x_2 s_\alpha\} \to z_2$, where $z_1 z_2$ are some points from $\overline{D_0 S} \cap X_{y_0} \subset D_0$. Since the semigroup S acts affinely on the fibres, $w = \frac{1}{2}(z_1 + z_2)$. We note that $z_1 \neq z_2$ because x_1 and x_2 are supposed to be distal. Thus we have shown that none of the points $w \in \overline{zS} \cap X_{y_0}$ is extremal for the set D_0. It is known (*cf.* Dunford and Schwartz [*1*]) that if Q is a compact subset of E and the closed convex hull $\overline{co}Q$ is compact, then the extremal points of $\overline{co}Q$ are contained in Q. Therefore the compact set $\overline{co}(\overline{zS} \cap X_{y_0}) \subset D_0$ does not contain extremal points of the set D_0. In fact, if x is an extremal point of D_0 and $x \in \overline{co}(\overline{zS} \cap X_{y_0})$, then x is an extremal point of $\overline{co}(\overline{zS} \cap X_{y_0})$, and hence $x \in \overline{zS} \cap X_{y_0}$, which is impossible, as was shown above. But D_0 has extremal points (see Dunford and Schwartz [*1*], p.476); hence $\overline{co}(\overline{zS} \cap X_{y_0})$ is a proper subset of D_0. Since the set F is compact and $z \in F(y_0)$, $Kp^{-1} \cap \overline{zS}$ is a closed subset of F. Hence we may apply Lemma 4·24.3 to the point z. We obtain $\overline{QS} \cap X_{y_0} \subset Q$, where $Q = \overline{co}(\overline{zS} \cap X_{y_0})$. Thus $Q \in \mathcal{A}$, $Q \subset D_0$ and $Q \neq D_0$. But this contradicts the minimality of D_0. ▮

4·24.6 COROLLARY: *Suppose that all the conditions of the preceding theorem are satisfied. Then either there exists a point* $z \in F(y_0)$ *such that the extension* $p:\overline{zS} \to \overline{y_0 S}$ *is almost automorphic or there exists a point* $a_0 \in E$, $a_0 \neq 0$, *such that for every neighbourhood* $V = V(0)$ *we can choose an element* $s \in S$ *with* $A(y_0, s)a_0 \in V(0)$.

Proof: If $z_1, z_2 \in \overline{zS} \cap X_{y_0}$ and $z_1 \neq z_2$, then set $a_0 = \eta(z_1, z_2)$. There exists a net $\{s_\alpha\}$, $s_\alpha \in S$, such that $\lim\{z_1 s_\alpha\} = \lim\{z_2 s_\alpha\}$. Therefore

$$\{A(y_0, s_\alpha)a_0\} = \{\eta(z_1 s_\alpha, z_2 s_\alpha)\} \to 0. \blacksquare$$

In the case of a skew product extension Theorem 4·24.5 can be reformulated as follows.

4·24.7 THEOREM: *Let E be a quasi-complete linear topological space, (Y,S,ρ) a dynamical system, $X = Y \times E$, $p:(X,S,\pi) \to (Y,S,\rho)$ an E-linear skew product extension,* $y_0 \in Y$ and $a_0 \in E$. *Suppose there exists a neighbourhood V_0 of the point y_0 such that the set $\overline{V_0 \cap \overline{y_0 S}}$ is compact. If, moreover, there exists a compact set $K \subset E$ with*

$$\{f(y_0,s) + A(y_0,s)a_0 \mid s \in S, y_0 s \in V_0\} \subset K,$$

then there is a point $b \in \overline{\text{co}}K$ such that the extension

$$p:(\overline{(y_0,b)S}, S, \pi) \to (\overline{y_0 S}, S, \rho)$$

is proximal in the fibre over y_0. If this fibre contains more than one point, then $\{A(y_0,s)c_0 \mid s \in S\} \cap V(0) \neq \emptyset$ for some element $c_0 \in E$, $c_0 \neq 0$, and every neighbourhood $V(0)$. $\blacksquare$

4·24.8 Now we shall assume that E is a uniformly convex Banach space and E^σ is the linear topological space obtained from the linear system E by endowing it with the weak topology $\sigma \equiv \sigma(E,E')$ (Dunford and Schwartz [*1*]). Let $X = (X,\tau)$ be a topological space and $p:((X,\tau),S,\pi) \to (Y,S,\rho)$ an E-linear extension. Suppose further that we have a topology τ^σ on X such that $p: ((X,\tau^\sigma),S,\pi) \to (Y,S,\rho)$ is an E^σ-linear extension. This situation occurs, for example, when we are considering E-linear skew product extensions $p:(Y \times E,S,\pi) \to (Y,S,\rho)$. In this case $(X,\tau) = Y \times E$ and $(X,\tau^\sigma) = Y \times E^\sigma$. Under the above circumstances the following theorem holds.

4·24.9 THEOREM: *Let $a \in X$ and $b = ap$. Suppose that for some neighbourhood V_0 of a, the set*

$$\{\eta(a_1,z) \mid a_1 \in V_0, z \in aS, (a_1,z) \in R\} \tag{24.15}$$

is bounded. Then there is a point c contained in the closed convex hull F of the set $\overline{aS}^\sigma \cap X_b$ (where $\overline{aS}^\sigma$ denotes the closure of aS in the space (X,τ^σ)) such that

$$\inf_{s \in S} \|\eta(x_1 \pi^s, x_2 \pi^s)\| = 0$$

for every $x_1, x_2 \in \overline{cS}^\sigma \cap F$.

Proof: Since the map $p:(X,\tau^\sigma) \to Y$ is open, the function $\eta:(R,\tau^\sigma \times \tau^\sigma) \to E^\sigma$ is continuous and the norm in E^σ is lower semi-continuous, we conclude that the set

$$\{\eta(a_1,z) \mid a_1 \in V_0, z \in \overline{aS^\sigma}, (a_1,z) \in R\} \tag{24.16}$$

is bounded. Recall that the fibre (X_b,τ^σ) is homeomorphic to E^σ. It follows from the boundedness of the set (24.16) that the set $\overline{aS^\sigma} \cap X_b$ is bounded (and weakly closed), hence $\overline{aS^\sigma} \cap X_b$ is weakly compact. By the Kreĭn-Šmulian theorem (see: Dunford and Schwartz [*1*]) the closed convex hull F of the set $\overline{aS^\sigma} \cap X_b$ is weakly compact. Let $x \in X$. Put

$$M(x) = \sup\{\|\eta(a_1,z)\| \mid a_1 \in V_0, z \in xS, (a_1,z) \in R\}$$

$$= \sup\{\|\eta(a_1,z)\| \mid a_1 \in V_0, z \in \overline{xS^\sigma}, (a_1,z) \in R\}, \tag{24.17}$$

$$M = \inf_{x \in F} M(x). \tag{24.18}$$

Since $a \in F$ and the set (24.16) is bounded, we have $M < \infty$.

We shall prove that there exists a point $c \in F$ with $M(c) = M$. Indeed, let $\{x_n\}$, ($x_n \in F$), be a minimizing sequence, *i.e.*,

$$M \leqslant \sup\{\|\eta(a_1,z)\| \mid a_1 \in V_0, z \in x_n S, (a_1,z) \in R\} \leqslant M + 1/n. \tag{24.19}$$

Since the set F is weakly compact, there exists a subsequence of $\{x_n\}$ which is weakly convergent to a point $c \in F$. We prove that $M(c) = M$. Suppose the contrary holds. Then there exists a number $\delta_0 > 0$ such that $M(c) > M + \delta_0$. We can choose an element $s^* \in S$ and a point $a^* \in V_0$ so that $\|\eta(a^*,cs^*)\| > M + \delta_0$. Since the map $p:X^\sigma \to Y$ is open, the function $\eta:(R,\tau^\sigma \times \tau^\sigma) \to E^\sigma$ is continuous and the norm is lower semi-continuous on E^σ, we can choose an arbitrarily large subscript n and a point $a_n \in V_0$ in such a way that $(a_n,x_n s^*) \in R$ and $\|\eta(a_n,x_n s^*)\| > M + \delta_0 > M + 1/n$. But this contradicts (24.19). Thus we have proved the existence of a point $c \in F$ for which

$$M = \sup\{\|\eta(a_1,z)\| \mid a_1 \in V_0, z \in \overline{cS^\sigma}, (a_1,z) \in R\}.$$

Let $x \in \overline{cS^\sigma} \cap F$. Then $M(x) \geqslant M$, by the definition of M. On the other hand, $x \in \overline{cS^\sigma}$. Hence $\overline{xS^\sigma} \subset \overline{cS^\sigma}$, and consequently $M(x) \leqslant M(c) = M$. Thus $M(x) = M$ for all $x \in \overline{cS^\sigma} \cap F$.

Let $x_1, x_2 \in \overline{cS^\sigma} \cap F$ and let $x_3 = \frac{1}{2}(x_1 + x_2)$. Then $x_3 \in F$ by the convexity of F. Hence $M(x_3) \geqslant M$, and thus for every positive integer n there exists a point $a_n \in V_0$ and an element $s_n \in S$ such that $(a_n, x_3 s_n) \in R$ and

$$\|\eta(a_n, x_3 s_n)\| > M - 1/n.$$

Since the semigroup S acts affinely on the fibres, we have

$$x_3\pi^{s_n} \equiv x_3s_n = \tfrac{1}{2}(x_1s_n + x_2s_n).$$

Therefore

$$\eta(a_n,x_3s_n) = \tfrac{1}{2}[\eta(a_n,x_1s_n) + \eta(a_n,x_2s_n)].$$

Let $p_n = \eta(a_n,x_1s_n)$, $q_n = \eta(a_n,x_2s_n)$. Then $\|p_n\| \leqslant M$, $\|q_n\| \leqslant M$, $\tfrac{1}{2}\|p_n + q_n\| = \|\eta(a_n,x_3s_n)\| \geqslant M - 1/n$.

It follows from the uniform convexity of E that for every $\varepsilon > 0$ there exists a $\delta > 0$ such that if $x,y \in E$, $\|x\| \leqslant M$, $\|y\| \leqslant M$ and $\tfrac{1}{2}\|x + y\| \geqslant M - \delta$, then $\|x - y\| \leqslant \varepsilon$. Thus

$$\|p_n - q_n\| = \|\eta(x_1\pi^{s_n},x_2\pi^{s_n})\| \to 0, \qquad (n \to +\infty). \blacksquare$$

4·24.10 THEOREM: *Let E be a uniformly convex Banach space, (Y,S,ρ) a dynamical system, Y compact, $X = Y \times E$, $p:(X,S,\pi) \to (Y,S,\rho)$ an E-linear skew product extension, $y_0 \in Y$ and $a_0 \in E$. Suppose that there is a neighbourhood V_0 of y_0 such that the set*

$$\{f(y_0,s) + A(y_0,s)a_0 \mid s \in S, y_0s \in V_0\}$$

is bounded. Then there is a point $b \in E$ such that either the extension $p:(\overline{(y_0,b)S},S,\pi) \to (\overline{y_0S},S,\rho)$ is almost automorphic or some point $c_0 \in E$, $c_0 \neq 0$, satisfies the condition

$$\inf_{s\in S} \|A(y_0,s)c_0\| = 0.$$

4·24.11 Let E be a Banach space and let $q:(Y \times E,S,\pi_1) \to (Y,S,\rho)$ be a homogeneous E-linear extension. We shall say that this extension satisfies the *condition on SEPARATION FROM ZERO* of compact (bounded) non-trivial motions in the fibre over $y \in Y$ if for each element $\ell \in E$, $\ell \neq 0$, the compactness (boundedness) of the set $\{A(y,s)\ell \mid s \in S\}$ implies

$$\inf_{s\in S} \|A(y,s)\ell\| > 0.$$

It is clear how one extends this definition to the case of a linear topological space E.

4·24.12 COROLLARY: *Let E be a quasi-complete linear topological space, (Y,S,ρ) a dynamical system, $q:(Y \times E,S,\pi_1) \to (Y,S,\rho)$ an E-linear skew product extension, $y_0 \in Y$* and *$a_0 \in E$. Suppose that the sets $\overline{y_0S}$ and $\overline{\{f(y_0,s) + A(y_0,s)a_0 \mid s \in S\}}$ are compact. If the non-trivial compact motions of the corresponding homogeneous extension are separated from zero in the fibre over the point y_0, then there is a point $b \in E$ such that the extension $q:(\overline{(y_0,b)S},S,\pi_1) \to (\overline{y_0S},S,\rho)$ is injective* *at the point* (y_0,b). *If, moreover, the set* $\overline{y_0S}$ *is minimal and the condition on separation from zero is satisfied for all points* $y \in \overline{y_0S}$, *then there is a point* $b \in E$ *such that the extension* $q:(\overline{(y_0,b)S},S,\pi_1) \to (\overline{y_0S},S,\rho)$ *is an isomorphism.*

Proof: The first assertion is an immediate consequence of Theorem 4·24.7. The second assertion follows from the first one by virtue of Corollary 3·12.8.∎

4·24.13 COROLLARY: *Let* E *be a uniformly convex Banach space,* (Y,S,ρ) *a dynamical system with* Y *compact,* $q:(Y \times E,S,\pi_1) \to (Y,S,\rho)$ *an* E*-linear skew product extension and let* $y_0 \in Y$ *and* $a_0 \in E$ *be such that the set* $\{f(y_0,s) + A(y_0,s)a_0 | s \in S\}$ *is bounded. If the non-trivial bounded motions of the corresponding homogeneous extension are separated from zero in the fibre over* y_0, *then there exists a point* $b \in E$ *such that the extension* $q:\overline{(y_0,b)S^\sigma} \to \overline{y_0S}$ *is one-to-one at the point* (y_0,b). *If, moreover,* $\overline{y_0S}$ *is a minimal set and the condition on separation from zero is satisfied for all points* $y \in \overline{y_0S}$, *then* $q:(\overline{(y_0,b)S^\sigma},S,\pi_1) \to (\overline{y_0S},S,\rho)$ *is an isomorphism.*

Proof: This follows directly from Theorem 4·24.9.∎

4·24.14 LEMMA: *Let* E *be a Banach space, let* $q:(Y \times E,S,\pi_1) \to (Y,S,\rho)$ *be a homogeneous* E*-linear extension. If the set* $E_1 = \{\ell | \ell \in E, \sup_{s \in S} \|A(b,s)\ell\| < \infty\}$ *is closed for some point* $b \in Y$, *then* $q:(\overline{(b,\ell)S},S,\pi_1) \to (Y,S,\rho)$ *is stable in the fibre over the point* $b (\ell \in E_1)$.

Proof: Clearly E_1 is a linear system and since E_1 is closed, it is a subspace of the Banach space E. By the Banach-Steinhaus theorem (see: Dunford and Schwartz [*1*] p.52), the family $\{A(b,s): E_1 \to E | s \in S\}$ is equicontinuous.∎

4·24.15 COROLLARY: *Let* E *be a Euclidean space and let* $q:(Y \times E,S,\pi_1) \to (Y,S,\rho)$ *be an* E*-linear homogeneous extension. If* $b \in Y$ *and* $\ell \in E$ *are such that* $\{A(b,s)\ell | s \in S\}$ *is bounded, then the extension* $q:(\overline{(b,\ell)S},S,\pi_1) \to (Y,S,\rho)$ *is stable in the fibre over the point* b.

4·24.16 Now we shall apply the above results to linear differential equations. Let E be a Banach space, $A \in C(\mathbb{R},L(E,E))$ and $f \in C(\mathbb{R},E)$. The differential equation

$$\frac{dx}{dt} = A(t)x + f(t) \qquad (24.20)$$

determines an E-linear skew product extension in the following manner. Let $Y = H(A,f)$ and let $(Y,\mathbb{R},\rho)$ be the shift dynamical system. The map $\pi:Y \times E \times \mathbb{R} \to Y \times E$ defined by $(B,g,x,t)\pi = (B_t,g_t,\varphi(t,x,B,g))((B,g) \in H(A,f), x \in E, t \in \mathbb{R})$, where $\varphi(t,x,B,g)$ denotes the solution of the equation

$$\frac{dx}{dt} = B(t)x + g(t), \tag{24.21}$$

with the initial condition $\varphi(0,x,B,g) = x$, determines a transformation group $(Y \times E, \mathbb{R}, \pi)$. The canonical projection $p: Y \times E \to Y$ is an E-linear skew product extension. The corresponding homogeneous extension is determined by the homogeneous equations

$$x' = B(t)x, \tag{24.22}$$

where $B \in H(A)$.

4•24.17 THEOREM: *Suppose that the functions A and f are Lagrange stable, there exists a compact solution $\varphi: \mathbb{R} \to E$ of* (24.20) *and each non-trivial compact solution ψ of the homogeneous equation corresponding to* (24.20) *is separated from zero (i.e., $\inf_{t \in \mathbb{R}} \|\psi(t)\| > 0$). Then Equation* (24.20) *has at least one synchronous solution. If the pair (A,f) is recurrent in the sense of Birkhoff, then the same is true even if the separation condition is replaced by the semiseparation condition (i.e., $\inf_{t \geq 0} \|\psi(t)\| > 0$). If, moreover, the condition on semiseparation of the compact solutions is satisfied for all equations of the form* (24.22) *with $B \in H(A)$, then* (24.20) *has a uniformly synchronous solution.*

Proof: This follows from Corollary 4•24.12 and Lemmas 4•22.10,4,12. ▌

4•24.18 A solution $\psi: \mathbb{R} \to E$ of Equation (24.20) is said to be a *WEAKLY SYNCHRONOUS SOLUTION* if for every neighbourhood U of $\psi \in C(\mathbb{R}, E^\sigma)$ (where E^σ denotes the linear topological space obtained from the linear system E by endowing it with the weak topology $\sigma(E,E')$) we can choose a number $\delta > 0$ and a segment $I \subset \mathbb{R}$ so that

$$\|A(t) - A(t+\tau)\| < \delta, \qquad \|f(t) - f(t+\tau)\| < \delta, \qquad (t \in I)$$

implies $\psi_\tau \in U$. The notion of a uniformly weakly synchronous solution can be introduced in a similar manner.

4•24.19 THEOREM: *If E is a uniformly convex Banach space, the functions A and f are Lagrange stable, $\varphi: \mathbb{R} \to E$ is a bounded solution of Equation* (24.20) *and each non-trivial bounded solution ψ of the homogeneous equation corresponding to* (24.20) *is separated from zero, then Equation* (24.20) *has a weakly synchronous solution. If the pair (A,f) is recurrent in the sense of Birkhoff, then the separation condition can be replaced by semiseparation. If the semiseparation condition is fulfilled for all Equations* (24.22) *with $B \in H(A)$, then there exists a weakly uniformly synchronous solution of* (24.20).

Proof: This is immediate from Corollary 4·24.13. ▮

4·24.20 It is easy to show that if a weakly synchronous (weakly uniformly synchronous) solution is compact, then it is synchronous (uniformly synchronous). An important question arises in this connection. Does the existence of a bounded solution imply the existence of a compact solution? This question is far from being completely solved (see Žikov [*9*]).

4·24.21 THEOREM: *Let E be a Euclidean space. Suppose that there is at least one bounded solution of Equation* (24.20). *If the pair (A,f) is recurrent in the sense of Birkhoff, then there is a set $H_0 \subset H(A,f)$ of the first category in $H(A,f)$ such that for every $(B,g) \in H(A,f) \setminus H_0$, the Equation* (24.21) *has a synchronous solution. If the function f is recurrent and $A(t) \equiv A$, then there exists a solution of* (24.20) *which is uniformly synchronous with f.*

Proof: In view of Theorem 4·24.5, we obtain a minimal extension with a proximal fibre. By Corollary 4·24.15, this extension has a stable fibre. Applying Corollary 3·13.19, we conclude that this extension is almost automorphic. In order to prove the first statement it suffices to refer to Lemmas 3·14.32 and 4·22.10. If $A(t) \equiv A$, then the extension is not only almost automorphic, but also distal because the non-trivial bounded solutions of the equation

$$\frac{dx}{dt} = Ax, \qquad (x \in \mathbb{R}^n, t \in \mathbb{R}),$$

are almost periodic and consequently separated from zero. Hence, the minimal extension under consideration is an isomorphism and the corresponding solution is uniformly synchronous with f, by Lemma 4·22.12. All other bounded solutions of the Equation (24.20) with $A(t) \equiv A$ are recurrent. ▮

4·24.22 Now we shall consider the question on the existence of synchronous solutions of Equation (24.20) in the case when $A(t) \equiv 0$. In other words, we shall investigate conditions under which the primitive

$$F(\varphi,t) = \int_0^t \varphi(\tau)d\tau, \qquad (24.23)$$

is synchronous with the function $\varphi \in C(\mathbb{R},E)$.

It is easy to verify that the function $F: C(\mathbb{R},E) \times \mathbb{R} \to E$ defined by (24.23) is continuous and satisfies the identity $F(\varphi,t+s) = F(\varphi,t) + F(\varphi\sigma^t,s)$ $(t,s \in \mathbb{R})$. This leads to the following generalization of the above problem. Let T be an additive topological group, X a topological space, (X,T,π) a topological transformation group and E a Banach space. A continuous function $F: X \times T \to E$

is called a (*TWO-DIMENSIONAL*) *COCYCLE over* (X,T,π) (see Subsection 3·16.6) if

$$F(x,t+s) = F(x,t) + F(x\pi^t,s), \qquad (x \in X; t,s \in T). \quad (24.24)$$

For example, if $\varphi: X \to E$ is a continuous function and $T = \mathbb{R}$, then the function $F: X \times \mathbb{R} \to E$ defined by

$$F(x,t) = \int_0^t \varphi(x\pi^\tau)d\tau, \qquad (x \in X, t \in \mathbb{R}),$$

is a cocycle. In particular, the integral (24.23) may be also considered as a cocycle over the shift dynamical system $(C(\mathbb{R},E), \mathbb{R},\sigma)$. Note that the space $C(\mathbb{R},E)$ is metrizable.

We shall say that a Banach space satisfies the *BOHL-BOHR CONDITION* if for every transformation group (X,T,π) with a metrizable phase space X, every point $x \in X$ and every cocycle F, the boundedness of the map $F_x: T \to E$, $F_x(t) = F(x,t)(t \in T)$ implies that the point $F_x \in C(T,E)$ is synchronous with the point $x \in X$.

Let $\mathbb{C}$ denote the Banach space of all convergent numerical sequences $a = \{a_n | n = 1,2,\dots\}$ with the norm $||a|| = \sup_n |a_n|$.

4·24.23 LEMMA: *If the space X is compact and the function $F: X \times T \to E$ is continuous, then for every $\varepsilon > 0$ and every compact subset $K \subset T$ there exists an index $\delta \in \mathfrak{U}[X]$ such that*

$$(x_1,x_2) \in \delta \Rightarrow ||F(x_1,t) - F(x_2,t)|| < \varepsilon, \qquad (t \in K). \quad (24.25)$$

Proof: Obvious.

4·24.24 THEOREM: *A Banach space satisfies the Bohl-Bohr condition iff it does not contain a subspace isomorphic to the space* $\mathbb{C}$.

Proof: Let X be a metrizable space, (X,T,π) a transformation group and E a Banach space, which does not contain a subspace isomorphic to $\mathbb{C}$. Further, let $F: X \times T \to E$ be a cocycle over (X,T,π) and let $||F(x,t)|| < L \leqslant +\infty$ for some point $x \in X$. Let us prove that the function F_x is synchronous with the point x. Suppose the contrary holds. Then there exists a sequence $\{t_n\}$, $t_n \in T$, such that $\{x\pi^{t_n}\} \to x$, but $\{F_x(t+t_n)\}$ does not converge to $F_x(t)$ in $C(T,E)$. Since $F(x,0) = 0$, we may assume without loss of generality that $||F_x(t_n)|| > a$ for some $a \geqslant 0$. By Lemma 4·24.23, we may also suppose (by passing to a subsequence, if necessary) that

$$||F(x\pi^{t_n},\xi) - F(x,\xi)|| < 1/2^n, \quad (24.26)$$

for all $\xi \in K_{n-1}$, where K_{n-1} denotes the set of all possible sums of the form $\delta_{n-1}\cdot t_{n-1} + \dots + \delta_1 \cdot t_1$, δ_i is equal to either 0 or 1 and $0 \cdot t = 0$, $1 \cdot t = t$.

Let $\tau_k = t_{n_k}$, $(k = 1,2,\ldots)$ denote any terms of the sequence $\{\tau_n\}$. We shall apply the following identity, which is a consequence of (24.24):

$$F(x,\tau_m + \ldots + \tau_1) - F(x,\tau_m) - \ldots - F(x,\tau_1)$$

$$= \sum_{k=1}^{m-1} \{F(x\pi^{\tau_{k+1}},\tau_k + \ldots + \tau_1) - F(x,\tau_k + \ldots + \tau_1)\}. \qquad (24.27)$$

Taking (24.27) and (24.26) into account, we conclude that

$$\left\| \sum_{k=1}^{m} F(x,\tau_k) \right\| \leqslant L + 1. \qquad (24.28)$$

Thus the series $\sum_n F(x,t_n)$ is divergent, but the equality (24.28) holds for every subsequence $\{F(x,\tau_k)\}$ and every positive integer m. It follows from a theorem due to Pełczyński [*1*,*2*] that E contains a subspace isomorphic to $\mathbb{C}$, but this contradicts the hypothesis.

Define an almost periodic function $\varphi:\mathbb{R} \to \mathbb{C}$ by:

$$\varphi(t) = \{(1/n)\cos(t/n) \,|\, n = 1,2,\ldots\}.$$

The primitive

$$F(t) = \int_0^t \varphi(\tau)d\tau = \{\sin(t/n) \,|\, n = 1,2,\ldots\}$$

of this function is bounded, but it is not even Poisson stable. Hence F is not synchronous with φ. ∎

4·24.25 COROLLARY: *If the space E does not contain a subspace isomorphic to $\mathbb{C}$, $\varphi:\mathbb{R} \to E$ is a uniformly continuous function and the set $\overline{\varphi(\mathbb{R})}$ is compact, then the boundedness of the primitive* (24.23) *implies that the primitive and the function φ are both isochronous.*

Proof: Let $\{t_n\}$ be a sequence such that

$$\left\{\int_0^{t+t_n} \varphi(\xi)d\xi\right\} \to \int_0^t \varphi(\xi)d\xi \qquad (24.29)$$

in the space $C(\mathbb{R},E)$. We shall prove that $\{\varphi(t + t_n)\} \to \varphi(t)$ in $C(\mathbb{R},E)$. Suppose the contrary holds. Since the family of all shifts of the function φ in the space $C(\mathbb{R},E)$ is compact, there

Let t_0 and t be arbitrary elements of the group T with $(x\pi^t, x\pi^{t_0}) \in \delta$. It follows from (24.30) that

$$x\pi^{t_0+k} \in x_0\beta \tag{24.33}$$

for some element $k \in K$. Since $(x\pi^t, x\pi^{t_0}) \in \delta \subset \gamma$, it follows from the choice of γ that $(x\pi^{t+k}, x\pi^{t_0+k}) \in \beta$. Hence

$$x\pi^{t+k} \in x\pi^{t_0+k}\beta \subset x_0\beta^2 \subset x_0\alpha. \tag{24.34}$$

It is easily seen from (24.24) that

$$F(x,t) - F(x,t_0) = \{F(x,t + k) - F(x,t_0 + k)\}$$

$$+ \{F(x\pi^{t_0},k) - F(x\pi^t,k)\}. \tag{24.35}$$

Since the oscillation of F^* in $x_0\alpha$ is less than $\varepsilon/2$, we get from (24.33) and (24.34) that

$$||F(x,t + k) - F(x,t_0 + k)|| = ||F^*(x\pi^{t+k})$$

$$- F^*(x\pi^{t_0+k})|| < \varepsilon/2. \tag{24.36}$$

But $(x\pi^t, x\pi^{t_0}) \in \delta$, $k \in K$ and (24.32) implies $||F(x\pi^{t_0},k) - F(x\pi^t,k)|| < \varepsilon/2$. Hence $||F^*(x\pi^t) - F^*(x\pi^{t_0})|| = ||F(x,t) - F(x,t_0)|| < \varepsilon$ whenever $(x\pi^t, x\pi^{t_0}) \in \delta$. Since the index $\delta \in \mathcal{U}[X]$ does not depend on t and t_0, the function $F^*: xT \to E$ is uniformly continuous. ▮

4·24.28 THEOREM: *Let X be a compact metrizable space, (X,T,π) a minimal transformation group and let E be a Banach space which does not contain a subspace isomorphic to* $\mathbb{C}$. *Let $F: X \times T \to E$ be a cocycle over (X,T,π) and let $x \in X$ be such that the motion $\pi_x: T \to xT$ is bijective. Suppose that $||F(x,t)|| \leqslant L < +\infty$ $(t \in T)$. Then the function $F_x \in C(T,E)$ is uniformly synchronous with the point x.*

Proof: It follows from Theorem 4·24.24 that the map p of the orbit xT onto the orbit of the point F_x in $(C(T,E),T,\sigma)$, where

$$xtp = F_x\sigma^t, \qquad (t \in T)$$

is well defined and continuous. Therefore also the map $F^*: xT \to E$ is continuous. By Lemma 4·24.27, the map F^* can be extended by continuity onto X. Denote this extension by $F_1^*: X \to E$ and define a map $f: X \to C(T,E)$ by exists a subsequence $\{\tau_k\} = \{t_{n_k}\}$ such that $\{\varphi(t + \tau_k)\} \to \psi(t)$ and $\varphi \neq \psi$. It follows from this and the equality:

$$\int_0^{t+\tau_k} \varphi(\xi)d\xi = \int_0^{\tau_k} \varphi(\xi)d\xi + \int_0^t \varphi(\xi + \tau_k)d\xi,$$

that the difference:

$$\int_0^t \varphi(\xi)d\xi - \int_0^t \psi(\xi)d\xi,$$

does not depend on t. Hence $\varphi = \psi$. This contradiction shows that φ is synchronous with the primitive. To conclude the proof we refer to Theorem 4·24.24.∎

4·24.26 Theorem 4·24.24 and Corollary 4·24.25 can be strengthened for minimal transformation groups (X,T,π) with a compact phase space and, consequently, for recurrent functions $\varphi: T \to E$.

Let (X,T,π) be a transformation group and let x be a fixed point of X such that the map $\pi_x : T \to X$ is bijective. Define a map $F^*: xT \to E$ by $F^*(x\pi^t) = F(x,t)$ $(t \in T)$. Although the function F^* is defined only on the orbit xT, it is meaningful to speak about the continuity of F^* at every point $y \in \overline{xT}$ and about the oscillation of F^* at such points.

We shall precede this by a proof of the following lemma.

4·24.27 LEMMA: *Let X be a compact space, (X,T,π) a minimal transformation group and $F: X \times T \to E$ a cocycle over (X,T,π). Suppose that the function $F^*: xT \to E$ is continuous at some point $x_0 \in X = \overline{xT}$. Then F^* is uniformly continuous on xT, and hence it can be extended by continuity onto X.*

Proof: Let ε be an arbitrary positive number and let the index $\alpha \in \mathcal{U}[X]$ be such that the oscillation of the function $F^*: xT \to E$ in the neighbourhood $x_0\alpha$ is smaller than $\varepsilon/2$. Let $\beta \in \mathcal{U}[X]$ be an index with $\beta^2 \subset \alpha$, $\beta = \beta^{-1}$. Since the transformation group (X,T,π) is minimal, there exists a compact subset $K \subset T$ such that

$$X \subset x_0\beta(-K) \equiv (x_0\beta, -K)\pi. \qquad (24.30)$$

Using the property of uniform integral continuity (see Subsection 1·1.5), we choose an index $\gamma \in \mathcal{U}[X]$ so that

$$(x_1,x_2) \in \gamma \Rightarrow (x_1\pi^t, x_2\pi^t) \in \beta, \qquad (t \in K). \qquad (24.31)$$

By Lemma 4·24.23 we can choose an index $\delta \in \mathcal{U}[X]$ such that $\delta \subset \gamma$ and

$$(x_1,x_2) \in \delta \Rightarrow \|F(x_1,t) - F(x_2,t)\| < \varepsilon/2, \qquad (t \in K). \qquad (24.32)$$

$$f_y(t) = F_1^*(y\pi^t), \qquad (y \in X, t \in T).$$

The function $f:X \to C(T,E)$ is also continuous in view of the uniform integral continuity property. Since (X,T,π) is a minimal transformation group and $f_x(t) = F^*(x\pi^t)$, the function $F^*(x\pi^t) = F_x(t)$ is recurrent in the sense of Birkhoff and f is a homomorphism of (X,T,π) onto the minimal transformation group (X^*,T,σ), where X^* is the orbit closure in $(C(T,E),T,\sigma)$ of the point $F_x:T \to E$. ∎

4·24.29 COROLLARY: *Let E be a Banach space which does not contain a subspace isomorphic to $\mathbb{C}$ and let $\varphi:\mathbb{R} \to E$ be a recurrent function in the sense of Birkhoff (in particular, a Bohr almost periodic function) whose primitive* (24.23) *is bounded. Then this primitive is a recurrent (respectively, almost periodic) function uniformly isochronous with φ.*

Proof: Since the function $\varphi \in C(\mathbb{R},E)$ is recurrent, the orbit closure X of the point φ in $(C(\mathbb{R},E),\mathbb{R},\sigma)$ is a compact minimal set. Let $\Phi:\mathbb{R} \to E$ denote the primitive (24.23) and let X^* denote the orbit closure of Φ in $(C(\mathbb{R},E),\mathbb{R},\sigma)$. It can be seen from the proof of Theorem 4·24.28 that there exists a continuous homomorphism $f:X \to X^*$. It remains to prove that the map f is bijective. Let $\psi_1,\psi_2 \in X$ and $f(\psi_1) = f(\psi_2)$. Since X is a minimal set, there exist sequences $\{t_n^{(1)}\}$ and $\{t_n^{(2)}\}$ such that $\{\varphi(t + t_n^{(i)})\} \to \psi_i(t)$ $(i = 1,2)$ in the space $C(\mathbb{R},E)$. Then

$$f\psi_i(t) = \lim_{n\to\infty} \int_0^{t_n^{(i)}+t} \varphi(\xi)d\xi = \lim_{n\to\infty} \left[\int_0^{t_n^{(i)}} \varphi(\xi)d\xi + \int_0^t \varphi(\xi + t_n^{(i)})d\xi\right]$$

$$= c^{(i)} + \int_0^t \psi_i(\xi)d\xi,$$

where $c^{(1)}$ and $c^{(2)}$ are elements of the space E. Hence it follows from the equality $f(\psi_1) = f(\psi_2)$ that $\psi_1 = \psi_2$. ∎

4·24.30 THEOREM: *Let (X,T,π) be a minimal transformation group with a compact metrizable phase space X, E a Banach space, $F:X \times T \to E$ a cocycle over (X,T,π) and let $x \in X$ be such that the set $\{F(x,t) | t \in T\}$ is weakly relatively compact. Then the point $F_x \in C(T,E)$, where $F_x(t) = F(x,t)$, is uniformly synchronous with x.*

Proof: Let f be a linear functional defined on E and let $\phi(x,t) = \langle f,F(x,t)\rangle$ $(t \in T)$. Clearly $\phi:X \times T \to \mathbb{R}$ is a cocycle. Therefore the function $\phi_x:T \to \mathbb{R}$, $\phi_x(t) = \phi(x,t)$ $(t \in T)$, is uniformly synchronous with the point x. Hence the map $\phi^*:xT \to \mathbb{R}$, where $\Phi^*(x\pi^t) = \Phi(x,t)$ $(t \in T)$, is uniformly continuous. Since the weak closure of the set $\{F(x,t) | t \in T\}$ is weakly compact, the map $F^*:xT \to E$ is weakly uniformly continuous, and therefore it can be extended continuously onto $X = \overline{xT}$. Denote this extension by $F_1^*: X \to E$. It follows from a result due to Gel'fand [*1*] that the function F_1 is strongly continuous at some point of X. By Lemma 4·24.27, $F_1^*: X \to E$ is a (strongly) continuous function. Hence it follows that the point F_x is uniformly synchronous with x.

4·24.31 COROLLARY: *Let E be an arbitrary Banach space and let $\varphi \in C(\mathbb{R},E)$ be a recurrent function such that the set of the values of the primitive* (24.23) *is weakly relatively compact. Then the function φ and its primitive* (24.23) *are uniformly isochronous.*

Proof: It follows from Theorem 4·24.30 that the primitive (24.23) is uniformly synchronous with the function φ. It remains to repeat the argument used in the proof of Corollary 4·24.29. ∎

4·24.32 COROLLARY: *Let E be a uniformly convex (or reflexive) Banach space and let $\varphi \in C(\mathbb{R},E)$ be a recurrent function. If the primitive* (24.23) *is bounded, then it is uniformly isochronous with φ.*

REMARKS AND BIBLIOGRAPHICAL NOTES

The question of the existence of almost periodic solutions of linear differential equations in a Euclidean space with a constant matrix A and an almost periodic free member $f(t)$ was first considered by Bohr and Neugebauer [*1*]. This case can be reduced to the famous Bohl-Bohr Theorem: *a bounded primitive of a real-valued almost periodic function is almost periodic*. Equations with almost periodic matrices $A(t)$ were investigated by Favard [*1,2*] by means of his minimax method. Favard's results have been extended to Levitan almost periodic functions (see: Levitan [*1*]). Almost periodic functions which take values in a Banach space were studied in the paper of Bochner [*1*] and in the book by Amerio and Prouse [*1*]. Favard's theory was extended to infinite-dimensional spaces in the papers by Amerio [*1-5*] and Žikov [*2*]. Theorems 4·24.5,9 are a further generalization of Favard's theory. They differ from Žikov's results in the following respects: (1) they consider not only skew products but also general linear extensions; (2) the conditions on compactness (or boundedness) are slightly relaxed. The proof of Theorem 4·24.5 seems to be new. The results of Ljubarskiĭ [*1,2*] and Dimitrov [*1*] can be deduced from Theorems 4·24.5,9. The first statement of Theorem 4·24.21 was proved by Žikov (in the almost periodic case). The second statement is a generalization of results due to Bohr and Neugebauer [*1*]. It was independently obtained by Ščerbakov [*6*]. Amerio [*4*] has extended the Bohl-Bohr Theorem to the case of almost periodic functions with range in a uniformly convex Banach space. A complete solution of the problem of the almost periodicity of the primitive of an almost periodic function was given by Kadets [*1*]. The results presented in 4•24.23-32 are an improvement and generalization of results due to Kadets [*1*]. Related results were obtained by Boles Basit [*1*], Bronšteĭn [*7*], Carrol [*1*], Doss [*1*], Günzler [*1*] and Žikov [*7*].

BIBLIOGRAPHY

ALEKSANDROV, P.S.

[*1*] *Introduction to the General Theory of Sets and Functions.* (Gostekhizdat Moscow), (1948),(Russian).

AL-KUTAIBI, S.H. AND RHODES, F.

*[*1*] Lifting Recursion Properties through Group Homomorphisms. *Proc. Amer. Math. Soc.*, **49**, (1975), 487-494; *M.R.*, **53**, # 4018.

ALLAUD, G. AND THOMAS, E.S. J

[*1*] Almost Periodic Minimal Sets. *J. Differential Equations.* **15**, (1974), 158-171; *M.R.*, **49**, #4049.

AMERIO, L.

[*1*] Soluzioni quasi-periodiche, o limitate, di sistemi differenziali quasi-periodichi, o limitati. *Ann. Mat. Pura Appl.*, **39** (1955), 97-119; *M.R.*, **18**, #128.

[*2*] Bounded or Almost Periodic Solutions of Non-Linear Differential Systems. *Proc. Conf. Diff. Equations*, (College Park, Md.), (1955), (Univ. Maryland Book Store), (1956), 179-182; *M.R.*, **18**, #738.

[*3*] Ancora sulle equazioni differenziali quasi-periodiche astratte. *Ricerche Mat.*, 10 (1961), 31-32; *M.R.*, **24**, #A3386.

[*4*] Sull'integrazione delle funzioni quasi-periodiche astratte. *Ann. Mat. Pura Appl.*, **53**, (1961), 371-382; *M.R.*, **24**, #A807.

[*5*] Su un teorema di minimax per le equazioni differenziali. *Rend. Accad. Naz. Lincei*, **34**, (1963), 409-416.

AMERIO, L. AND PROUSE, G.

[*1*] *Almost Periodic Functions and Functional Equations*. (Van Nostrand Reinhold, New York); *M.R.*, **43**, #819.

ANDERSON, R.D.

[*1*] Minimal Sets under Flows on Tori and Derived Spaces. *Notices Amer. Math. Soc.*, **5**, (1958), 843.

ANOSOV, D.V. AND KATOK, A.B.

[*1*] New Examples in the Smooth Ergodic Theory. Ergodic Diffeomorphisms. *Trudy Moskov. Mat. Obsc.*, 23, (1970), 3-36, (Russian); *M.R.*, **43**, #6947.

AUSLANDER, J

[*1*] Mean-L-Stable Systems. *Illinois J. Math* 3, (1959), 566-579; *M.R.*, **26**, #6950.
[*2*] On the Proximal Relation in Topological Dynamics. *Proc. Amer. Math Soc*,**11**, (1960), 890-895; *M.R.*, **29**, #1632.
[*3*] Endomorphisms of Minimal sets. *Duke Math. J.*, **30**, (1963), 605-614; *M.R.*, **27**, #5245.
[*4*] Regular Minimal Sets (I). *Trans. Amer. Math. Soc.*,**123**, (1966), 469-479.
[*5*] Homomorphisms of Minimal Transformation Groups. *Topology*, **9**, (1970), 195-203; *M.R.*, **42**, #2455.
[*6*] *Structure and homomorphisms of minimal sets*. (Lecture Notes in Mathematics), (Springer Berlin), **235**, (1971), 23-30.

AUSLANDER, J. AND HAHN, F.

[*1*] Point Transitive Flows, Algebras of Functions and the Bebutov System. *Fund Math.*, **60**, (1967), 117-137; *M.R.*, **36**, #4541.

AUSLANDER, J. AND HORELICK, B.

[*1*] Regular Minimal Sets. II. The Proximally Equicontinuous case. *Compositio Math.*, **22**, (1970), 203-214; *M.R.*, **42**, #2456.

AUSLANDER, L., GREEN, L. AND HAHN, F.

[*1*] *Flows on Homogeneous Spaces*. (Annals of Mathematics Studies, **53**),(Princeton), (1963); *M.R.*, **29**, #4841. (Russian edn., Mir), (1966); *M.R.*, **35**, #4332.

AUSLANDER, L. AND HAHN, F.

[*1*] Real Functions Coming from Flows and Concepts of Almost Periodicity. *Trans. Amer. Math. Soc.*, **106**, (1963), 415-426; *M.R.*, **26**, #1871.

BARBALAT, I.

[*1*] Solutii aproape-periodiche ale ecuatiilor differentiale nelineare. *Com. Acad. Rep. Pop. Romîne*, **11**, (1961), 155-159; *M.R.*, **24**, #A292.

BARBASIN, E.A.

[*1*] On Homorphisms of Dynamical Systems. *Mat Sbornik*, **27**, (1950), 455-470, (Russian); *M.R.*, **12**, #422.
[*2*] On Homomorphisms of Dynamical Systems. II. *Mat Sbornik*, **29**, (1951), 501-518, (Russian); *M.R.*, **13**, #473.
[*3*] *Introduction to Stability Theory.* ("Nauka", Moscow), (1967), (Russian).

BAUM, J.D.

[*1*] An Equicontinuity Condition in Topological Dynamics. *Proc. Amer. Math. Soc.*, **12**, (1961), 30-32; *M.R.*, **22**, #11381.

BEBUTOV, M.V.

[*1*] On Dynamical Systems in the Space of Continuous Functions *Bull. Moskov. Univ., Matematika*, **2**, (1941), 1-52, (Russian).

BENDER, P.R.

[*1*] Recurrent Solutions to Systems of Ordinary Differential Equations. *J. Differential Equations*, **5**, (1969), 271-282;*M.R.* **38**, #1357.

BHATIA, N.P. AND HAJEK, O.

[*1*] *Local Semi-Dynamical Systems.* (Lecture Notes in Mathematics), (Springer, Berlin-New York), **90**, (1969); *M.R.*, **40**, #4559.

BIRKHOFF, G.D.

[*1*] *Dynamical Systems.* AMS Colloquium Publications, **9**, (1927) (Russian edn), (GITTL, Moscow), (1964)
[*2*] Some Unsolved Problems of Theoretical Dynamics. *Science*, **94**, (1941), 598-600; *M.R.*, **3**, #279.

BIRJUK, G.I.

[*1*] A Theorem on Existence of Almost Periodic Solutions of Some Systems of Non-Linear Differential Equations with a Small Parameter. *Dokl. Akad. Nauk SSSR*, **96**, (1954), 5-7, (Russian).

BOCHNER, S

[*1*] Abstrakte fastperiodische Funkionen. *Acta Math.*, **61**,

BOCHNER, S. (Contd)

(1933), 149-184.
[*2*] A New Approach to Almost Periodicity. *Proc. Nat. Acad. Sci. USA.*,**48**, (1962), 2039-2043; *M.R.*, **26**, #2816.
[*3*] Continuous Mappings of Almost Automorphic and Almost Periodic Functions. *Proc. Nat. Acad. Sci. USA,* **52**,(1964), 907-910; *M.R.*, **29**, #6252.

BOHL, P.

[*1*] Uber eine Differentialgleichung der Storungstheorie. *J. Reine Angew. Math.,* **131**, (1906), 268-321.

BOHR, H.

[*1*] Zur Theorie der fastperiodischen Funktionen. I Teil: Eine Verallgemeinerung der Theorie der Fourierreihen. *Acta Math.* **45**, (1925), 29-127.
[*2*] Kleinere Beitrage zur Theorie der fastperiodischen Functionen. VIII. Uber den Logarithmus einer positiven fastperiodischen Functionen. *Danske Vid. Selsk. Mat.-Fys. Medd.*, **14**, (1936), 17-24.

BOHR, H. AND FLANDERS, D.A.

[*1*] Algebraic Equations with Almost Periodic Coefficients. *Danske Vid. Selsk. Mat.-Fys. Medd.*, **15**, (1937), 1-49.

BOHR, H. AND NEUGEBAUER, O.

[*1*] Uber lineare Differentialgleichungen mit konstanten Koeffizienten und fastperiodischer rechter Seite. *Nachr. Ges. Wiss. Gottingen, Math.-Phys. Klasse,* (1926), 8-22.

BOLES BASIT, R.

[*1*] A Generalization of Two-Theorems Due to M.I. Kadets on the Primitive of Abstract Almost Periodic Functions. *Mat. Zametki.*, **9**, (1971), 311-321, (Russian).

BORUKHOV, L.E.

[*1*] On Almost Periodic Solutions of Some Differential Equations. *Učen. Zap. Saratov. Gos. Univ.,* **70**, (1961), 11-24.

BOURBAKI, N.

[*1*] *Topological Vector Spaces.* ("Inostr. Lit.", Moscow), (1959), (Russian);(Hermann et Cie, Paris), (1953); *M.R.*, **14**, #880.
[*2*] *General Topology. Fundamental Structures.* ("Nauka", Moscow), (1968), (Russian); *M.R.*, **39**, #6238; (Hermann et Cie, Paris), (1940,1951,1961); *M.R.*, **25**, #4480;

(English edn), (Addison-Wesley, Reading), I, II, (1966); *M.R.*, **34**, #5044a.

[*3*] *General Topology. Topological Groups. Numbers and the groups and Spaces Related to Them.* ("Nauka", Moscow), (1969), (Russian); *M.R.*, **41**, #984; *M.R.*, **22**, #2503; (Hermann et Cie, Paris), (1942,1947,1960); *M.R.*, **25**, #4021, (English edn), (Addison-Wesley, Reading); *M.R.*, **34**, #5044b.

BRONŠTEĬN, I.U.

[*1*] On Homogeneous Minimal Sets. *Papers on Algebra and Mathematical Analysis, Kishinev,* (1965), 113-115, (Russian); *M.R.*, **36**, #4542.

[*2*] On a Class of Distal Minimal Sets. *Sibirsk. Mat. Ž.*, 7 (1966), 746-756, (Russian), *M.R.***36**, # 4543.

[*3*] On Levitan Almost Periodic Motions in Compact Dynamical Systems. *Mat. Issled., Kishinev,* 2, (1967), 3-13, (Russian).

[*4*] A Contribution to the Theory of Distal Minimal Sets and Distal Functions. *Dokl. Akad. Nauk SSSR,* **172**, (1967), 255-257, (Russian).

[*5*] A Generalization of a Theorem Due to Green. *Izv. Akad. Nauk. MSSR, Kishinev,* **8**, (1967), 67-71, (Russian).

[*6*] Equivalence Relations in Distal Minimal Transformation Groups. *Izv. Akad. Nauk MSSR, Kishinev,***11**, (1968), 51-59, (Russian); *M.R.*, **38**, #701.

[*7*] Extensions of Transformation Groups. *Sibirsk. Matt. Ž.*,**9** (1968), 13-20, (Russian).

[*8*] Group Extensions of Minimal Sets. *Sibirsk. Mat. Ž.*, **10** (1969), 507-521, (Russian)

[*9*] Locally Trivial Extensions of Transformation Groups. *Sibirsk. Mat. Ž.*, **10**, (1969), 820-832, (Russian); *M.R.*, **40**, #4954.

[*10*] A Criterion for Distallity of Transitive Transformation Groups. *Mat. Zametki,* **5** (1969), 77-80, (Russian).

[*11*] *Minimal Transformation Groups.* (Red.-Izdat. otdel Akad. Nauk MSSR, Kishinev), (1969), (Russian).

[*12*] On Disjointness of Minimal Sets. *Mat. Issled., Kishinev,* **5**, (1970), 70-78; *M.R.*, **43**, #6900.

[*13*] Distal Extensions of Minimal Transformation Groups. *Sibirsk. Mat. Ž.*, **11**, (1970), 1215-1235 (Russian); *M.R.*, **43**, #413.

[*14*] On the Structure of Distal Extensions and Finite-Dimensional Distal Minimal Sets. *Mat. Issled., Kishinev,* **6**, (1971), 41-62, (Russian); *M.R.*, **45**, #2687.

[*15*] A Structure Theorem for Almost Distal Extensions of Minimal Sets. *Mat. Issled., Kishinev,* **6**, (1971), 22-32 (Russian).

[*16*] On a General Appraoch to Extensions of Minimal Sets. *Mat. Issled., Kishinev,* **7**, (1972), 28-61, (Russian); *M.R.***47**, #9585.

BRONŠTEĬN, I.U. (Contd)

[*17*] On the Galois Theory of Distal Minimal Extensions. *Mat. Issled., Kishinev*, **7**, (1972), 28-61, (Russian); *M.R.*, **47**, #9586.

[*18*] Stable and Equicontinuous Extensions of Minimal Sets. *Mat. Issled., Kishinev*, **8** (1973), 3-11, (Russian); *M.R.*, **50**, #5764.

[*19*] On Birkhoff's Recurrent Solutions of Algebraic Equations. *Mat. Zametki*, **13**, (1973), 619-625, (Russian).

*[*20*] Regular Extensions and the Structure of Minimal Sets. *Izv. Akad. Nauk. MSSR*, **1**, 5-14, (1976), 94, (Russian); *M.R.*, **54**, #1187.

BRONŠTEĬN, I.U. AND ČERNIĬ, V.F.

[*1*] On Extensions of Dynamical Systems with Uniformly Asymptotically Stable Points. *Differencial 'nye Uravneniya.* **10**, (1974), 1225-1230, (Russian); *M.R.*, **51**, #1788.

*[*2*] Linear Extensions Satisfying the Conditions on Exponential Dichotomy. *Izv. Akad. Nauk. MSSR*, **3**, (1976) 12-16, (Russian).

*[*3*] Linear Extensions of Dynamical Systems. *Mat. Issled., Kishinev, 'Štiintsa'*, **44**, (1977), 42-48, (Russian).

BRONŠTEĬN, I.U. AND KHOLODENKO, M.S.

[*1*] On Homogeneity of Minimal Sets. *Differencial'nye U Uravneniya*, **8** (1972), 406-414; *M.R.*, **46**, #8199.

CARROL, F.W.

[*1*] On Bounded Functions with Almost Periodic Differences. *Proc. Amer. Math. Soc.*, **15**, (1964), 241-243.

CARTWRIGHT, M.L.

[*1*] Almost Periodic Solutions of Systems of two Periodic Equations. *Trudy Meždunar. Simposiuma Nelineĭnykh Kolebanii, Kiev*, (1963), 256-263.

[*2*] Almost Periodic Differential Equations and Almost Periodic Flows. *J. Differential Equations.* **5**, (1969), 167-181.

[*3*] *Almost Periodic Solutions of Differential Equations and Flows.* (Lecture Noes in Mathematics), (Springer, Berlin), **235**, (1971), 35-43.

CERESIZ, V.M.

[*1*] V-Monotone Systems and Almost Periodic Solutions. *Sibirsk. Mat. Ž.*, **13**, (1972), 921-932, (Russian); *M.R.*, **47**, #8981.

[*2*] Uniformly V-Monotone Systems. Almost Periodic Solutions *Sibirsk. Mat. Ž.*, **13**, (1972), 1107-1123, (Russian).

CHU, H.

[*1*] On Universal Transformation Groups. *Illinois J. Math.*, 6, (1962), 317-326; *M.R.*, **25**, #1545.
[*2*] On Totally Minimal Sets. *Proc. Amer. Math. Soc.*, **13**, (1962), 457-458; *M.R.*, **25**, #1544.
[*3*] Algebraic Topology Criteria for Minimal Sets. *Proc. Amer. Math. Soc.* **13**, (1962), 503-508; *M.R.*, **25**, #1533.
[*4*] On the Structure of Almost Periodic Transformation Groups. *Pacific J. Math.*, 38, (1971), 359-364; *M.R.*, **46**, #4555.

CHU, H. AND GERAGHTY, M.A.

[*1*] The Fundamental Group and the First Cohomology Group of a Minimal Set. *Bull. Amer. Math. Soc.*, **69**, (1963), 377-381; *M.R.*, **26**, #4334.

CHU , J.P.

[*1*] Proximity Relations in Transformation Groups. *Trans. Amer. Math. Soc.*, **108**, (1963), 88-96; *M.R.*, **27**, #4218.
[*2*] Variations on Equicontinuity. *Duke Math. J.*, **30**, (1963) 423-431; *M.R.*, **27**, #4219.

CONNER, P. AND MONTGOMERY, D.

[*1*] An Example for *SO*(3) Action. *Proc. Nat. Acad. Sci. USA*, **48**, (1962), 1918-1922; *M.R.*, **26**, #6300.

DALETSKIĬ, Ju. L. AND KREIN, M.G.

[*1*] *Stability of Solutions of Differential Equations in a Banach Space.* ("Nauka", Moscow), (1970), (Russian).

DEMIDOVIČ, B.P.

[*1*] On the Existence of an Almost Periodic Solution of an Ordinary Differential Equation of the First Order. *Uspehi Mat. Nauk*, **8**, (1953), 103-106, (Russian).
[*2*] *Lectures on the Mathematical Theory of Stability.* (Fizmatgiz, Moscow), (1967), (Russian)

DEYSACH, L. AND SELL, G.R.

[*1*] On the Existence of Almost Periodic Motions. *Michigan Math. J.*, **12**, (1965), 87-95; *M.R.*, **30**, #3279.

DIEUDONNÉ, J.

[*1*] *Foundations of Modern Analysis.* ("Mir", Moscow), (1964), (Russian); (Academic Press, New York), (1960,1968); *M.R.*, 22, #11074; (Gauthier-Villars, Paris),(1968); *M.R.*, **38**, #4246.

DIMITROV, D.B.

[*1*] On Almost Periodic and N-Almost Periodic Solutions of a Difference Equation. *Ukrain. Mat. Ž.*, **25**, (1973), 193-199, (Russian).

DOBRYNSKIĬ, V.A. AND ŠARKOVSKIĬ, A.N.

[*1*] On Orbitally Stable Trajectories. *Differencial'nye Uravneniya*, **9**, (1973), 558-559, (Russian).

[*2*] The Genericity of Dynamical Systems Almost All of Whose Trajectories Are Stable Under Constantly Acting Perturbations. *Dokl. Akad. Nauk SSSR*, **211**, (1973), 273-276, (Russian); *M.R.*, **48**, #9762.

DOSS, R.

[*1*] On Bounded Functions with Almost Periodic Differences. *Proc. Amer. Math. Soc.* **12**, (1961), 488-489; *M.R.*, **23**, #A3424.

DUBOLAR', V.K.

[*1*] On Recurrent Solutions of Differential-Functional Equations in a Banach Space. *Differencial'nye Uravneniya*, **6** (1970), 1395-1401, (Russian).

DUNFORD, N. AND SCHWARTZ, J.T.

[*1*] *Linear Operators. General Theory.* (Wiley, New York), (1957); (Russian edn), ("Inostr. Lit.", Moscow), (1962); *M.R.***22**, # 8302.

EISENBERG, M.

[*1*] Maximally Almost Periodic and Universal Equicontinuous Minimal Sets. *Michigan Math. J.*, **14**, (1967), 101-105.

[*2*] *A Theorem on Extensions of Minimal Sets.* (Lecture NOtes in Mathematics), (Springer, Berlin), **235**, (1971), 61-64.

ELLIS, R.

[*1*] Locally Compact Transformation Groups. *Duke Math. J.*, **24**, (1957), 119-125; *M.R.*, **19**, 561.

[*2*] Distal Transformation Groups. *Pacific J. Math.*, **8**, (1958) 401-405; *M.R.*, **21**, #96.

[*3*] Equicontinuity and Almost Periodic Functions. *Proc. Amer. Math. Soc.*, **10**, (1959), 637-643; *M.R.*, **21**, #5950.

[*4*] Universal Minimal Sets. *Proc. Amer. Math. Soc.*, **11**, (1960) 540-543; *M.R.*, **22**, #8491.

[*5*] A Semigroup Associated with a Transformation Group. *Trans. Amer. Math. Soc.*, **94**, (1960), 272-281; *M.R.*, **23**, #A961.

[*6*] Point Transitive Transformation Groups. *Trans. Amer. Math. Soc.*, **101**, (1961), 384-395; *M.R.*, **24**, #A1119.

[*7*] Locally Coherent Minimal Sets. *Michigan Math. J.*, **10**, (1963), 97-104; *M.R.*, **27**, #4217.

[*8*] Global Sections of Transformation Groups. *Illinois J. Math.*, **8**, (1964), 384-394; *M.R.*, **29**, #5234.

[*9*] The Construction of Minimal Discrete Flows. *Amer. J. Math.*, **87**, (1965), 564-574.

[*10*] Group-Like Extensions of Minimal Sets. *Trans. Amer. Math. Soc.*, **127**, (1967), 125-135; *M.R.*, **36**, #4544.

[*11*] The Structure of Group-Like Extensions of Minimal Sets. *Trans. Amer. Math. Soc.*, **134**, (1968), 261-287; *M.R.*, **38**, #6569.

[*12*] The Beginnings of an Algebraic Theory of Minimal Sets. *Topological Dynamics, An International Symposium, (Fort Collings, 1967)*, (Benjamin, New York), (1968), 165-184.

[*13*] *Lectures on Topological Dynamics.* (Benjamin, New York), (1969); *M.R.*, **42**, #2463.

[*14*] The Veech Structure Theorem. *Trans. Amer. Math. Soc.*, **186**, (1973), 203-218.

*[*15*] A Group Associated with an Extension. *Recent Advances In Topological Dynamics.* (Springer, New York), (1973), 86-93

ELLIS, R., GLASNER, S. AND SHAPIRO, L.

*[*1*] Proximal-Isometric (PI) Flows. *Advances in Math.*, **17**, (1975), 213-260; *M.R.*, **52**, #1652.

ELLIS, R. AND GOTTSCHALK, W.H.

[*1*] Homomorphisms of Transformation Groups. *Trans. Amer. Math. Soc.*, **94**, (1960), 258-271; *M.R.*, **23**, #A960.

ELLIS, R. AND KEYNES, H.B.

[*1*] A Characterization of the Equicontinuous Structure Relation. *Trans. Amer. Math. Soc.*, **161**, (1971), 171-184.

[*2*] Bohr Compactifications and a Result of Følner. *Israel J. Math.*, **12**, (1972), 314-330; *M.R.*, **47**, #1992.

ENGLAND, J.W.

[*1*] Stability in Topological Dynamics. *Pacific J. Math.*, **21** (1967), 479-485; *M.R.*, **35**, #3648.

FAVARD, J.

[*1*] Sur les équations différentielles linéaires à coefficients presque-périodiques. *Acta Math.*, **51**, (1927), 31-81.

[*2*] *Lecons sur les fonctions presque-périodiques.* (Gauthier-Villars, Paris), (1933).

FINK, A.M.

[*1*] Almost Automorphic and Almost Periodic Solutions which Minimize Functionals. *Tôhoku Math. J.*, **20**, (1968), 323-332.
[*2*] Semi-Separated Conditions for Almost Periodic Solutions. *J. Differential Equations*, **11**, (1972), 245-251.
[*3*] *Almost Periodic Differential Equations.* (Lecture Notes in Mathematics), (Springer, Berlin), 377, (1974).

FINK, A.M. AND FREDERICKSON, P.O.

[*1*] Ultimate Boundedness Does not Imply Almost Periodicity. *J. Differential Equations*, **9**, (1971), 280-284.

FINK, A.M. AND SEIFERT, G.

[*1*] Liapunov Functions and Almost-Periodic Solutions for Almost-Periodic Systems. *J. Differential Equations*, **5**, (1969), 307-313.

FLOR, P.

[*1*] Rhythmische Abbildungen abelscher Gruppen. II. *A. Wahrscheinlichkeitstheorie verw. Geb.*, **7**, (1967), 17-28.

FLOYD, E.E.

[*1*] A Nonhomogeneous Minimal Set. *Bull. Amer. Math. Soc.*, **55**, (1949), 957-960; *M.R.*, **11**, #453.

FRÉCHET, M.

[*1*] Les fonctions asymptotiquement presque-periodiques. *Rev. Scientifique*, **79**, (1941), 341-354; *M.R.*, **7**, #127.

FURSTENBERG, H.

[*1*] The Structure of Distal Flows. *Amer. J. Math.*, **85**, (1963) 477-515; *M.R.*, **28**, #602.
[*2*] Disjointness in Ergodic Theory, Minimal Sets, and a Problem in Diophantine Approximation. *Math. Systems Theory*, **1**, (1967), 1-50.

FURSTENBERG, H., KEYNES, H. AND SHAPIRO, L.

[*1*] Prime Flows in Topological Dynamics. *Israel J. Math.*, **14**, (1973), 26-38; *M.R.*, **47**, #9588.

GEL'FAND, I.M.

[*1*] Abstrake Functionen und lineare Operatoren. *Mat. Sbornik*, **4**, (1938), 235-286.

GERKO, A.I.

[*1*] Recursiveness, Minimal Sets,and Liapunov Stability in Transformation Semigroups. *Mat. Issled., Kishinev,* **7**, (1972), 62-80, (Russian); *M.R.*, **47**, #9589.
[*2*] On the Structure of Distal Extensions of Minimal Transformation Semigroups. *Mat. Issled.,,Kishinev* **7**, (1972). 69-86, 269, (Russian); *M.R.*, **47**, #9590.
*[*3*] Liapunov Stable and Regionally Distal Transformation Semigroups. *Mat. Issled., Kishinev,* **9**, (1974), 19-27, (Russian).
*[*4*] Regionally Distal and Group Extensions of Transformation Semigroups. *Mat. Issled., Kishinev,* **10**, (1975), 94-106, (Russian).

GHEORGHIU, N.

[*1*] Asupra soluţiilor aproape-periodice si aşimptotic aproape-periodice ale ecuţiilor differenţiale neliniare de primul ordin. *Anal. Şt. Univ. Iaşi,* **1**, (1955), 17-20; *M.R.*, **18**, #899.

GLASNER, S.

*[*1*] Topological Dynamics and Group Theory. *Trans. Amer. Math. Soc.,* **187**, (1974), 327-334; *M.R.*, **49**, #1496.
*[*2*] Compressibility Properties in Topological Dynamics. *Amer. J. Math.,* **97**, (1975), 148-171.
*[*3*] A Metric Minimal Flow Whose Enveloping Semigroup Contains Finitely Many Minimal Ideals is PI. *Israel J. Math.*, **22**, (1975), 87-92.
*[*4*] *Proximal Flows.* (Lecture N tes in Mathematics), (Springer, Berlin), **517**, (1976).

GORIN, E.A. AND LIN, V.J .

[*1*] Algebraic Equations with Continuous Coefficients and some Questions of the Algebraic Theory of Braids. *Mat. Sbornik,* **78**, (1969), 579-610, (Russian).

GOTTSCHALK, W.H.

[*1*] Almost Periodicity, Equi-Continuity and Total Boundedness. *Bull. Amer. Math. Soc.,* **52**, (1946), 633-636; *M.R.*, **8**, #34.
[*2*] Transitivity and Equicontinuity. *Bull. Amer. Math. Soc.,* **54**, (1948), 982-984; *M.R.*, **10**, 199.
[*3*] Characterizations of Almost Periodic Transformation Groups. *Proc. Amer. Math. Soc.,* **7**, (1956), 709-712; *M.R.*, **18**, 141.
[*4*] Minimal Sets: An Introduction to Topological Dynamics *Bull. Amer. Math. Soc.*, **64**, (1958), 336-351; *M.R.*, **20**, #6484.
[*5*] Universal Curve of Sierpiński Is Not a Minimal Set. *Notices Amer. Math. Soc.,* **6**,(1959), 257.

GOTTSCHALK, W.H. (Contd)

[*6*] An Irreversible Minimal Set. *Ergodic Theory (Proc. International Symposium) Tulane University.*, (1961), 135-150; *M.R.*, **28**, #341.
[*7*] Substitution Mimal Sets.*Trans. Amer. Math. Soc.*, **109**, (1963), 467-491.
[*8*] Minimal Sets Occur Maximally. *Trans. N.Y. Acad. Sci.*, **26**, (1963-64), 348-353; *M.R.*, **28**, #5438.
[*9*] A Survey of Minimal Sets. *Ann. Inst. Fourier*, **14**, (1964) 53-60; *M.R.*, **29**, #4044.
[*10*] *Bibliography for Dynamical Topology.* (Wesleyan University), (1972),(Fifth edn); *M.R.*, **41**, #2655.
[*11*] *Some General Dynamical Notions.*(Lecture Notes in Mathematics),(Springer, Berlin),**318**, (1973), 120-125; *M.R.*, **53**, #11591.

GOTTSCHALK, W.H. AND HEDLUND, G.A.

[*1*] Recursive Properties of Transformation Groups. *Bull. Amer. Math. Soc.*, **52**, (1946), 637-641);*M.R.***8**, #34.
[*2*] *Topological Dynamics. (AMS Colloguium Publications)*, **36**,1955; *M.R.*, **17**, #650.

GRABAR', M.I.

[*1*] On Homogeneity of Dynamical Systems and a Problem of Birkhoff, *Učen. Zapiski Moskov. Univ.*, **186**, (1959), 161-177, (Russian).

GÜNZLER, H.

[*1*] Integration of Almost Periodic Functions. *Math. Zeitschrift*, **102**, (1967), 253-287.

HÁJEK, O.

[*1*] *Dynamical Systems in the Plane.* (Academic Press, London, New York), (1968); *M.R.*, **39**, #1767.

HALANAY, A.

[*1*] Soluţii aproape-periodice ale ecuaţiei Riccaty. *Studii şi Cercetări Mat.*, **4**, (1953), 345-354; *M.R.*, **16**, #475.

HARTMAN, P.

[*1*] Ordinary Differential Equations. (Wiley, New York), 1964; (Russian edn);("Mir", Moscow), (1970).

HEDLUND, G.A.

[*1*] Fuchsian Groups and Transitive Horocycles. *Duke Math. J.*, **2**, (1936), 530-542.

HILMY, G.F.

[*1*] On a Property of Minimal Sets. *Dokl. Akad. Nauk SSSR*, **14**, (1937), 261-262.

HOFMANN, K.H. AND MOSTERT, P.S.

[*1*] *Elements of Compact Semigroups*. (Merill, Columbus), (1966); *M.R.*, **35**, #285.
[*2*] Compact Groups Acting with $(n - 1)$-Dimensional Orbits on Subspaces of n-Manifolds. *Math. Ann.*, **167**, (1966), 224-239.

HORELICK, B.

[*1*] Group-Like Extensions and Similar Algebras. *Math. Systems Theory*, **3**, (1969), 139-145.
[*2*] Pointed Minimal Sets and S-Regularity. *Proc. Amer. Math. Soc.*, **20**, (1969), 150-156.

HUREWICZ, W. AND WALLMAN, H.

[*1*] *Dimension Theory*. (Princeton University Press, Princeton), (1941); (Russian edn),("Inostr. Lit.", Moscow), (1948).

HUSEMOLLER, D.

[*1*] *Fibre Bundles* (McGraw-Hill, New York), (1966);(Russian edn), ("Mir", Moscow), (1970).

HU, SZE-TSEN

[*1*] *Homotopy Theory*. (Academic Press, New York-London) (1959); *M.R.*, **21**, #5186; (Russian edn), ('Mir', Moscow), (1964).

ISHII, IPPEI

[*1*] On a Non-Homogeneous Flow on the 3-Dimensional Tours. *Funkcial. Ekvacioj*, **17**, (1974), 231-248.

JEWETT, R.I.

[*1*] The Prevalence of Uniquely Ergodic Systems. *J. Math. Mech.*, **19**, (1970), 717-729.

KADEC, M.I.

[*1*] On Integration of Almost Periodic Functions with Values in a Banach Space. *Funkcional. Anal. i Priložen*, **3**, (1969), 71-74, (Russian).

KAHN, P.J. AND KNAPP, A.W.

[*1*] Equivalent Maps on to Minimal Flows. *Math. Systems Theory*, **2**, (1968), 319-324.

KATOK, A.B.

[*1*] Minimal Diffeomorphisms on Principal S^1-Fiber Spaces. *Tezisy VI Vsesojuznoi Topol. Konf. Tbilisi*, (1972), 63, (Russian).

KATO, J. AND NAKAJIMA, F.

[*1*] On Sacker-Sell's Theorem for a Linear Skew Product Flow. *Tohoku Math. J.*, **28**, (1976), 79-88.

KATO, J. AND YOSHIZAWA, T.

[*1*] A Relationship between Uniformly Asymptotic Stability and Total Stability. *Funkcial. Ekvacioj*, **12**, (1970), 233-238.

KELLEY, J.L.

[*1*] *General Topology*. (D. Van Nostrand, Princeton), (1955); *M.R.*, **16**, #1136; (Russian edn), ("Nauka", Moscow), (1968).

KENT, J.F.

[*1*] *Locally Connected Almost Periodic Minimal Sets.* (Springer, Berlin), (Lecture Notes in Mathematics), **318**, (1973), 150-166.

KEYNES, H.B.

[*1*] Topological Dynamics in Coset Transformation Groups. *Bull. Amer. Math Soc.*, **72**, (1966), 1033-1035.

[*2*] A Study of the Proximal Relation in Coset Transformation Groups. *Trans. Amer. Math. Soc.*, **128**, (1967), 389-402; *M.R.*, **35**, #4333.

[*3*] On the Proximal Relation being Closed. *Proc. Amer. Math. Soc.*, **18**, (1967), 518-522; *M.R.*, **35**, #3645.

[*4*] The Proximal Relation in a Class Substitution Minimal Sets. *Math. Systems Theory*, **1**, (1967), 165-174; *M.R.*, **35**, #997

[*5*] The Structure of Weakly Mixing Minimal Transformation Groups. *Illinois J. Math.*, **15**, (1971), 475-489; *M.R.*, **44**, #3306.

[*6*] Disjointness in Transformation Groups. *Proc. Amer. Math. Soc.*, **36**, (1972), 253-259.

KEYNES, H.B. AND ROBERTSON, J.B.

[*1*] On Ergodicity and Mixing in Topological Transformation Groups. *Duke Math. J.*, **35**, (1968), 809-819; *M.R.*, **38**, #2758.

[*2*] Eigenvalue Theorems in Topological Transformation Groups. *Trans. Amer. Math. Soc.*, **139**, (1969), 359-369; *M.R.*, **38**, #6029.

KHARASAKHAL, V.K .

[*1*] *Almost Periodic Solutions of Ordinary Differential Equations.* ("Nauka", Alma Ata), (1970), (Russian).

KNAPP, A.W.

[*1*] Distal Functions. *Proc. Nat. Acad. Sci. USA*, **52**, (1964) 1409-1412.
[*2*] Distal Functions on Groups. *Trans. Amer. Math. Soc.*, **128**, (1967), 1-40; *M.R.*, **36**, #5923.
[*3*] Functions Behaving like Almost Automorphic Functions. *Topological Dynamics (An International Symposium).* (Benjamin, New York), (1968), 299-317; *M.R.*, **38**, #6570.

KRASNOSEL'SKIĬ, M.A., BURD, V.S. AND KOLESOV, JU.S.

[*1*] *Non-Linear Almost Periodic Vibrations.* ("Nauka", Moscow) (1970), (Russian).

KURATOWSKI, K.

[*1*] *Topologie, Vol.I*, (PWN, Warsaw, (1958); *Vol. II*, (PWN, Warsaw), (1961); (Russian edn), ("Mir", Moscow), *Vol. I*, (1966); *Vol. II*, (1969).

KUROSH, A.G.

[*1*] *Group Theory.* (New York, Chelsea), (1955-56); (Russian edn), ("Nauka", Moscow), (1967).

LAM, PING-FUN

[*1*] Inverses of Recurrent and Periodic Points under Homomorphisms of Dynamical Systems. *Math. Systems. Theory*, **6**, (1972), 26-36; *M.R.*, **46**, #673.
[*2*] Almost Equicontinuous Transformation Groups. *Trans. Amer. Math. Soc.*, **195**, (1974), 165-169.

LASOTA, A. AND YORKE, J.A.

[*1*] The Generic Property of Existence of Solution of Differential Equations in Banach Spaces. *J. Differential Equations*, **13**, (1973), 1-12.

LANG, S.

[*1*] *Algebra.* (Addison-Wesley, Reading), (1965); (Russian edn), ("Mir", Moscow), (1968).

LIAPIN, E.S.

[*1*] *Semigroups.* (GIFML, Moscow), (1960, (Russian); *M.R.*, **22**, #11054; (English edn), (A.M.S., Providence), (1963).

LIUBARSKIĬ, M.G.

[*1*] On the Primitive of a Levitan Almost Periodic Function. *Teorija Funktsii, Funkcional. Anal. i ikh Priložen., Izd. Kharkov. Univ.*, **16**, (1972), 139-149, (Russian).

[*2*] Extension of Favard's Theory to the Case of Systems of Linear Differential Equations with Unbounded Levitan Almost Periodic Coefficients. *Dokl. Akad. Nauk. SSSR*, **206**, (1972), 808-810, (Russian).

MACLANE, S.

[*1*] *Homology*. (Springer-Verlag, New York-Berlin-Heidelberg), (1963); *M.R.*, **28**, #122; (Russian edn), ("Mir", Moscow), (1966).

MAL'CEV, A.I.

[*1*] On a Class of Homogeneous Spaces. *Izv. Akad. Nauk SSSR, (Ser. Mat.)*, **13**, (1949), 8-32, (Russian); *M.R.*, **10**, #507.

MARKLEY, N.G.

[*1*] *Non-Trivial Minimal Sets — A Survey*. (Lecture Notes in Mathematics), (Springer, Berlin), **144**, (1970), 153-159.

[*2*] *F*-Minimal Sets. *Trans. Amer. Math. Soc.*, **163**, (1972), 85-100.

[*3*] Locally Circular Minimal Sets. *Pacific J. Math.*, **47**, (1973), 177-197; *M.R.*, **48**, #12491.

MARKOV, A.A.

[*1*] Sur une propriété générale des ensembles minimaux de Birkhoff. *Compt. Rend. Acad. Sci., Paris*, **193**, (1931), 823-825.

[*2*] Stabilitat in Liapounoffschen sinne und Fastperiodizitat. *Math. Zeitschrift*, **36**, (1933) 708-738.

[*3*] Almost Periodicity and Harmonizability. *Trudy II Vsesojusnogo s'jesda, Leningrad*, (1936), 227-231,(Russian).

MARTIN, J.C.

[*1*] Substitution Minimal Flows. *Amer. J. Math.*, **93**, (1971), 503-526; *M.R.*, **43**, #2691.

MASSERA, J.L. AND SCHÄFFER, J.J.

[*1*] *Linear Differential Equations and Function Spaces*. (Academic Press, New York),(1966); *M.R.*, **35**, #3197; (Russian edn), ("Mir", Moscow), (1970); *M.R.*, **42**, #331.

MCMAHON, D.

*[*1*] Relativized Problems with Abelian Phase Group in Topo-

logical Dynamics. *Proc. Nat. Acad. Sci. USA*, **73**, (1976) 1007.

*[2] On the Role of an Abelian Phase Group in Relativized Problems in Topological Dynamics. *Pacific J. Math.*, **64**, (1976), 493-504.

MCMAHON, D. AND WU, T.S.

[*1*] On Weak Mixing and Local Almost Periodicity. *Duke Math. J.*, **39**, (1972), 333-343; *M.R.*, **49**, #1497.

*[2] *Relative Equicontinuity and Its Variations.* (Lecture Notes in Mathematics),(Springer,Berling), **318**, (1973), 201-205.

*[3] Distal and Proximal Extensions of Minimal Sets. *Bull. Inst. Math. Acad. Sinica.* **2**, (1974), 93-107.

*[4] On the Connectedness of Homomorphisms in Topological Dynamics. *Trans. Amer. Math. Soc.*, **217**, (1976), 257-270; *M.R.*, **54**, #1188.

MILLER, R. K.

[*1*] Almost Periodic Differential Equations as Dynamical Systems with Applications to the Existence of Almost Periodic Solutions. *J. Differential Equations*, **1**, (1965), 337-345; *M.R.*, **32**, #2690.

MILLER, R.K. AND SELL, G.R.

[*1*] *Volterra Integral Equations and Topological Dynamics.* (AMS Memoirs), **102**, (1970).

MILLIONŠČIKOV, V.M.

[*1*] Recurrent and Almost Periodic Limiting Trajectories of Non-Autonomous Systems of Differential Equations. *Dokl. Akad. Nauk SSSR*, **161**, (1965), 43-44, (Russian); *M.R.*, **30**, #4043.

[*2*] On Recurrent and Almost Periodic Limiting Solutions of Non-Autonomous Systems. *Differencial'nye Uravneniya*, **4**, (1968), 1555-1559, (Russian).

MONTGOMERY, D. AND ZIPPIN, L.

[*1*] *Topological Transformation Groups.* (Interscience, New York), (1955); *M.R.*, **17**, #383.

NAMIOKA, I.

[*1*] Right Topological Groups, Distal Flows, and a Fixed Point Theorem. *Math. Systems Theory*, **6**, (1972), 193-209.

NEMYTSKII, V.V.

[*1*] General Dynamical Systems. *Dokl. Akad. Nauk. SSSR*, **53**,

NĔMYTSKIĬ, V.V. (Contd).

(1946), 491-494, (French), 495-498, (Russian); *M.R.*, **8**, #280.

[*2*] Topological Problems of the Theory of Dynamical Systems. *Uspehi Mat. Nauk*, **4**, (1949), 91-153, (Russian); *M.R.*, **11**, #526.

NEMYTSKIĬ, V.V. AND STEPANOV, V.V.

[*1*] *Qualitative Theory of Differential Equations.* (GITTL, Moscow), (1949),(Russian); *M.R.*, **10**, 612; (English edn), (Princeton University Press, Princeton), (1960); *M.R.*, **22**, #12258.

OPIAL, Z.

[*1*] Sur les solutions presques-periodiques des equations differentielles du premier et du second ordre. *Ann. Polon. Math.*, **7**, (1959), 51-61; *M.R.*, **22**, #799.

[*2*] Sur une equation differentie-le presque-periodique sans solution presque-periodique. *Bull. Acad. Polon. Sci., Ser. Sci. Math. Astron. Phys.*, **9**, (1961), 673-676; *M.R.*, **24**, #A2707.

PALAIS, R.S.

[*1*] Local Triviality of the Restriction Map for Embedding. *Comment. Math. Helv.*, **34**, (1960), 305-312; *M.R.*, **23**, #A666.

PARRY, W.

*[*1*] *Class-Properties of Dynamical Systems.*(Lecture Noes in Mathematics), (Springer, Berlin),**318**, (1973), 218-225.

PARRY, W. AND WALTERS, P.

[*1*] Minimal Skew-Product Homeomorphisms and Coalescence. *Compositio Math.*, **22**, (1970), 283-288.

PEŁCZYŃSKI, A.

[*1*] On *B*-Spaces Containing Subspaces Isomorphic to the Space C_0. *Bull. Acad. Polon. Sci., Ser. Math. Astron. Phys.* **5**, (1957), 797-798; *M.R.*, **19**, 565.

[*2*] Projections on Certain Banach Spaces. *Studia Math.*, **19**, (1960), 209-228; *M.R.*, **23**, #A3441.

PELEG, R.

[*1*] Some Extensions of Weakly Mixing Flows. *Israel J. Math.*, **9**, (1971), 330-336; *M.R.*, **43**, #6903.

[*2*] Weak Disjointness of Transformation Groups. *Proc. Amer. Math. Soc.*, **33**, (1972), 165-170; *M.R.*, **45**, #7694.

PETERSEN, K.E.

[*1*] Disjointness and Weak Mixing of Minimal Sets. *Proc. Amer. Math. Soc.*, **24**, (1970), 278-280; *M.R.*, **40**, #3522.
[*2*] Extension of Minimal Transformation Groups. *Math. Systems Theory*, **5**, (1971), 365-375; *M.R.*, **46**, #2699.

PLISS, V.A.

[*1*] *Non-Local Problems in Vibration Theory*, ('Nauka', Moscow-Leningrad), (1964), (Russian).

PONTRIAGIN, L.S.

[*1*] *Continuous Groups*, ('Nauka', Moscow), (1973); *Topological Groups*, (English edn), (Princeton University Press), (1939); *M.R.*, **1**, #44; *Topological Groups*, (English 2nd edn), (Gordon and Breach, New York), (1976).

RHODES, F.

*[*1*] Lifting Recursion Properties. *Math. Systems Theory*, **6**, (1973), 302-307.

SACKER, R.J. AND SELL, G.R.

*[*1*] Skew-Product Flows, Finite Extensions of Minimal Transformation Gropus and Almost Periodic Differential Equations. *Bull. Amer. Math. Soc.*, **79**, (1973), 802-805; *M.R.*, **51**, #9040.
*[*2*] Finite Extensions of Minimal Transformation Groups. *Trans. Amer. Math. Soc.*, **190**, (1974), 325-334; *M.R.*, **50**, #3207.
*[*3*] Existence of Dichotomies and Invariant Splittings for Linear Differential Systems, I. *J. Diff. Equations*, **15**, (1974), 429-458; *M.R.*, **49**, #6209.
*[*4*] Existence of Dichotomies and Invariant Splittings for Linear Differential Systems, II. *J. Diff. Equations*, **22**, (1976), 478-496.
*[*5*] Existence of Dichotomies and Invariant Splittings for Linear Differential Systems, III. *J. Diff. Equations*, **22**, (1976), 497-522.

ŠČERBAKOV, B.A.

[*1*] Recurrent Solutions of Differential Equations. *Dokl. Akad. Nauk USSR*, 167, (1966), 1004-1007, (Russian); *M.R.*, **34**, #4642.
[*2*] Recurrent Solutions of Differential Equations and the General Theory of Dynamical Systems. *Differencial'nye Uravneniya*, 3, (1967), 1450-1460, (Russian); *M.R.*, **36**, #6728.

ŠČERBAKOV, B.A. (Contd)

[*3*] On a Certain Class of Poisson Stable Solutions of Differential Equations. *Differencial'nye Uravneniya*, **4**, (1968), 238-243, (Russian).

[*4*] Poisson Stable Solutions of Differential Equations and Topological Dynamics. *Differencial'nye Uravneniya*, **5**, (1969), 2144-2155, (Russian).

[*5*] The Method of Limiting Transformations in the Problem of Existence of Poisson Stable Solutions of Differential Equations. *Dokl. Akad. Nauk USSR*, **190**, (1970), 796-799, (Russian).

[*6*] *Topological Dynamics and Stability, in the Sense of Poisson, of Solutions of Differential Equations*, ('Ŝtiintsa', Kishinev), (1972), (Russian).

*[*7*] On Synchronous Recurrence of Bounded Solutions of Differential Equations of the First Order. *Differencial'nye Uravneniya*, **10**, (1974), 270-275, (Russian).

SEIFERT, G.

[*1*] Stability Conditions for Separation and Almost Periodicity of Solutions of Differential Equations. *Contrib. Diff. Equations*, **1**, (1963), 483-487; *M.R.*, **36**, #6501.

[*2*] Uniform Stability of Almost Periodic Solutions of Almost Periodic Systems of Differential Equations. *Contrib. Diff. Equations*, **2**, (1963), 269-276; *M.R.*, **27**, #5987.

[*3*] Stability Condition for the Existence of Almost Periodic Solutions of Almost Periodic Systems. *J. Math. Anal. Appl.*, **10**, (1965), 393-408.

[*4*] A Condition for Almost Periodicity with Some Applications to Functional Differential Equations. *J. Diff. Equations*, **1**, (1965), 393-408.

[*5*] Almost Periodic Solutions for Almost Periodic Systems of Ordinary Differential Equations. *J. Diff. Equations*, **2**, (1966), 305-319; *M.R.*, **34**, #4613.

[*6*] Almost Periodic Solutions and Asymptotic Stability. *J. Math. Anal. Appl.*, **21**, (1968), 136-149; *M.R.*, **36**, #1774.

SELGRADE, J.F.

*[*1*] Isolated Invariant Sets for Flows on Vector Bundles. *Trans. Amer. Math. Soc.*, **203**, (1975), 359-390; *M.R.*, **51**, #4322; *M.R.*, **54**, #1309.

SELL, G.R.

[*1*] Non-Autonomous Differential Equations as Dynamical Systems. *Differential Equations and Dynamical Systems, (Symposium, Puerto Rico, 1965)*, (Academic Press, New York), (1967), 531-536.

[*2*] Non-Autonomous Differential Equations and Topological Dynamics. I. The Basic Theory. *Trans. Amer. Math. Soc.*, **127**, (1967), 241-262.

[3] Non-Autonomous Differential Equations and Topological Dynamics. II. Limiting Equations. *Trans. Amer. Math. Soc.*, **127**, (1967), 263-283.

[4] *Topological Dynamics and Ordinary Differential Equations*, (Van Nostrand Reinhold, New York), (1971).

[5] Topological Dynamics Techniques for Differential and Integral Equations. *Ordinary Differential Equations, (1971 NRL-MRC Conference)*, (Academic Press, New York), (1972), 287-304.

*[6] Differential Equations without Uniqueness and Classical Topological Dynamics. *J. Diff. Equations*, **14**, (1973), 42-56; *M.R.*, **48**, #4430.

*[7] A Note on Almost Periodic Solutions of Linear Partial Differential Equations. *Bull. Amer. Math. Soc.*, **79**, (1973), 428-430; *M.R.*, **47**, #3804.

*[8] Almost Periodic Solutions of Linear Partial Differential Equations. *J. Math. Anal. Appl.*, **42**, (1973), 302-312; *M.R.*, **48**, #4463.

SHAPIRO, L.

[1] Proximality in Minimai Transformation Groups. *Proc. Amer. Math. Soc.*, **26**, (1970, 521-525; *M.R.*, **42**, #1091.

[2] Distal and Proximal Extensions of Minimal Flows. *Math. Systems Theory*, **5**, (1971), 76-88; *M.R.*, **46**, #877.

[3] *The Structure of H-Cascades*. (Lecture Notes in Mathematics), (Springer-Verlag, Berlin), **235**, (1971), 117-122;

[4] *On the Structure of Minimal Flows*. (Lecture Notes in Mathematics), (Springer-Verlag, Berlin), **235**, (1971), 123-128.

SIBIRSKIĬ, K.S.

[1] *Introduction to Topological Dynamics*. (Red.-Izdat. otdel Akad. Nauk MSSR, Kishinev), (1970), (Russian); *M.R.*, **43**, #651; (English edn), (Noordhoff, Leyden), (1975); *M.R.*, **50**, #10452.

SMALE, S.

[1] Differentiable Dynamical Systems. *Bull. Amer. Math. Soc.*, **73**, (1967), 747-817; *M.R.*, **37**, #3598.

STEFANITSE, I.M.

[1] Poisson Stable Solutions of Linear Systems of Difference Equations. *Differencial'nye Uravneniya*, **8**, (1972), 2062-2072, (Russian).

STEPANOV, V.V. AND TIKHONOV, A.N.

[1] On the Space of Almost Periodic Functions. *Mat. Sbornik*, **41**, (1934), 166-178, (Russian).

TERRAS, R.

*[1] Almost Automorphic Functions on Topological Groups. *Math. J. Indiana Univ.*, **21**, (1972), 759-774.

*[2] On Almost Periodic and Almost Automorphic Differences of Functions. *Duke Math. J.*, **40**, (1973), 81-92; *M.R.*, 47, #719.

VEECH, W.A.

[1] Almost Automorphic Functions. *Proc. Nat. Acad. Sci. USA*, **49**, (1963), 462-464.

[2] Almost Automorphic Functions on Groups. *Amer. J. Math.*, **87**, (1965), 719-751; *M.R.*, **32**, #4469.

[3] The E uicontinuous Structure Relation for Minimal Abelian Transformation Groups. *Amer. J. Math.*, **90**, (1968), 723-732.

[4] Minimal Transformation Groups with Distal Points. *Bull. Amer. Math. Soc.*, **75**, (1969), 481-486; *M.R.*, **41**, #4508.

[5] Point-Distal Flows. *Amer. J. Math.*, **92**, (1970), 205-242.

*[6] A Fixed Point Theorem-Free Approach to Weak Almost Periodicity. *Trans. Amer. Math. Soc.*, **177**, (1973), 353-362; *M.R.*, **49**, #7998.

*[7] *Minimal Sets and Souslin Sets*. (Lecture Notes in Mathematics), (Springer-Verlag, Berlin), **318**, (1973), 253-265.

WALTER, A.

[1] Algebraische Functionen von fastperiodischen Functionen. *Monatschr. Math. Phys.*, **40**, (1933), 444-457.

WESTERBECK, K.E. AND WU, T.S.

*[1] The Equicontinuous Structure Relation for Ergodic Abelian Transformation Groups. *Illinois Math. J.*, **17**, (1973), 421-441

WU, T.S.

[1] Proximal Relations in Topological Dynamics. *Proc. Amer. Math. Soc.*, **16**, (1965), 513-514.

[2] Two Homomorphic but Non-Isomorphic Minimal Sets. *Michigan Math. J.*, **14**, (1967), 401-404.

[3] Construction of Locally Almost-Periodic Minimal Transformation Groups. *Math. Systems Theory*, **1**, (1968), 157-163.

[4] Notes on Coset Transformation Groups. *Topological Dynamics, An International Symposium, (Fort Collins, 1967)*, (Benjamin, New York), (1968), 507-512.

[5] Disjointness of Minimal Sets. *Michigan Math. J.*, **15**, (1968), 369-371.

[6] *Disjointness of Minimal Sets*. (Lecture Notes in Mathematics), (Springer-Verlag, Berlin), **235**, (1971), 134-140.

[*7*] A Note on the Minimality of Certain Bitransformation Groups. *Pacific J. Math.*, **36**, (1971), 553-556; *M.R.*, **44**, #1009.

*[*8*] Notes on Topological Dynamics. I. Relative Disjointness, Relative Regularity and Homomorphisms. *Bull. Inst. Math. Acad. Sinica*, **2**, (1974), 343-356; *M.R.*, **51**, #11465.

*[*9*] Notes on Topological Dynamics, II. Distal Extensions with Discrete Fibers and Prime Flows. *Bull. Inst. Math. Acad. Sinica*, **3**, (1975), 49-60; *M.R.*, **53**, #9182.

YOSHIZAWA, T.

[*1*] Extreme Stability and Almost Periodic Solutions of Functional-Differential Equations. *Arch. Rat. Mech. Anal.*, 17, (1964), 148-170.

[*2*] Stability and Existence of Periodic and Almost-Periodic Solutions. *Proceedings of the United States-Japan Seminar on Differential and Functional Equations*, (1967), 411-427.

[*3*] *Stability for Almost Periodic Systems.* (Lecture Notes in Mathematics), (Springer-Verlag, Berlin), **243**, (1971), 29-39.

*[*4*] *Stability Theory and the Existence of Periodic Solutions and Almost Periodic Solutions.* (Applied Mathematical Sciences), (Springer-Verlag, Berlin), **14**, (1975).

ŽIDKOV, N.P.

[*1*] Certain Properties of Discrete Dynamical Systems. *Uč. Zap. Moskov. Gos. Univ. Mat.*, **163**, (1952), 31-59, (Russian); *M.R.*, **17**, #394.

ŽIKOV, V.V.

[*1*] Dini's Theorem for Monotone Sequences of Almost Periodic, Almost Automorphic and Recurrent Functions and Related Questions. *Vestnik Moskov. Univ. Matematika, Mekhanika*, **4**, (1967), 12-16, (Russian).

[*2*] A Contribution to the Problem of Existence of Almost Periodic Solutions of Differential and Operator Equations. *Naučnye Trudy Vladimir. Več. Politekhn. Instituta, Matematika, Vladimir*, **8**, (1969), 94-188, (Russian).

[*3*] On a Supplement to the Classical Favard Theory. *Mat. Zametki*, **7**, (1970), 239-246, (Russian).

[*4*] Existence of Levitan Almost Periodic Solutions of Linear Systems. (Second Supplement to the Classical Theory of Favard). *Mat. Zametki*, **9**, (1970), 409-414, (Russian); *M.R.*, **43**, #7720.

[*5*] On the Problem of Existence of Almost Periodic Solutions of Differential and Operator Equations. *Dokl. Akad. Nauk Azerb. USSR*, **26**, (1970), 3-7, (Russian).

[*6*] Almost Periodic Solutions of Linear and Non-Linear Equations in a Banach Space. *Dokl. Akad. Nauk USSR*, **195**, (1970), 278-281, (Russian); *M.R.*, **43**, #667.

ŽIKOV, V.V. (Contd)

[*7*] Some Remarks on the Compactness Condition in Connection with the Paper by M.I. Kadec on the Integration of Almost Periodic Functions. *Funkcional. Anal. i Priložen.*, **5**, (1971), 30-36, (Russian).

[*8*] On Certain Functional Methods in the Theory of Almost Periodic Solutions. I. *Differencial'nye Uravneniya*, **7**, (1971), 215-225, (Russian); *M.R.*, **43**, #7719.

[*9*] Monotonicity in the Theory of Almost Periodic Solutions of Non-Linear Operator Equations. *Mat. Sbornik*, **90**, (1973), 214-228, (Russian).

*[*10*] Some New Results in the Abstract Favard Theory. *Mat. Zametki*, **17**, (1975), 33-40, (Russian); *M.R.*, **51**, #1060.

*[*11*] On the Solvability of Linear Equations in Classes of Besicovitch and Bohr Almost Periodic Functions. *Mat. Zametki*, **18**, (1975), 553-560; *M.R.*, **52**, #14524.

ŽIKOV, V.V. AND LEVITAN, B.M.

*[*1*] On Favard's Theory. *Uspehi Mat. Nauk*, **32**, (1977), 122-171, (Russian).

ZIMMER, R.J.

*[*1*] Distal Transformation Groups and Fibre Bundles. *Bull. Amer. Math. Soc.*, **81**, (1975), 959-960; *M.R.*, **51**, #11462.

SUBJECT INDEX

Typesetter: *Academic Industrial Epistemology*
Printer: *Samsom Sijthoff Grafische Bedrijven*
Binder: *Abbringh*